From Sight to Light

From Sight to Light

THE PASSAGE FROM ANCIENT
TO MODERN OPTICS

A. Mark Smith

The University of Chicago Press   CHICAGO & LONDON

The University of Chicago Press, Chicago 60637
The University of Chicago Press, Ltd., London
© 2015 by The University of Chicago
All rights reserved. Published 2015.
Paperback edition 2017
Printed in the United States of America

23 22 21 20 19 18 17   2 3 4 5 6

ISBN-13: 978-0-226-17476-1 (cloth)
ISBN-13: 978-0-226-52857-1 (paper)
ISBN-13: 978-0-226-17493-8 (e-book)
DOI: 10.7208/chicago/9780226174938.001.0001

Library of Congress Cataloging-in-Publication Data

Smith, A. Mark, author.
From sight to light : the passage from ancient to modern optics / A. Mark Smith.
    pages cm
Includes bibliographical references and index.
ISBN 978-0-226-17476-1 (cloth : alkaline paper) — ISBN 978-0-226-17493-8 (e-book)
1. Optics—History. I. Title.
QC352.S57 2014
535'.209—dc23
2014015929

♾ This paper meets the requirements of ANSI/NISO Z39.48-1992 (Permanence of Paper).

CONTENTS

Preface ix

CHAPTER 1   Introduction / 1

CHAPTER 2   The Emergence of Optics as a Science: The Greek and Early
Greco-Roman Background / 23

1	Early Intimations	25
2	Physical and Psychological Theories of Vision	29
3	The Anatomical and Physiological Grounds of Vision	36
4	Theories of Color and Color Perception	43
5	The Euclidean Visual Ray Theory	47
6	Euclidean Catoptrics	55
7	Burning Mirrors and the Analysis of Focal Properties	68
8	Conclusion	72

CHAPTER 3   Ptolemy and the Flowering of Greek Optics / 76

1	The Ptolemaic Account of Visual Perception	80
2	The Ptolemaic Account of Reflection	92
3	The Ptolemaic Account of Refraction	108
4	Atmospheric Refraction and the Moon Illusion	121
5	Conclusion	127

CHAPTER 4   Greco-Roman and Early Arabic Developments / 130

1	Plotinus's Theory of Visual Perception	134
2	The Later *De anima* Commentators	143
3	Saint Augustine's Psychological Model: The Inward Ascent	150
4	The Arabic Transition: The *De anima* Tradition	155

5   The Arabic Transition: Geometrical Optics   166
6   Conclusion   178

CHAPTER 5   Alhacen and the Grand Synthesis / 181

1   The Elements of Alhacen's Analysis   184
2   Visual Discrimination, Perception, and Conception   189
3   Reflection and Its Visual Manifestations   195
4   Refraction and Its Visual Manifestations   206
5   Conclusion   224

CHAPTER 6   Developments in the Medieval Latin West / 228

1   Background to the Translation Movement   232
2   The Translation Movement and the Inroads of Aristotelianism   242
3   The Scholastic Analysis of Perception and Cognition   245
4   Geometrical Optics and the Evolving Science of *Perspectiva*   256
5   Conclusion   275

CHAPTER 7   The Assimilation of Perspectivist Optics during the Later Middle Ages and Renaissance / 278

1   Optics as a Quadrivial Pursuit in the Arts Curriculum   280
2   Theology and the Emergence of Optical Literacy   287
3   Optical Motifs in Literature   291
4   Renaissance Art, Naturalism, and Optics   298
5   Conclusion   316

CHAPTER 8   The Keplerian Turn and Its Technical Background / 322

1   Technological, Social, and Cultural Changes: 1450–1600   323
2   Rethinking Concave Mirrors and Convex Lenses   333
3   Rethinking the Eye   350
4   Kepler's Analysis of Retinal Imaging   353
5   The Analytic Turn   363
6   The Epistemological Turn   367
7   Conclusion   370

CHAPTER 9   The Seventeenth-Century Response / 373

1   The Conceptual and Cultural Context for the Keplerian Turn   376
2   Extending Vision in Both Directions   381

3   New Theories of Light   391
4   Recasting Color   400
5   The Epistemological Consequences   408
6   Conclusion   415

Bibliography 417
Index 441

The merest glance at any modern optics textbook leaves no doubt that, as currently understood, the science of optics is about light, about its fundamental properties and how they determine such physical behavior as reflection, refraction, and diffraction. But this understanding of optics and its appropriate purview is relatively new. For the vast majority of its history, the science of optics was aimed primarily at explaining not light and its physical manifestations, but sight in all its aspects from physical and physiological causes to perceptual and cognitive effects. Consequently, light theory was not only regarded as subsidiary to sight theory but was actually accommodated to it. And so it remained until the seventeenth century, when the analytic focus of optics shifted rather suddenly, and definitively, from sight to light. Marking the turn from ancient toward modern optics, this shift of focus evoked an equivalent shift in the order of analytic priority. Henceforth, sight theory would become increasingly subsidiary to light theory, the former now accommodated to the latter.

Why this particular turn at this particular time? That is the formative question for this study, and in response I will argue that Johannes Kepler's theory of retinal imaging, which was published in 1604, was instrumental in prompting the turn, as well as in giving it direction and shape, hence my later characterization of the turn as Keplerian. In support of this argument, I will show how Kepler's theory of retinal imaging not only fit within, but also transcended a long-evolving optical tradition that traced back to Greek antiquity via the Muslim Middle Ages. More to the point, I will show how the optical analysis that emerged within this tradition was driven by the need to explain how the visual process can yield a faithful mental picture of physical reality from the initial sensation of light and color. Accordingly, as it developed up to the late Renaissance, the science of optics was expressly designed to vindicate what amounts, quite literally, to a worldview, based on the assumption that objective reality somehow truly conforms to the abstract "image" we have of it in our minds.

After explaining in some detail how this optical tradition evolved between classical antiquity and roughly 1600, I will turn to a close examination of the ways in which Kepler both drew on and—far more importantly—flouted it in advancing his theory of retinal imaging. In the course of this examination, I will argue that it is in this latter respect, as a reaction against tradition and, moreover, a major step toward undermining it, that Kepler's new theory of sight takes on true historical significance. It is also in this respect that by treating the eye as a mere light-focusing device his account of retinal imaging helped spur the shift in analytic focus that marked the turn toward modern optics. Not just a transformation of optics, though, the Keplerian turn entailed an even more profound transformation of worldview, grounded in deep skepticism about whether objective reality is anything like the mental picture we have of it. Well on the way to completion by the end of the seventeenth century, this latter transformation is the subject of the book's concluding chapter, where I will show how it came about in response to certain implications of Kepler's new visual model.

As is evident from the preceding synopsis, this study covers an extraordinarily long chronological span, some two millennia in all. It also covers a broad spectrum of topics that include the physics of radiation, ray geometry, ocular anatomy, neuroanatomy and physiology, psychology, and epistemology. In certain instances, my treatment of these topics is moderately technical. In particular, I have had to deal in some detail with ray geometry and its theoretical underpinnings in order, eventually, to make proper, contextual sense of Kepler's analysis of convex lenses, which forms the linchpin for his account of retinal imaging. I realize that in delving into the intricacies of ray geometry I risk losing readers who find mathematics uncongenial, if not downright nettlesome. Like the devil, however, Clio is sometimes in the details, and in this case the details are too historically significant to avoid. Still, I have done my best to minimize them while explaining them in the simplest terms possible, all in the hope of making them understandable to any educated and willing reader, no matter how skimpy his or her technical or mathematical background. I am, in short, aiming as much at interested generalists as I am at specialists in the history of science.

Given the chronological and topical scope of this study, I have had to venture into nooks and crannies of scholarship that lie well outside my comfort zone. Who am I, a historian of medieval science, after all, to pontificate on classical Greek and late antique philosophy, early Muslim educational ideology, medieval Christian speculative theology, or Renaissance art, all of which crop up in the course of this study? Fortunately, while I was in the throes of writing this book, I managed to persuade several suitably knowledgeable friends and colleagues to give relevant chapters a close, critical scrutiny and thus, I trust, keep me from straying too far from the acceptable bounds of fact or interpretation. For this service I warmly thank, in alphabetical order: Alan Bowen, Sven Dupré, Jeremiah "Jerry" Hackett, George Hand, Jon McGinnis, Mary Quinlan-McGrath, and Yaakov Zik. Special thanks, finally, to Robert Hatch not only for carefully vetting the portions of the study that deal with the late Renaissance and early modern periods, but also for casting a gimlet eye over the entire manuscript to root out mistakes of fact or interpretation, as well as infelicities of style.

CHAPTER 1

Introduction

First published by the University of Chicago Press in 1976 and reissued in 1981, David Lindberg's *Theories of Vision from al-Kindi to Kepler* has stood to this day as the definitive study of pre-Keplerian optics and its historical evolution. The reasons are not far to seek. For one thing, that study is based on impeccable scholarship; Lindberg delved deeply and broadly into the appropriate primary and secondary sources available to him at the time. Crucial among these sources was a small group of late thirteenth-century "perspectivist" optical writers who drew heavily upon the Latin version of Ibn al-Haytham's *Kitāb al-Manāẓir* ("Book of Optics"). Probably completed by 1030, this treatise was rendered from Arabic into Latin sometime around 1200 under the title *De aspectibus* and attributed to "Alhacen."[1]

Another reason for the continuing success of Lindberg's study is his gift for clarity and precision in both thought and articulation, a gift that is evident in the deceptively simple, linear way he framed his narrative. Picking up the threads of his analysis in Greek antiquity, Lindberg showed in admirable detail how they were unraveled, modified, and augmented over the succeeding centuries until Johannes Kepler finally managed to weave them together into a coherent whole at the turn of the seventeenth century. Moreover, the narrative

1. In the actual manuscripts of the Latin version of the *Kitāb al-Manāẓir*, Ibn al-Haytham is most often referred to by the Latin transliteration "Alhacen" of his given name "al-Ḥasan" (Abū 'Alī al-Ḥasan ibn al-Ḥasan ibn al-Haytham). The work itself is most often titled *De aspectibus*. The forms "Alhazen" of his name and *Perspectiva* for the title of the work are later accretions. For discussion of these points, see A. Mark Smith, *Alhacen's Theory of Visual Perception* (Philadelphia: American Philosophical Society Press, 2001), xxi.

itself is compellingly thesis driven; Lindberg argued clearly and forcefully that Kepler's theory of retinal imaging in the *Ad Vitellionem Paralipomena* ("Supplement to Witelo") of 1604 represented not a fundamental break with, but rather a continuation of the perspectivist optical tradition as it evolved over the later Middle Ages and Renaissance. Lindberg made no bones about it: "Kepler was the culminating figure in the perspectivist tradition"; and Lindberg's analysis leads us systematically and inexorably to concur with this point.[2]

While thus emphasizing the centrality of perspectivist optics in the development of Kepler's visual model, Lindberg also showed that the theoretical foundations of those optics are as sophisticated as they are systematic and coherent. In other words, *pace* T. S. Kuhn, perspectivist optics provided what amounts to a scientific paradigm during the later Middle Ages and Renaissance.[3] Furthermore, Lindberg made it eminently clear that pre-Keplerian optics was not focused narrowly on the physical analysis of light and color but dealt more broadly with issues of visual perception. In order to be properly understood, then, pre-Keplerian optics had to be analyzed within this broader context because light theory and sight theory were interdependent before the seventeenth century—hence the focus on "vision" rather than "light" in the title of Lindberg's study.[4]

That no one has yet offered a compelling alternative to Lindberg's account is a testament to its coherence and persuasiveness.[5] Yet over the past three de-

2. Quotation from David Lindberg, *Theories of Vision from al-Kindi to Kepler* (Chicago: University of Chicago Press, 1976), 207. It is worth noting that Lindberg was arguing against the prevailing view of the time, championed most notably by Alexandre Koyré, that early modern science marked a revolutionary departure from its medieval forebear. Today, of course, the notion of an early modern scientific revolution has been, if not discredited, then modified considerably; see, e.g., Stephen Shapin, *The Scientific Revolution* (Chicago: University of Chicago Press, 1996), for one of the more radical reactions to this notion.

3. For Thomas S. Kuhn's claim that there was no optical paradigm before the end of the seventeenth century, see *The Structure of Scientific Revolutions*, 2nd ed. (Chicago: University of Chicago Press, 1970), 12–13.

4. In this regard Lindberg was reacting against Vasco Ronchi's *Storia della luce* ("History of Light") of 1939, which was translated into English by V. Barocas as *The Nature of Light* (Cambridge, MA: Harvard University Press, 1970). Lindberg intended his *Theories* as a corrective to Ronchi's study, which Lindberg considered to be riddled with factual errors and misconceptions; see his review: "New Light on an Old Story" (essay review of Vasco Ronchi, *The Nature of Light: An Historical Survey*, trans. V. Barocas [Cambridge, MA: Harvard University Press, 1970]), *Isis* 62 (1971): 522–24.

5. To be sure, books dealing with the development of optics have appeared since Lindberg published his *Theories*. Two fairly recent examples are David Park, *The Fire within the Eye:*

cades a considerable amount of work, some of it revisionary, has been done on pre-Keplerian optics. New texts have been brought to light, Roshdi Rashed's studies of previously unknown or little-known medieval Arabic sources being especially noteworthy in this regard.[6] Old texts have been critically edited or reedited, translated, and closely analyzed. Abdelhamid I. Sabra's work on the Arabic text of Ibn al-Haytham's *Kitāb al-Manāẓir* and mine on its medieval Latin counterpart, Alhacen's *De aspectibus*, serve as related examples.[7] New interpretive avenues have also been opened. Katherine Tachau, for instance, has shown how deeply implicated perspectivist theory was in later medieval discussions of epistemology, and recent work on "practical" optics in the sixteenth century, particularly the study of mirrors and lenses, has shed light not only on the scientific milieu within which Kepler conducted his optical

---

A Historical Essay on the Nature and Meaning of Light (Princeton, NJ: Princeton University Press, 1997); and Mark Pendergrast, *Mirror Mirror: A History of the Human Love Affair with Reflection* (New York: Basic Books, 2003). It is no derogation to these books, however, to say that neither is, or purports to be, a work of serious historical scholarship. Both are written by nonspecialists in the field (Park is a physicist and Pendergrast a journalist), both are pitched at a fairly popular level, and neither engages with historiographical or interpretive issues; see, e.g., my review of Park in *Physics Today* 51 (1998): 62–63. Even more recent, and more scholarly in its focus, is Olivier Darrigol's *A History of Optics from Greek Antiquity to the Nineteenth Century* (Oxford: Oxford University Press, 2012), but he devotes only thirty-six pages to the development of optics from Greek Antiquity to Kepler, and his treatment of pre-Keplerian optics is mostly derivative.

6. See, e.g., Roshdi Rashed, *Géométrie et dioptrique au Xᵉ siècle* (Paris: Les Belles Lettres, 1993).

7. So far Sabra has published critical Arabic editions of the first five of the seven books comprising the *Kitāb al-Manāẓir* in *The Optics of Ibn al-Haytham. Books I-II-III: On Direct Vision* (Kuwait: National Council for Culture, Arts, and Letters, 1983) and *The Optics of Ibn al-Haytham. An Edition of the Arabic Text of Books IV–V: On Reflection and Images Seen by Reflection* (Kuwait: National Council for Culture, Arts, and Letters, 2002). In addition, he has published a two-volume English translation of the first three books in *The Optics of Ibn al-Haytham: Books I-III on Direct Vision* (London: Warburg Institute, 1989). For my critical editions and translations of all seven books of the *De aspectibus*, see *Alhacen's Theory of Visual Perception* (Philadelphia: American Philosophical Society Press, 2001); *Alhacen on the Principles of Reflection* (Philadelphia: American Philosophical Society Press, 2006); *Alhacen on Image-Formation and Image-Distortion in Mirrors* (Philadelphia: American Philosophical Society Press, 2008); and *Alhacen on Refraction* (Philadelphia: American Philosophical Society Press, 2010). For a recent edition and French translation of book 7, see Paul Pietquin, *Le septième livre du traité* De aspectibus *d'Alhazen, traduction latine médiévale de* l'Optique *d'Ibn al-Haytham* (Louvain: Académie royale de Belgique, 2010).

research but also on the conceptual and methodological basis of that research.[8] In addition, the relationship between Renaissance art and perspectivist optics has been reexamined and, in some cases, reconfigured since Lindberg's day.[9]

Although none of these developments, singly or collectively, calls for an outright rejection of Lindberg's account, they do call for a significant revamping of it. That is what I propose to do in this study. I say "revamping" because Lindberg's account will serve as the backbone of my own in terms of factual detail (or most of it), as well as basic lineaments. Accordingly, the cast of main characters will remain essentially the same, although the roles of some will change. Ptolemy, for example, will play a far more important part in my narrative, and Alhacen's place in that narrative will be significantly altered by my linking him more tightly to Ptolemy and more loosely to Kepler than Lindberg did.

There will be some changes in emphasis, as well. One such change centers on Lindberg's eschewal of psychological and epistemological issues in order that his "investigation [not] get out of hand." To be sure, he acknowledged, such issues were "often raised within the context of visual theory," but they "were never its central concerns."[10] Long at odds with Lindberg over this point, I see these concerns as, if not absolutely central, then certainly integral to the formation of visual theory, and thus optics in general, from antiquity right up to the time of Kepler.[11] In fact, I will argue that perspectivist optics was expressly designed with these concerns in mind. Thus, whereas Lindberg's analysis effectively stops at the back of the eye, mine will trace the entire perceptual process from eye to brain, from the lowest-level apprehension of visible radiation to the conceptual and intellectual grasp of its object sources. Another change in emphasis involves reflection and refraction. I think Lindberg gave these two phenomena far shorter shrift than they merit because I am convinced that sixteenth-century efforts to understand reflection from concave mirrors and refraction through convex lenses played a crucial, perhaps determinative role in Kepler's model of retinal imaging. I will therefore pay closer attention

8. See, e.g., Katherine Tachau, *Vision and Certitude in the Age of Ockham* (Leiden: Brill, 1988); Sven Dupré, "Ausonio's Mirrors and Galileo's Lenses: The Telescope and Sixteenth-Century Practical Optical Knowledge," *Galileiana* 2 (2005): 145–80; and Eileen Reeves, *Galileo's Glassworks* (Cambridge, MA: Harvard University Press, 2008).

9. See, e.g., David Summers, *The Judgement of Sense* (Cambridge: Cambridge University Press, 1987).

10. Lindberg, *Theories of Vision*, x.

11. See, e.g., A. Mark Smith, "Getting the Big Picture in Perspectivist Optics," *Isis* 72 (1981): 568–89.

than Lindberg did to how these two phenomena, especially refraction, were understood and analyzed from antiquity to the seventeenth century.

This increased emphasis on reflection and refraction requires a somewhat more extensive discussion of ray theory and its mathematical underpinnings than Lindberg offered. Consequently, I have devoted significant portions of chapters 2 and 3 to showing how the foundations of that theory were laid in classical antiquity and how the ray geometry at its heart was used to explain a spectrum of optical phenomena from size perception to image formation in variously shaped mirrors. I have, however, tried to keep the ray-theoretical analysis in those chapters to a focused minimum aimed at providing the necessary background, and no more, for the discussion of spatial perception, reflection, and refraction in later chapters. I have also simplified that analysis as much as possible by taking a relatively superficial, descriptive approach to the mathematics on which it is based so as not to get entangled in the details of proof. I have therefore done my best to make at least the gist, if not the technical details, of that analysis accessible to any patient and attentive reader, no matter how math averse.

These are just some of the more salient revisions I will be making to Lindberg's account. But the most marked difference between my account and his resides in our respective views of Kepler's visual model and its relationship to perspectivist theory. Lindberg stressed continuity in that relationship. I, on the other hand, will emphasize discontinuity by arguing that Kepler's theory of retinal imaging did represent a break, a radical break, with the perspectivist tradition. This break, I will show, occurred at two levels. First, Kepler's theory of retinal imaging put the perspectivist visual model in jeopardy by severing the perceptual and epistemological link between eye and brain the perspectivists so carefully forged. As a result, the study of light was increasingly dissociated from the study of sight in post-Keplerian optics, as it evolved over the seventeenth century. The consequent shift of analytic focus from sight to light between roughly 1600 and 1700 constitutes what I call the Keplerian turn, and it is this turn, not Kepler himself, that is the ultimate focus of this book. Second, Kepler's analysis of convex lenses, which was central to his theory of retinal imaging, depended on theoretical and methodological concepts that were nowhere to be found in the perspectivist sources available to him. In certain key respects, in fact, he succeeded in that analysis despite rather than because of what he could have gleaned from those sources. I am not, I hasten to add, suggesting that Kepler's model of retinal imaging was a *creatio ex nihilo*. There is no question that he constructed that model on perspectivist foundations, but in the process he undermined those foundations in radical and ultimately lethal ways. In short, perspectivist optics served more as a foil than as a springboard for Kepler.

Significant though they are, these differences should not mask the fundamental points of agreement between Lindberg's account and my own. Both have as an ultimate goal to explain Kepler's theory of retinal imaging in proper historical context, although I will follow some of the ramifications of that theory into the seventeenth century, as they eventuated in the Keplerian turn. Both take perspectivist optics as a central component of that context. Both follow similar narrative paths, tracing the evolution of pre-Keplerian optics from Greek antiquity, through the Arabic Middle Ages, into the Latin Middle Ages and Renaissance. Both acknowledge Alhacen as a pivotal figure in the transition from classical Greek to medieval optics. Both recognize that pre-Keplerian optical theory was oriented toward the analysis of light *and* vision, not light alone. And both take a decidedly thematic or conceptual approach. At bottom, then, my aim here is not so much to supplant as to supplement Lindberg's account.

At this point, I should briefly explain what I intend to do in the following chapters and how I propose to do it. To start with, I make no claims to writing a comprehensive, global history of optics from antiquity to the seventeenth century.[12] My approach will be considerably more focused and thematic than that because, like Lindberg's, my narrative path leads more or less directly to and through Kepler and is thus pretty restrictively goal oriented. Hence, I will pay scant attention, if any, to various bypaths along the way that do not loop back to the main track leading toward Kepler. This is not to say that such bypaths are uninteresting or insignificant; they are simply irrelevant to my purposes. The medieval "Arabic" optical tradition serves as a prime

---

12. The recently revived push toward a global approach to the history of science is reflected in the five essays by Marwa Elshakry, Helen Tilley, Shruti Kapila, Neil Safier, and Sujit Sivasundaram in the "Focus" section of *Isis* 101 (2010): 95–158. See especially Marwa Elshakry, "When Science Became Western: Historiographical Reflections," 98–109. This global approach calls for a radical loosening of the definitional constraints on "science" posed by rigid "Western" scientific norms in order to give non-Western cultures their due in the development of scientific thought. Presumably, this should eventuate in a grand, global synthesis, but in adverting to "the fantasy of a singular global history," Sivasundaram seems to doubt that it will; see p. 95. For a fairly recent example of such a "global" approach, see Arun Bala, *The Dialogue of Civilizations in the Birth of Modern Science* (New York: Palgrave Macmillan, 2006); cf. Edward Grant, "Reflections of a Troglodyte Historian of Science," *Osiris* 27 (2012): 133–55, for a critical evaluation of Bala's approach.

example.[13] That it, like its medieval Latin counterpart, was firmly rooted in Greek sources is beyond question, as is the fact that several key Arabic thinkers—Alhacen foremost among them—were instrumental in the development of optics in the medieval Latin West. But to concentrate on these figures alone is to ignore some truly interesting and innovative optical work carried out in the medieval Muslim world.

Take the tenth-century mathematician Ibn Sahl (d. ca. 1000) associated with the ʿAbbāsid court in Baghdad. In the course of analyzing curved mirrors and lenses in his *On Burning Instruments*, he provided an elegant and ingenious mathematical demonstration of the focusing property of hyperboloidal lenses based on what amounts to the sine law of refraction.[14] As far as we know, Ibn Sahl's version of that law predates the earliest European version, generally attributed to Willibrord Snel, by more than six centuries. So too, as far as we know, it would take well over six centuries before anyone in Europe undertook an equivalent analysis of hyperboloidal lenses.[15] Why, then, not include Ibn Sahl as an integral part of my narrative? Because, not having been translated into a European language until very recently (by Roshdi Rashed), his *On Burning Instruments* remained unknown in the Latin West. There is, in short, no evidence whatever that Ibn Sahl had anything to do with the sine law arrived at by Snel. Examples like this abound of brilliant and innovative Arabic thinkers, such as Kamāl al-Dīn al-Fārisī (d. 1318), whose optical work

13. An abiding problem for anyone dealing with science in the Muslim world is how best to denominate it according to kind. I have chosen the linguistic characterization "Arabic" in order to avoid not only the religious overtones of "Muslim" or "Islamic" but also the ethnic overtones of "Arabian." Many early contributors to Arabic science were neither Muslim nor Arab, yet Arabic did serve as the lingua franca for scholarship throughout the medieval Muslim world. Marshall Hodgson's "Islamicate" might perhaps be better in view of its cultural connotations, but I decided against it because it is somewhat clumsy and has never achieved widespread usage.

14. See Roshdi Rashed, "A Pioneer in Anaclastics: Ibn Sahl on Burning Mirrors and Lenses," *Isis* 81 (1990): 464–91. Simply put, the sine law states that when light is refracted in passing from one transparent medium to another, the sines of the angles of incidence and refraction will be in constant proportion, which is to say that for all angles of incidence, sine i / sine r = constant. Accordingly, in figure 9.2 (see chapter 9), if angles of incidence KBM and ABH yield angles of refraction NBL and NBI, respectively, then KBM / NBL = ABH / NBL = constant.

15. Although Snel is generally credited with having "discovered" the sine law sometime around 1620, the Englishman Thomas Harriot arrived at it by no later than 1602 and perhaps as early as the late 1590s. René Descartes was the first to publish the sine law, along with a specious proof, in his *Dioptrique* of 1637, where he also demonstrated the focusing property of hyperboloidal lenses.

developed along byways that never looped back to my main track and to whom I will therefore give little more than lip service.[16]

One unavoidable consequence of my taking this tack is that the resulting narrative will have a definite "Western" slant. That such a slant smacks of Edward Said's "Orientalism" hardly needs saying.[17] After all, as far as the development of modern optics goes (via Kepler), some of the most innovative and forward-looking Arabic thinkers play little or no part whatever in my narrative. I can thus be accused of treating the development of modern optics Eurocentrically, as a uniquely "Western" phenomenon to which the "West" can claim exclusive proprietary rights.[18] One of the most prominent recent critics of such a culturally isolationist view of science is George Saliba, who poses the rhetorical question, "Whose science [is] it . . . anyway?"[19] His an-

16. Author of the *Tanqīḥ al-Manāẓir* ("Revision of [Alhacen's] Optics"), Kamāl al-Dīn al-Fārisī made significant adjustments to and elaborations on Alhacen's optical work. He was a student of the renowned mathematician and astronomer, Quṭb al-Dīn al-Shīrāzī, who in turn was a student of Naṣīr al-Dīn al-Ṭūsī, founder of the Maragha observatory in 1259 under the Mongol il-khan Hulagu.

17. Edward Said, *Orientalism* (New York: Parthenon, 1978). In light of Said's scathing attack on Eurocentric views of the "East" as undifferentiated and backward, I will bracket "East" with quotation marks in recognition of its problematic conceptual status. After all, the medieval Muslim "East" extended through North Africa into Spain at the far western reaches of Europe. I will do the same with "West," which, ironically enough, Said presents as monolithic in its arrogance and its push for dominance, all the while demanding that "Westerners" develop a more nuanced and sympathetic approach to the "East."

18. As Roshdi Rashed puts it, "scientific activity outside of Europe, poorly integrated into the history of science, is the object of an anthropology of science whose academic translation is nothing more than Orientalism," in *The Development of Arabic Mathematics: Between Arithmetic and Algebra* (Dordrecht: Kluwer Academic, 1994), 333.

19. Saliba poses this question at the end of a long essay titled "Whose Science Is Arabic Science in the Renaissance" on his website (http://www.columbia.edu/~gas1/project/visions/case1/sci.5.html). His main point is that to isolate science according to cultural categories, such as Islamic, Western, Greek, is both futile and hegemonic in the context of the "grand narrative" of triumphal modern Western science. In his recent *Islamic Science and the Making of the European Renaissance* (Cambridge, MA: MIT Press, 2007), Saliba attempts to demonstrate this point by showing that in order to account for technical details of planetary motion in the *De revolutionibus*, Copernicus borrowed certain analytic devices from a trio of medieval Muslim astronomers, two of whom had been ensconced at the observatory of Maragha. That being the case, we are forced to ask whether the so-called Copernican revolution was actually "Copernican" and, therefore, whether it was meaningfully "Western." Cf., however, the reviews by Toby Huff, in *Middle East Quarterly* 15 (2008): 77–79, and Owen Gingerich, in *Journal of Interdisciplinary History* 39 (2008): 310–11.

swer, of course, is that science belongs to everyone, to every culture; any claim to ownership within a particular culture is thus hegemonic. In principle I agree with Saliba; it would be difficult for me not to, having spent over a quarter of a century closely studying Alhacen's optics. But in historical practice I find his open stance problematic because it is based on what I view as an unwarranted absolutism according to which a given "science" or scientific concept remains temporally and culturally constant or atomic.

Let us go back to the sine law for a moment. If we take that law as factually determinate, and if we think in terms of temporal priority alone, then the sine law obviously belongs to Ibn Sahl and, by extension, the Arabic "East." But to take the sine law in that way, as a brute scientific fact, or at least a declaration of scientific fact, is naively reductionist. As the French phrase *les faits sont faits* sums it up nicely, facts may be facts, but they are also constructed. Indeed, as recent sociologists of scientific knowledge would have it, they are *socially* constructed.[20] Consequently, if we place Ibn Sahl's and Snel's versions of the sine law in their respective "marketplace of ideas," they take on an entirely different cast. Whereas there appear to have been no buyers in Ibn Sahl's marketplace, there was a brisk trade in Snel's.[21] It was therefore in the "West," not the "East," that the sine law became historically significant and meaningful because it was there that it became communal and therefore fruitful.

The same holds for the evolution of modern optics over the sixteenth and seventeenth centuries. It may well be that certain key ideas, laws, and concepts that contributed to that evolution were anticipated by Arabic or, for that matter, Indian, Chinese, or Mesoamerican thinkers. And it is certainly the case that there was a lively cross-cultural marketplace of commodities and ideas between the Latin "West" and Arabic "East" throughout the Middle Ages and Renaissance. The fact remains, though, that it was in Europe that those ideas, laws, and concepts were eventually assimilated, refined, channeled, and combined in such a way as to form the basis for what most of us today would characterize as modern optics. Any claim to the contrary strikes me as historically perverse. Furthermore, to contend that the evolution of modern optics over the

20. For an extreme defense of social construction, see Stephen Shapin, *A Social History of Truth* (Chicago: University of Chicago Press, 1994); for a more moderate one, see Stephen Shapin and Simon Schaffer, *Leviathan and the Air Pump* (Princeton, NJ: Princeton University Press, 1989).

21. That Alhacen did not adopt Ibn Sahl's law is particularly telling as an indication of its failure to influence Arabic optics; for a discussion of this point, see Smith, *Alhacen on Refraction*, lxxxii–lxxxiv. That the sine law became a staple of optical analysis in seventeenth-century Europe is so obvious it needs no belaboring.

sixteenth and seventeenth centuries happened in Europe is not to give Europe proprietary rights to that science or to accord Europe cultural exceptionalism or superiority for having developed it. I therefore strongly resist any charge of being trapped, whether wittingly or unwittingly, in some grand, master narrative or of engaging in hegemonic discourse. I do, however, freely acknowledge that I will be telling a particular story, not necessarily the whole story, nor the "true" story, nor the only story, nor even the best story. I also acknowledge that my story is framed within a specific historiographic or metahistorical tradition in which historical narratives have a beginning, middle, and end.[22]

No less selective than the narrative path I intend to follow is the set of landmarks I will use to define it. Those landmarks will be primarily textual because, as I mentioned earlier, my approach will be thematic or conceptual rather than *événementielle* (to borrow from Fernand Braudel) or prosopographical. Its primary focus will therefore not be on events or personalities. As might be expected, the core of the textual sources used in this study will comprise works devoted wholly or in great part to the systematic analysis of visual theory, light and color, or both. Euclid's *Optics* and *Catoptrics*, Alhacen's *De aspectibus*, Roger Bacon's *Perspectiva*, and Kepler's *Paralipomena* are obvious examples. Aristotle's *De anima*, Galen's *De usu partium*, and Avicenna's *Shifā* ("Healing") are perhaps less obvious but no less representative examples.

Secondary, but by no means marginal, will be artifactual sources that provide relevant technological information. Glass looms large in this respect because a variety of optical theorists from Ptolemy and Alhacen onward based, or at least claim to have based, empirical studies of refraction on the passage of visual rays or light rays through it. The type and quality of the glass available to those theorists has an obvious bearing on the feasibility or accuracy of these studies and their purported results. Lenses are more problematic. Lens-like objects dating back well before 1500 BC have been uncovered and closely analyzed for optical properties, and it is tempting on that basis to suppose they were used as magnifying devices for close work or viewing.[23] Unfortunately, there is no way of determining whether they were actually intended for such use or any way of knowing whether they were produced with any optical principles, either pragmatic or theoretical, in mind. For that reason, these artifacts

22. See Hayden White, "The Value of Narrativity in the Representation of Reality," in *The Content of the Form* (Baltimore: Johns Hopkins Press, 1987), 1–25.

23. See, e.g., Jay M. Enoch, "The Enigma of Early Lens Use," *Technology and Culture* 39 (1988): 273–91.

cannot be meaningfully included in a history of optics or, for that matter, a history of lenses. From my narrative perspective, which is admittedly conservative, that history will only begin to take shape during the Greco-Roman period.

I am well aware of the potential shortcomings and pitfalls of such a conceptual, text-based approach. It is all too easy, for instance, to treat concepts or ideas as having a life of their own without due consideration of how they were assimilated and modified over time and place. Thus one might assume incorrectly that "ray" meant the same thing for Euclid as it did for Alhacen or Descartes insofar as all three took it as rectilinear and therefore geometrically representable by a straight line. It is also easy to view similarity of concepts or ideas as an indication of influence. On that basis one might be led to suppose, again incorrectly, that Snel somehow got the sine law from Ibn Sahl either directly or through connected, intermediary sources as yet unknown. Moreover, the kinds of texts upon which my study is based are finished products whose actual manufacture we know little or nothing about. As a result, these texts provide almost no information about the scholarly networks—the marketplace of ideas, if you will—within which they were produced and disseminated. Early in the *De aspectibus*, for instance, Alhacen divides contemporary and antecedent optical theorists into two main groups: mathematicians (*mathematici*) and natural philosophers (*naturales*). Precisely whom he had in mind is impossible to say with any certainty because he gave no names.[24] The best we can do is make educated guesses, and tentative ones at that. Nor is Alhacen unique in this regard; most of the pre-Keplerian optical theorists I will deal with in the course of this study (Roger Bacon being a salient exception) were equally reticent about citing sources. As a result, the social dimension of my narrative will be severely but necessarily limited.

A further limitation is imposed according to the time period my study spans and the relative scarcity of sources and types of sources available for most of it. Especially acute for antiquity, this scarcity has a variety of causes. First and most obvious is loss over time due to natural deterioration, fire, earthquake, flooding, and so on. The fate of the Alexandrian "library" in successive holocausts between the first century BC and the seventh century AD is a clear case in point.[25] Second is supersession. Certain works, such as Euclid's *Elements* and Ptolemy's *Almagest*, assumed such canonical status that earlier works

24. For a brief discussion of this point, see Smith, *Alhacen's Theory*, xxv.

25. See Diana Delia, "From Romance to Rhetoric: The Alexandrian Library in Classical and Islamic Traditions," *American Historical Review* 97 (1992): 1449–67.

treating the same subjects simply disappeared through obsolescence and consequent disuse.[26] In such cases we have the culminating works but not the various stages leading to them. Third is a lack of working texts. Rough copies did not survive because they were done on media such as wax tablets that were designed for easy erasure and reuse. Finally, private, off-the-cuff correspondence among classical scholars and *littérateurs* seems to have been unusual insofar as letter writing was considered to be a rhetorical exercise strictly bound by formulaic conventions.[27] Much of the correspondence that has survived, though nominally private, was intended either for circulation among compeers or for "publication" through copying.[28] Classical letter writers were thus aiming to perfect their thoughts and the articulation of them, not to reveal the process by which they perfected them. Additionally, although we can be fairly certain that such correspondence passed through networks of readers, we have little or no indication of who those readers were, apart from the person to whom the letter was addressed. It was only in the late Renaissance that private, scholarly correspondence became private in the modern sense, and only then that we can begin to delineate the networks through which such correspondence passed on the basis of surviving letter collections, some of them quite extensive.

Things get significantly better for the medieval period, if only because the ravages of time have had less time to take effect. Also, the shift from papyrus to parchment during late antiquity meant that texts were being written on a far more durable, climate-resistant medium. In the Muslim world the equivalent shift—from papyrus or parchment to paper—occurred somewhat later.[29]

26. On this point see Gerald J. Toomer, trans., *Ptolemy's Almagest* (Princeton, NJ: Princeton University Press, 1998), 1–2; and Glenn R. Morrow, trans., *Proclus: A Commentary on the First Book of Euclid's* Elements, 2nd ed. (Princeton, NJ: Princeton University Press, 1992), xlv–lxvi; see also A. Mark Smith, *Ptolemy and the Foundations of Ancient Mathematical Optics* (Philadelphia: American Philosophical Society Press, 1999), 13–16.

27. See, e.g., Carol Poster, "A Conversation Halved: Epistolary Theory in Greco-Roman Antiquity," in *Letter-Writing Manuals and Instruction from Antiquity to the Present*, ed. Carol Poster and Linda Mitchell (Columbia: University of South Carolina Press, 2007), 21–51.

28. A clear example is the preface to Ptolemy's *Almagest*, which is addressed to a certain Syrus, to whom Ptolemy addressed other works as well; see Toomer, *Ptolemy's Almagest*, 35.

29. See Jonathan Bloom, *Paper before Print: The History and Impact of Paper in the Islamic World* (New Haven, CT: Yale University Press, 2001). Paper was in widespread use in the Muslim world by the ninth century. Manufactured in Muslim Spain by the late eleventh century, it was produced in Italy by the thirteenth. Until the early fifteenth century, it was not widely used in Europe, in part because it was not trusted and in part because it was no less expensive than parchment. It is estimated that by the time Gutenberg published his Bible in the mid-fifteenth century, paper was around six times less expensive than parchment.

Consequently, the shortage of relevant texts is not nearly as acute as it is for the antique period. But time and changes in writing materials are not the only factors. Institutional support for learning played a critical role in how knowledge was created and disseminated during this period, and in this regard there were crucial differences between medieval Europe and the Muslim world. These differences are worth pondering because they go a long way toward explaining why modern optics evolved in the Latin West rather than in the Arabic world.

Between roughly 500 and 1150 in Europe, institutional support for education came primarily from monastic and cathedral schools. Within this context the focus of education was primarily religious. However, from at least the sixth century, the curricular structure of Christian education had coalesced around the pagan liberal arts, which were theoretically reducible to the *trivium* (grammar, logic, and rhetoric) and *quadrivium* (arithmetic, geometry, music, and astronomy). Adopting this curricular model as the basis for religious education was problematic in several ways. Because most of the educated elite in the western portion of the Roman Empire had lost facility in Greek, they had no direct access to the textual sources of Hellenic and Hellenistic learning. Instead, they had to depend on Latin translations. Fairly few and far between, these included a smattering of Plato and Aristotle as well as some basic philosophical and scientific commentaries. Also, because those sources were pagan, they had to be bowdlerized in order to suit Christian norms. And finally, they had to be conserved and reproduced in what became an increasingly rough-and-ready environment.

By the second half of the twelfth century this situation had changed dramatically. Emerging universities had begun to supplant the old monastic and cathedral schools in response to a complex set of demographic, commercial, political, and intellectual circumstances.[30] Crucial among these latter was the flood of new texts translated into Latin during the twelfth and early thirteenth centuries, primarily from Arabic and primarily in Spain, that washed over northern Europe.[31] These texts included long-lost Greek sources, Aristotle

30. Although dated, John W. Baldwin's *The Scholastic Culture of the Middle Ages* (Lexington, MA: D. C. Heath, 1971) is still quite useful for a basic overview of the development of medieval universities and the educational system within them. His focus, however, is mostly limited to the two "archetypal" universities of Paris and Bologna; see especially 1–44.

31. While Spain was the primary source of translations during this period, Norman Sicily was of some importance. It was there, for instance, that Ptolemy's *Optics* was translated from Arabic in the mid-twelfth century, and Michael Scot was responsible for translations of Aristotelian works from Arabic in the early thirteenth century under the patronage of Frederick II. Not all translations were from Arabic; James of Venice translated a number of Aristotelian treatises

in particular, that had been rendered into Arabic during the ninth and tenth centuries, but they also included a wealth of commentaries and original works by Arabic thinkers. The inroads of these texts led to a standardization of learning based on a much-expanded liberal arts program that formed the foundation for all university education, from the lowest level (bachelor of arts) to the highest (doctor of theology).[32] Every student, whatever his final academic destination, was expected to gain an appropriate grounding in the liberal arts, so it was critical that this grounding be common. Therefore, the texts to be taught had to be common, and as far as the "scientific" disciplines are concerned, Aristotle eventually won out, his works on natural philosophy providing the core of a much-expanded arts curriculum throughout most of the later Middle Ages and Renaissance.[33] Other works fell within the gravitational pull of that core and were studied as exemplifications of its underlying principles.[34] Among these was Ibn al-Haytham's *Kitāb al-Manāẓir* ("Book of Optics"), which was rendered into Latin under the title *De aspectibus* and attributed to Alhacen, the Latin transliteration of Ibn al-Haytham's given name, al-Ḥasan.

---

from Greek in the early twelfth century, and later in the thirteenth century translations from Greek to Latin were made from manuscripts that became available after the conquest of Constantinople during the Fourth Crusade and the subsequent establishment of the Latin Empire of Constantinople. For a list of Arabic-to-Latin translations between roughly 1130 and 1300, see Charles Burnett, "Arabic Philosophical Works Translated into Latin," in *The Cambridge History of Medieval Philosophy*, vol. 2, ed. Robert Pasnau and Christina Van Dyke (Cambridge: Cambridge University Press, 2010), 814–22.

32. There were two levels of matriculation in the arts, the lower one leading to the bachelor's degree, the higher one to the master's degree. Beyond the master's degree one could go on to the advanced study of medicine, Roman law, canon law, or theology, this latter being by far the most prestigious of the advanced disciplines. Each of these subdivisions was represented by a faculty in the given university, although not all these faculties were represented at every given university. For a brief overview of this system and its implications, see Baldwin, *Scholastic Culture*, 44–97.

33. The incorporation of Aristotle into the arts curriculum was neither uniform nor smooth, particularly at Paris, where instruction in his works on natural philosophy, perhaps only select ones, was forbidden between 1210 and 1231. By 1254, however, the Aristotelian works prescribed for study at Paris included not only his *Organon* and *Ethics* but also his *Physics*, *Metaphysics*, *On the Soul*, *History of Animals*, *On the Heavens*, *Meteorology*, *On Generation and Corruption*, *On Sense and Sensibles*, *On Sleep and Waking*, *On Memory*, *On Death and Life*, *On Plants*, and the pseudo-Aristotelian *Book of Causes*; see Hastings Rashdall, *The Universities of Europe in the Middle Ages*, vol. 1 (Oxford: Clarendon Press, 1895), 435–36.

34. Far and away the most popular of these satellite works was Johannes de Sacrobosco's *De sphera*, which was used to teach arts students the rudiments of Aristotelian cosmology and Ptolemaic astronomy.

The production and subsequent fate of this translated text is instructive. To begin with, there is the question of when and by whom it was Latinized. Gerard of Cremona has long been suggested, and if so, then the *De aspectibus* must have been rendered into Latin by no later than 1187, the year of his death. But this attribution is problematic because we now know that at least two, and probably three, different translators were involved, only one of whom could have been Gerard.[35] Still, we can be fairly certain that the translation was completed by the very beginning of the thirteenth century at the latest and probably in Spain. That leads automatically to the next question: why was the *De aspectibus* translated at this particular time? This question arises because the lion's share of the treatise consists of highly technical and sophisticated geometrical demonstrations based not only on Euclid's *Elements* but also on Apollonius of Perga's *Conics*. Very few, if any, Europeans at the time had the appropriate mathematical background to fully grasp the Euclidean steps of Alhacen's demonstrations, much less the Apollonian steps, which are based on certain properties of hyperbolic sections.[36] So the very fact that the text was translated despite these impediments has two important implications. First, there were translators available who were adequate to the task of rendering the text, with all its complexity, in a reasonably faithful manner.[37] Second, there must have been a potential readership; otherwise, why undertake the laborious process of translating the text in the first place?[38] We can thus assume that by no later than the beginning of the thirteenth century there was already a community of scholars in Europe prepared to assimilate or at least attempt to assimilate the *De aspectibus*.

The subsequent fate of the treatise can be charted in two ways: by its manu-

35. For a full discussion of these points, see Smith, *Alhacen's Theory*, xix–xxi; *Alhacen on Image-Formation*, xlv–xlvii; and *Alhacen on Refraction*, cxxv–cxxvi.

36. See Smith, *Alhacen on the Principles*, xxxvii–lxxxii.

37. There are of course some significant deviations from the Arabic original, but overall the Latin translation seems to be relatively faithful to it, one indication being that the geometrical proofs in the Latin version are logically valid throughout. It is not clear, however, whether the deviations in the Latin version reflect the Arabic text or texts the translators used and whether that text might have deviated from the ones currently extant and used by Sabra and Rashed.

38. There is considerable evidence that translators, generally working in teams, were commissioned and thus paid to translate particular texts or types of texts. It is therefore not unlikely that the *De aspectibus* was produced to order. For an interesting discussion of the Spanish translation movement and its general context, see Thomas Glick, "Science in Medieval Spain: The Jewish Contribution in the Context of *Convivencia*," in *Convivencia: Jews, Muslims, and Christians in Medieval Spain*, ed. Vivian Mann, Thomas Glick, and Jerrilynn Dodds (New York: Georges Braziller and the Jewish Museum, 1992), 83–111.

script diffusion and by its influence. As to the first, we are currently aware of eighteen complete, or nearly complete, and five fragmentary manuscript copies. Of the eighteen complete ones, the vast majority (fifteen) dates from the second half of the thirteenth century to the end of the fourteenth century.[39] From this we can infer that the period during which interest in the *De aspectibus* was liveliest lasted from around 1250 to 1400 at the outside. It was during this same period that European universities expanded most rapidly in both size and number in the Middle Ages and Renaissance. As to influence, the clearest indicator is the set of perspectivist optical texts produced between roughly 1260 and 1280 by Roger Bacon, Witelo, and John Pecham—all of whom drew heavily upon Alhacen for theoretical and analytic support. These texts, too, were diffused widely during the later Middle Ages and Renaissance.[40]

The Arabic original of Alhacen's *De aspectibus* presents a sharp contrast. Whereas eighteen complete, or nearly complete, and five fragmentary versions of the Latin translation are known to exist, only one complete version of the Arabic original has so far been uncovered, the four additional extant versions all being incomplete, sometimes woefully so.[41] It is possible, of course, that as-yet-undiscovered manuscripts of the Arabic text are squirreled away in badly cataloged collections, but it is highly unlikely that these will add significantly to the current total. As far as the *De aspectibus* is concerned, then, there was a significantly larger community of interest in medieval Europe than in the medieval Muslim world.[42] This community expands exponentially when we

39. See Smith, *Alhacen's Theory*, clv–clx, for a list and description of seventeen of the eighteen complete versions. Since the publication of that list, a new version of the *De aspectibus* has been uncovered—Biblioteca Casanatense, Rome, MS 1393—probably dating from the fourteenth century. This manuscript has been digitized under the direction of Klaus Werner and Andreas Thielemann at the Bibliotheca Hertziana in Rome and can be accessed at http://db.biblhertz.it/alhacen/.

40. The core set of perspectivist texts consists of Roger Bacon's *Perspectiva* (ca. 1265), Witelo's *Perspectiva* (ca. 1275), and John Pecham's *Perspectiva communis* (ca. 1280). Peripheral are Bacon's *De multiplicatione specierum* and *De speculis comburentibus*, the latter partially based on Alhacen's *De speculis comburentibus*, and Pecham's *Tractatus de perspectiva*, a brief treatise that predates his *Perspectiva communis*. One might also include Robert Grosseteste's various works on optics, especially the *De luce* and the *De lineis, angulis, et figuris*, but whether Grosseteste ever read Alhacen's *De aspectibus* is open to question.

41. See Sabra, *Optics* (1989), lxxx–lxxxiii, for a list and description of these manuscripts.

42. Although Ibn al-Haytham's actual treatise is extant in very few manuscript copies, Kamāl al-Dīn al-Fārisī included much of that treatise's contents in his early fourteenth-century *Tanqīḥ al-Manāẓir*. Currently extant in fifteen manuscripts, of which at least nine date from the seven-

take into account the perspectivist works based on the *De aspectibus*. Altogether, with the *De aspectibus* included, these are represented by nearly 190 manuscripts, and this is surely the tip of a textual iceberg whose size is a matter of conjecture.[43]

One obvious reason for the relatively sparse dissemination of the Arabic original of the *De aspectibus* is that institutional support for the study of natural philosophy was relatively decentralized and sporadic in the medieval Muslim world. Also, the focus of Muslim education was on understanding proper religious observance and behavioral standards through close reading of the Qur'ān and intense study of *ḥadīth*, the set of maxims attributed to Muḥammed and, to most, the basis of *sunnah*.[44] Suffice it to say that Aristo-

---

teenth century, this work was instrumental in preserving Ibn al-Haytham's optical thought in Arabic, particularly during the early modern period; see A. I. Sabra, "The 'Commentary' That Saved the Text: The Hazardous Journey of Ibn al-Haytham's Arabic 'Optics,'" *Early Science and Medicine* 12 (2007): 117–33. Nonetheless, even with the addition of Kamāl al-Dīn's *Tanqīḥ*, the dissemination of Ibn al-Haytham's optical thought in the medieval Muslim world appears to have been comparatively limited.

43. For a list of the manuscripts of the core perspectivist texts, see David C. Lindberg, *A Catalogue of Medieval and Renaissance Optical Manuscripts* (Toronto: University of Toronto Press, 1975), 40–42, 68–72, and 77–79. Since Lindberg compiled this catalog, other manuscripts have been uncovered, and more will undoubtedly come to light in the future. Precisely how many have been permanently lost or destroyed since the Middle Ages is impossible to determine, but the number is probably large. Aside from fire, flood, war, careless storage, and the like, the advent of print in the early modern period was a crucial factor in the disappearance of manuscripts because it rendered them obsolete. In a recent statistical analysis of survival rates for medieval Latin manuscripts, Eltjo Buringh estimates that 9,508,080 manuscripts were produced between 1200 and 1500, out of which 615,774 are currently extant. This yields a gross survival rate of roughly one out of fifteen, which at best might give us a rough idea of the survival rate of the core perspectivist texts. See *Medieval Manuscript Production in the Latin West: Explorations with a Global Database* (Leiden: Brill, 2011), especially table 5.4, p. 261.

44. See Fazlur Rahman, *Islam*, 2nd ed. (Chicago: University of Chicago Press, 1979), 181–92; see also Toby Huff, *The Rise of Early Modern Science: Islam, China, and the West*, 2nd ed. (Cambridge: Cambridge University Press, 2003), 47–117 and 149–79. For a fascinating account of this educational regime and its ramifications as seen through the eyes of a late fourteenth-century Muslim scholar, see Ibn Khaldūn, *The Muqaddimah*, trans. Franz Rosenthal (Princeton, NJ: Princeton University Press, 1969), 333–459. As Huff points out, there was no set curriculum in this regime, nor was there a corporate standard by which a student was judged to have mastered it. A crucial distinction between medieval Muslim and Christian education is that the former had its wellsprings in a close study of the Qur'ān, so all other knowledge had to flow from and be properly attuned to it. In medieval Europe, on the other hand, the study of

telian natural philosophy did not hold the same privileged place within this educational regime as it did in the Latin West. Not that natural philosophy was rejected outright by the Muslim intelligentsia, which consisted primarily of legal experts. Quite the contrary, certain disciplines within the broad spectrum of natural philosophy were pursued quite vigorously and brought to an extraordinary level of sophistication. Such disciplines were considered to be acceptably "Islamic" and were thus favored because they answered the needs of Muslim law and observance.[45]

Astronomy is paradigmatic in this regard. Establishing the exact times of prayer and the precise direction of Mecca (the *qibla*) toward which prayer should be offered was, and still is, crucial to Muslim observance, and both require precise observational data, accurate astronomical and geodesic parameters, and advanced mathematical tools, such as plane and spherical trigonometry.[46] Mathematics, too, constituted a properly Islamic discipline and at least partly on that account was brought to an exceptionally high level in both geometry and arithmetic.[47] On the other hand, disciplines within natural philosophy that had no clear bearing on Islamic law or observance were considered to be either non-Islamic—and thus not favored—or un-Islamic—and thus actively disfavored.[48] Optics evidently fell within the first category, not

---

the Bible was reserved for those who had the appropriate training and preparation to read and interpret it in an appropriately orthodox manner.

45. Strictly speaking, the Islamic sciences comprised those disciplines needed for a mastery of the classical Arabic language and its proper use in reading and understanding the Qur'ān. In contradistinction to these were the "foreign" or "ancient" disciplines that derived primarily, though not exclusively, from Greek sources. This distinction comes through quite clearly in Ibn Khaldūn's discussion of the various sciences and their utility or value. Being a *qāḍī*, or judge, he was of course keenly aware of the centrality of Islamic jurisprudence (*fiqh*) in the proper conduct of social and religious affairs. For a somewhat elementary but nuanced, intelligent, and balanced account of the development of Arabic scientific thought in its proper cultural and religious context, see Danielle Jacquart, *L'Épopée de la science arabe* (Paris: Gallimard, 2005).

46. See Saliba, *Islamic Science*, 78–90 and 186–87. I am taking "Islamic" broadly here to include those disciplines that were considered to be of fundamental service to the Muslim religion even though they were "foreign."

47. The standard justification for including mathematics among the acceptable "Islamic" sciences is the need to apply it to inheritance law.

48. An example of an actively disfavored "science" is sorcery, which, according to Ibn Khaldūn, "consists of directing oneself to the spheres, the stars, the higher worlds, or to the devils by means of various kinds of veneration and worship"; for this, Ibn Khaldūn concludes, "sorcerers should be killed" (Rosenthal, *Muqaddimah*, 392). Because social utility had a pro-

favored because it had no clear, direct bearing on Islamic law or observance, but not actively disfavored because it conflicted with neither.

Princely patronage provided a crucial outlet for the pursuit of philosophical disciplines at, and sometimes beyond, the margins of accepted Islamic learning. From at least the late eighth century, various caliphs, emirs, and viziers adopted the role of patron, inviting renowned scholars to reside at their courts as an earnest of their liberality and erudition. It was under such patronage that Greek philosophical and scientific works were translated in the ninth and tenth centuries. The motivation for patronage was not necessarily intellectual, though. Having a noteworthy scholar at one's court—or, better yet, a covey of such scholars—redounded to the patron's credit and reputation in much the same way as it did in the medieval and Renaissance courts of Europe. There were practical motivations as well. Many court scholars served as personal physicians or astrologers (shades, again, of the European Middle Ages and Renaissance), and as long as they fulfilled their core duties according to need and expectation, they were free to carry out research and writing on whatever subject they pleased, provided that it did not meet with the patron's disapproval. In order to snag a patron, though, a scholar had to establish a reputation, and he had to maintain it while supported by his patron. Consequently, scholarship was a competitive, edgy, sometimes cutthroat business in which practitioners were encouraged to outdo one another in order to find and keep patrons.[49]

It was also Darwinian. Only the fit survived, and in order to survive, scholars had to be abreast of whatever discipline or disciplines they staked their reputation on. They thus formed a community—a relatively small one, to be

---

found effect on the Muslim attitude toward science, medicine and pharmacology were approved and encouraged even though neither discipline is directly related to religious observance or sociopolitical behavior.

49. "While our contemporaries witness the fields of knowledge multiply," the famed astronomer and mathematician al-Bīrūnī (d. 1048) remarks, "and while they are naturally inclined to seek perfection in every science, and while they even succeed, by accrued merits, where the most illustrious of the ancients had failed, we nonetheless find among them behavior that contrasts with what we have just said. A bitter rivalry puts those who are competing in opposition with one another. Quarrels and disputes take them to the point where each envies the other and takes boastful credit for what is not his. Such a person steals the discoveries of others, attributes them to himself, and profits from them, and he would still like one to pretend not to notice; better, whoever denounces his imposture is quickly taken apart and faced with his punishment. That is what we have seen in the midst of an élite group of our contemporaries"; quoted from Jacquart, *Épopée*, 102–3 (my translation).

sure—of intense interest in various subjects, optics included, bound in a network of communication through which texts and ideas passed more or less freely according to political circumstances. It was an ever-shifting community, though. Scholars could easily fall into patronal disfavor, so they had to be ready to pick up stakes at a moment's notice in order to seek preferment elsewhere while fleeing punishment. As a result, scholarship became truly international in the medieval Muslim world because individual thinkers were often forced to wander hither and thither in search of support. In the process, they came into contact with other scholars in far-flung reaches, sharing ideas through discussion and argumentation.

This was an extraordinarily fluid system that encouraged deep learning, agile wits, strength of intellectual purpose, and a powerful drive toward professional and physical self-preservation. It also demanded creativity and innovation in thought. Accordingly, although the study of natural philosophy in the medieval Muslim world occupied fewer scholars than it did in later medieval Europe, it was commensurately deeper, more creative, and more sophisticated. And the same holds by extension for optics. The relative dearth of Arabic texts in that discipline at its highest level reflects not a fundamental lack of interest in or aversion to optics during the Muslim Middle Ages, but the selectively small size of the community of scholars actively engaged in it. It should therefore come as no surprise that the Arabic text of Alhacen's *De aspectibus* is represented by so few manuscripts. What is surprising, perhaps, is that it survives in so many, even if in fragmentary form.

This brief excursus serves to illustrate several major points. First, the difference in educational norms between the Latin West and the Muslim world is not symptomatic of a fundamental, ideological difference between Christianity and Islam at the time. Christianity was no more intrinsically open to freewheeling rationalism than Islam. In both religions unbridled philosophical and scientific curiosity was considered vain and was thus frowned upon. That Christian education eventually found a basis in the liberal arts is due at least as much to historical accident as it is to tolerance for secular learning. Second, at the very moment universities began to develop, there already existed a community of scholars eager for new texts and new learning. Those scholars, moreover, had achieved an adequately high level of intellectual sophistication to absorb that new learning effectively, something that could probably not have been done a century or so earlier.[50] Third, with the subsequent growth of

50. That the twelfth century was a period of intense intellectual and cultural development in Europe was firmly established by Charles Homer Haskins in his seminal *The Renaissance*

universities came an increased demand not only for new texts but also for more copies of them. In response, there soon evolved a thriving publication industry centered on university communities and based on manuscript production rather than print production.[51] In all likelihood, many or most of the extant manuscript copies of Alhacen's *De aspectibus* and its perspectivist offshoots were produced within this university-based industry and were thus destined for use by teachers and students.

All of this points to the fact that by the second half of the thirteenth century the educational system in the Latin West had created a much larger, more centralized marketplace of philosophical and scientific ideas than its equivalent in the Arabic East. The two marketplaces were interconnected through the marketplace of trade goods, and for much of the period between 1150 and 1250, the flow of ideas was predominately from East to West. As time wore on, however, the European marketplace of ideas became increasingly independent and self-sustaining. This move toward intellectual independence was accelerated over the fifteenth and sixteenth centuries by such things as the transition from manuscript to print, the production of relatively cheap paper, improvements in glassmaking, the growth of lay literacy and the concomitant acceptance of vernaculars, new techniques in painting and spatial representation, and the emergence of extrauniversity scientific academies and societies.

Over the course of the study that follows, I will revisit and elaborate on many of the points made in this rough background sketch. For now, though, it is enough to reiterate (perhaps unnecessarily) that modern optics did not evolve between antiquity and the seventeenth century in a social or cultural vacuum. Ideas were at play throughout, to be sure, but they were proposed, defended, adopted, rejected, and modified by people who, being thoroughly human, acted and reacted on the basis not just of the internal logic of those ideas but also of cultural, social, and personal predilections. Habit, intellec-

---

of the Twelfth Century (Cambridge, MA: Harvard University Press, 1927). Drawing in part on this idea, the French medievalist Sylvain Gouguenheim has recently argued that intellectual development in Europe during the thirteenth and succeeding centuries was so firmly rooted in this twelfth-century renaissance as to render the Arabic learning introduced through translated texts irrelevant; see *Aristote au Mont-Saint-Michel* (Paris: Seuil, 2008). As bold as it is misguided, this thesis has aroused considerable controversy in academic circles, above all in France.

51. For a relatively detailed study of this industry at Paris, see Richard Rouse, "The Book Trade at the University of Paris, ca. 1250–ca. 1350," in *Authentic Witnesses: Approaches to Medieval Texts and Manuscripts*, ed. Mary A. and Richard H. Rouse (Notre Dame, IN: Notre Dame University Press, 1991), 259–338.

tual inertia, obduracy, uncritical acceptance, and amour propre were no less integral to that evolution than intellectual curiosity, open-mindedness, and dispassionate reasoning. Yet for all that, the result of this somewhat chaotic and malformed evolutionary process was not a genetic freak but something truly compelling in its elegance and coherence.

CHAPTER 2

The Emergence of Optics as a Science
The Greek and Early Greco-Roman Background

Reconstructing or, more properly, constructing a history of Greek and Greco-Roman optics from extant textual evidence is essentially an exercise in archaeology because most of that evidence consists of fragments, not complete textual artifacts. Worse, virtually all of these fragments have reached us not in original form but in quotations or paraphrases later interpreters provided. Aristotle, for instance, has much to tell us piecemeal about Plato and his pre-Socratic antecedents, but he generally cites them out of context in order to rebut them or to use them in bolstering his own case. Since we are "seeing" those sources through his eyes and from his interpretive perspective, therefore, Aristotle's reliability as a dispassionate witness is clearly suspect. The problem of reliability is compounded by the fact that many of the witnesses who cite earlier thinkers postdate them not, as in Aristotle's case, by decades but by centuries and may therefore be inadvertently misquoting or misreporting. Casting about for corroborating sources or witnesses is of little help because these in turn may, and most likely do, rely on questionable reports or sources.[1]

Then there are the few actual texts on optics or optical matters that have come down to us in more or less complete form from that period. Among them, the pertinent works of Plato, Aristotle, and Galen are relatively but by no means entirely unproblematic. Far more problematic are Euclid's *Optics* and *Catoptrics*, Hero of Alexandria's *Catoptrics*, and Ptolemy's *Optics*, all of them

---

1. For a brief discussion of these source problems, see A. Mark Smith, *Ptolemy and the Foundations of Ancient Mathematical Optics* (Philadelphia: American Philosophical Society Press, 1999), 11–19.

of questionable authenticity and/or ascription. The *Catoptrics* now attributed to Hero, for example, exists only in a thirteenth-century Latin translation and is ascribed there not to Hero but to Ptolemy.[2] Even if we suppose that current attributions are correct, we have no way of determining whether the text itself is original or a redaction. Euclid's *Optics* may thus be Euclidean, but it may not be Euclid's. This certainly seems to be the case with his *Catoptrics*, which has been subject to considerable debate over how much of it, if any, is genuinely Euclid's or even Euclidean.

Forming a coherent historical narrative from such a hodgepodge of textual snippets is therefore much like piecing a few scattered pottery shards together into some recognizable shape, with most of them either missing or known only through description. Depending on how the pieces are assumed to fit according to interpreted shape and size, the result may be an amphora, a krater, or whatever other form of pottery seems suitable. The possibilities are virtually endless, and, so it seems, are the efforts to interpretively reshape the various pieces through close analysis of their minute linguistic features and their possible interrelated patterns. No less endless, therefore, are scholarly reinterpretations of these textual fragments and debates over the fine points and broad implications of those reinterpretations.

Insofar as possible, I will skirt these debates by wielding Ockham's razor when it comes to the finer points of attribution, dating, and interpretation. Hence, while mindful of these points, I will accept traditional attributions and dates unless there is compelling reason to do otherwise. Admittedly, in taking this tack, I will sacrifice nuance for simplicity and clarity, but I do so in order to keep my narrative within reasonable bounds. The primary focus of that narrative, moreover, will be on conceptual rather than chronological development, not that the two are necessarily mutually exclusive. Whether actually Euclid's or even Euclidean, the *Catoptrics* attributed to him manifests a lower level of sophistication than the *Catoptrics* attributed to Hero, which is in turn manifestly less sophisticated than Ptolemy's *Optics*. It is not unreasonable, then, to suppose that these works represent stages of development that are ordered in rough chronological sequence, if not in exact temporal succession according to the traditional dating for Euclid (early third century BC), Hero (ca. AD 10–ca. AD 70), and Ptolemy (ca. AD 100–ca. AD 180), this latter serving as a conceptual terminus ad quem for the entire account and thereby giving it shape. The same holds for the rest of the sources upon which I will draw, most of them of equally questionable authenticity and attribution. Still, tenta-

2. See note 92 below.

tive though it may be, the resulting narrative does have a few relatively secure chronological and interpretive anchor points in Plato, Aristotle, and Galen.

1. EARLY INTIMATIONS

There is a solid kernel of truth in the idea that naming something brings it into existence. This, of course, is an epistemological rather than an ontological creation, yet it is a creation nonetheless. By naming things we make them definite and, in the process, fit them into taxonomic niches that distinguish them from everything else, be they concrete or abstract. But the very act of naming also imposes the task of determining and defining what it is that makes the named thing what we take it to be. This task is crucial because the effort to define something as clearly and thoroughly as possible invariably transmutes it into something else that is genetically related to, yet different—sometimes radically different—from what it was before.

Taken in a general sense, this book is about "optics," which is as much a name as a thing. More specifically, it is about how that thing was transformed, as its definition was refined and modified over the course of some two millennia. Etymologically, the name "optics" comes from the Greek *optika*, which in turn derives from the verb *opteuō* ("to see"), both words related to the noun *ōps* ("eye"). In its original Greek sense, then, optics has to do with seeing. But we can be more specific than that. In the second book of the *Physics*, Aristotle (384–322 BC) adverts to "the more natural (*ta phusikōtera* = 'physical') of the branches of mathematics, such as optics, harmonics, and astronomy." These mathematical disciplines, he explains, "are in a way the converse of geometry [for] while geometry investigates natural lines but not *qua* natural (*phusikē*), optics investigates natural lines, but *qua* natural, not *qua* mathematical."[3] This statement can be taken in at least two ways. On the one hand, Aristotle may be drawing a distinction between a purely imaginary line—a one-dimensional "breadthless length" in Euclidean parlance—and a physical line in "space" with at least two dimensions (breadth and length) or at most three (breadth, depth, and length). In this interpretation, the mathematical line would be an idealization of its natural counterpart and would thus be an analytic fiction when applied to physical reality. On the other hand, Aristotle may mean that mathematical lines really do exist in physical reality and that we abstract them

3. *Physics*, 2, 2, 194a7–12, trans. R. P. Hardie and R. K. Gaye, in *The Complete Works of Aristotle*, ed. Jonathan Barnes (Princeton, NJ: Princeton University Press, 1984), 331. See also *Posterior Analytics*, 1, 13, 78b34–79a15, trans. Jonathan Barnes, in Barnes, *Complete Works*, 114–66.

intellectually. In this interpretation, the mathematical line is not so much an analytic fiction as an actual representation of what obtains in physical reality. We will return to this issue somewhat later.

A brief look at Aristotle's explanation of the rainbow in chapters 2–6 of the third book of the *Meteorology* gives further insight into his conception of optics. The gist of this explanation is as follows. Let semicircle SAP in figure 2.1a lie on a great circle of a hemisphere in the plane of the paper, let SP be a diameter of that hemisphere, and let K be its center. Locate point D on that diameter between S and K and point B on the extension of that diameter such that SD and DK are in the same ratio as SB and BK, that is, SD:DK = SB:BK (or, taken quite loosely, SD / DK = SB / BK). Find point M on semicircle SAP and connect lines SM and MK such that SM:MK = SB:BK. Aristotle says nothing about the actual lengths of those lines, but he does tell us that in order to locate M, we must find midpoint C of line DB and construct a semicircle on that point with DB as a diameter. M will be the point of intersection of that semicircle with semicircle SAP. Now assume that M is a point on a raincloud whose surface is defined by the hemisphere that contains semicircle SAP and that SB lies in the plane of the horizon. Assume, further, that an observer stands at point K and that the sun lies at point S. Accordingly, Aristotle invites us to "let KM be reflected to [S]" along M[S].[4] Drop a perpendicular from point M to point O on horizon line SB, and form semicircle NMQ on O with radius OM. This is equivalent to the arc that would be described by point M on the hemisphere that contains semicircle SAP if triangle SMK were rotated about SB as an axis.[5]

Arc NMQ, or rather its equivalent on the hemisphere, will therefore be an arc on a rainbow seen by the observer at K, and so "it is clear that the rainbow is a reflection of sight (*opseōs*) to the sun."[6] The rainbow itself is the composite image of the sun seen in individual water droplets within a band on the hemispheric cloud containing that arc, and because the droplets are so small, the resulting image in each is not of the sun itself but of its apparent color at various points within that band.[7] Both the color and its brightness depend on certain conditions, such as the relative darkness of the cloud in which the

4. *Meteorology*, 3, 5, 375b30, trans. E. W. Webster, in Barnes, *Complete Works*, 604. I have adjusted the order of Aristotle's analysis in order to make it simpler and clearer.

5. For the complete text of this geometrical account, see *Meteorology*, 3, 5, 375b16–376b22, in Barnes, *Complete Works*, 604–5.

6. Ibid., 3, 4, 373b34, in Barnes, *Complete Works*, 602. The term *opseōs* is the genitive singular of *opsis*, meaning "sight"; it is also the term used later for "visual rays."

7. See ibid., 373a32–374a3, in Barnes, *Complete Works*, 602–3.

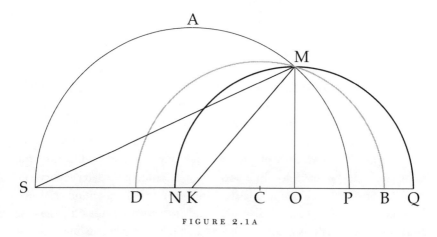

FIGURE 2.1A

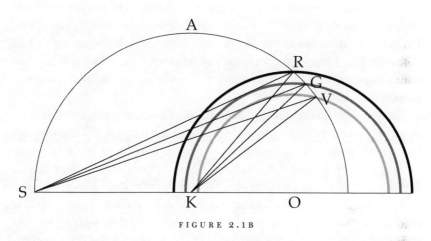

FIGURE 2.1B

rainbow appears. But there is a constant pattern: in the primary rainbow, the color bands range from red at the top, to green in the middle, and to violet at the bottom. In the secondary rainbow, which lies above the primary one and is considerably fainter, the colors are reversed. The threefold pattern of colors seen in the primary rainbow depends on the ratio of the incident and reflected rays.

In figure 2.1b, for instance, red will appear at point R according to the ratio KR:RS, green according to the ratio KG:GS, and violet according to the ratio KV:VS, the result being three concentric arcs as seen straight on from K along horizon line SKO. Another factor determining the apparent colors within the rainbow is the weakening of sight caused by viewing the sun reflected in the cloud. Strong sight will yield red, weaker sight green, and weakest violet. Since sight weakens with distance, ray couple KRS in figure 2.1b will yield red because it is the shortest distance for sight to follow, whereas ray couple RVS will yield violet because it is the longest.[8]

Neither the technical ramifications of this explanation nor its manifold shortcomings need detain us. Its importance lies in what it tells us about the state and nature of optics as conceived around the middle of the fourth century BC and about the line of development this science would follow for a long time afterward. First of all, it strongly suggests that by this time the science of optics was, if not fully born, then certainly in its later embryonic stages. Aristotle's explanation is fairly complex and involves a number of analytic and empirical factors. It also tells us that the mathematical lines at the heart of optical analysis are rectilinear and, in that form, define lines of sight passing outward from the eye. In short, they are visual rays, and they rebound from reflecting surfaces, such as water droplets in clouds, in a particular way. Furthermore, it tells us that such reflection is mathematically determined by a specific ratio between the incident and reflected rays, which is to say that the process of reflection follows a precise law, albeit a speciously precise and patently invalid one.[9]

As of the mid-fourth century, then, the science of optics, as inferred from Aristotle's account of the rainbow, had as its primary goal to explain why and how we see certain things as we do, and it appears to be based on the following three principles: (1) that sight is due to visual rays reaching out in perfectly straight lines from the eye to visible objects; (2) that these rays are broken and diverted by certain kinds of objects that can be defined as reflective; and (3) that the visual process is mathematically determined not only by the rectilinearity of the visual rays themselves but also by certain specific rules governing their interaction with physical bodies.[10] Implicit in Aristotle's account of the rainbow, these principles will be explicitly articulated and refined in

---

8. See ibid., 374a4–375b8, in Barnes, *Complete Works*, 601–4.

9. As was recognized by no later than the end of the fourth century BC, reflection is governed by the equal-angles law, which holds that the angle of incidence is equal to the angle of reflection. In Aristotle's account, visual ray KM is incident to the semicircle at 90° no matter where M might be situated, since KM is a radius of that semicircle, so it should reflect back along itself at an equal angle rather than along MS.

10. In "Archytas and Optics," *Science in Context* 18 (2005): 35–53, Miles Burnyeat argues that mathematical optics, as a formal science based on these principles (including the equal-angles law of reflection), can be traced back to Archytas of Tarentum in the early fourth century BC. This, however, is a highly controversial claim based not on textual but contextual evidence.

the Euclidean *Optics* and *Catoptrics*. But before turning to those works, we should take a brief look at various nonmathematical theories of vision that were proposed in Greek antiquity and that may provide a conceptual context for the visual ray theory articulated by Euclid and subsequently brought to culmination by Ptolemy.

2. PHYSICAL AND PSYCHOLOGICAL THEORIES OF VISION

Let us start with a few preliminary generalizations. First, as mentioned earlier, any effort to trace the development of visual theory among ancient Greek thinkers from the pre-Socratics to their Greco-Roman successors is hamstrung by a lack of primary sources. As a result, we are forced to rely on secondary or tertiary sources that are often centuries removed from the events, personalities, and ideas they describe. Second, on the basis of the sources we do have, it is fairly safe to say that Greek visual theorists all accepted that the visual act requires mediation. Convinced that vision occurs neither by direct contact between eye and object nor by action at a distance, they assumed that something has to intervene in order to link the two and thus initiate the visual act. Third, such intervention was accounted for in one of two fundamental and contradictory ways: according to either an extramission of something from within the eye out to visible objects or an intromission of something from visible objects into the eye.

The visual ray theory upon which Aristotle based his explanation of the rainbow obviously falls within the extramissionist camp, and it probably originated in the notion that in order to see, the eye has to emit its own light. In the *De sensu et sensibilibus* ("On Sense and the Sensibles"), for instance, Aristotle tells us that Empedocles (499–432 BC) likened the eye to a lantern, "a gleam of fire blazing through the stormy night, adjusting thereto . . . the transparent sides, which scatter the breath of the winds . . . while, out through them leaping the fire." Just so, "the primaeval fire, fenced within the membranes and delicate tissues [of the eye] fended off the deep surrounding flood, while letting through the fire" that makes the surroundings visible.[11] Similarly, the interlocutor Timaeus of the eponymous Platonic dialogue describes a fire within the eye that does not burn but yields a "gentle light." The eye itself, according to him, "its middle part in particular, [is] close-textured, smooth, and dense, to enable [it] to keep out all the other, coarser stuff, and let that kind of fire

---

11. *De sensu*, 2, 437b26–438a 3, trans. J. I. Beare, in Barnes, *Complete Works*, 695.

pass through pure by itself."[12] However, as Aristotle points out, Empedocles was not consistent in his extramissionism; "sometimes he accounts for vision thus," Aristotle cautions, "but at other times he explains it by emanations from the visible objects."[13] Nor is the theory described in the *Timaeus* unequivocally extramissionist inasmuch as Timaeus argues that "whenever daylight surrounds the visual stream, like makes contact with like and coalesces with it to make up a single, homogeneous body aligned with the direction of the eyes."[14] The phrase "aligned with the direction of the eyes" suggests that the emanation of visual fire from the eye was understood to be rectilinear.

Perhaps the clearest example of the intromissionist alternative is to be found in the theory of Epicurus (341–270 BC). An atomist within the Democritean tradition, Epicurus reportedly argued that "particles are continually streaming off from the surface of bodies. . . . And those given off for a long time retain the position and arrangement which their particles had when they formed part of the solid bodies."[15] The resulting atom-thick shells of the object form replicas (*eidōla*) that enter the eye to be visually apprehended. The Roman, Lucretius (ca. 94–55? BC), followed Epicurus's lead in claiming that these replicas (*simulacra*) are "carried about and thrown off scattered abroad in all directions [so that] when we turn our sight, there all things strike it with shape and colour."[16] According to Aristotle's student, Theophrastus (ca. 371 BC–ca. 287 BC), Empedocles had something similar in mind when he asserted that "perception occurs because something fits into the passages of the particular [sense organs]," which in the eye "are arranged alternately of fire and of water: by the passages of fire we perceive white objects; by those of water, things black; for in each of these cases [the objects] fit into the given [passages]."[17] Interestingly enough, the theory of Democritus (b. ca. 460 BC), who is considered to be the co-founder of atomism with Leucippus (fifth century BC), is less crassly particulate than that of his disciples Epicurus and Lucretius. If Theophrastus is to be credited, Democritus believed that the object compresses the surrounding air so that it "becomes imprinted (*tupousthai*), since there is always an efflu-

12. *Timaeus*, 45b–c, trans. Donald Zeyl, *Plato*: Timaeus (Indianapolis: Hackett, 2000), 33.

13. *De sensu*, 2, 438a 4–5, in Barnes, *Complete Works*, 695.

14. *Timaeus*, 45c, in Zeyl, *Plato*, 33.

15. Diogenes Laertius, *Lives of the Eminent Philosophers*, vol. 2, trans. R. D. Hicks (Cambridge, MA: Harvard University Press, 1959), 577–79.

16. *De rerum natura*, 4, 239, in William H. D. Rouse and M. F. Smith, trans., *Lucretius: De rerum natura* (Cambridge, MA: Harvard University Press, 1982), 297.

17. Theophrastus, *On the Senses*, 7, in George M. Stratton, trans., *Theophrastus and the Greek Physiological Psychology before Aristotle* (London: Allen and Unwin, 1917), 71–73.

ence of some kind arising from everything." The resulting impression in the air "is reflected in the eyes," and as Theophrastus sums it up, "Democritus himself, in illustrating the character of the 'impression,' says that 'it is as if one were to take a mould in wax.'"[18] For Democritus, then, the mediating entity in vision seems not to be a particulate replica of the object, but an impression in the air that replicates that replica.

Certainly the most extensive, and perhaps the most coherent and penetrating, account of vision in antiquity was given by Aristotle in the *De anima* ("On the Soul") and the *De sensu*. In certain key respects, that account bears a distinctly Democritean imprint. Like Democritus, Aristotle placed the stress on air as the primary mediating entity in the visual process; like Democritus, he fell back on the notion of imprinting; and like Democritus, he took a decidedly intromissionist stance. This stance, of course, is utterly inconsistent with his extramissionist account of the rainbow.[19] Unlike Democritus, however, Aristotle based his theory on a formal or qualitative transformation of, rather than a physical impulse through, the aerial medium. Perhaps the best way to approach Aristotle's theory of sight is according to his notion of potency and act and his fourfold distinction of final, material, efficient, and formal causes. Not surprisingly, Aristotle locates the final cause of vision—the "that for the sake of which"—in completion of the act itself, so the entire visual process is driven toward that goal. The formal cause of seeing is color. This, Aristotle insists, is the sole proper sensible for vision and is thus the only thing that is seen per se, not shape, size, distance, or any other associated characteristic that we reflexively take as visible.[20]

Color, for its part, is an intrinsic quality of visible objects and, as such, renders them potentially visible. In order to see color, however, the eye must be linked to visible objects by a continuous transparent medium, such as air or water, through which the color can be transmitted to it. Such media are not actually transparent by their nature, though; they are only potentially so. In order to become actually transparent, they must be exposed to light, so light is the formal cause of transparency in these media. Likewise, the eye has the potential to see, but it can fulfill that potential only if it faces a colored object

18. Theophrastus, *On the Senses*, 50, in Stratton, *Theophrastus*, 108–11.

19. The clear contradiction between Aristotle's extramissionist account of the rainbow and his intromissionist account of vision in the *De anima* and *De sensu* is puzzling, but his rejection of extramissionism in *De sensu*, 2, 438a26–28, is unequivocal: "It is, to state the matter generally, an irrational notion that the eye should see in virtue of something issuing from it; that the visual ray should extend itself all the way to the stars . . . as some say" (Barnes, *Complete Works*, 696).

20. See *De anima*, 2, 6, 418a13, trans. J. A. Smith, in Barnes, *Complete Works*, 665.

lying in an actually transparent medium. The eye's capacity to see is also dependent on its being properly integrated into a living, ensouled body. If the eye were an animal, Aristotle explains, "sight would be its soul, for sight is the substance or essence of the eye [and so] when seeing is removed the eye is no longer an eye, except in name—it is no more a real eye than the eye of a statue or a painted figure."[21]

So far we have identified the final cause of seeing in the act itself, the formal cause in color, and the material cause in an appropriate succession of continuous physical bodies from the colored object, through the transparent medium, to the eye. Light, as we have seen, is a secondary formal cause serving as a sort of catalyst to transform potential transparency to actual transparency. In that regard, it can be thought of as "the proper colour of what is transparent"; as such, it "exists whenever the transparent is excited to actuality by the influence of fire or something resembling 'the uppermost body.'"[22] But if color is the formal cause of sight, it is also the efficient cause, for by its very nature it transforms any continuous transparent medium in contact with it from transparent to colored. And it does so instantaneously. Both the visible object and the transparent medium thus provide a material substrate in which the color can be formally instantiated, yet the color's status in each is radically different. In the object it is intrinsic and thus integral; in the medium it is extrinsic and thus evanescent.[23] Likewise, in the eye the color is "accidental" in two respects: according to its presence on the cornea as a reflected image and according to its presence in the eye as a proper object for vision. In this latter guise, it makes an impression that can be understood by analogy to the imprint created in very soft wax by a seal, so it affects the eye "just as if the impression on the wax were transmitted as far as the wax extends."[24] It is important to remember that this is just an analogy and that Aristotle does not mean to imply that the impression is really mechanical or incised into the surface of the eye the way an imprint in wax is.

For Aristotle the visual process is not limited to the apprehension of color

<hr/>

21. Ibid., 2, 7, 412b18–22, in Barnes, *Complete Works*, 657.

22. Ibid., 418b11–13, in Barnes, *Complete Works*, 666. The uppermost body to which Aristotle adverts is the aither, of which the celestial (i.e., supralunar) realm is composed. In associating light with it, Aristotle evidently meant to emphasize the quintessential and thus transcendent nature of light.

23. By intrinsic and integral I mean that it is a definitive property of that particular visible object as *that* object and no other. By extrinsic and evanescent I mean that it has no definitive bearing on the transparent medium as transparent.

24. *De anima*, 3, 12, 435a10, in Barnes, *Complete Works*, 691.

by the eye because sight forms part of a more general system of sensation that includes touch, taste, hearing, and smell. Like sight, each of these senses has its unique special sensible to which the other senses are impervious; we obviously do not see sweetness or hear redness. Accordingly, in eating an apple, we may sense its redness separately through sight, its sweetness through taste, and its hardness and smoothness through touch. Nevertheless, as we eat the apple, we associate all these sensations to get a combined impression of red, sweet, hard, smooth, and so on. This associative process is carried out in the so-called common sensibility (*aisthēsis koinē*), which filters the sense impressions unique to each sense and associates them with the object from which they originate. It also filters sense impressions that are common to more than one sense. These include such things as shape, size, motion, rest, and number, which are common to sight and touch—as well as to hearing in the case of motion, rest, and number.[25] Still, these are not properly sensible in the way that the special sensibles are because they are perceptually dependent upon them. The common sensibility must therefore abstract them from the primary sense impressions that convey them. Out of all these special and common sensibles is formed a single representation or image (*phantasma*) of the object in the imagination (*phantasia*). There it stands for the object when it is no longer present, so the imagination serves as a sort of mnemonic storehouse of images that represent particular objects that have been previously sensed.[26]

It is also at the level of the imagination that we make judgments about what we perceive, concluding, for instance, that what we see is actually red, round, hard, smooth, and so forth. In these judgments we can be mistaken, even though the senses themselves never or very rarely lie.[27] Hence, the imaginative faculty functions not only in perceiving the sense data but also in judging what they are and what objects they belong to. Aristotle offers as an example our perception of the sun as "a foot in diameter," which is not merely a perception but also a judgment. Intellectually, however, we know this judgment to be false

25. See ibid., 3, 1, 425a14–425b4, in Barnes, *Complete Works*, 676.

26. See ibid., 3, 3, 427b23–429a9, in Barnes, *Complete Works*, 680–82. In distinguishing sensation of the special sensible from perception of the common sensibles based on that sensation, I am following what Pavel Gregoric calls "the standard view," which views the common sense as of a higher order of sensibility than the individual senses. This, however, is by no means an unproblematic interpretation of Aristotle's conception of the common sense and the common sensibles. For some discussion of this and other interpretations, see Gregoric, *Aristotle on the Common Sense* (Oxford: Oxford University Press, 2007); see also Joseph Owens, "Aristotle on Common Sensibles and Incidental Perception," *Phoenix* 36 (1982): 215–36.

27. *De anima* 3, 3, 428b18–24, in Barnes, *Complete Works*, 681.

because we understand that the sun is actually "larger than the inhabited part of the earth."[28] But perceptual judgment does not end here; in apprehending the special and common sensibles, we also apprehend incidental things, such as the fact that the white thing before us is Cleon's son.[29] Of this sort, apparently, is our perception that the bright thing facing us in the sky, which looks to be a foot in diameter, is the sun rather than the moon. This kind of perceptual judgment is the most subject to error.

Perception, however, is not thinking or knowing, even though it seems to involve judgments, for these judgments are reflexive and involuntary. We cannot help perceiving the sun as a foot in diameter, even though we know it to be much larger than that. Thinking, on the other hand, is voluntary, and although it is based on the particular perceptual representations in the imagination, the conclusions it draws from them are perfectly abstract and general. When correctly drawn, such conclusions constitute actual knowledge for Aristotle, and upon reaching such a state of knowledge, the mind "becomes" what it knows by a sort of assimilation.[30] In that respect, its potential for knowing is actualized, and having the potential to know everything knowable, the mind is theoretically capable of becoming everything knowable. The knowable itself is the form (*eidos*), and as Aristotle puts it, "the faculty of thinking . . . thinks the forms (*eidē*) in the images (*phantasmata*)."[31] Abstracted conceptually from the physical or perceptual particulars represented in the imagination, these forms constitute what Aristotle calls "universals" (*ta kathalou*). Thus it is that "the mind when it is thinking the objects of mathematics thinks of them as separate [from physical objects] though they are not separate at all."[32] Still,

28. Ibid., 428b3, in Barnes, *Complete Works*, 681.

29. See ibid., 3, 1, 425a25, in Barnes, *Complete Works*, 676. Aristotle makes this same point earlier in *De anima*, 2, 6, 418a20–22, when he says that "we speak of an incidental object of sense where e.g. the white object which we see is the son of Diares; here because being the son of Diares is incidental to the white which is perceived, we speak of the son of Diares as being incidentally perceived" (Barnes, *Complete Works*, 674).

30. See *De anima*, 3, 4, 429b6–9 and 3, 4, 430a14, in Barnes, *Complete Works*, 682–84. This process of assimilation also occurs in sensation: "What has the power of sensation is potentially like what the perceived object is actually; that is, while at the beginning of the process of its being acted upon the two interacting factors are dissimilar, at the end the one acted upon is assimilated to the other and identical in quality to it" (ibid., 3, 5, 418a3–6, in Barnes, *Complete Works*, 665).

31. Ibid., 3, 7, 431b2, in Barnes, *Complete Works*, 686.

32. Ibid., 431b15–16, in Barnes, *Complete Works*, 686.

transcendent though it may be, thinking is not autonomous or unconnected from perception. "To the thinking soul," Aristotle explains, "images serve as if they were contents of perception [which] is why the soul never thinks without an image."[33] Hence, as Aristotle sums it up, "no one can learn or understand anything in the absence of sense, and when the mind is actively aware of anything it is actively aware of it along with an image; for images are like sensuous contents except in that they contain no matter."[34]

From this brief account it should be obvious that Aristotle did not conceive of vision as a single, simple act. Instead, he viewed it as a complex process that can be analyzed according to four basic phases. The first of these involves the physical transmission of color through an appropriately transparent and continuous medium to the eye. The second involves the sensitive apprehension of that color by the eye, which is capable of apprehending it that way by virtue of its integration into a living, ensouled body. Otherwise, it would be physically but not sensibly affected by the color, just as the eye of a statue can be physically colored without seeing. The third stage is perceptual.[35] It is at this stage that the special and common sensibles passed from each sense to the common sensibility are combined into a unified representation of the object and remanded to the imagination for mnemonic storage. It is also here that we recall the images associated with previous thinking through the process of recollection (*anamnēsis*).[36] During the fourth and final stage of perception, which is intellectual or conceptual, the mind generalizes from the perceptual representations in the imagination to reach a perfectly abstract, formal grasp of what they represent "deep down." At this level we acquire a generalized understanding of, for example, "human" from the imagined representation of a specific man, such as Cleon's son, which in turn arises from our perception of a particular white thing. At bottom, then, seeing is only an interim final cause of vision for humankind. The true final cause—the final, final cause, so to speak—is knowing.

33. Ibid., 3, 7, 431a16–17, in Barnes, *Complete Works*, 685.

34. Ibid., 3, 8, 432a6–9, in Barnes, *Complete Works*, 687.

35. It should be noted that Aristotle does not have a special term for *perception* to distinguish it from *sensation*; I am using it here simply to indicate that this phase of sensation is more abstract and complex than the antecedent phase during which only the special sensible is apprehended.

36. For Aristotle's discussion of recollection, see *De memoria et reminiscentia* ("On Memory and Recollection"), trans. J. I. Beare, in Barnes, *Complete Works*, 14–20.

3. THE ANATOMICAL AND PHYSIOLOGICAL
GROUNDS OF VISION

One feature of all the visual theories discussed to this point is that they have little or no basis in the anatomical or physiological structure of the eye or the optic system as a whole. To be sure, issues were raised about what makes the eye percipient according to either its elemental composition or its physical structure. Empedocles, as we saw earlier, is said to have traced its percipience to the fiery and watery composition of the passages in the eye that allow certain kinds of particles to enter it and stimulate vision. Democritus focused on the eye's watery nature, which makes its front surface highly reflective and thus susceptible to visual impression. Aristotle agreed with Democritus that water is a crucial component of the eye but argued that its importance lay in the fact that water is not only transparent but also "more easily confined and more easily condensed than air; [so] the pupil (*korē*), that is, the eye proper, consists of water."[37] As Theophrastus reported, Diogenes of Apollonia (fifth century BC) thought that air was the critical elemental component because "he [connected] it with both life and thought [and thus opined that] sight arises when objects are reflected in the pupil, but it [i.e., the reflection or rebound] occasions perception only when mingled with the internal air [for] if the ducts become inflamed, there is no union with the internal [air], and sight is impossible."[38] These ducts would seem to be the optic nerves, in which case the internal air with which union had to be made would have been in the brain. It was apparently to these ducts that Aristotle was adverting when he observed that "it is a matter of experience that soldiers wounded in battle by a sword slash to the temple, so inflicted as to sever the passages of the eye, feel a sudden onset of darkness, as if a lamp had gone out."[39] Finally, there were those, like Plato, who sought the ultimate explanation in a gentle fire suffusing the eye from within.

In order to get beyond these vague hints to a truly systematic anatomical and physiological account of vision equivalent in depth and sophistication to Aristotle's psychological account, we need to leap ahead several centuries to Galen (AD 129–216?). Opinionated, argumentative, and blessed with a

---

37. *De sensu*, 2, 438a15–16, in Barnes, *Complete Works*, 696. In normal usage among ancient Greek optical thinkers, the term "pupil" (*korē*) denotes the central portion of the cornea rather than the aperture in the iris.

38. Theophrastus, *On the Senses*, 39–40, in Stratton, *Theophrastus*, 101.

39. *De sensu*, 2, 438a11–13, in Barnes, *Complete Works*, 696.

probing intellect, Galen had a strong philosophical predisposition coupled with an equally strong empirical bent. Together these made him a formidable medical theoretician and an equally formidable anatomist who insisted on testing theory with anatomical fact. Convinced that anatomical form and physiological function are perfectly symbiotic, Galen was equally convinced that the human body and its functions are ordered as rationally and efficiently as possible. Time and again he marvels at Nature's foresight in crafting things for the best. Galen, in short, was a firm believer in intelligent design. This conviction informed his functionalist account of the optic system in books 8–10 of the *De usu partium corporis humani* ("On the Usefulness of the Parts of the Human Body") and the seventh book of the *De placitis Hippocratis et Platonis* ("On the Opinions of Hippocrates and Plato").[40]

Key to that account is the linkage between the eye and the brain. Structurally, the Galenic brain is encompassed by two membranes, the harder of which, the *dura mater*, serves both to contain the brain and to provide a layer of protection within the skull itself. Inside the dura mater, the softer, thinner *pia mater* provides a secondary layer of protection. Aside from safeguarding the brain, the dura mater serves as the source of the motor nerves, starting at the brain stem, whereas the softer sensory nerves have their source in the pia mater.[41] The body of the brain contains four hollow, interconnected cells or ventricles ranged from front (forehead) to back (occiput). Two of these are paired laterally in the anterior portion of the brain, where the sense nerves are all channeled into it.[42] Emerging from these laterally paired ventricles, the optic nerves form hollow tubes sheathed on the outside by a tougher, outer layer springing from the dura mater and on the inside by a softer, inner layer springing from the pia mater. Continuing out from the brain and toward each other, the two optic nerves merge at the optic chiasma and then branch out again toward the eye sockets in the shape of an X with the optic chiasm at its intersection.[43] After penetrating the openings in the eye sockets, the two nerves

40. For a wide-ranging, though at times muddled and anachronistic, account of Galen's theory of vision and its historical context, see Rudolph E. Siegel, *Galen on Sense Perception* (Basel; New York: Karger, 1970), especially 40–126.

41. See *De usu partium*, 8, 9, trans., Margaret Tallmadge May, *Galen on the Usefulness of the Parts of the Body* (Ithaca, NY: Cornell University Press, 1968), 409–12.

42. See *De usu partium*, 8, 10–13, in May, *Usefulness*, 412–18. See also book 9 of Galen's *On Anatomical Procedures* in Charles Singer, trans., *Galen on Anatomical Procedures* (London: Oxford University Press, 1956). For a detailed overview of Galen's anatomical and physiological understanding of the brain, see Julius Rocca, *Galen on the Brain* (Leiden: Brill, 2003).

43. See *De usu partium*, 10, 12, in May, *Usefulness*, 491.

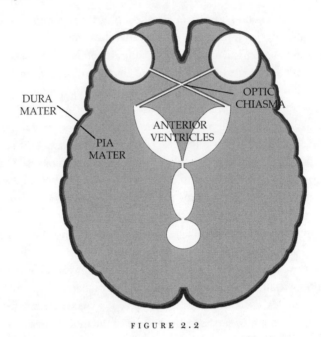

FIGURE 2.2

funnel out to form the eyeballs, which are paired so that, if one is injured or destroyed, the other will still function.[44] This sort of duplication, Galen points out time and again, is a mark of the intelligence with which the human body is designed.[45] Figure 2.2 offers a very rough schematic representation of the cerebral system described to this point, as viewed in cross section from below.

Like the optic nerves, the eyes that grow out of them consist of nesting sheaths or tunics. The outermost sheath, which contains the eyeball as a whole and emerges from the outer sheath of the nerve, constitutes the scleral or "hard" tunic (*chitōn sklēros*). Nesting inside this tunic is the choroid or "afterbirth-like" tunic (*chitōn choroeidēs*), inside of which is couched a netlike membrane (*amphiblēstroeidēs*)—the retina—which Galen does not consider to be a true tunic.[46] Along with the retina, the scleral and choroid tunics are attached to the lens along its "equator" so that half of it protrudes out in front. The lens itself forms an oblate sphere.[47] From the point where the sclera intersects the lens, the thin, hard, and transparent cornea (*keratoeidēs*) extends over

44. See ibid., 8, 10, in May, *Usefulness*, 413.
45. See ibid., 10, 1, in May, *Usefulness*, 463.
46. See ibid., 10, 2, in May, *Usefulness*, 465–69.
47. See ibid., 10, 6, in May, *Usefulness*, 479.

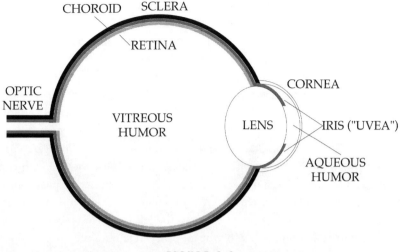

CHOROID   SCLERA

RETINA

OPTIC
NERVE

VITREOUS
HUMOR

CORNEA

LENS

IRIS ("UVEA")

AQUEOUS
HUMOR

FIGURE 2.3

the front portion of the lens to protect it from external irritants.[48] A second, flimsier protection is provided by a partial extension of the choroid tunic over the front of the lens that forms what we now refer to as the iris. This protective extension is perforated at the center by the pupil.[49] The lens itself consists of a "moderately hard," highly transparent gel called crystalline humor (*krustalloeidēs*) because of its ice-like, crystal clarity. The large cavity behind the lens is filled with vitreous humor (*hualoeidēs*), so called because of its glass-like appearance, and the small basin between the lens and the cornea is filled with a clear liquid (*ōoeidēs*) that has the consistency and transparency of egg white. Today we refer to this as aqueous humor.[50] So formed, the eyeball is encased in the conjunctiva, which is attached to fat that holds the eye in place in its socket and muscles that move it up and down and from side to side. The lids and lashes provide yet another layer of protection.[51] Figure 2.3 gives a cross-sectional view of the eyeball and its main components as described by Galen.

Although it is physically positioned toward the front of the eye, the lens is functionally central because, according to Galen, it is the seat of visual sensation. The rest of the eye, with all its tunics and humors, is therefore designed to

48. See ibid., 10, 3, in May, *Usefulness*, 469–72.
49. See ibid., 10, 3–4, in May, *Usefulness*, 472–75.
50. See ibid., 10, 1 and 4, in May, *Usefulness*, 464–65 and 475.
51. See ibid., 10, 6, in May, *Usefulness*, 480.

facilitate this function by protecting and nourishing the lens, which is uniquely suited to its role not only because of its "clear, radiant, gleaming, and pure" nature, but also because of its location in the optic system as a whole.[52] As we have seen, that system originates anatomically in the brain with its ventricles. Physiologically, it is based on a vitalizing mixture of fire and air called pneuma (*pneuma*), which is the animating principle of all living things. The cerebral ventricles are continually suffused with a highly refined and rarefied modification of this principle in the form of psychic pneuma (*pneuma psuchikon*), which is distilled from blood flowing from the left ventricle of the heart to the brain through the carotid arteries.[53] This blood is already infused with vital pneuma (*pneuma zōtikon*) from the inhaled air that supposedly passes from the lung into the left ventricle through the pulmonary vein and mixes with blood that has seeped from the right into the left ventricle via microscopic channels in the septum. Reaching the so-called rete mirabile, the "wonderful network" of tiny blood vessel at the base of the brain, the vitalized blood conveyed through the carotid arteries circulates through that network and is thereby distilled into psychic pneuma.[54] This is the medium through which the ruling principle (*hēgemonikon*), the motivating "self" of the body, exercises its perceptual, intellectual, and motor functions.[55]

In the case of vision, the ruling principle exercises its function by means of the psychic pneuma that flows through the hollow optic nerves from the forefront of the brain through the eye to the anterior surface of the cornea. There it comes into contact with the surrounding air. If vision is actually to occur, that air must be appropriately pneumatized by light so as to become "an instrument of vision of the same description as the pneuma coming to it from the brain." Hence, when the psychic pneuma from the eye makes contact with the pneumatized air, the resulting medium becomes "a sympathetic instrument by virtue of the change effected in it by the outflow of pneuma."[56] In essence, then, the transformed air becomes a percipient extension of the ruling principle, putting it in visual contact with external objects. The actual visibility of those objects is a function of their inherent color, "for colors are the first thing [the

52. Ibid., 10, 1, in May, *Usefulness*, 464.

53. See ibid., 10, 4, in May, *Usefulness*, 430–34.

54. See *De placitis*, 7, 3.19–29, trans. Phillip De Lacy, *Galen: On the Doctrines of Hippocrates and Plato*, Corpus Medicorum Graecorum, 4, 1, 2 (Berlin: Akademie Verlag, 1980), 443–47.

55. See ibid., 8, 1.1–24, in De Lacy, *Galen*, 481–87.

56. Ibid., 7, 7.19, in De Lacy, *Galen*, 475.

eye] perceives . . . and it alone of all sense organs perceives them."[57] Color, in other words, is the sole proper sensible for sight, and in order to be perceived, it must be transmitted almost instantaneously back through the pneumatized air to the crystalline lens, which senses it visually.[58] Along with color, moreover, we perceive such things as shape, size, position, and distance, but we do so inferentially, because these things "are incidental [insofar as] they require reasoning and memory, not merely sensation."[59]

With some apparent (and perhaps disingenuous) reluctance, Galen ends his account of vision in the tenth book of the *De usu partium* by appealing to those of us—evidently few in number—who "know what circles, cones, axes, and other things of that sort are."[60] Galen then goes on to describe a cone with its vertex in the pupil of the eye and its base on a circle that defines the field of view for that eye. He urges us to conceive of that cone as formed by filaments as thin as the strands of a cobweb extending in straight lines from the vertex to every point on the circumference of the circle at the base as well as to the center of that circle along the cone's axis.[61] Accordingly, we can imagine that a small object placed inside the cone on the axis will block some of the filaments from reaching the circle at the base. All we need do now is to "call . . . those thin filaments extending from the pupil to the circumference of the circle . . . visual rays" and imagine every point within that circle to be connected by such rays.[62]

Bearing this visual model in mind, Galen asks us to imagine both eyes staring at some object EF in figure 2.4 from vertices A and B of their respective cones. Place some object CD in front of EF so that it forms cone CAD with the eye at A and cone CBD with the eye at B and so that the axes of both cones intersect at center point O. When viewed from A, object CD will occlude portion KL of the distant object, whereas when viewed from B it will occlude portion GH of that object. Consequently, against the background of object EF

57. Ibid., 7, 5.33, in De Lacy, *Galen*, 461.

58. See ibid., 7, 7.1–4, in De Lacy, *Galen*, 467.

59. Ibid., 7, 6.24, in De Lacy, *Galen*, 467.

60. *De usu partium*, 10, 12, in May, *Usefulness*, 492. "Most people pretending to some education," Galen laments at the beginning of *De usu partium*, 10, 12, "not only are ignorant of [geometry] but also avoid those who do understand it and are annoyed with them" (May, *Usefulness*, 490).

61. See ibid., in May, *Usefulness*, 492–93.

62. Ibid., in May, *Usefulness*, 493.

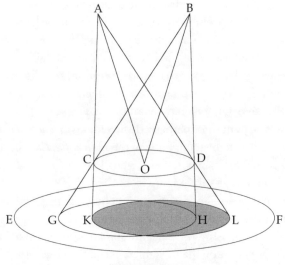

FIGURE 2.4

the image of CD will be seen at disparate locations by the two eyes.[63] For the most part, though, we see things singly rather than double, and the reason is that the axis of each eye passes through the center of the optic nerve feeding into it. Since the two optic nerves in their entirety lie in the same plane, the disparate images from both eyes will be fused at the optic chiasma where they cross.[64] This fused image represents a sort of average of the two disparate ones. That, in brief, is why we normally perceive a given object as precisely the same with both eyes.

In its individual elements there is nothing particularly original in Galen's theory of vision, not even in his account of image fusion. Most of the anatomical details upon which he based that theory were already noted centuries earlier by the great Alexandrian anatomist Herophilus (ca. 325–255 BC) and his younger contemporary Erasistratus (ca. 304–250 BC). Indeed, Galen cites Herophilus and Erasistratus copiously throughout his writing, Herophilus in particular meeting with his approval.[65] Nor was the idea that sensation is pneumatic new to Galen. It was developed most notably by the Stoics in the early third century BC, and pneuma figures prominently in the physiological

63. Ibid., in May, *Usefulness*, 493–97.

64. See ibid., in May, *Usefulness*, 497–502.

65. See Heinrich von Staden, *Herophilus: The Art of Medicine in Early Alexandria* (Cambridge: Cambridge University Press, 1989), especially 195–208 and 313–18.

theory of both Herophilus and Erasistratus. There are clear Stoic resonances in Galen's notion of pneumatized air as a "sympathetic instrument" and in his appeal to a "ruling principle" that exercises perceptual and intellectual control from its seat in the brain, although the Stoics located it in the heart.

On the other hand, Galen's insistence that color is the sole proper sensible for sight and his emphasis on the virtual instantaneity with which the air is pneumatized and the color is transmitted back through it to the eye are strongly reminiscent of Aristotle. And so is his assertion that although we perceive such "incidental" things as shape, size, position, and distance along with color, these things are not actually sensed but inferred. They are, after all, among Aristotle's common sensibles. Yet Galen sided with Plato against Aristotle not only by arguing in favor of extramissionism and vehemently opposing intromissionism, but also by locating the perceptual faculties in the brain rather than in the heart, where Aristotle placed it.[66] Furthermore, as we have just seen, Galen used ray geometry to explain binocular vision and the need for image fusion at the optic chiasma. To sum it up, then, Galen was a shrewdly syncretic thinker whose originality lay not so much, perhaps, in the specific concepts upon which he based his visual theory as in the remarkably cogent and systematic way he combined those concepts into a coherent whole.

4. THEORIES OF COLOR AND COLOR PERCEPTION

It is clear from their analyses of vision that both Aristotle and Galen assumed color to have objective existence as an inherent quality in visible things, independent of any observer. As such, it is only potentially visible. In order to become actually visible, the color must be in contact with a properly receptive medium, such as air rendered transparent or appropriately pneumatized by light, and it must face a properly receptive eye that is also in contact with that medium. On the face of it, Aristotle's account of the rainbow seems inconsistent with this model because the color appears not in the object but in water droplets, where it exists incidentally rather than inherently. But even in this instance it is the color of the visible object itself—the sun—that is actually seen by means of visual rays reflected to it from the individual water droplets. The resulting solar images, reduced to color by the smallness of the droplets, may *appear* to be in the cloud those droplets form, but it is the sun itself that is actually seen by the visual rays reflected to it.

66. For the background to Aristotle's "cardiocentric" theory of perception and Galen's "encephalocentric" alternative, see Rocca, *Galen on the Brain*, 17–47.

Not every Greek thinker shared this understanding of color as objectively determined. Some believed it to be subjectively determined. We already noted, for instance, Empedocles's idea that visual perception is based on passages in the eye that accommodate particles emitted from external objects. These passages, he explains, "are arranged alternately of fire and of water" so that "by the passages of fire we perceive white objects; by those of water, things black; for in each of these cases [the objects] fit into the given [passages]."[67] Evidently, then, the perception of white or black depends on specific kinds and sizes of particles that fit into specific kinds and sizes of passages in order to stimulate the eye to see white or black. White and black, in short, exist subjectively in the perceiver, not objectively in the thing seen. Plato is even clearer on this point. As he puts it in the *Timaeus*:

> Now the parts that move from the other objects and impinge on the ray of sight are in some cases smaller, in others larger than, and in still other cases equal in size to, the parts of the ray of sight itself. Those that are equal are imperceptible, and these we naturally call *transparent*. Those that are larger contract the rays of sight, while those that are smaller, on the other hand, dilate it, and so are "cousin" to what is cold or hot in the case of the flesh, and, in the case of the tongue, to what is sour, or to all those things that generate heat and that we have therefore called "pungent." So *black* and *white*, it turns out, are properties of contraction and dilation, and are really the same as these other properties, though in a different class, which is why they present a different appearance. This, then, is how we should speak of them: *white* is what dilates the ray of sight, and *black* is what does the opposite.[68]

Several things are worth noting about this account. First, it is essentially corpuscular; both the visual ray and what affects it "objectively" consist of particles. Second, black and white—or, rather, the perception of them—is dependent upon the size of those particles. And third, those two colors (taking black loosely to be a color rather than absence of all color) are ultimately due to the mechanical interaction between particles, although precisely what the resulting dilation and contraction of the visual ray means in this context is

67. See note 17 above. I should point out that my use of "objective" and "subjective" here and later in this book reflects a post-Kantian conception of the terms and does not reflect the way "object" and "subject" were understood before the late eighteenth century. Nonetheless, I think the distinction is implicit in pre-Kantian thought and is most easily grasped by a modern audience in such anachronistic terms.

68. *Timaeus*, 67d–e, in Zeyl, *Plato*, 61.

unclear to the point of opacity. One thing seems clear, however: as qualities, in the Aristotelian sense, black and white are not necessarily objective for Plato; they exist in the perceiver, not in the thing perceived.

Disagreement there may have been about whether color is in the eye of the beholder, so to speak, or in the thing beheld, but there was almost no disagreement about what color is. The consensus among Greek thinkers, at least from the time of Aristotle, was that the color spectrum is bounded by pure white and pure black at the extremes and that every intermediate color is a blend of the two in a specific proportion. The more white in that blend, the brighter and more vivid the color, and conversely, the more black in the blend, the darker it is. Thus, as Aristotle remarks in *De sensu*,

> It is conceivable that the white and the black should be juxtaposed in quantities so minute that either separately would be invisible, though the joint product would be visible; and that they should thus have the other colours for resultants. Their product . . . must be of a mixed character—in fact, a species of colour different from either. Such, then, is a possible way of conceiving the existence of a plurality of colours besides the white and black; and we may suppose that many are the result of a ratio. [Another possibility] is that the black and the white appear the one through the medium of the other, giving an effect like that sometimes produced by painters overlaying a less vivid upon a more vivid colour, as when they desire to represent an object appearing under water or enveloped in a haze, and like that produced by the sun, which in itself appears white, but takes a crimson hue when beheld through a fog or a cloud of smoke.[69]

According to both explanations, color is objectively absolute. Red is red by virtue of the precise blend or overlay of black and white that makes it so, and the same holds for every other color.

Things are not quite that simple, though. As we saw earlier, Aristotle reduced the rainbow to three primary colors in order from the most vivid (red) to the least vivid (violet), with green standing intermediate. "These," Aristotle asserts, "are almost the only colours which painters cannot manufacture, for there are colours which they create by mixing, but no mixing will give red, green, or purple."[70] The notion of subdividing the spectrum into primary colors was not new to Aristotle, nor was there clear consensus among

69. *De sensu*, 3, 439b20–440a12, in Barnes, *Complete Works*, 698.
70. *Meteorology*, 3, 2, 372a6–9, in Barnes, *Complete Works*, 600.

his predecessors on precisely what those colors must be.[71] Moreover, Aristotle was aware that in the rainbow there is a band of yellow between the red and green. The appearance of this band, according to him, "is due to contrast; for the red is whitened by its juxtaposition with green. We can see this from the fact that the rainbow is purest when the cloud is blackest; and then the red shows more yellow. . . . So the whole of the red shows white by contrast with the blackness of the cloud around."[72] The yellow is thus created by a contrast between, rather than a mixture of, the red and green—which is to say that the resulting effect of "yellow" is psychological rather than physical in origin. In addition, the distance the visual ray has to travel in order to establish contact with the colored body affects the vividness of the resulting impression. Altogether, then, "we must recognize . . . first, that white color on a black surface or seen through a black medium gives red; second, that sight when strained to a distance becomes weaker; [and] third, that black is in a sort the negation of sight."[73]

The Pseudo-Aristotelian *On Colours*, variously ascribed to Theophrastus or Strato of Lampsacus (ca. 335–269 BC), is even more explicit about the circumstantial nature of color perception:

> We never see a colour in absolute purity: it is always blent, if not with another colour, then with rays of light or with shadows, and so it assumes a tint other than its own. That is why objects assume different tints when seen in shade and in light and sunshine, and according as the rays of light are strong or weak. . . . Again they vary when seen by firelight or moonlight or torchlight, because the colours of those lights differ somewhat. . . . Thus all hues represent a threefold mixture of light, a translucent medium (e.g., water or air), and underlying colours from which the light is reflected.[74]

There is an obvious tension here between what a given color actually is, according to a physical blend of black and white, and how it appears, according to a variety of circumstantial factors. The resulting ambiguity will be explored and exploited by Renaissance artists in their search for ways to represent physical reality as naturalistically as possible in visual space.

71. See J. L. Benson, *Greek Color Theory and the Four Elements* (ScholarWorks@UMass Amherst, 2000—http://scholarworks.umass.edu/art_jbgc/1/).

72. *Meteorology*, 3, 4, 375a7–13, in Barnes, *Complete Works*, 603.

73. Ibid., 374b10–12, in Barnes, *Complete Works*, 603.

74. *On Colours*, 3, 793b14–21, trans. T. Loveday and E. S. Forster, in Barnes, *Complete Works*, 1221.

5 . THE EUCLIDEAN VISUAL RAY THEORY

Up to this point we have touched only briefly and obliquely on the visual ray theory, noting how it was applied to specific problems Aristotle and Galen raised. It is now time to backtrack, presumably to the turn of the third century BC, in order to examine the earliest known formal articulation of that theory in Euclid's *Optics*.[75] Although the analysis in the *Optics* is astonishingly primitive in comparison to Euclid's masterful treatment of plane and solid geometry in the *Elements*, the two works seem to have a couple of key things in common. First, there is some reason to believe that Euclid contributed little original content in either of the two; instead, he may have drawn for the most part on ideas and theorems earlier thinkers had developed. His primary contribution in both works, therefore, may have been to organize and refine those ideas and theorems systematically.[76] For another thing, Euclid follows the same methodology in both works, laying down the necessary axiomatic foundations and then building his analysis in logical order, theorem by theorem, upon those foundations.

The axiomatic foundations of the *Optics* consist of seven postulates that conspire to define the grounds of spatial perception.[77] The first two establish that discrete rays disperse outward in bundles from a point inside the eye to form a cone with its vertex at that point and its base on a circle at the limit of visibility. This is Galen's visual cone with the vertex transplanted from the surface to the interior of the eye. The third postulate asserts that whatever

75. Here, of course, I am assuming (provisionally) that the *Optics* is authentically Euclidean and therefore was composed sometime around 300 BC.

76. See the introduction to Thomas L. Heath, trans., *The Thirteen Books of Euclid's Elements*, 2nd ed. (New York: Dover Press, 1956), especially 1–6. In his commentary on the first book of the *Elements*, the fifth-century Neoplatonist thinker Proclus admired Euclid for having "brought together the *Elements*, systematizing many of the theorems of Eudoxus, perfecting many of those of Theaetetus, and putting in demonstrable form propositions that had been rather loosely established by his predecessors" (Glenn R. Morrow, trans., *Proclus: A Commentary on the First Book of Euclid's* Elements, 2nd ed. [Princeton, NJ: Princeton University Press, 1992], 56).

77. For the critical Greek text with an accompanying Latin translation, see I. L. Heiberg, *Euclidis Opera Omnia*, vol. 7 (Leipzig: Teubner, 1895), 1–121; for an English translation, see Harry E. Burton, trans., "The Optics of Euclid," *Journal of the Optical Society of America* 35 (1945): 357–72; and for a French translation, see Paul ver Eecke, *Euclide: L'Optique et la Catoptrique* (Paris: Blanchard, 1959), 1–51. In the following analysis I will refer to specific propositions by number rather than cite pages in these works, since those propositions can be easily found without page references.

the visual rays touch within the visual field is seen and that whatever they do not touch remains invisible. The fourth maintains that the apparent size of any viewed object depends on the visual angle at the vertex of the cone under which it is seen; the fifth that objects seen by higher or lower rays appear higher or lower, respectively, in the field of view; and the sixth that whatever is seen by leftward or rightward rays appears to the left or right, respectively, in the field of view. The final postulate claims, in a rather awkward way, that when an object is touched by more rays, it will be seen more clearly than when it is touched by fewer.[78]

After the seven postulates come fifty-eight theorems, of which we will limit ourselves to a few representative samples. Based on the first three postulates, for example, the first theorem demonstrates that we can never see an entire object at once because, being discrete, the visual rays strike it at separate points.

Thus, as illustrated in figure 2.5, object line A is seen only at the points where the rays emanating from E strike it, the spaces between remaining invisible. Things seem to be perceived as a whole, the theorem concludes, "because the visual rays shift rapidly," presumably because of visual scanning from side to side and top to bottom.[79] The second theorem appeals to postulate seven in demonstrating that if two objects of the same size face the eye directly at different distances, and if they are parallel, the closer one will be seen more clearly because it will intercept more rays. This point is illustrated in figure 2.5, where (despite the visual illusion) object lines A and B are equal and parallel. Being closer, object line A intercepts more rays, so it is seen more clearly. Proposition three maintains that for every object there is a distance at which it will no longer be seen because it falls between visual rays, as they disperse outward with distance. Accordingly, object line C in figure 2.5 will not be seen at distance EC because it is untouched by the rays on each side. In proposition four we are told that if equal segments of the same line are viewed from a slant, the one lying at a greater distance will appear smaller. In figure 2.5, segments DF, FG, and GH of line DH are equal, but it is easily demonstrated that visual

78. As Burton translates it, the seventh postulate assumes that "things seen within several angles appear to be more clear." What this means is that the more rays the object subtends at the base of the visual cone, the more angles will be formed at the cone's vertex (and the smaller they will be). Consequently, more rays will strike the object's surface. It is worth noting that Euclid uses both *opsis* and *aktis* to denote "visual ray" in these postulates and continues to use them interchangeably throughout the *Optics* and *Catoptrics*. Unlike *opsis*, which applies specifically to sight, the term *aktis* carries a more general sense of "ray" or "beam," and can be applied to sight as well as to light.

79. Translation adapted from Burton.

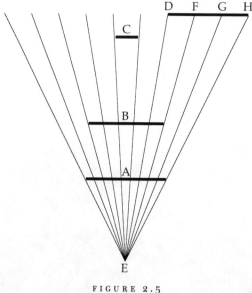

FIGURE 2.5

angle GEH subtended by farther segment GH is smaller than visual angle FEG subtended by nearer segment FG, which is smaller yet than angle DEF subtended by nearest segment DF. According to postulate four, then, segment GH will appear smallest, and segment DF will appear largest. Finally, proposition five demonstrates that when two equal objects, such as A and B in figure 2.5, are parallel and lie at unequal distances from the eye, the one subtending the smaller visual angle will appear more distant.

Propositions four and five set the pattern for the rest of the *Optics* by establishing the relationship between size perception and distance perception on the basis of the visual angle subtended by the object. Take as an example proposition six, which claims that "parallel lines, when seen from a distance, appear not to be equidistant from one another."[80] The proof is based on figure 2.6, where E is the eye facing equal and parallel lines BD, ZL, and TK, whose endpoints are joined by parallel lines BA and DC. From viewpoint E, TK will appear smaller than ZL, which will appear smaller than BD (by proposition five). Thus, since the apparent distance between corresponding points T and K, Z and L, and B and D on the parallel lines AB and CD diminishes continually, those lines will not appear parallel. Returning to this apparent nonparallelism in proposition twelve, Euclid draws on postulate six to conclude that point T

80. Translation adapted from Burton.

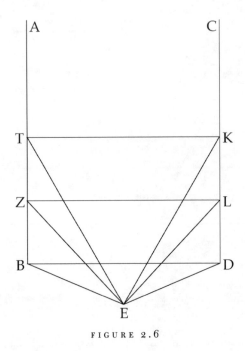

FIGURE 2.6

in figure 2.6 will appear to the right of point Z, and point Z will appear to the right of point B, so the entire line BA will appear to incline inward toward the right. Conversely, the entire line DC will appear to incline inward toward the left, so the two parallel lines will appear to converge as they get farther from E. The same sort of analysis can be used to explain the apparent convergence of ceilings and floors in hallways, which Euclid explains in propositions ten and eleven. By implication, then, Euclid has provided the geometrical basis for artistic perspective, as it may have been practiced in antiquity and was certainly, and explicitly, practiced in the early fifteenth century.[81]

Propositions eighteen to twenty-one address some simple surveying problems. In proposition eighteen, for example, Euclid poses the problem of determining the height of line AB in figure 2.7a. Let D be the viewpoint, and let the sun shine over AB along GD to cast a shadow of measured length DB. Place some object ZE of known height so that it fits into triangle ADB. Measure DE. Then, since lengths DE, ZE, and DB are known, length AB can be cal-

81. For a recent, highly speculative study of artistic perspective in antiquity, see Rocco Sinisgalli, *Perspective in the Visual Culture of Classical Antiquity* (Cambridge: Cambridge University Press, 2012). On the development of Renaissance artistic perspective, see chapter 7, section 4.

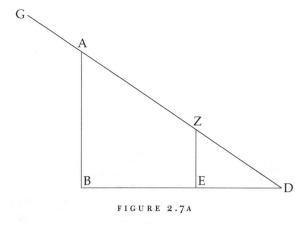

FIGURE 2.7A

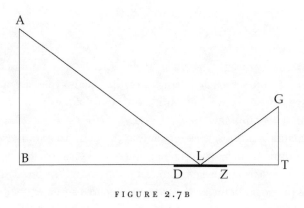

FIGURE 2.7B

culated according to the similarity of triangles ABD and ZED: that is, ZE:ED = AB:BD. If the sun is not shining, Euclid continues in proposition nineteen, then let AB in figure 2.7b be the height to be determined and GT the height of the observer viewing from point T, and let the foot-to-foot distance between them be measured by length BT. Place mirror DZ flush with BT, and let ray GL pass from the eye to the mirror so as to reflect to A along LA. Since we know from the *Catoptrics* (which will be examined later) that the angles of incidence and reflection are equal, it follows that angles GLT and ALB are equal, so triangles GLT and ALB will be similar.[82] Therefore, given that we know the lengths of LT, LB, and GT, we can determine the length of AB according to

82. As we will see shortly, Euclid establishes the equal-angles law of reflection in proposition one of his *Catoptrics*.

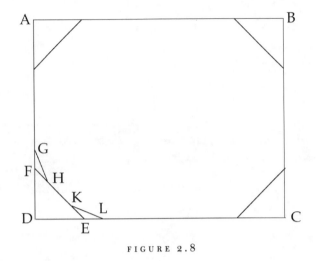

FIGURE 2.8

the proportionality of sides in similar triangles. The main point to note about this theorem, of course, is that it depends on the equal-angles law of reflection, which, as we saw earlier, was either unknown to Aristotle or flouted by him in his account of the rainbow.

The remaining propositions range from the pedestrian to the absurd. In this latter category is proposition nine, which demonstrates that "rectangular objects, when seen from a distance, appear rounded."[83] The proof, such as it is, is based on proposition three. Accordingly, Euclid's argument with reference to figure 2.8 is that the angles at A, B, C, and D of rectangle ABCD will disappear at a great distance because they fall between rays. Having "lost" those angles, the figure then appears octagonal rather than rectangular, according to the apparent replacement of the angles with sides, as illustrated by the apparent replacement of the angle at D with the side FE. "In the same way also in the case of the other angles, this will happen," Euclid concludes, "so that the whole thing will appear rounded."[84] In other words, just as the angles at A, B, C, and D are lost to sight, so are the angles at E and F, which are replaced by sides KL and GH, and so forth.

For the most part, though, the pedestrian far outweighs the absurd. Thus in proposition twenty-three Euclid shows that if a sphere is viewed by one eye, less than half the sphere will be seen, whereas in proposition twenty-four

83. Translation adapted from Burton.
84. Translation adapted from Burton.

he shows that as the eye approaches the sphere, the amount seen will appear to get larger, even though a smaller portion of the sphere is actually seen. On the other hand, if both eyes view the sphere, precisely half will be seen if the distance between the eyes is equal to the sphere's diameter (proposition twenty-five), more than half will be seen if that distance is greater than the sphere's diameter (proposition twenty-six), and less than half will be seen if that distance is smaller than the sphere's diameter (proposition twenty-seven). The same applies mutatis mutandis to cylinders and cones (propositions thirty-one and thirty-two). All of these propositions are based on dropping tangents from the eye or eyes to the curved surfaces. Euclid also addresses the apparent shape distortion of objects viewed at a slant. The paradigm case is the way circles look ellipsoidal at a slant because the diameters appear to be of different lengths according to the obliquity of the viewpoint with respect to the circle's center. That is why "the wheels of chariots sometimes appear circular and sometimes distorted" (proposition thirty-six).[85] Euclid also gives a brief account of apparent motion, an example being the perceived backward movement of a stationary object with respect to other objects perceived to travel in a forward direction (proposition fifty-two).

Cursory though it is, this summary gives an adequate overview of Euclid's approach to optics, as well as of the phenomena susceptible to that approach. The resulting analytic focus is almost exclusively upon size perception, distance perception, and shape perception as dictated by the geometrical structure of the visual cone with its constituent rays. The virtue of this approach is its simplicity. By transforming rectilinear lines of sight into rays that funnel out from the eye to form cones, Euclid was able to account for spatial perception in a limited but relatively effective way according to the principles of surveying.

More telling than what Euclid succeeded in doing with his visual ray theory, however, is what he failed to do. For instance, if vision is really due to discrete rays that pick out certain spots on visible surfaces, then why do we see those surfaces as continuous? Appealing to the scanning process, as Euclid apparently does in proposition one, fails to address this problem adequately because that process ought to yield an annoyingly stroboscopic view of the object as new points come into view while previous ones disappear in succession.[86] The same holds for objects that disappear at certain distances when they fall

85. Translation directly from Burton.

86. Furthermore, if the visual rays strike the object's surface at true points, then nothing should be seen because Euclidean points have absolutely no spatial dimension.

between visual rays, for as we move our eyes from side to side or up and down, those objects ought continually to come into and disappear from view.

Another problem is Euclid's failure to account for variations in visual acuity among different observers. Do some emit denser bundles of visual rays than others who see less clearly? Yet another problem is Euclid's assumption in postulate four that apparent size is determined by visual angle alone. If that is so, then why, when we view two objects lying at different distances but subtending the same visual angle, do we perceive the farther one as larger? The problematic nature of Euclid's analysis becomes even clearer when we take into account his failure to mention light as a factor in visual perception (apart from the oblique reference in proposition eighteen), his complete disregard for color perception, his refusal to discuss how and from what the visual rays are formed in the eye, and his silence on the psychological aspects of sight. As we will see in the next chapter, which deals with Ptolemy's *Optics*, the evolution of visual ray theory after Euclid was determined to a great extent by the need to resolve these issues.

It is difficult to believe that someone of Euclid's mental acuity in geometry was blind to the analytic shortcomings of this minimalist approach to optics. Why, then, did he choose to take it anyway? The main reason, I suggested long ago, was Euclid's commitment to a methodology that was eventually articulated as "saving the appearances."[87] This methodology is predicated on the assumption of a single, overarching principle according to which a complex set of apparently irregular phenomena can be reduced to regularity. In mathematical astronomy, the assumed principle was that celestial motion is perfectly uniform and circular, from which it follows that all apparent irregularities—or anomalies—in that motion can be resolved by combining uniform circular movements in various ways. The optical counterpart of this principle is that visual contact occurs along perfectly straight lines, from which it follows that all anomalies in visual appearances can be resolved according to ray geometry. However, the choice of overarching explanatory principle also dictates the choice of anomalies to which it can be appropriately applied. In the case of optics, for instance, ray geometry cannot explain such visual anomalies as the apparent difference in length between lines A and B in figure 2.5, even though the two lines are equal, because that anomaly is psychological in origin. For this reason, I suggest, Euclid stripped both the world of visual appearances and the visual act itself to their barest geometrical bones so that the appear-

87. See A. Mark Smith, "Saving the Appearances of the Appearances: The Foundations of Classical Geometrical Optics," *Archive for History of Exact Sciences* 24 (1981): 73–100.

ances and the act of seeing them would dovetail as perfectly as possible. The limited scope of his analysis was therefore due to a self-imposed limit on the phenomena Euclid intended to address.

6. EUCLIDEAN CATOPTRICS

In his commentary on the first book of Euclid's *Elements*, Proclus (410/412–485), citing Geminus (first century BC), tells us that optics, taken in the general sense, "uses visual lines and the angles made by them [and] is divided into a part specifically called optics, which explains the illusory appearances [the anomalies] presented by objects seen at a distance, such as the converging of parallel lines or the rounded appearance of square towers." A second part, he continues, consists of "catoptrics (*katoptrikē*) in general, which is concerned with the various ways in which reflection [occurs]."[88] According to Proclus's description of catoptrics, then, its subject matter is limited to reflection, that is, "bending back" or "flexure" (*anaklasis*), from mirrors (*katoptra*). As we will soon see, this limitation is somewhat looser than it may appear to be at first blush.

For the most part, the analysis in Euclid's *Catoptrics* reflects Proclus's description; that is, it is concerned almost exclusively with reflection from mirrors whose surfaces take three specific forms: plane, convex spherical, and concave spherical.[89] As in the *Optics*, so in the *Catoptrics*, Euclid opens with

88. Morrow, *Proclus*, 33 (translation modified slightly by me). In his *Apology*, the second-century writer Apuleius of Madaura offers a more expansive account of catoptrics according to its need to address such anomalies as image magnification in concave mirrors, image reduction in convex mirrors, and image reversal in plane, convex, and concave mirrors; see Apuleius, *Apologia*, chapter 16, in Vincent Hunink, trans., *Apuleius of Madauros: Pro Se De Magia (Apologia)* (Amsterdam: J. C. Gieben, 1997), 47; for an English translation, see H. E. Butler, *The Apologia and Florida of Apuleius of Madaura* (Oxford: Clarendon Press, 1909), 42.

89. For the critical Greek text with an accompanying Latin translation, see I. L. Heiberg, *Euclidis Opera Omnia*, vol. 7 (Leipzig: Teubner, 1895), 285–343; and for a French translation, see Paul ver Eecke, *Euclide*, 99–123; as before, I will refer to specific propositions by number rather than cite pages. The authenticity of this work has long been in dispute. Euclid's late nineteenth-century editor, I. L. Heiberg, decided that it is actually a recension by the fourth-century thinker Theon of Alexandria, and most scholars since Heiberg have accepted his verdict; see, especially, Albert Lejeune, *Recherches sur la catoptrique grecque*, Mémoires de l'Académie Royale de Belgique: Classe des lettres et des sciences morales et politiques 52, no. 2 (Brussels: Palais des Académies, 1957), 112–36. Not everyone has concurred, though. As far as I know, the most recent scholar to take issue with Heiberg's judgment is Ken'ichi Takahashi, who argues forcefully that the *Catoptrics* is authentically Euclidean in its entirety; see Taka-

a set of postulates, this time numbering six rather than seven. The first re-establishes that visual rays form perfectly straight lines and the second that all visible objects are seen along such straight lines. In a profoundly elliptical way, the third postulate maintains that if AB in figure 2.9a is a plane mirror, EB a viewer standing upright on it, and O an object point at perpendicular height OA above the mirror, then if visual ray ER reflects to O from R, EB:OA = BR:AR. In other words, triangles ORA and ERB will be similar. In the fourth and fifth postulates Euclid claims that in plane, spherical convex, and spherical concave mirrors, if a perpendicular (the so-called cathetus) is dropped from an object to the mirror's surface, and if the point at which it meets that surface is covered, the object will no longer be seen.

The apparent sense of these two postulates can be understood by means of figures 2.9a–c, representing a plane, convex, and concave mirror, respectively. In all three mirrors, EB represents the viewer standing upright on the reflecting surface, according to the tangent at point B, and OA the perpendicular distance of the object from that surface, according to the tangent at point A. ER is the incident visual ray and RO the reflected ray. According to postulates four and five, then, if point A on the mirror is blocked by something, the object will no longer be visible. Suffice it for now to observe that these two postulates are problematic in the extreme. Problematic, as well, is the sixth postulate because, in dealing with refraction rather than reflection, it is anomalous. It asserts that if an object placed in a vessel disappears from sight just below the rim when seen from a particular viewpoint, then if water is poured into the vessel while the viewpoint is maintained, the object will come back into view as if by floating upward.

Of the thirty propositions that follow these postulates, a few are worth discussing in moderate detail. The first three, for instance, are concerned with the equal-angles law and its immediate ramifications. Proposition one offers a proof of that law on the basis of the third postulate, which establishes the similarity of triangles ORA and ERB in figure 2.9a. From this it follows that the respective angles of those triangles will be equal, including angle ERB formed with the mirror by the incident ray and angle ORB formed by the reflected ray. For the two curved mirrors in figures 2.9b and c, Euclid has us imagine plane

---

hashi, *The Medieval Latin Traditions of Euclid's* Catoptrica (Kyushu: Kyushu University Press, 1992), 13–37. Although perfectly cogent, Takahashi's argument does not strike me as entirely convincing. Nonetheless, I am inclined (somewhat inertially, perhaps) to agree that, except for a handful of non-Euclidean accretions, such as the sixth postulate dealing with refraction, the *Catoptrics* is authentically Euclidean.

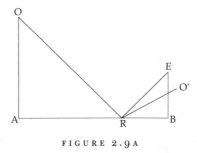

FIGURE 2.9A

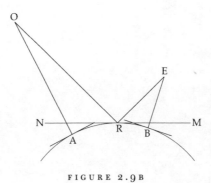

FIGURE 2.9B

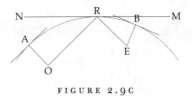

FIGURE 2.9C

mirror MN tangent to the reflecting surface at point R. Since it has just been demonstrated that angles ERM and ORN of incidence and reflection in plane mirror MRN must be equal, and since the angles of tangency MRB and NRA are equal, it follows that angles ERB and ERA of incidence and reflection are equal. The next two propositions demonstrate that in all three mirrors a perpendicular ray will reflect back on itself, whereas a nonperpendicular ray will reflect neither back on itself nor at an angle smaller than the angle of incidence, as measured with respect to the mirror. In other words, ray ER in figure 2.9a will not reflect back along ER or along RO′ at angle O′RB smaller than angle of incidence ERB.

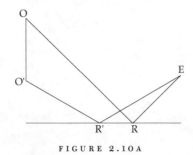

FIGURE 2.10A

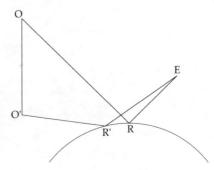

FIGURE 2.10B

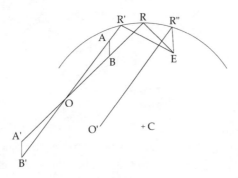

FIGURE 2.10C

Proposition four demonstrates that in plane and convex mirrors, rays incident at two points on the mirror from the same eye will never reflect to the same point. Thus, as illustrated in figures 2.10a and b, reflected ray R′O′ will never intersect reflected ray RO beyond the mirror. Propositions five and six, on the other hand, demonstrate that in concave mirrors the reflected rays can converge, depending on where the eye is placed. In figure 2.10c, for instance, if the eye is located at the center of curvature C, every ray emitted from that point to every point on the mirror's surface will reflect back on itself because, being a radius, it will be perpendicular to the reflecting surface. If the eye is located at E, the rays reflected from R and R´ will converge on O. On the other hand, ray R˝O´ reflected from R˝ will obviously not intersect reflected ray RO.

Implicit in these three propositions is that the number of possible images of any object point seen at any one time in a mirror is a function of the number of reflected rays converging on that object point. Hence, in plane and convex mirrors only one image of a given object point can ever be seen, whereas in concave mirrors more than one image of an object point may be seen. Euclid goes on to analyze image inversion for all three kinds of mirrors in propositions seven, eight, and eleven. Accordingly, in the first two of these propositions, he shows that upright object lines always appear inverted in plane and convex mirrors. In figures 2.10a and b, therefore, object line OO´ will appear upside down in the mirror. On the other hand, as he shows in proposition eleven, object lines may appear upright or inverted in concave mirrors, depending on where they stand in relation to the converging reflected rays. In figure 2.10c, for instance, object line AB between point of convergence O and points R and R´ of reflection will appear inverted, whereas object line A´B´ beyond O will appear upright.

Propositions sixteen to eighteen signal a major transition in Euclid's analysis because it is in these three propositions that he establishes the cathetus rule of image location, which was considered inviolable until the beginning of the seventeenth century. This rule stipulates that in all three kinds of mirrors the image of any object point lies where the cathetus dropped from that object point to the mirror intersects the extension of the incident ray. As illustrated in figures 2.11a–c, image point I in all three kinds of mirrors will lie at the intersection of cathetus OA, extended to I, and rectilinear extension RI of incident ray ER. If we take into account that the cathetus is simply an analytic convention independent of the actual lines of radiation, then the problematic nature of postulates four and five becomes clear: blocking point A, where the cathetus meets the reflecting surface, should have no effect whatever on the visibility of the image viewed from point E. What, then, was Euclid trying to establish with these two postulates? Among several possible explanations, the most plausible is that they represent a maladroit attempt to show on empirical grounds that object points are always seen along the cathetus. Interpreted thus, the intended point of the two postulates is that, if O in figures 2.9a–c is a small object, and if the eye continually moves in its vicinity, the object's image will continue to show in the mirror until the eye passes directly above the object to peer along

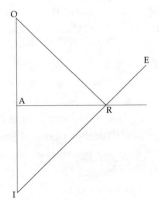

FIGURE 2.11A

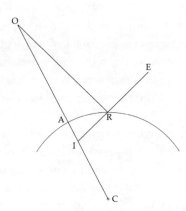

FIGURE 2.11B

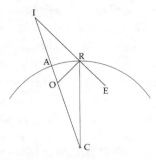

FIGURE 2.11C

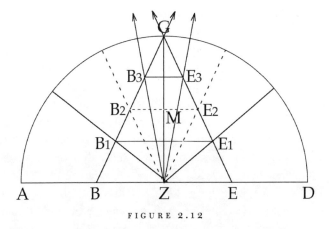

FIGURE 2.12

OA. At that point, the image will disappear because the object itself will block it.[90] This interpretation is borne out by the way Euclid proves the cathetus rule in proposition sixteen, appealing first to the supposition in postulates four and five that object O in figures 2.11a–c is no longer seen when point A is blocked and then to the supposition in postulate two that the object is seen along a straight line, that is, incident visual ray ERI extended to the cathetus.

Armed with the cathetus rule, Euclid is now prepared to address a range of anomalies related to mirror images. In proposition nineteen, for instance, he demonstrates that although they suffer left–right reversal, images in plane mirrors appear to be the same size as their objects and also appear to lie the same distance from the reflecting surface as their objects. Image reversal also occurs in convex mirrors, but in these the images appear smaller and closer to the reflecting surface than their objects (propositions twenty and twenty-one). In addition, the smaller the convex mirror is, the smaller the resulting images of the same objects look (proposition twenty-two). Finally, all images look convex in convex mirrors (proposition twenty-three).

Turning to concave mirrors, Euclid shows in proposition twenty-seven that if AGD in figure 2.12 is a hemispherical section of a concave mirror with center of curvature Z on diameter AZD, and if the two eyes B and E are placed on this diameter equidistant from center point Z, then neither eye will see itself or its mate in the mirror. To start with, if it were possible for the eyes to see themselves, then they could only do so by reflection from point G because that is the only point on the hemisphere at which the angles of incidence BGZ or

90. See Takahashi, *Medieval Traditions*, 20–26.

EGZ and the resulting angles of reflection ZGE or ZGB, respectively, are equal. That neither eye E nor eye B could see itself is clear from the fact that incident rays BG and EG are not perpendicular to the reflecting surface. Thus, eye E would have to see eye B, and vice versa. But diameter AD is the cathetus for both E and B because it is the only line through them that is perpendicular to the reflecting surface. Therefore, by the cathetus rule, if B were to see E, it would have to do so at the intersection of cathetus AD with incident ray BG, and there is no such intersection in the direction of reflecting surface AGD. The same holds mutatis mutandis for E, which can therefore not see B in the direction of the reflecting surface.

Euclid caps his analysis of image formation in concave mirrors with propositions twenty-eight and twenty-nine as follows. Let radius GZ of mirror AGD in figure 2.12 be bisected at point M. If the two eyes are placed at the endpoints of line B_1E_1 between M and diameter AD such that B_1E_1 is perpendicular to radius GZ and is bisected by it, then neither eye will see the other because catheti ZB_1 and ZE_1 will not intersect their respective incident rays E_1G and B_1G. Likewise, if the two eyes are placed on line B_2ME_2 passing through the midpoint of radius GZ, the incident rays B_2G and E_2G will be parallel to their respective catheti ZE_2 and ZB_2, so there will be no intersection. If, however the two eyes are placed on a line B_3E_3 between midpoint M of the radius and the reflecting surface, there will be an image because catheti ZB_3 and ZE_3 will intersect the respective incident rays E_3G and B_3G extended beyond the mirror. Since, therefore, line B_3E_3 lies on the viewer's face, its image will appear upright and reversed, and it will appear to lie at a greater distance from the reflecting surface than that between B_3E_3 and the surface. It will also be magnified. Furthermore, as line B_3E_3 draws toward line B_2E_2, the apparent size of the image continually increases along with its apparent distance from the reflecting surface.

Finally, if the two eyes E and B are placed beyond diameter A, as in figure 2.13, then image B′E′ of line BE will appear unreversed in left–right orientation, smaller than its object, and closer to the reflecting surface. The farther away line BE draws from the diameter, moreover, the smaller the image will be, and the closer to the reflecting surface it will appear to get. Oddly enough, Euclid fails to mention that in this situation the image will be inverted.

Like his *Optics*, Euclid's *Catoptrics* is virtually bereft of physical or psychological content because it, too, focuses on anomalies that can be explained solely by ray geometry. His analysis of mirrors thus shares the shortcomings of his analysis of direct vision in the *Optics*. It has its own shortcomings, as

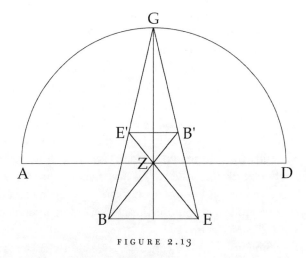

FIGURE 2.13

well. Surely the most salient of these involves the lack of image when line BE between the eyes in figure 2.12 is placed on radius GZ between diameter AD and midpoint M of the radius. After all, according to the third postulate of the *Optics*, whatever a visual ray touches will be seen, and ray GB reflected from incident ray EG certainly touches B. B ought therefore to be seen by E, and conversely, E ought to be seen by B. Still, despite its shortcomings, Euclid's analysis of reflection goes a long way toward explaining key anomalies, such as image displacement (that is, that the object appears at a location other than its real one), distortions in image size and shape, and image reversal and inversion. Particularly significant in this regard is his account of concave spherical mirrors because the anomalies created by these are so diverse and contradictory. Depending on the placement of eye and object, the resulting images may not be formed at all; or, when formed, they may be magnified or diminished; or they may appear behind or in front of the mirror; and they may be upright and reversed or inverted and unreversed. Crucial for its long-term implications is Euclid's association of all these anomalies with midpoint M on the mirror's radius.

Mirrors of excellent reflective quality and of every size and shape are so commonplace today that we tend to take the visual anomalies they create for granted. Such was not the case in antiquity, when mirrors were relatively uncommon, and mirrors of reasonably good reflective quality were rarer still. Glass mirrors, for instance, were problematic for a variety of reasons, and it was only in the Middle Ages that lead-backed glass mirrors were manufactured.

Even then, the quality of imaging in these mirrors was fairly low.[91] Consequently, most mirrors from antiquity to the sixteenth century were formed from metal, the best consisting of polished silver because it is both highly reflective and relatively colorless. Mirrors fashioned from bronze were most common in antiquity because bronze was relatively inexpensive, and, being fairly malleable, it could be polished to a fairly high sheen. Given these constraints, ancient thinkers were more alive than most of us today to the marvelous visual effects mirrors can produce. For them, mirrors were a sort of funhouse attraction that could be manipulated in various ways to create startling illusions. Even Euclid felt compelled to discuss such illusions in propositions thirteen to fifteen and twenty-nine of the *Catoptrics*, where he describes a few simple ways to manipulate or multiply images according to certain arrangements of plane and curved mirrors.

This zest for funhouse illusions comes through clearly in Hero of Alexandria's *Catoptrics*, which was presumably written some 250 years after Euclid's treatise on reflection.[92] Of the eighteen propositions contained in this work, the concluding eight are concerned with how to arrange plane mirrors to produce startling visual effects. The last of these eight propositions is a prime example. It promises to show how "to put a mirror in a given place so that everyone who approaches will see neither himself nor someone else, but only whatever picture someone has chosen in advance."[93]

Illustrated in figure 2.14, the setup requires that mirror AB be posed against a wall at an angle of 30° and that some plaque CD with a picture on it face mirror AB at the same angle. A barrier is set up in front of the plaque to hide it from a viewer whose eye is at E. Accordingly, when the viewer looks straight ahead from E, he will see image C′D′ of the picture straight ahead and behind mirror AB. This apparition is due to the reflection of the visual rays EA and EB along AC and BD, respectively. The top of the picture appears at C′ along the continuation of incident ray EA and the bottom at D′ along the continua-

91. See Sarah J. Schechner, "Between Knowing and Doing: Mirrors and Their Imperfections in the Renaissance," *Early Science and Medicine* 10 (2005): 137–62.

92. Unfortunately, this treatise is extant only in the Latin translation produced by William of Moerbeke, probably in 1269; no Greek version has yet to come to light. For a recent critical edition and English translation, see Alexander Jones, "Pseudo-Ptolemy *De Speculis*," SCIAMVS 2 (2001): 145–86. An older edition of the Latin text along with a German translation can be found in W. Schmidt, ed. and trans., *Heronis Alexandrini Opera Quae Supersunt Omnia*, vol. 2.1 (Leipzig: Teubner, 1900), 317–65. Because Jones's edition is so difficult to find, I will cite both editions in what follows.

93. Jones, "Pseudo-Ptolemy," 178; Schmidt, *Opera*, 358.

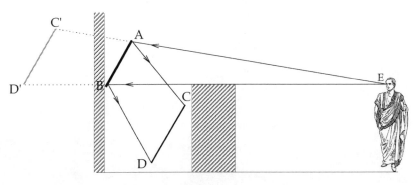

FIGURE 2.14

tion of incident ray EB skimming the top of the barrier. In proposition twelve, Hero describes an arrangement of two hinged mirrors, one of whose effects is to cause an approaching onlooker to see an image of Athena emerging from the head of Zeus, the point presumably being to inspire wonder and awe in an onlooker entering a temple.[94]

Not all of Hero's *Catoptrics* is devoted to such "practical" matters. Before taking them up, Hero offers a fairly brief theoretical account of reflection in order to lay the analytical foundations necessary for understanding how and why the mirror arrangements he will eventually describe work, although his explanations of those workings are rather minimal. He starts with an analogy between visual radiation and projectile motion. Like arrows shot at high velocity, visual radiation follows rectilinear trajectories because, being incredibly swift, it is inclined to travel the shortest possible distance.[95] In reflection, this radiation rebounds from certain bodies whose surfaces are perfectly smooth and dense (*spissa*) in the same way that stones hurled with great force bounce back from a wall because it is hard (*durus*). When such radiation encounters soft (*molli*) bodies, it does not rebound because its force is absorbed in much the same way as the force of a hurled stone is absorbed by wool.[96] Although reflective, water and glass also allow visual radiation to pass through because their surfaces are not uniformly dense and resistive but are interrupted by portions that are less dense (*rara*) and thus yield to it.[97]

The way visual radiation rebounds in reflection is dictated by its inclina-

94. See Jones, "Pseudo-Ptolemy," 173–74; Schmidt, *Opera*, 342–44.

95. Jones, "Pseudo-Ptolemy," 168; Schmidt, *Opera*, 320.

96. Jones, "Pseudo-Ptolemy," 168; Schmidt, *Opera*, 322–24.

97. See Jones, "Pseudo-Ptolemy," 168; Schmidt, *Opera*, 324.

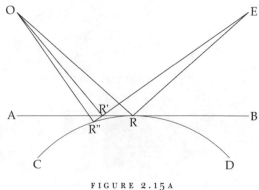

FIGURE 2.15A

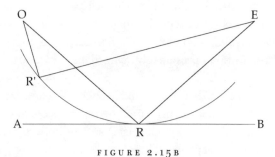

FIGURE 2.15B

tion to travel the shortest possible distance, and that in turn necessitates its following the equal-angles law. Hero proceeds to demonstrate this point in the first two propositions of the *Catoptrics*, which can be summarized as follows.[98] Suppose that CRD in figure 2.15a is a segment of a convex spherical mirror and ARB a plane mirror tangent to it at point R. Let E be an eye emitting radiation along ER such that it reflects along RO according to the equal-angles law. Now suppose that it reflects from some other point R´ on the plane mirror, or R˝ on the convex one, so as to form unequal angles. In either case, it can be easily demonstrated that ER´ + R´O or ER˝ + R˝O > ER + RO. In short, just as direct radiation follows the shortest possible single trajectory, reflected radiation follows the shortest possible double trajectory. The equal-angles law is therefore a special case of the least-distance principle governing direct visual radiation, which is thereby compelled to follow perfectly straight lines.

With the equal-angles law established according to the least-distance

98. See Jones, "Pseudo-Ptolemy," 169–79; Schmidt, *Opera*, 324–30.

principle, Hero turns in the third proposition to an oblique demonstration of the cathetus rule in plane mirrors based on the fact that if R in figure 2.9a is blocked (not A, as Euclid seemed to argue in the fourth postulate of his *Catoptrics*), the image of O will no longer be seen. The reason, according to Hero, is that triangles OAR and ERB are similar, so it presumably follows that the image of O must be seen directly through R. Implicit in this argument is that the image of O will be seen along the rectilinear extension of incident ray ER and that it will lie on cathetus OA, as in figure 2.11a. In the fourth and fifth propositions, Hero demonstrates that two rays from the same eye reflected at different points on plane and convex mirrors will never intersect, and in the seventh proposition he shows that such reflected rays can converge in concave mirrors.[99] These propositions simply recapitulate what Euclid demonstrated in propositions four to six of his *Catoptrics*.

The Euclidean basis of Hero's theory of reflection is evident in both his use of the visual ray and his choice of theorems dealing with the nonconvergence of reflected rays in plane and convex mirrors and their convergence in concave mirrors. Equally evident is that Hero imported some physical content into Euclid's theory and, moreover, attempted to justify the resulting model on metaphysical grounds. Accordingly, the "what" of visual radiation seems now to consist of particles shot at incredible speed from the eye and interacting with reflective and nonreflective surfaces in ways that are, if not identical with, at least analogous to the way physical projectiles interact with physical surfaces according to their texture and hardness. The "why" of visual radiation and its interaction with reflective bodies is given by the least-distance principle that governs both. The radiation itself must occur along straight lines because those are the most efficient possible trajectories for it to follow. Likewise, reflection must occur at equal angles because it is at those angles, and those angles only, that the visual radiation will complete its travel most efficiently.

That reflection from concave spherical mirrors does not follow this principle, however, is obvious from figure 2.15b, where ER + RO is obviously *longer* than ER′ + R′O. Despite this exception, Hero's attempt to explain direct radiation and reflection according to maximum efficiency has a certain logical force because it appeals to a principle of natural economy popularly expressed in the notion that "nature does nothing in vain." Commonplace

---

99. For the third, fourth, fifth, and seventh propositions of Hero's *Catoptrics*, see Jones, "Pseudo-Ptolemy," 170–72; Schmidt, *Opera*, 330–36. The sixth proposition simply demonstrates that when the eye lies at the center of curvature of a concave spherical mirror, all the rays emitted from it will reflect back upon themselves.

in Greek philosophical and scientific thought, Aristotle and Galen used this principle to especially good effect in their efforts to explain the physical world at both the macrocosmic and microcosmic (human) level.[100]

7. BURNING MIRRORS AND THE ANALYSIS OF FOCAL PROPERTIES

In the thirtieth and final proposition of his *Catoptrics*, Euclid undertakes to demonstrate that "fire is ignited from concave [spherical] mirrors facing the sun."[101] The demonstration is bipartite. First, Euclid assumes that line FTB in figure 2.16a is a ray emanating from the sun through center of curvature T of mirror ABC. Let FC be another ray striking the mirror but not passing through T, and let it be reflected at equal angles to K, which will lie between the mirror's surface and center of curvature T. That ray will be matched on the other side by some ray FA that will also reflect at equal angles to K—assuming, that is, that ray FTB is perpendicular to the sun's surface, a point that Euclid passes over in silence. The same will hold for all other rays from F striking the mirror at angles equal to those at C and A, so an infinitude of solar rays will be reflected to K from a circle that passes through C and A on the mirror's surface. This circle can be produced by rotating the entire figure about FTB as an axis. On the other hand, let rays GTD and HTE pass through T. Since these are all perpendicular to the mirror's surface, they will all reflect back to T, so the infinitude of solar rays projected through T to the circle passing through E and D on the mirror's surface will be augmented by the infinitude of rays reflecting back through that point. Consequently, T will be the point at which enough rays converge to cause kindling.

Exactly what Euclid is getting at in the first part of this proposition is difficult to determine, but it is plausible to assume that he means us to understand that the congregation of rays at points like K on axis FTB is inadequate to cause kindling because those rays originate at only one point on the sun and reflect from only one circle on the mirror. On the other hand, rays from every point on the solar surface between G and H will pass through center point T, so the addition of their reflected rays will be enough to cause kindling at T. Whatever the case, Euclid's analysis is flawed in two fundamental respects. First, it is simply incorrect; as we will see in a moment, the center of curvature

100. See A. Mark Smith, "Extremal Principles in Ancient and Medieval Optics," *Physics* 31 (1994): 113–40.

101. Euclid, *Catoptrics*, in Heiberg, *Opera*, 340–41.

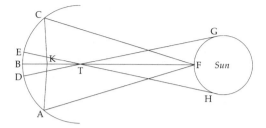

FIGURE 2.16A

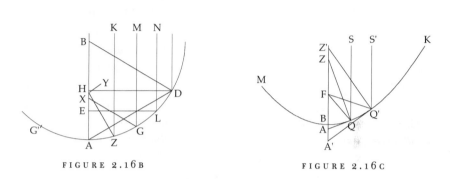

FIGURE 2.16B   FIGURE 2.16C

is not the focal point of the mirror. As we will also see in a moment, the rays responsible for kindling in concave spherical mirrors do not converge at a single point; instead, they aggregate in a spot near the focal point. Second, the proof assumes that the sun and the mirror are of roughly commensurate size and fairly close to one another. Consequently, the rays from the sun will be oblique to both the sun's surface and that of the mirror, and rays from every point on the sun will pass through the mirror's center. In fact, the sun's distance from the mirror is so great as to be virtually infinite, so all the solar rays reaching the mirror's surface are virtually parallel, which means that only the axial ray will strike the mirror's surface along the perpendicular. The rest will strike it at an angle so as to be reflected.

These points are taken into proper account in Diocles's *On Burning Mirrors*, which was presumably written around 200 BC.[102] Diocles takes up the

102. Like Hero's *Catoptrics*, Diocles's *On Burning Mirrors* is not extant in Greek; it currently exists only in a medieval Arabic version. For the Arabic text and an English translation, see Gerald J. Toomer, *Diocles on Burning Mirrors* (New York: Springer, 1976). We know nothing definitive about Diocles apart from what we can glean from this source, but it seems likely that he was more or less contemporary with Apollonius of Perga, whose landmark study of

analysis of concave spherical mirrors in propositions two and three of that work, operating on the assumption that all the solar rays reaching the mirrors are parallel to the axial ray.[103] Accordingly, let DAG′ in figure 2.16b be the mirror and BHA the axial ray passing through center of curvature B. Let arc AD be 60°, drop DH perpendicular to axial ray BHA, and connect BD and DA. Point H will be the midpoint of radius BA because triangle BDA is equiangular, and so perpendicular DH bisects side BA of that triangle. Let point G bisect arc DA so that arc GA = 30°. Let solar ray MG be parallel to axis BHA, and let it be reflected at equal angles so that its reflected ray intersects axis BHA at point X. Let solar ray KZ, also parallel to axis BHA, strike the mirror between G and A. Its reflected ray will strike BHA at point Y between X and H. On the other hand, ray NL striking the mirror between D and G will reflect to point E below H. Any ray, such as KZ, striking the mirror between G and A will thus reflect to some point between X and H, whereas any ray between G and D will reflect to some point below X. No matter how close to A the solar ray strikes the mirror, however, it will reflect to some point below H. H is therefore the limiting point to which no reflection can occur, and it constitutes the focal point of the mirror. The failure of all the reflected rays to coalesce at a single point constitutes spherical aberration, and it is because of such aberration that the actual "focusing" of those rays will occur in a small area just below the focal point, where the individual intersection points are most densely packed. The mirror itself can be formed by imagining the entire figure rotated 360° about axis BHA. It follows, then, that the effective surface of the mirror for the purposes of burning is quite small, certainly no larger than the sector bounded by arc GAG′ of 60°. In order to maximize the amount of solar radiation brought toward focus, then, it is necessary to increase the mirror's radius of curvature rather than expose more of the mirror to the sun.

Even better for focusing solar rays, however, is a paraboloidal mirror, as Diocles demonstrates in proposition one.[104] Let MBK in figure 2.16c be a parabolic section on such a mirror, and let F be its focal point. What Diocles sets out to prove is that all parallel solar rays reaching MBK will be reflected to F. The proof itself is based on erecting tangent AQ to the parabolic section at some arbitrary point Q and then dropping perpendicular ZQ to that tangent. On that basis, Diocles demonstrates that angle SQZ of incidence is equal to

---

parabolic, elliptical, and hyperbolic sections in the *Conics* had an enormous impact on the development of mathematics after the second century BC.

103. For the actual text of these two propositions, see Toomer, *Diocles*, 54–62.

104. See ibid., 44–48.

angle ZQF of reflection, and the proof will hold for any other point, such as Q´, on the parabolic section. Consequently, all the solar rays striking the parabolic section will be reflected to F, and the larger that section, the more rays will reach F. What holds for the parabolic section holds a fortiori for the paraboloidal mirror formed by rotating the entire section MBK about axis ZFB. Such a mirror is obviously more efficient as a focal device than a concave spherical mirror, and as such, it can be used to "produce fire in temples and at sacrifices and immolations, so that the fire is clearly seen to burn the sacrificial victims [as happens] in certain remote cities, especially on the days of great celebration, [causing] the people of those cities to marvel."[105] This, of course, is the same sort of pragmatic rationale Hero gives for the arrangement of mirrors described in proposition eighteen of his *Catoptrics*.

The vast superiority of Diocles's analysis of burning mirrors over that of Euclid is a testament to major advances in the mathematics of conic sections between the time of Menaechmus (fourth century BC?) and Apollonius of Perga (late third century BC) rather than to any improvement in the technology of mirrors or in their practical application. Particularly important in the study of conic sections just before the time of Apollonius was Archimedes (ca. 287 BC–ca. 212 BC), who is also credited with a now-lost work on optics. It is probably no accident that his reputed interest in both conic sections and optics inspired the legendary account of his using concave mirrors to burn a Roman fleet besieging his native Syracuse.[106] That Diocles was at least indirectly linked to Archimedes is indicated by his citing Conon of Samos and Dositheus of Pelusium, both of whom were closely connected to Archimedes. Diocles thus seems to have been part of a network of late third- and early second-century scholars who shared a deep interest in mathematics and, consequently, a fascination with burning mirrors. Harking back to the time

---

105. Ibid., 44.

106. In chapter 16 of his *Apologia*, Apuleius refers to Archimedes's "massive volume" (*volumen ingens*) on optics; see Hunink, *Apuleius*, 47; and Butler, *Apologia*, 42. The legend of Archimedes's use of mirrors to burn the Roman fleet evolved over several centuries between Lucian (fl. ca. AD 160), who simply claimed that Archimedes burned the Roman fleet without mentioning mirrors, and Anthemius of Tralles (fl. ca. AD 500), who claimed that Archimedes deployed several hexagonal, plane mirrors in a parabolic array to burn the fleet; see E. J. Dijksterhuis, *Archimedes* (Princeton, NJ: Princeton University Press, 1987), 28–29; for Anthemius, see George L. Huxley, *Anthemius of Tralles: A Study in Later Greek Geometry* (Cambridge, MA: Easton Press, 1959), 12–15. For a thorough but skeptical examination of the controversy surrounding the Archimedean mirror legend, see D. L. Simms, "Archimedes and the Burning Mirrors of Syracuse," *Technology and Culture* 18 (1977): 1–24.

of Euclid, and probably even earlier, this particular fascination was part of a more general passion for mirrors and the various optical effects they produced. Nonetheless, the study of burning mirrors was anomalous in one key respect: instead of focusing on visual rays as vehicles of sight, it focused on light rays as vehicles of heat.

8. CONCLUSION

There is no question that between the time of the later pre-Socratics and that of Galen, mathematical optics and visual theory had achieved a remarkably high level of sophistication. Especially salient in this regard are the contributions of Aristotle, who broke the visual act into a succession of psychological stages, and Galen, who embodied that act in an anatomical and physiological system extending from the brain to the anterior surface of the cornea.

Unfortunately, the dearth of relevant primary sources and questions about the authenticity of the ones we do have make it difficult, if not impossible, to reconstruct the incremental stages that led to this achievement.[107] What we have to settle on, instead, is a few snapshot glimpses at sporadic intervals between roughly 400 BC and AD 180. Consequently, any generalizations we might make about the development of visual theory and mathematical optics over that period are bound to be as tentative as they are limited. But there are a few things we can be fairly sure about.

For a start, ancient thinkers were ambivalent about the ontological status of light and its role in the visual process. Most visual theorists, whether intromissionist or extramissionist, viewed light as a catalytic agent, not as a visible entity

---

107. The issue of authenticity has nagged historians of ancient optics for many decades. As mentioned earlier in note 89 above, Heiberg cast doubt on the authorship of Euclid's *Catoptrics*, ascribing it instead to Theon of Alexandria. More recently, Takahashi attempted to reestablish Euclid's authorship. Part of the problem lies in long-standing misascriptions of texts. For several centuries, for instance, Hero's *Catoptrics* was falsely ascribed to Ptolemy. But part of the problem also lies in the sorts of evidence, much of it subjective, used to establish the case for or against authenticity. A particularly good example of such reasoning is to be found in Wilbur Knorr, "Archimedes and the Pseudo-Euclidean *Catoptrics*: Stages in the Ancient Geometrical Theory of Mirrors," *Archives Internationales d'Histoire des Sciences* 35 (1985): 28–104. Among other things, Knorr argues that the work ascribed by Apuleius to Archimedes is actually Euclid's *Catoptrics* and that Ptolemy's *Optics* is not authentically Ptolemaic but may have been composed by a contemporary, Sosigenes. If nothing else, this erudite and wide-ranging study goes a long way toward demonstrating that the issue of authenticity will probably never be definitively resolved for any ancient text.

in its own right. Aristotle, for instance, argued that light contributes to sight by actualizing the potential transparency of the aerial medium between the eye and visible objects. In much the same vein, Galen held that light pneumatizes this medium so that once cotransformed by the psychic pneuma reaching it from the eye, the air will form a percipient bridge between eye and object. Even earlier, Plato had emphasized the need for external light to mingle with the eye's light in order to initiate the visual process. Euclid and Hero went so far as to give light no effective role whatever in their accounts, preferring instead to emphasize visual radiation as the sole means of sight.

But not everyone treated light so cavalierly. As we saw earlier, the author of the Pseudo-Aristotelian *On Colours* explicitly mentioned illumination by external light sources as a factor in color perception, which is why colors look different in sunlight than they do in firelight. Illumination, in the form of solar rays, was also at play in the analysis of burning mirrors. Furthermore, although there seems to have been a general consensus among ancient thinkers that color is produced by mixing the polar opposites of black and white, there was considerable disagreement about how color is perceived, as well as about which colors, if any, are primary. The very idea that certain colors are primary because they cannot be produced by mixture is in fact inconsistent with the notion that all colors are a blend of black and white in specific proportions.

It also seems clear that there was a fundamental dichotomy between mathematical optics, as exemplified by Euclid and Hero, and "philosophical" optics, as exemplified by Plato and Aristotle. The mathematicians were perfectly content to strip the visual act down to a geometrical minimum, thus disregarding all psychological and physical considerations that were not susceptible to geometrical analysis. The philosophers, on the other hand, focused on precisely those considerations with little or no regard for the geometry of sight. I have suggested elsewhere that this dichotomy arose from a fundamental difference in theoretical approach, and Miles Burnyeat has made the same point more recently.[108] Mathematical opticians were concerned only with the geometrical properties of the ray and the visual phenomena that could be "saved" accordingly. The physical or psychological cause and quality of the visual act, and whether it was due to something transmitted from the eye to external objects or from external objects into the eye, were irrelevant. At bottom, all that mattered for them was that the act be subject to geometrical analysis according

108. See A. Mark Smith, "The Physiological and Psychological Grounds of Ptolemy's Visual Theory: Some Methodological Considerations," *Journal of the History of the Behavioral Sciences* 34 (1998): 231–46, especially 231–34.

to straight lines and angles with little or no regard to what those lines might represent in physical reality.[109] In short, the mathematical opticians, Euclid in particular, took an essentially instrumentalist approach to visual analysis. Plato and Aristotle, by contrast, took a realist approach to that analysis, attempting to explain vision according to actual physical and psychological principles. Burnyeat, in fact, suggests that Plato framed his account of vision in the *Ti-maeus* as "a polemic against the idea of a mathematics of visual appearances" and, therefore, as a repudiation of the instrumentalist approach the ray theorists took.[110] If so, then we are left to scratch our heads in puzzlement over Plato's supposedly wholehearted approval of a mathematics of astronomical appearances. Why the one and not the other?

From all this we are pretty safe in concluding that by the mid-second century AD, the science of optics, taken in its broadest sense, was fragmented along several blurred lines. At a gross level, there seems to have been a fundamental rift between those who favored extramissionism and those who favored intromissionism. At a deeper level, though, there were apparent divisions within each of these camps. One such division was between mathematical opticians, who were relatively indifferent to the physical and psychological nature of the visual act, and philosophers, who were relatively indifferent to the formal, geometrical analysis of that act. But even here there were divisions. Galen, for instance, would seem to fall squarely into the philosophical camp, focused as he was on the physical and physiological grounds of vision, yet he gave a nod to ray theory when accounting for image fusion. Aristotle, too, was willing to apply the visual ray model in explaining the rainbow—perhaps taking a self-consciously instrumentalist tack—but as a philosophical realist, he argued vehemently against that model. Hero of Alexandria, on the other hand, would seem to fall squarely into the mathematical camp, yet he imported physical content into his analysis by transforming the ray from a mere line into a trajectory.

To complicate matters further, there were divisions over the nature of light. Some, like Aristotle and Galen, treated it as a mere catalytic agent in sight. Others accorded it a more direct role as a source of illumination that affects the visual apprehension of color. And others yet viewed it as a fully existent and

109. According to this interpretation, mathematical opticians, such as Euclid and Hero, took an extramissionist position out of convenience rather than conviction. It should be noted, however, that, unlike Euclid, Hero offered both a physical and metaphysical explanation of the visual ray as a trajectory.

110. Burnyeat, "Archytas," 46–48.

independent entity governed by the same principles as visual radiation. Color theory, as well, witnessed a rift between those who viewed color as objectively determined and those who viewed it as subjectively determined. Overall, the lines of division among ancient opticians and visual theorists were both manifold and permeable. No matter their fundamental theoretical allegiance, ancient optical thinkers were not averse to incorporating ideas from rival thinkers when it suited their purposes. Manifest in Galen's account of vision, this eclectic attitude is equally manifest in the account Ptolemy provided in his *Optics*. As we will see in the next chapter, Ptolemy's principal goal in framing that account was to accommodate ray geometry to the physics and psychology of sight while, in the process, resolving various problems the Euclidean analysis of direct and reflected vision raised.

Ptolemy and the Flowering of Greek Optics

Ptolemy's *Optics* presents an unusual cluster of problems.[1] First and most obvious is that the treatise exists only in a twelfth-century Latin translation based on an Arabic version that is no longer extant, so it lies at least two linguistic removes from the lost Greek original and at least one from the lost Arabic intermediary. The translator, Eugene of Sicily, was a Byzantine Greek who served under a succession of Sicilian rulers from the Norman king William I (d. 1166) to the Hohenstaufen emperor Henry VI (d. 1197). In recognition of this service, he was raised to the exalted rank of *amiratus* ("admiral" or "emir") a decade or so before his death in 1203. Well placed in the Norman and Hohenstaufen administration of Sicily, Eugene was also noted for his learning. "A man as expert in Greek as in Arabic, and not ignorant in Latin," and part of a tight group of court scholars, he was involved in the effort, spearheaded by Henry Aristippus, to translate Ptolemy's *Almagest* into Latin.[2] This project was completed around 1160, and Eugene may have finished translating the *Optics* at roughly the same time. His choice as translator of Ptolemy's *Optics*,

---

1. For a detailed analysis of the points made in this preliminary sketch, see the introduction to A. Mark Smith, trans., *Ptolemy's Theory of Visual Perception: An English Translation of the* Optics *with Introduction and Commentary* (Philadelphia: American Philosophical Society Press, 1996), 1–21. See also Albert Lejeune, ed. and trans., *L'Optique de Claude Ptolémée dans la version latine d'après l'arabe de l'émir Eugène de Sicile* (Leiden: Brill, 1989), 7*–28*.

2. The characterization of Eugene's linguistic talents comes from the preface to the 1160 translation of the *Almagest* (*Virum tam grece quam arabice peritissimum, latine quoque non ignarum*), quoted in Charles Homer Haskins, *Studies in the History of Mediaeval Science* (Cambridge, MA: Harvard University Press, 1927), 191.

whether self-imposed or thrust on him, was doubtless due to his expertise in Arabic. But as a Latinist he was significantly less proficient, the result being that his translation is somewhat rough and idiosyncratic in style: "assez bar-bare," as his modern editor, Albert Lejeune, characterizes it.[3] Consequently, it is sometimes difficult to make full sense of the text, a problem made all the more acute by the apparent misplacement of certain passages in it. Whether all or some of these problems stem from deficiencies in the Arabic version from which the translation was drawn or in the translation itself is uncertain.[4]

A more serious problem is that the text is incomplete, reflecting the Arabic text upon which it was based, according to Eugene's testimony.[5] Not only is the entire first book of the treatise missing, but also the fifth book comes to an abrupt halt in the middle of a proposition dealing with image distortion in refraction. How much of book five has been lost and whether it was followed by additional books in the original Greek version are matters of guesswork. We do, however, get a hint of the content of the lost first book from the prologue of the second, where Ptolemy offers a brief résumé of what was covered in book one. There, he assures us, we learned "everything that one can gather about what enables light and visual flux to interact, how they assimilate to one an-other, how they differ in their powers and operations, what kind of essential difference characterizes each of them, and what sort of effect they undergo."[6] In its current form, then, Ptolemy's *Optics* consists of a whiff of book one, books two, three, and four in their entirety, and a significant portion of book five.

The topical structure of the treatise follows the order of books. Accord-ingly, in book two, Ptolemy deals with visual perception in general, accord-ing to straight, unimpeded visual radiation—the subject of optics proper by Proclus's account. Books three and four in turn are devoted to reflection and image formation in plane, convex spherical, and concave spherical mirrors. These topics, of course, belong within the domain of catoptrics, as Proclus described. In the fifth and final book, Ptolemy passes from mirrors to an anal-ysis of refraction according to the penetration of visual rays from air into water, air into glass, and water into glass. Unprecedented in scope and sophistication, this analysis is nonetheless foreshadowed in the sixth and concluding postu-

3. Lejeune, *L'Optique*, 7*.

4. At the beginning of book two, Eugene tells us that he had two Arabic texts at hand and that he relied on the "more recent one" because it was more accurate; see Smith, *Ptolemy's Theory*, 70.

5. See ibid.

6. Ibid.

late of Euclid's *Catoptrics*, which is the source of Ptolemy's "floating coin" experiment in book five.[7] Refraction, in short, was not accorded its own subdomain by ancient opticians. Rather, it was lumped with reflection under the broad head of catoptrics, presumably because both involve the "flexure" (*anaklasis* or *flexio*) of rays.[8]

Of all Ptolemy's works, which run a topical gamut from geography, through astrology, to music, surely the best known is his massive astronomical compendium, the *Mathēmatikē Suntaxis* ("Mathematical Systematization"), which eventually came to be known as the *Almagest*.[9] This work provides a useful benchmark for situating the *Optics* within Ptolemy's oeuvre because of certain methodological and topical connections between the two. These we will explore in further detail at the appropriate time, but for now it will be enough to signal their existence and to point out that in regard to these connections, the *Almagest* appears to have been a source of inspiration for the *Optics*, rather than the reverse. It follows, then, that the *Optics* was written later than the *Almagest*, which was probably completed by around AD 150, perhaps a little later. How much later than that the *Optics* was composed is open to speculation, but it is reasonable to suppose that it was finished sometime between AD

---

7. See Ptolemy, *Optics*, V, 5, in ibid., 230–31. Ptolemy's experiment involves placing a coin at the bottom of a vessel, having the viewer step back from the vessel until the coin disappears below the rim, and then pouring water into the vessel until the coin appears to float back up into view.

8. According to Hero of Alexandria, "Negotium autem quod circa visus dividitur in opticam, id est visivam, et dioptricam, id est perspectivam, et katoptricum, id est inspectivum negotium"; Jones, "Pseudo-Ptolemy," 153. Loosely translated, the division described is according to "optics, i.e., the study of seeing; dioptrics, i.e., the study of perception; and catoptrics, i.e., the study of reflection." Proclus offers some insight into the meaning of "dioptrics," as used by Hero, when he tells us that catoptrics is "closely bound up with the art of representation and studies what is called scenography, showing how objects can be represented by images that will not seem disproportionate or shapeless when seen at a distance"; Proclus, *Commentary*, in Morrow, *Proclus*, 33 (translation slightly modified by me). Another clue to the meaning of "dioptrics" in antiquity is the word itself, which derives from *dioptra*, a sighting instrument used for surveying. It is apparent, then, that in antiquity, unlike today, refraction was not the subject matter of dioptrics.

9. Aside from the *Almagest*, the surviving works of Ptolemy include the *Geography*, the *Tetrabiblos* (on astrology), the *Harmonics*, the *Canon* (a list of Egyptian rulers), the *Handy Tables*, and the *Planetary Hypotheses*. Ptolemy is also credited with a work titled *Peri kritēriou kai hēgemonikou* ("On the Criterion [of Truth] and the Governing Faculty"), which shows strong Stoic leanings. There is some doubt about the authenticity of this work.

160 and 170. Ptolemy would thus have composed it relatively late in his life, which probably spanned the period between roughly AD 100 and AD 180.[10]

Naturally, the *Optics* has not escaped the issue of authenticity. Having never seen print until the late 1800s, it fell into oblivion from the early seventeenth century to the very end of the eighteenth, when a trio of scholars unearthed manuscript copies. During the next two decades or so, the text was examined fairly closely, and certain scholars, the astronomer and historian of astronomy Jean Baptiste Delambre (d. 1822) notable among them, doubted the Ptolemaic authorship of the *Optics* on the grounds of its inferiority to the *Almagest* at both the mathematical and methodological level. During the early 1870s, however, the question of authenticity was largely laid to rest, and the modern editor of the *Optics*, Albert Lejeune, drove a stake through its heart in the late 1950s. Despite Wilbur Knorr's effort to revive the issue in the mid-1980s, the consensus in favor of authenticity remains firm, and I see no reason to go against it.[11]

All indications are that Ptolemy spent his working life at Canobus (or Canopus) in the Nile delta, within a stone's throw of the city of Alexandria. Although Alexandria had lost some of its original, Hellenistic luster by the second century AD, it was still an important and vibrant commercial and intellectual center, and Ptolemy seems to have benefited from the various intellectual currents feeding into it at that time. In his astronomical work, for example, he clearly drew on observational and tabular methods developed in the "Babylonian" Middle East and brought to fruition during the Seleucid period.[12] Ptolemy's syncretism in astronomy is fully matched by his syncretism in optics. This is evident in his interdisciplinary approach to visual theory. As mentioned earlier, for example, Ptolemy addressed certain problems related to astronomy in his *Optics*, and he applied analytic methods from astronomy to his tabulations for refraction, a point to be discussed later at some length.

Surely the most fundamental convergence between Ptolemy, the astronomer, and Ptolemy, the optician, lies in his use of geometry as an analytical tool. Just as in astronomy, so in optics, Ptolemy was bent on "saving" the appearances through mathematics. Euclid, of course, was his primary source for that enterprise, but unlike Euclid, Ptolemy was mindful of the limitations of ray geometry and the attendant limitations of Euclid's instrumentalist ap-

10. See Smith, *Ptolemy's Theory*, 1–3.

11. For Knorr's argument, see "Archimedes and Pseudo-Euclid," 96–104 (previously cited in note 107, chapter 2).

12. For the classic account of these methods and Ptolemy's use of them, see Otto Neugebauer, *The Exact Sciences in Antiquity*, 2nd ed. (New York: Dover, 1969), especially 97–190.

proach to vision. In order, therefore, to overcome those limitations, Ptolemy looked to sources and ideas that would provide appropriate physical and psychological support for his ray-based analysis. The resulting account of visual radiation and perception bears evident Stoic and Aristotelian traces, and it is possible but unlikely that Ptolemy found inspiration in Hero's *Catoptrics* for his account of the physics of reflection and refraction. Significant, as well, are the sources against which Ptolemy argued, the "Pythagorean" theory of particulate interaction in Plato's *Timaeus* being an example. On the whole, then, Ptolemy's *Optics* represents a sophisticated integration of concepts and methods drawn from a spectrum of technical and philosophical sources. As such, it reveals the same sort of wide-ranging erudition and spirit of accommodation that we saw before in Ptolemy's younger contemporary, Galen.

1. THE PTOLEMAIC ACCOUNT OF VISUAL PERCEPTION

Like Euclid's account of sight, Ptolemy's is predicated on radiation emitted from a point in the eye and forming a visual cone (*piramis visibilis*) whose base defines the field of view. Unlike Euclid's, however, Ptolemy's visual radiation is absolutely continuous in all dimensions and, therefore, not actually composed of discrete visual rays (*radii visibiles*). For Ptolemy, then, visual rays are not real lines in space; they are virtual lines and, as such, mathematical abstractions that allow us to treat certain visual phenomena according to the geometry of straight lines and angles.[13] Given Ptolemy's conception of visual radiation, it follows that things look continuous to us not because of visual scanning, as Euclid evidently supposed, but because the visual flux by means of which we see them is continuous. About the physical nature of this flux, we can only speculate because whatever details Ptolemy might have provided are lost with the first book. In all likelihood, he conceived of it as pneumatic and generically related to light, this latter point indicated by his claim that the two "interact [and] assimilate to one another."[14]

According to Ptolemy, this flux radiates physically out from the eye to visible objects rather than simply establishing a sympathetic link between the two

13. See *Optics*, II, 50, in Smith, *Ptolemy's Theory*, 91; for the Latin, see Lejeune, *L'Optique*, 37. Henceforth, these works will be cited by author only.

14. *Optics*, II, 1, in Smith, 70, and Lejeune, 11. Later, in *Optics*, II, 23, Ptolemy asserts that the visual flux is affected by light and color "because it shares their genus"; Smith, 80, and Lejeune, 23. It is important to note that this analysis of Ptolemy is based on a Latin translation that is at least two removes from the Greek original, so it may not adequately reflect Ptolemy's intentions in that original.

through the surrounding medium. Each visual ray thus represents a trajectory of sorts, and on that basis Ptolemy vests the flux with dynamic qualities. Although Ptolemy gives no details about the anatomical or physiological structure of the eye or of the optic system as a whole, there are hints in the *Optics* that he regards the eye, or at least the visually functioning part of it—which he calls the "viewing" portion (*aspiciens*)—as perfectly spherical. There are also hints in the *Optics* that Ptolemy locates the vertex of the visual cone at the center of curvature of this sphere.[15] The term *aspiciens* most likely refers to the cornea, not the anterior surface of the crystalline lens, and it may imply that this surface is visually sensitive, although such an interpretation is tentative at best.[16]

Objectively, or as Ptolemy phrases it, "according to the disposition of visible objects," the visible world presents itself at three levels.[17] At the most fundamental level, visibility is a function of the luminosity and physical compactness, or opacity, of a given object, the one allowing the visual flux to be properly affected by the object's surface when it comes into contact with it and the other preventing the flux from penetrating that surface without "feeling" it. Hence, to put it in Ptolemy's own words, "luminous compactness (*lucida spissa*: literally, "luminous, dense things") is . . . intrinsically visible (*vere videntur*), for objects that are subject to vision must somehow be luminous, either in and of themselves or from elsewhere, for that is essential to [the functioning of] the visual sense." They must also be compact "in order to impede the visual flux, so that its power may enter into them rather than pass through without incident effect."[18] Notice, then, that for Ptolemy light has become more than a mere catalytic agent; it plays a direct, even palpable role in vision by rendering

15. See Smith, 28–29. Especially telling is Theorem IV.1 in *Optics*, IV, 3–5 (ibid., 175–76), where Ptolemy demonstrates that if the vertex of the visual cone lies at the center of a concave spherical mirror, all the rays emanating from that point to the mirror will reflect back to that center point through the concentric *aspiciens*. See also *Optics*, III, 16, in ibid., 137–38, and Lejeune, 96–97.

16. At several points in the *Optics* Ptolemy refers to the "pupil" (*pupilla*), which translates to *korē* in Greek and, as we saw earlier in chapter 2, refers not to the opening in the iris (as we take it today), but to the portion of the anterior surface of the cornea in front of that opening. Some took this portion to be the sensitive part of the eye, and Ptolemy may have concurred, although he does not say so explicitly. On the other hand, if Ptolemy took the *aspiciens* and *pupilla* to be equivalent, he may have understood this portion of the eye to be a sort of window through which external objects are viewed from the perspective of the visual cone's vertex at the eye's center.

17. Quoted passage from *Optics*, II, 3, in Smith, 71, and Lejeune, 12.

18. *Optics*, II, 4, in Smith, 71, and Lejeune, 12.

the objects themselves visible rather than by permeating the medium to make it effectively transparent or percipient.

At the next level of visibility come colors, which are "primarily visible (*primo videntur*), because nothing, besides light, that does not have color is seen." Color, in short, is the sole, proper object for sight. Nonetheless, Ptolemy continues, "colors are not intrinsically visible, since [they] are somehow contingent on the compactness of bodies and are not visible per se without light."[19] The remaining visible characteristics, are "secondarily visible" (*sequenter videntur*) because their perception depends on the apprehension of color. As listed by Ptolemy, these secondarily visible characteristics range from corporeity (*corpus*, that is, the fact that the thing is a physical body), through size, shape, and place, to activity, or motion (*motus*), and rest, all of which actually are, or are equivalent in kind to, Aristotle's common sensibles.[20]

Corporeity is sensed visually as soon as the flux emanating from the eye meets the visual resistance of a luminous, opaque surface, just as it is sensed tactilely when a probing finger encounters the physical resistance of an outlying, solid surface. The rest of the secondarily visible characteristics are grasped "by means of . . . illuminated colors, not insofar as they are colors but only insofar as they have boundaries."[21] It is thus by defining the surfaces of visible objects that colors allow us to infer the size, shape, place, motion, or rest of such objects according to the boundaries delineated by color contrasts among those surfaces within the field of view. In his insistence that color is an inherent quality of physical objects, that it coincides with and defines their surfaces, that it is the proper object of sight, and that it is the mediating entity through which the visual faculty apprehends spatial characteristics, Ptolemy to this point is perfectly consistent with Aristotle.[22]

For Ptolemy, as to some extent for Galen, the process of visual perception is initiated when the visual flux makes contact with the surface of a visible object and undergoes the effect (*passio*) of illumination and coloring. Purely qualitative rather than material, the resulting alteration of the visual flux (in the form of *coloratio*) is transmitted back to the eye, the flux serving as a continuous me-

19. *Optics*, II, 5, in Smith, 71, and Lejeune, 13.

20. Aristotle lists common sensibles in several places among his works, and the lists do not always correspond, which suggests that he was providing examples rather than a definitive and limited catalog.

21. *Optics*, II, 7, in Smith, 72, and Lejeune, 13.

22. For further discussion of the Aristotelian aspects of Ptolemy's theory, see A. Mark Smith, "The Psychology of Visual Perception in Ptolemy's *Optics*," *Isis* 79 (1988): 189–207, especially 200–203.

dium to support that transmission. Ptolemy explicitly denies that this process involves any material or mechanical interaction that would imply a subjective determination of color. Those, he says, who "have thought that whiteness is perceived through a spreading out of the visual rays, while blackness is perceived through a constriction of them" are wrong—the "those" presumably being thinkers who accept the "Pythagorean" account of color perception in Plato's *Timaeus*.[23] The capacity of the visual flux to undergo the qualitative change of "coloring" (*coloratio*) is due to its being "perspicuous," for it "should have no qualification but should be pure and should suffer the qualification [passed to it] by light and color."[24] This qualification is eventually conveyed through the visual flux to the governing faculty (*virtus regitiva*), which makes visual sense of it according to both the color itself and the secondary, spatial characteristics contingent on that color.[25] Clearly reminiscent of the Stoic *hēgemonikon*, to which Galen also appealed in his theory of perception, Ptolemy's governing faculty judges and controls all sense perception while providing a sense of self-location that enables the perceiver to discern where things are in the space defined by the visual cone.[26]

The perception of where things are in visual space and how much of it they occupy is determined from the vertex of the visual cone, which Ptolemy characterizes as a "source" (*principium*) and which provides the ultimate analytic viewpoint for Ptolemaic optics. Right or left and up or down are thus perceived according to where the visual ray, or set of visual rays, that touch an object at the base of the visual cone lie in relation to the axis. From this the visual faculty gets a sense of where the endpoints of those rays lie with respect to the center of the cone's base. This is essentially Euclid's explanation of spatial position within the base of the visual cone. Unlike Euclid, however, Ptolemy supposes that the visual faculty senses distance along the ray according to its length; the longer the ray, the farther away what it sees is "felt" to lie.

Consequently, the perception of size is a function not only of the visual angle subtended by a given object but also of its perceived distance, as well as of its slant. According to figure 3.1, of the four object lines AB, CD, CG, and FG within visual cone AED, AB will be perceived as largest from viewpoint E because of its linear distance along axis EX of the cone and its slant, which

23. Quotation from *Optics*, II, 24, in Smith, 80, and Lejeune, 24.

24. *Optics*, II, 23, in Smith, 80, and Lejeune, 23.

25. See *Optics*, II, 23–24, in Smith, 79–80, and Lejeune, 22–24.

26. See *Optics*, II, 76, in Smith, 103, and Lejeune, 51–52. Ptolemy does not specify the location of the governing faculty, so there is no way of determining whether he thought it lay in the brain or in the heart.

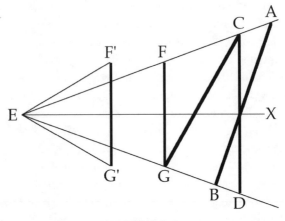

FIGURE 3.1

is perceived according to the difference in length between ray EA and ray EB. Object line FG will therefore be perceived as smallest, and CD will be perceived as larger than CG because its greater axial distance outweighs the slant of CG. Hence, by taking the perception of distance and obliquity into account, Ptolemy resolves the Euclidean problem of why objects subtending the same visual angle do not necessarily look the same size.

The ability of the visual faculty to sense distance according to ray length is limited, though. When it becomes too great, the separation between eye and object can no longer be properly perceived, and judgments based on it can no longer be properly made. Take the perception of shape as an example. "The visual faculty perceives shape," according to Ptolemy, "by means of the shapes of the bases upon which the visual rays fall."[27] It does so by perceiving the boundary defined by the object's color within the field of view, as well as by the convexity or concavity of the surface in contact with the visual cone, "just as . . . objects are perceived by touch, convex ones being apprehended through the concavity of the encircling hand, and concave ones being apprehended through the convexity of the encircling hand."[28] At a great enough distance, though, concavity or convexity will become imperceptible because the relative difference in ray lengths touching more prominent or more sunken points on the surfaces of concave or convex bodies will be undetectable. That is why the sun and moon appear as flat disks rather than as spheres; they lie so far away

27. *Optics*, II, 64, in Smith, 98, and Lejeune, 46.
28. *Optics*, II, 67, in Smith, 99, and Lejeune, 47.

that the visual faculty can no longer sense their convexity. For the same reason, we might not perceive that what looks like an ellipse within the field of view is actually a circle seen obliquely at a great distance or that what looks rounded from far away is actually angular in shape.[29] The disappearance from view of a given object is also a function of distance. The farther the object recedes from the viewpoint, the smaller the visual angle becomes, until, finally, it becomes imperceptibly small. At that point, the object becomes imperceptible as well, not because it disappears between discrete rays, as Euclid argues, but because it becomes insensibly tiny.

The perception of motion and rest is complicated somewhat not only by the fact that motion can be lateral to the visual axis or to and fro along it, but also by the fact that it is affected by whether and how the vertex of the visual cone moves. It is for that reason, Ptolemy warns us, that "visual perception of the phenomena associated with location depends primarily on the visual faculty itself [rather than on the object undergoing the change]."[30] For instance, when the viewpoint at the vertex of the visual cone is stationary, an object is perceived not to move toward or away from it along the visual axis if the distance between the object and the viewpoint remains constant along the axis over a perceptible span of time. On the other hand, when the viewpoint does move along the axis, the object will be perceived as stationary if the distance it appears to move toward or away from the viewpoint matches the distance the viewpoint is perceived to move in the given direction during the same amount of time.[31] Likewise, the perception of lateral motion is dependent on the relationship between the viewpoint and object. If the viewer senses that he is stationary with his eye held steady and perceives that an object occupies successive spots within his field of view, he will perceive the object to move. By the same token, if the viewer perceives that his eye is moving horizontally or vertically and perceives an object to move in the opposite direction at the same rate, he will perceive the object as stationary. Otherwise, he will perceive it to move, the direction and rate of its motion visually determined according to how far ahead or behind the axis it appears to fall in a given span of time.[32] The

29. See *Optics*, II, 70–72, in Smith, 100–101, and Lejeune, 48–49. Euclid had already dealt with the rounding of polygonal shapes seen at a distance in proposition nine of his *Catoptrics*, and it would remain a standard trope for optical theorists for centuries to come.

30. *Optics*, II, 76, in Smith, 103, and Lejeune, 51.

31. See *Optics*, II, 78–79, in Smith, 103–4, and Lejeune, 52–53.

32. See *Optics*, II, 80–81, in Smith, 104–5, and Lejeune, 53–54.

perception of motion is therefore heavily dependent on the viewer's spatial self-awareness, which comes to ground in the governing faculty.[33]

Visual perception, or the quality of it, is ultimately dependent on the clarity with which things are seen, and that diminishes according both to distance from the vertex of the visual cone and to distance from the visual axis within the cone. Visual clarity decreases with distance from the cone's vertex, according to Ptolemy, because visual radiation loses power or intensity as it recedes from its source at the center of the eye. This sort of loss is seen in physical projection, as well as in the radiation of heat and light: the farther from the source, the weaker the effect.[34] Accordingly, the closer a given object lies to the viewpoint at the vertex of the visual cone, the more intensely it will be "illuminated" by the visual flux reaching it. The greater clarity of a closer view is due in part to the visual angle, which allows more flux to strike the object's surface, as illustrated in figure 3.1, where the object represented by F´G´ is the same size as that represented by FG but subtends a larger visual angle and therefore presents a larger surface to the oncoming flux. In addition, the power or intensity of the radiation along lines EF and EG, as well as along all the other lines reaching to FG from E, will be less than that along lines EF´, EG´, etc., reaching F´G´. On that basis, FG will be seen more clearly than CD, even though they subtend the same visual angle and are thus "illuminated" by the same number of rays. Within the visual cone itself, the axial ray yields the clearest view because the farther from the axis toward the edge of the visual cone the rays get, the weaker their power. Ptolemy's rationale for this claim is unclear, but there is no question that the claim itself is empirically sound; things do appear clearer at the center of the visual field than they do at its edges.[35] A final factor in visual acuity is the doubling of the eyes, which "nature" has contrived "so that we may see more clearly and so that our vision may be

33. As Ptolemy phrases it in *Optics*, II, 76, "The motion and rest of the visual flux . . . is apprehended not by the visual faculty but by the sense of touch that extends to the Governing Faculty, in the same way that we do not discern the motion of our hands by sight when our eyes are closed but by means of a continuous [sense link] that reaches to the Governing Faculty"; Smith, 103, and Lejeune, 52.

34. See *Optics*, II, 20, in Smith, 78, and Lejeune, 14.

35. In *Optics*, II, 28, Ptolemy claims that the weakening of vision toward the edge of the field of view is due to the increasing absence (*privatio*) of rays toward the edge of the visual cone; see Smith, 77–78, and Lejeune, 19–20. What he means by "absence" is unclear at best. If he had in mind that the base of the visual cone is a circle and thus planar, then he might have been referring to the fact that the flux becomes increasingly dispersed over the plane of the circle from center to edge.

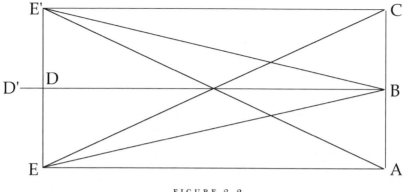

FIGURE 3.2

regular and definite."[36] This appeal to "nature" as providential in designing the optic system for efficiency is reminiscent of Galen, and Ptolemy refers to a provident nature more than once in the *Optics*.[37]

Having established the basic grounds of visual perception, Ptolemy sets out upon a fairly detailed analysis of how the visual process can go awry in ways that lead to misperception. He provides a typology of such misperceptions or illusions according to cause. One type is due to abnormal conditions in the physical circumstances under which sight occurs. Diplopia, or double vision, is an example. In binocular vision, Ptolemy explains, we normally see a single object as single rather than double because "we are naturally disposed to turn our raised eyes unconsciously in various directions with a remarkable and accurate motion, until both axes converge on the middle of a visible object" so as to form a single base "composed of all the correspondingly arranged rays [within the separate visual cones]."[38] What this means is illustrated in figure 3.2, where E and E´ are the right and left eyes, respectively, and EB and E´B the visual axes meeting at point B on object line AC. Rays EA and E´A on the right side of their respective visual cones AEC and AE´C are corresponding rays, as are EC and E´C on the left side. Consequently, points A, B, and C will appear single, and so will the entire object line AC.

The way that "nature . . . joins [the two visual axes] according to the location of the visible object" is by keeping them focused at the point where the

36. *Optics*, II, 28, in Smith, 83, and Lejeune, 27.

37. See *Optics*, II, 45; III, 35; and V, 35, in Smith, 90, 144, and 244, and Lejeune, 34, 106, and 246.

38. *Optics*, II, 28, in Smith, 83, and Lejeune, 27.

"common axis" intersects the object.[39] This axis runs from its own *principium*, or "apex," to the midpoint on the object, as represented by DB in figure 3.2, which bisects line EE′ and is perpendicular to it. When the three axes are arranged in this way, with the object facing the viewer frontally, the object is seen as clearly as possible. Exactly where the apex lies in relation to the eyes is open to debate, but Ptolemy does say that it "is where the vertices of the visual cones ought to intersect."[40] On that basis, I have suggested very tentatively that he may have had the optic chiasma in mind, in which case it would lie at some point D′ beyond D. Against this suggestion, Gérard Simon has argued in favor of Aristotle's common sensibility as the controlling *principium*, locating it at point D between the eyes.[41] Of one thing, however, there is no doubt: it is from the apex that the governing faculty controls the focusing of the two visual axes in order to ensure that both eyes yield a single image.[42]

Diplopia, or double vision, occurs when the natural tendency of the axes to be brought to focus is disturbed in such a way that noncorresponding rays see an object point. As a result, the object point appears displaced. In order both to test and explain this point empirically, Ptolemy offers some simple experiments based on a short plaque and two pegs, one white, the other black. The first three of these experiments are illustrated in figure 3.3, where the rectangle represents the plaque with the right and left eyes placed at E and E′, respectively. The white peg is placed at A and the black one at B, both of them on the midline of the plaque, which corresponds to the common visual axis. If the two axes EA and E′A are brought to convergence on peg A, then both eyes will see it as single. But E will see peg B with ray EB, which lies to the right of axis EA within the visual cone emanating from E, and E′ will see peg B with ray E′B, which lies to the left of axis E′A in the visual cone emanating from E′. Consequently, from the perspective of E, peg B will appear to the right of peg A, and from the perspective of E′, peg B will appear to the left of peg A at an equal distance. The resulting view is represented at the lower left of the figure. Now let the two axes be brought to convergence on peg B. In that case, E′A will be a rightward ray with respect to axis E′B and EA a leftward ray with

39. *Optics*, III, 35, in Smith, 144, and Lejeune, 106.

40. *Optics*, III, 35, in Smith, 144, and Lejeune, 106.

41. See Smith, 28–29; cf. Gérard Simon, "La vision binoculaire chez Ptolémée et Ibn al-Haytham," in *Archéologie de la vision* (Paris: Seuil, 2003), 131–63, especially 141–58.

42. "It has thus been demonstrated," says Ptolemy in *Optics*, III, 61, "how the visual cones attain a perceptual grasp [that] leads to a single, primal impression in terms of both sensible effect and location . . . when both axes . . . are brought to convergence from the Apex toward the visible object by the Governing Faculty"; Smith, 152, and Lejeune, 117.

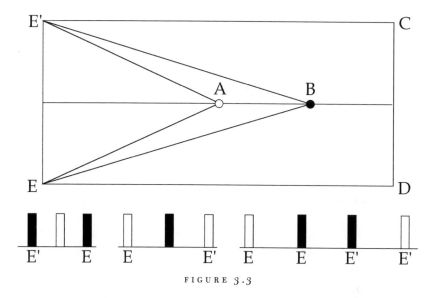

FIGURE 3.3

respect to axis EB. Hence, peg B will be seen single with both eyes, whereas peg A will appear to the left of it from the perspective of E and to the right of it from the perspective of E´, as represented in the middle arrangement below the figure. Finally, if the two eyes stare off to infinity, so that their respective visual axes coincide with ED and E´C, the two pegs will appear double in the arrangement illustrated at the lower left of the figure.[43]

Diplopia involves a misperception of place (and number), and it is due to a circumstantial abnormality forced on the natural system of image fusion for binocular vision. Another example of such a circumstantial abnormality is farsightedness, which is due to excessive moisture in the eye that affects the visual flux as it passes outward.[44] Color perception can also be affected by circumstantial abnormalities. A brightly illuminated flower, for instance, can tint surrounding objects with its color so as to cause their actual colors to be misperceived.[45] Daubs of individual colors on a rapidly spinning potter's wheel can appear as a muddled blend because there is not enough time for the visual faculty to distinguish each color from the other. Or, if a parti-colored object is seen at too great a distance, it may appear to be of a uniform, blended

43. For Ptolemy's description of these three experiments, see *Optics*, II, 33–36, in Smith, 84–86, and Lejeune, 29–30.

44. See *Optics*, II, 87, in Smith, 107, and Lejeune, 57.

45. See *Optics*, II, 94, in Smith, 109, and Lejeune, 59.

color because the individual spots of color are too small to be properly discerned at that distance.[46] Too much distance can also affect the perception of size and shape, as we saw in the case of the sun and moon, which appear as flat disks because their convexity is imperceptible. Moreover, because the planets are so distant, their motion is imperceptible over short to moderate amounts of time. On the other hand, the motion of rapidly spinning disks is sometimes imperceptible for lack of adequate time to discern the motion of the disk's individual parts.[47]

The two remaining types of illusion involve abnormalities that affect the visual faculty itself and those that affect perceptual judgment. As to the first of these, the most salient abnormality involves "breaking" of the visual flux at reflective or refractive interfaces.[48] As a result, objects appear displaced (for example, "behind" plane or convex mirrors), and they may look distorted in size, shape, and color, and nearer or farther away than they ought to look. Coloring of the visual flux when it passes through a tinted, translucent body is another example of an abnormal effect on the visual flux because it causes a misperception of the colors behind the body.[49] Afterimage is similar in that being abnormally affected by an intensely luminous color, the visual flux retains that color for a while, even after looking away.[50] Yet another example of a misperception affecting the visual faculty itself is the oculogyral illusion, which occurs when a rapidly spinning viewer stops suddenly and perceives his surroundings to spin about him. This misperception arises, Ptolemy explains, because the visual flux continues to be roiled by the original spinning motion.[51]

Illusions involving perceptual judgment can arise in a variety of ways. Of two objects in the same vicinity, for instance, the brighter one will appear closer, which is why "mural painters use weak and tenuous colors to render things that they want to represent as distant."[52] In the same vein, the dimmer of two objects that subtend equal visual angles and that lie the same distance from the viewer will be perceived as larger because its relative dimness will indicate greater distance.[53] The illusion that a boat anchored in a smoothly but swiftly

46. See *Optics*, II, 95–96, in Smith, 109–10, and Lejeune, 59–61.

47. See *Optics*, II, 98, in Smith, 110–11, and Lejeune, 62.

48. See *Optics*, II, 104–5, in Smith, 113, and Lejeune, 65–66.

49. See *Optics*, II, 98, in Smith, 110–11, and Lejeune, 62.

50. See *Optics*, II, 108, in Smith, 114–15, and Lejeune, 66–67.

51. See *Optics*, II, 121, in Smith, 119–20, and Lejeune, 73–74.

52. *Optics*, II, 124, in Smith, 120, and Lejeune, 74.

53. See *Optics*, II, 126, in Smith, 121, and Lejeune, 75–76.

moving river is speeding upstream through that river is another example of a perceptual misjudgment.[54] So too, perceptual misjudgment of shape can arise from a shading of colors that makes things look convex or concave. For that reason, "a painter who wishes to represent these two shapes by means of colors paints the part he wants to appear higher a bright color, whereas the part he wants to appear concave he paints with a weaker and darker color."[55]

Implicit throughout this summary account of illusions is the assumption that in order for visual perception to occur correctly, certain normative conditions must be met. If the distance is too great, the object too small, its surface inadequately luminous or opaque, the visual flux inordinately weakened by old age or excessive ocular moisture, the time needed for proper perception too short or too long, the two visual axes unable to converge properly on the object, the flux broken at reflecting or refracting interfaces—in short, if any or all such abnormalities obtain—misperception of corporeity, color, size, shape, place, motion, or rest is bound to result. Meantime, Ptolemy's typology of visual illusions according to abnormalities in physical circumstances, visual apprehension, and perceptual judgment bespeaks his understanding of vision as a complex process, much of it psychological in nature. Moreover, visual illusions are not monocausally fixed for Ptolemy; they can cross typological lines. Diplopia, for instance, involves an abnormal physical circumstance in that it occurs when the axes of the visual cones are forced from proper convergence, but it also involves an abnormality in visual apprehension because it stems from an effect wrought on the visual flux. Likewise, the illusion that a painted sphere in a mural landscape is convex involves an abnormality in visual apprehension because it is based, quite literally, on a trompe l'oeil, yet it also involves the perceptual misjudgment of shape, place, and size. Most important, however, many of the visual illusions Ptolemy discusses in the *Optics* are not susceptible, or at least not fully susceptible, to geometrical explanation. That a dimmer object looks larger and farther away than a brighter one subtending an equal visual angle not only resists but actually defies such an explanation. And so does the fact that the room keeps spinning after I stop twirling.

With Ptolemy's theory of vision we are obviously a far cry, both chronologically and conceptually, from Euclid. For a start, Ptolemy's approach to vision is considerably more comprehensive than Euclid's because he takes the issue of perception seriously, following it well beyond the limits of mere geometri-

54. See *Optics*, II, 131, in Smith, 123–24, and Lejeune, 62.

55. *Optics*, II, 127, in Smith, 122, and Lejeune, 78–79.

cal analysis. Ptolemy therefore deals with a much broader spectrum of visual phenomena, particularly those involving misperception or illusion, than did Euclid. In order to do so, he has to strike a balance between empirical observation and theoretical explanation. A case in point is Ptolemy's observation that in order to represent things as farther away, painters render them in dimmer colors, according to what is now called aerial perspective. Theoretically, the reason this representational technique works is that in reality, according to Ptolemy, closer objects appear more clearly and vividly than distant ones because they are struck by visual rays of greater intensity. Ptolemy's empirical bent is thus manifested in the numerous types of visual illusions he provided in relatively systematic order, but it is especially salient in his experiments to test binocular vision according to a theoretical model based on the convergence of axial rays. As we will see in the next two sections, Ptolemy applies his empirical bent in a particularly effective way to the analysis of reflection and refraction.

2. THE PTOLEMAIC ACCOUNT OF REFLECTION

Ptolemy opens book three of the *Optics* by noting that in addition to the illusions arising in direct, unbroken radiation, which have been adequately covered in book two, there are those that arise when the visual rays are broken at refractive or reflective interfaces. "It is to this phenomenon," Ptolemy informs us, "that we have devoted the greater part [of this treatise]."[56] In the case of refraction, he continues, this breaking is only partial, allowing the visual flux to continue through the surface after striking it. In reflection, on the other hand, the breaking is complete because the reflecting surface blocks penetration entirely. It is to this latter kind of breaking, Ptolemy says, that he will now turn, focusing his analysis on plane, convex spherical, and concave spherical mirrors, as well as on combinations of them.

As with any scientific investigation, according to Ptolemy, we must start this one with fundamental principles. In the case of reflection, these are three in number, and they are all empirically verifiable.[57] The first principle is that objects viewed in mirrors are seen along the extension of the incident visual ray. This principle can be confirmed by the simple expedient of looking into any of the three types of mirrors and noting that, "if we mark on [their] surface . . .

---

56. *Optics*, III, 1, in Smith, 131, and Lejeune, 60.

57. For Ptolemy's listing of these three principles, see *Optics*, III, 3, in Smith, 131–32, and Lejeune, 88–89.

the spots where visible objects appear and then cover them, the image of the visible object will no longer appear." Subsequently, "if we uncover one spot after another . . . the designated spots will appear together with the image of the visible object along the line of sight projected from the vertex of the visual cone."[58] This verification has a distinctly Heronian ring in that it establishes much the same thing that Hero does in the third proposition of his *Catoptrics* (see chapter 2, section 6 above), and it does so in much the same way.

Ptolemy's second principle is that individual spots on a visible object are seen along the perpendicular dropped from them to the mirror's surface, that is, the cathetus. In order to confirm this, we need only stand a moderately long, straight object upright on the surface of any mirror and notice that its image lies directly in line with it.[59] In tandem, these two principles establish the cathetus rule of image location, according to which the image of any object point seen in a mirror will lie at the intersection of the cathetus dropped orthogonally from the object point to the mirror's surface and the extension of the incident ray. The third principle Ptolemy adduced states that no matter the shape of the reflecting surface, the angles of incidence and reflection will always be equal, whether incidence is oblique or along the perpendicular. Taken as a whole, these three principles establish that the incident and reflected rays, as well as the cathetus, will lie in a single plane of reflection perpendicular to the mirror's surface, and the normal or perpendicular dropped to the point of reflection will form the common section of all possible planes of reflection passing through that reflection point.[60]

In order to verify the equal angles relationship expressed in the third principle, Ptolemy suggests two ways, the first of which we will ignore.[61] The second, however, merits close scrutiny, because it involves a simple, yet relatively ingenious, experimental confirmation.[62] The setup for this confirmation is presented in slightly adapted form in figure 3.4a, where the circle centered on point A represents a bronze disk "of moderate size" (*moderate quantitatis*), divided into four quadrants. Each quadrant is meant to be subdivided into 1° increments, the diagram showing the subdivision in 5° increments because of size limitations. Line LA is drawn in one of the quadrants, and a sighting device, called *dioptra* in Latin, is set up on the disk, perhaps on an arm that

58. *Optics*, III, 4, in Smith, 133, and Lejeune, 89.

59. See *Optics*, III, 4, in Smith, 132, and Lejeune, 89.

60. See *Optics*, III, 5, in Smith, 132, and Lejeune, 89.

61. See *Optics*, III, 6, in Smith, 133, and Lejeune, 90–91.

62. For the full account of this experiment, see *Optics*, III, 8–11, in Smith, 134–35, and Lejeune, 91–94.

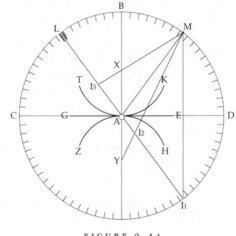

FIGURE 3.4A

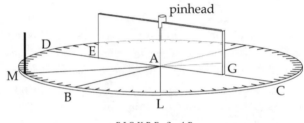

FIGURE 3.4B

pivots around point A.[63] The dioptra is then rotated along the edge of arc BC until the line of sight through it is fixed along LA. Meantime, a colored peg M is set up on the disk, perhaps also on a rotating arm, so that it will slide along the disk's edge within arc BD. Finally, three mirrors are fashioned from highly polished iron strips, one of them plane, the other two formed into circular arcs. When stood upright on the bronze disk at A, the plane mirror GAE will be flush with diameter CD, and the reflecting surfaces of convex mirror ZAH and concave mirror TAK will be tangent to that diameter at point A. Point X on line BA is the center of curvature for the concave mirror, point Y on the

---

63. Ptolemy gives no description of the dioptra. It may be a sighting slit or a tube, but its purpose is obviously to constrict the field of view as much as possible in order to keep the eye focused along a single line of sight.

extension of line BA the center of curvature for the convex mirror. When each mirror is properly situated, a pin is inserted into the bronze disk at point A to hold the mirror in place.

With plane mirror GAE so arranged, we are to fix a line of sight through the dioptra along LA. We are then to slide the colored peg M along the edge of quadrant BD until it reaches line AM, at which point its image in the mirror will coincide with the pinhead at A. This situation is represented in figure 3.4b, which shows what should be seen along line of sight LA in the plane mirror when things are properly arranged. Finding arcs BL and BM equal, we will have thus confirmed that angle of incidence LAB is equal to angle of reflection BAM. The same thing holds for convex mirror ZAH and concave mirror TAK: when the line of sight is established along LA and the peg is moved to point M, its image will coincide with the pinhead, and angles LAB and BAM will be equal. Within certain limitations, this test can be made for any line of sight chosen within arc BC; no matter the angle of incidence, the resulting angle of reflection will be equal to it.

The rationale behind this experiment is dependent on the cathetus rule of image location. Hence, in the case of plane mirror GAE in figure 3.4a, the image of peg M seen in the mirror will lie at I_1, where cathetus MI_1 intersects the extension of incident ray LA, so it will be directly in line with and behind the pin along LA. In the case of convex mirror ZAH, the image I_2 of peg M will lie at the intersection of cathetus MY and the extension of incident ray LA. It, too, will be directly in line with and behind the pin along LA. In the concave mirror, the image I_3 of peg M will lie at the intersection of incident ray LA and the extension of cathetus MX. In this case, the image will lie directly in line with the pin but in front of rather than behind the mirror.

As actually described in the text, the setup for this experiment is somewhat more nebulous and confused than I have presented it, but I am confident that I have captured its intent. Nonetheless, the very vagueness of the textual description raises the issue of whether Ptolemy actually conducted, or could have successfully conducted, the experiment as described. For one thing, the accuracy of the experiment depends upon the size of the bronze disk and the exactitude of the degrees into which it is subdivided. Ptolemy's instruction that it should be of moderate size leaves a great deal to be desired in specificity. Fortunately, we have some indication from his use of that disk later in book five that it should be large enough to be accurate to within roughly half a degree, which would probably make it somewhere around 20 cm in diameter. Given the technological capabilities of Ptolemy's day there is no doubt that a bronze disk this size could be produced or that it could be subdivided into degrees with fair accuracy.[64]

Moreover, although Ptolemy's choice of iron for the mirrors is somewhat puzzling because of the difficulty of working that metal cold, fashioning mirrors from thin, polished iron strips would have been relatively easy. The drawback is that the curved mirrors fashioned from those strips would have been cylindrical rather than spherical, so his test for spherical mirrors would have been somewhat compromised by his use of cylindrical ones. Also, the size of the angles to be effectively tested for the concave mirrors would have depended to some extent on the radius of their curvature, about which the text gives us no useful information. Nevertheless, there is little reason to doubt that Ptolemy could have conducted the experiments within these limitations and that his results would have been good enough if not to perfectly confirm, then at least not to disconfirm the equal-angles relationship in reflection.[65]

After describing his experimental confirmation of the equal-angles law, Ptolemy takes up the issue of why that law should obtain. One reason, he suggests, is the determinate nature of the image according to both the incident ray and the cathetus. Since both lines are unique for any given object point and point of reflection, their intersection is unique, and this intersection mandates that the angles of incidence and reflection be equal.[66] Another reason is that reflection "follows a natural course" in the same way that physical rebound does, "for projectiles are scarcely obstructed by objects they strike at tangents, whereas they are obstructed to a considerable extent by objects that resist them [directly] along the line of projection."[67] By extension, "the action of the visual ray itself must . . . follow this rule, and any of the rays that approaches a

64. Ptolemy's description of the circular sighting device in *Almagest*, I, 12, makes it eminently clear that by his time it was possible to manufacture quite precise instruments for astronomical measurement and to subdivide them accurately not only into degrees but also into fractions of a degree; see Toomer, *Ptolemy's Almagest*, 61–63. At the beginning of book 5, Ptolemy describes the manufacture of an even more complex armillary sphere with two interlocking, graduated rings; see Toomer, *Ptolemy's Almagest*, 216–19. It is almost certain that the bronze disk Ptolemy describes in the *Optics* was originally intended for astronomical measurement or was modeled after an equivalent device intended for astronomical measurement.

65. As will become clear when we deal with his refraction experiments, Ptolemy apparently believed that observed and true values can be discrepant and that the observed ones must be adjusted accordingly. Hence, a very near equality between the observed angles of incidence and reflection could be taken to indicate exact equality.

66. See *Optics*, III, 16–18, in Smith, 137–39, and Lejeune, 96–98.

67. *Optics*, III, 19, in Smith, 139, and Lejeune, 98.

mirror and then bounces back . . . must maintain the disposition that occurs in the paradigm case,"[68] namely, that the angles of incidence and reflection be equally disposed with respect to the point of reflection. The paradigm case par excellence, it would therefore seem, is projection and rebound along the normal itself. Since, moreover, visual radiation acts according to the principles of physical projection, it follows that the very act of reflection saps it of some of its intensity, just as the act of rebound causes a physical projectile to lose some of its force. That, in a nutshell, is why things look dimmer in mirrors than they would if seen straight on, without reflection, at the same apparent distance.[69]

Ptolemy's reliance here on the analogy between visual radiation and projectile motion raises the specter of Hero yet again, suggesting that Ptolemy borrowed the analogy from him. In favor of this suggestion is the similarity between Ptolemy's and Hero's empirical verifications of the cathetus rule for plane mirrors (see chapter 2, section 6 above). Two things, however, argue rather forcefully against this suggestion. First, although they are certainly similar in a general way, Hero's and Ptolemy's analogies differ in certain key details. Most telling of these is that, unlike Hero, Ptolemy does not relate reflection or refraction to the surface structure of the bodies struck by visual flux. Ptolemy never refers to the relative softness, hardness, or porosity of such surfaces in order to explain how impinging visual rays interact with them. Instead, as we will see in short order, he appeals to the relative density of the reflecting or refracting body. A second, more compelling, reason to doubt that Ptolemy borrowed from Hero is his failure to offer anything resembling Hero's least-distance justification of the equal-angles law. It is difficult to believe that he would not have availed himself of that justification had he known of it. On balance, then, it seems unlikely that Ptolemy was influenced by Hero or, for that matter, that he ever read his *Catoptrics*.

Having validated the cathetus rule to his satisfaction, Ptolemy returns to his empirical analysis of binocular vision and diplopia on the basis of object lines rather than object points in the field of view. Not surprisingly, he shows that whether such object lines are seen as single or double depends upon whether and how the two visual axes and the common axis converge on the given object line. The resulting account of diplopia therefore adds nothing of moment to what Ptolemy has already shown in book two. Nor does the

68. *Optics*, III, 20, in Smith, 140, and Lejeune, 99.

69. See *Optics*, III, 22, in Smith, 140, and Lejeune, 100–101. In Ptolemy's day, the dimming effect of reflection was due primarily to the relatively poor reflective quality of the mirrors available to him.

analysis of plane and convex mirrors that immediately follows add anything of moment to what Euclid already established in his *Catoptrics*. Accordingly, Ptolemy demonstrates that in reflection from both kinds of mirrors, there can be only one image of any given point seen from any given viewpoint, and it will lie behind the reflecting surface.[70] He then demonstrates that in plane mirrors images always look the same size and shape as their objects and appear to lie as far behind the reflecting surface as their objects do above it. In convex mirrors, on the other hand, images are always smaller and more convex than their objects, and they appear to lie closer to the reflecting surface than their objects actually do. Finally, he shows that in both types of reflection the images suffer left–right reversal.[71]

In his analysis of image formation and distortion in plane and convex mirrors, which occupies less than half of book three, Ptolemy has done little more than plow Euclidean ruts, albeit with somewhat greater mathematical rigor than Euclid. This changes dramatically when he takes up the analysis of concave spherical mirrors in book four of the *Optics*. Here Ptolemy breaks entirely new ground in ways whose importance will become evident later on, when we deal with Alhacen. As we saw earlier, Euclid demonstrated in a rather vague way that an eye and an object point can be situated with respect to a concave spherical mirror in such a way that there can be more than one reflection between the two. How many possible reflections there might be, and precisely how the number of reflections relates to the disposition of the eye and the object point, Euclid leaves unanswered. Ptolemy's analysis of concave spherical mirrors represents a sophisticated and systematic response to both of these questions. It also provided the key to Alhacen's method for determining precisely where the points of reflection will be according to that disposition (see chapter 5, section 3).

Ptolemy starts with the most trivial case, in which the eye is located at the center of the mirror to form its own object, the result being that all the radiation from the eye will be reflected back upon itself from the entire surface of the mirror. Accordingly, the eye will see itself at every point on that surface, just as Euclid showed in proposition five of his *Catoptrics*. Next, Ptolemy locates the eye and object point on a diameter of the mirror at points equidistant from the center of curvature, as illustrated in figure 3.5a, where H is the eye, E the

70. See *Optics*, III, 68–71 and 99–109, in Smith, 154–55 and 162–65, and Lejeune, 120–22 and 133–37.

71. See *Optics*, III, 79–96 and 110–30, in Smith, 156–61 and 165–71, and Lejeune, 124–31 and 137–45.

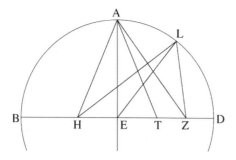

FIGURE 3.5A

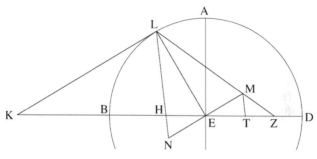

FIGURE 3.5B

center of curvature, and T the object point, all three on diameter BD. Let EA be a radial segment of the diameter normal to BD, and draw HA and AT. From the equality of triangles HAE and TAE, it is obvious that incident visual ray HA will reflect along AT because of the equality of angles HAE and TAE. Moreover, if some point Z is chosen beyond T on diameter BD, it is obvious that there cannot be a second reflection to it from A or from any point, such as L, between A and D on arc AD because the resulting angles of incidence and reflection will be unequal. Conversely, if Z is the eye, there can be no reflection to H from any point between D and A on arc AD. From this it follows that there can be only one reflection in the plane of the circle, and it will be from point A. If we rotate the entire figure around diameter DB as an axis, point A will describe a great circle on the mirror's surface, so visual rays from H will reflect to T from every point on that circle.[72]

Ptolemy has so far demonstrated that only one ray from H will reflect to radius ED from arc AD, and it will do so from point A and from no other point

72. See *Optics*, IV, 3–10, in Smith, 175–77, and Lejeune, 148–51.

on that arc. There can, however, be a reflection from arc AB, as illustrated in figure 3.5b. The actual point of reflection can be found by marking off ET equal to HE and extending diameter DB to point K such that the ratio of line segment ZT to line segment ET (= EH) is the same as the ratio of line segment ZH to line segment KH: that is, ZT:ET = ZH:KH. Drop tangent KL to arc AB, and connect HL and LE. It turns out under these conditions that angle of incidence HLE is equal to angle of reflection ELM.[73] Hence, there will be a reflection to point Z from point L on arc AB within the plane of the circle, and this will be the only possible reflection to point Z within that plane. If, as before, we rotate the figure about diameter DB as an axis, point L will describe a circle smaller than a great circle on the mirror's surface. Visual rays from H will therefore reflect to Z from every point on that smaller circle.

Now let the eye and the object lie on a chord of the sphere, as in figure 3.6a, where D is the sphere's center of curvature and DEA is on a diameter normal to chord BG, which contains the eye at H and the object point at Z. As before, Ptolemy starts with the most trivial case, in which H and Z are equidistant from normal DEA passing through the chord, that is, HE = ZE. Produce a circle through H, Z, and center point D of the sphere, and draw lines DHL and DZM. Let circle HDZ intersect the mirror's surface at points T and K. Accordingly, there will be three reflections within the plane. One of them will occur from point A, as is obvious from the equality of triangles HAE and ZAE. The other two will be from T and K because in both cases the angles of incidence and reflection are subtended by equal arcs HD and HZ within circle HDZ. Consequently, by Euclid, *Elements*, III, 21, the angles themselves must be equal. That there can be no reflection from some point T′ outside circle HDZ follows from the fact that although the angles of incidence and reflection are subtended by equal arcs in that circle, the point of reflection lies outside it.

Nor can there be reflection from any point, such as T′, on arc TK, as represented in figure 3.6b, because, although the angles of incidence and reflection subtend equal arcs within circle HDZ, the point of reflection lies inside that circle. Furthermore, no reflection can occur from a point such as T″ at or below point L because the angle formed by the incident and reflected rays will no longer be split by a normal from D. The same holds by symmetry for

73. See *Optics*, IV, 11–19, in Smith, 177–81, and Lejeune, 151–55. In order to understand why the ratio ZT:ET = ZH:KH dictates the equality of angles HLE and MLE, draw line NEM parallel to KL and line MT parallel to LH. Accordingly, triangles KLZ and EMZ will be similar. Since LH and MT are parallel and cut sides KZ and EZ proportionally, it follows that ZT:ET = ZH:KH. It also follows that triangle NLM is isosceles, so LE bisects base NM because it is perpendicular to it. Consequently, angles NLE and MLE must be equal.

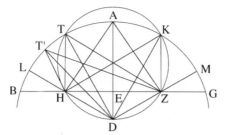

FIGURE 3.6A

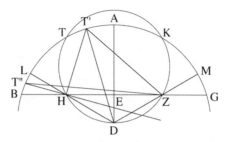

FIGURE 3.6B

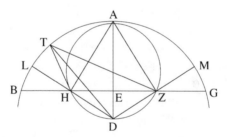

FIGURE 3.6C

arc KM; no reflection can occur from any point on it or from any point below M. Finally, if circle HDZ fails to intersect the mirror, as in figure 3.6c, then the only possible reflection will be the one occurring at A because, as we just saw, no reflection can occur at or below L or M, and no reflection can occur from any point lying inside or outside circle HDZ when the two angles formed on it are subtended by equal arcs within that circle.[74]

74. See *Optics*, IV, 26–33 and 41–45, in Smith, 182–84 and 187–88, and Lejeune, 157–60 and 163–64.

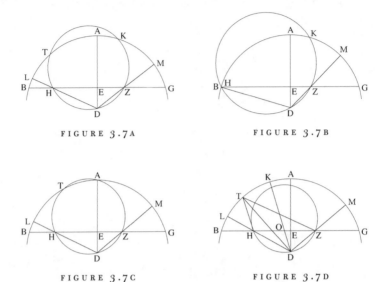

FIGURE 3.7A   FIGURE 3.7B

FIGURE 3.7C   FIGURE 3.7D

Taking this relatively trivial case as a paradigm, Ptolemy turns to the situation in which the eye and the object point on chord BG are not equidistant from normal DAE, as represented in figure 3.7a, where distance HE between the eye and the normal is greater than distance ZE between the object and the normal. As before, we draw circle HDZ. If it intersects the mirror at points T and K flanking A, then there will be three reflections. Two of these reflections, Ptolemy goes on to prove, will be from arc AK and one from arc TL. But if H and Z are situated as in figure 3.7b, where point H coincides with point B and circle HDZ cuts arc AM at point K, then reflection on the side of arc AB will be preempted because there will be no arc between B and circle KZDB from which it can occur. Hence, only two reflections will occur, both from points on arc AK.

By the same token, if circle HDZ passes through point A on the normal, as in figure 3.7c, reflection on the side of arc AG will be preempted, leaving only one reflection from arc TL. Finally, if circle HDZ fails to intersect the mirror at all, as illustrated in figure 3.7d, and if line DOK is drawn through point O, which bisects HZ, there will be one reflection only, and it will occur from some point T between K and L.[75]

The last situation with which Ptolemy deals is the one in which the eye and the object lie on a line beyond the center of curvature, as illustrated in figure 3.8, where H is the center of sight on line HZ´, and EDA is normal to that line.

75. See *Optics*, IV, 34–40 and 46–49, in Smith, 184–89, and Lejeune, 160–66.

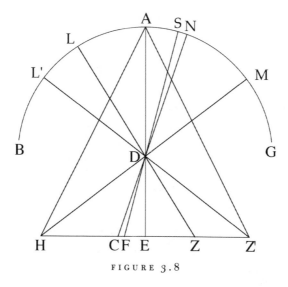

FIGURE 3.8

Obviously, if Z´ is the object point, and if EZ´ = HE, then ray HA will reflect to it at equal angles along AZ´. If we draw normals HDM and Z´DL´, then it is equally obvious that no reflection can occur from M or L or any point below them toward G or B. Nor, as Ptolemy goes on to demonstrate, can there be any reflection from arcs AL´ or AM. Hence, the reflection from A is the only possible one. Likewise, if the object is Z, and if HE > ZE, there can be one and only one reflection. Where it will occur is determined by bisecting HZ at C and drawing line CDN to the mirror and then bisecting angle HDZ with line FDS. Accordingly, the reflection will occur somewhere between N and S.[76] Consequently, when the eye and the object lie on a line beyond the center of curvature, there will be only one reflection, no matter whether the eye and object are equidistant from normal EDA or not.

To generalize, Ptolemy has broken the problem of reflection from concave spherical mirrors into three primary cases, depending upon whether the eye and the object lie on the diameter of the sphere containing the mirror, on a chord within that sphere between the mirror's surface and its center of curvature, or on a line lying beyond the mirror's center of curvature. Each of these primary cases is then divided into two classes according to whether the eye and the object are equidistant or at unequal distances from the point at which the normal from the center of curvature to that line intersects it.

76. See *Optics*, IV, 81–96, in Smith, 98–201, and Lejeune, 178–84.

In all three primary cases, if the eye and object are equidistant from the point at which the normal through the sphere's diameter intersects the line on which they lie, reflection will occur from the point at which that normal intersects the mirror's surface. In one case only will there be more than one reflection, and it occurs when the eye and object lie on a chord at equal distances from the normal and the circle passing through the eye, the object, and the center of curvature intersects the mirror's surface. Under these conditions there will be three reflections, one from the point where the normal intersects the mirror and one from each point at which the circle intersects the mirror's surface.

Similarly, if the eye and object are not equidistant from the point where the normal intersects the line upon which they lie, there can be one and only one reflection when they lie on a diameter of the mirror or on a line beyond the center of curvature. When the eye and object lie on the diameter, the reflection will occur from the arc on the mirror on the side of the eye, whereas when the eye and object lie on a line beyond the center of curvature, the reflection will occur from the arc on the mirror on the other side of the eye. If, however, the eye and object lie on a chord, there can be several possible reflections ranging from none to three, depending upon whether and where the circle passing through the eye, the object, and the center of curvature intersects the mirror. The key to all these cases involving the chord is thus the circle that passes through the eye, the object, and the center of curvature. Somewhat complex in geometrical detail, this analytic expedient is truly elegant in its generality and conceptual simplicity, and Alhacen used it as a basis for determining the actual points of reflection for any pair of points facing a concave spherical mirror from any position (see chapter 5, section 3).

After this painstaking, case-by-case analysis of possible reflection points in concave spherical mirrors, Ptolemy takes up the problem of image formation for single object points. This problem is complicated by the fact that according to the cathetus rule, images in concave mirrors can be located behind the mirror, between the mirror and the eye, at the eye itself, or behind the eye. In fact, there may be no image location at all.

This can be explained geometrically by reference to figure 3.9. Let the eye at E face the mirror, and let incident ray EA reflect along AB. Let a range of object points from O_1 to O_6 be placed on that reflected ray. According to the cathetus rule, the image location for object O_1 is determined by the intersection of cathetus DO_1 and incident ray EA, both of which converge at I_1, when extended. Image I_1 will therefore appear behind the mirror. The same holds for object point O_2, whose image I_2 also lies behind the mirror, but at a vastly increased distance. On the other hand, object point O_3 is situated such that

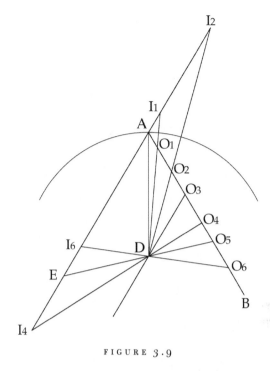

FIGURE 3.9

cathetus DO_3 is parallel to incident ray EA, so the two will never intersect to yield an image. It follows, therefore, that if the object point is located anywhere between A and O_3, its image location will be behind the mirror, and the farther it gets from A, the exponentially farther behind the mirror its image location will be until it reaches point O_3, where no image is formed. As soon as the object point is moved beyond O_3, eventually reaching some location O_4, its image will lie behind the eye, as is the case with I_4, and the image will continue to lie behind the eye until the object reaches O_5, at which point cathetus O_5D, when extended, intersects incident ray EA at the eye itself. Finally, the image location of every point beyond O_5 on AE will lie between the eye and the mirror, as illustrated by point O_6 and its image I_6.[77]

All of this, Ptolemy assures us, can be empirically verified with the apparatus used earlier to confirm the equal-angles law for the concave mirror. Simply install the concave mirror TAK on the disk pictured in figure 3.10, place the eye at E with the line of sight fixed along EA through the dioptra, and then

77. This analysis is based on a far more cursory discussion in *Optics*, IV, 63–65, in Smith, 192–93, and Lejeune, 170–71.

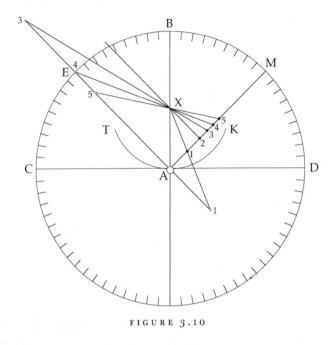

FIGURE 3.10

plant the colored peg at various points along line AM. When the peg is placed close to A at position 1, it will appear behind and relatively close to the reflecting surface. As it is moved farther back from A, it will appear increasingly farther behind the reflecting surface until, suddenly, the peg disappears when it reaches position 2. As it continues to be moved beyond that point to positions 3 and 4, the peg should remain invisible, but soon as it is moved beyond position 4, it should reappear. From then on it should appear in front of the mirror, as is the case when it is placed at position 5.[78] That the object can no longer be properly seen in the mirror when it lies between positions 2 and 4 in figure 3.9 is easily explained according to the intersection of the cathetus and the incident ray. When the object is at position 2, there is no such intersection, so there is no image to be seen. At position 3, on the other hand, there is an intersection, but it lies behind the eye at location 3, so the image at that point is effectively masked from view. This is also the case when the peg is placed at position 4 because its image at location 4 coincides with the eye. Once moved beyond position 4, though, the peg comes back into view, but this time in front of the mirror, as is the case when it stands at position 5 with its image at corresponding image location 5.

78. See *Optics*, IV, 71–73, in Smith, 194–95, and Lejeune, 174–75.

This is essentially how Euclid explained the phenomenon (see chapter 2, section 6), and the problem with this explanation is that it conflicts with the core assumption that any luminous, opaque object the visual flux touches will be seen. Therefore, since the object is reached by the visual flux at every point on reflected ray AB, it ought to be seen. Rather than duck this problem, as did Euclid, Ptolemy attempts to resolve it by appealing to the psychology of visual perception. In essence, Ptolemy argues that although the image may be out of sight, it is not out of mind. For instance, "because [the image] is difficult to see" when it is located in the eye itself, "the visual sensitivity inclines toward the surface from which the reflection occurs" and transfers the image to that location. "Consequently, the colors of such images . . . are either indistinguishable or barely distinguishable from the colors of the mirrors."[79] In other words, the visual faculty (under control of the governing faculty) transposes the image to the mirror's surface so that it coalesces with it and takes on its color, becoming a mere ghost of itself. The visual faculty does the same thing when the cathetus and incident ray fail to intersect at all, but if they intersect behind the eye, the visual faculty transposes the image to a point somewhere between the mirror and the eye rather than to the mirror's surface.[80] That explains why, when we look into a concave mirror from a close distance and slowly withdraw from it, the image of our face spreads out momentarily in a nebulous, chaotic blur on the reflecting surface before coming back into view inverted.

Compared to his analysis of reflection and image formation for object points in concave spherical mirrors, Ptolemy's account of image formation for object lines is surprisingly cursory and lackluster. It is also severely limited by the stipulation that the object line be "directly facing," so that the common axis passes orthogonally through its center and the two axial rays converge on that point. Accordingly, Ptolemy shows that when the eye lies between the reflecting surface and the center of curvature with the object line located in front of it, the image will appear behind the mirror and will look larger than the object itself. It will also look more distant from the mirror than the object, and it will maintain the same right-to-left orientation as the object. By this he means that the left and right sides of the image will correspond with the left and right sides of the object, even though the two would be reversed according to a face-to-face disposition.[81] On the other hand, Ptolemy continues, if the

---

79. *Optics*, IV, 25, in Smith, 182, and Lejeune, 157.

80. See *Optics*, IV, 69–70, in Smith, 184–89, and Lejeune, 160–66.

81. See *Optics*, IV, 120–22 and 142–46, in Smith, 207–8 and 214–15, and Lejeune, 192–94 and 202–4.

image appears between the mirror and the eye, it may sometimes look larger, sometimes the same size, and sometimes smaller than the object. Whatever the case, it will suffer both inversion and left and right reversal insofar as the left and right sides of the object will appear on the right and left sides of the inverted image.[82] Also, whether they appear behind or in front of the mirror, straight and convex lines will look concave, whereas concave lines will look convex under certain conditions and concave under others.[83] With these points demonstrated, Ptolemy concludes book four with a brief account of reflection from composite mirrors, including cylindrical and conical mirrors, which are formed from circular and rectilinear sections.

As haphazard and banal as it sometime is, Ptolemy's overall analysis of reflection is nonetheless incomparably more systematic, sophisticated, and comprehensive than Euclid's. Particularly noteworthy is his effort to determine the number of possible reflections in concave spherical mirrors according to various dispositions of the eye and the object point vis-à-vis the reflecting surface. Indeed, his demonstration of roughly where, within specific arcs on the mirror, the points of reflection will lie when the eye and the object point are not equidistant from the normal is something of a tour de force. Left unanswered is the question of how to locate those points within their appropriate arcs. That problem will be resolved nearly a millennium later by Alhacen, but the very raising of the problem, albeit tacitly, and the terms in which it was posed by Ptolemy were crucial to Alhacen's success in resolving it. Noteworthy, too, is Ptolemy's effort to explain on psychological grounds why there seems to be no image, or at least no definite image, in concave mirrors when the image location is at or behind the eye because it bespeaks his willingness to ignore the imperatives of ray geometry when it leads to conclusions that contradict theoretical or empirical expectations. Thus Ptolemy was perfectly content to shift the burden of explanation from mathematics to psychology in order to "save the appearances" (or, rather, the apparent disappearances) of images under certain conditions of reflection from concave mirrors.

3. THE PTOLEMAIC ACCOUNT OF REFRACTION

Because we know so little about the development of mathematical optics and visual theory before Ptolemy, it is difficult to tell how much he owed to earlier

---

82. See *Optics*, IV, 125–29 and 147–51, in Smith, 208–10, and Lejeune, 195–97 and 204–5.

83. See *Optics*, IV, 130–41, in Smith, 210–14, and Lejeune, 197–201.

theorists and how much was original to him. Definite, identifiable sources for his overall account of visual perception are thin on the ground. There is absolutely no doubt that Euclid was one of them. Not only can we identify his *Optics* and *Catoptrics* as textual sources, but we can even link specific propositions in them to Ptolemy. That Ptolemy also drew upon Aristotelian, Platonic, and Stoic sources is virtually beyond doubt, although we cannot be certain about what specific texts among those sources he may have used. Possible sources, on the other hand, are as legion as they are vague. Least vague among them, perhaps, are Hero's *Catoptrics* and the *Catoptrics* attributed to Archimedes. For reasons given earlier, it is highly unlikely that Ptolemy was influenced by or even knew of Hero's *Catoptrics*. Archimedes's *Catoptrics* is even more problematic as a source because, although it is mentioned by several post-Ptolemaic authors, it is now lost.[84] Consequently, without any direct evidence of its contents, we can only speculate about what Ptolemy may have borrowed from it, if in fact he even had access to it.

The issue of originality becomes especially prominent when we turn to Ptolemy's analysis of refraction because, as far as we can tell from direct evidence, it is totally unprecedented in scope and sophistication. To be sure, Ptolemy was not the first to bring refraction into the domain of mathematical optics. After all, the sixth and final postulate of Euclid's *Catoptrics* deals explicitly with refraction and was presumably meant to set the stage for an analysis that, for whatever reason, never eventuated.[85] There is thus no telling what form that analysis might have taken, but given the primitive nature of the postulate, it is unlikely that any theoretical account flowing from it would have been much less primitive. For lack of any meaningful evidence to the contrary, then, we are driven to conclude that Ptolemy's study of refraction was almost entirely original. Indeed, it may well be his most original contribution to the science of mathematical optics.

Ptolemy opens his study in book five by noting the fundamental similarities between refraction and reflection. For a start, both involve breaking of the visual ray. Physically, this breaking is due to the resistance of certain bodies to penetration by impinging visual rays. When the body's surface is reflective, its surface resists penetration completely, so the visual ray is fully broken by

84. For instance, as we saw earlier, in note 106 to chapter 2, Apuleius adverts to Archimedes's "massive volume" on optics.

85. Whether such an analysis was part of the *Catoptrics* and has since been lost, or whether the sixth postulate was added later, is a matter of debate that is unlikely ever to be resolved; see Takahashi, *Medieval Traditions*, 17–18.

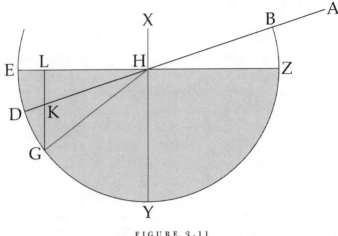

FIGURE 3.11

rebound (*reverberatio*).[86] When the body is transparent, the resistance at its surface is only enough to deflect the visual ray as it passes through, so it is only partially broken. Refraction is thus a special case of reflection, in which the rebound is incomplete.[87] The amount of deflection the visual ray undergoes in refraction depends on how much resistance it encounters, and that is a function of the relative "density" of the refractive body. Accordingly, when it passes from a rarer (*subtilior et tenuior*) body, such as air, into a denser (*grossior*) one, such as water, the visual ray will be deflected in a particular way. Likewise, when it passes from a denser into a rarer body, the visual ray will be deflected in a particular way.[88]

Reflection and refraction also share three fundamental principles. In both cases, the image is seen along the continuation of the incident ray. In both cases, that image is located at the juncture of the cathetus and the continuation of the incident ray. And in both cases, the incident ray and its broken counterpart lie in a plane perpendicular to the interface of the resisting surface.[89]

Offering the "floating coin" experiment as a verification of these principles, Ptolemy explains the result of that experiment geometrically on the basis of figure 3.11. Let the circular section represent a vessel inside of which a coin is

86. Only once is the Latin term *reflexio* used to denote "reflection" in the *Optics*. In all other cases the term *reverberatio* is used.

87. See *Optics*, V, 36, in Smith, 244–45, and Lejeune, 246.

88. See *Optics*, V, 2, in Smith, 229, and Lejeune, 223–24.

89. See *Optics*, V, 3, in Smith, 230, and Lejeune, 224–25.

placed at point G. Let ABHD be a visual ray passing over the rim of the vessel and over the coin so that it cannot be seen. When the vessel is filled with water to line EHZ, however, the coin floats into view at K, directly above G. The reason for this phenomenon is that the incident ray AH is refracted from the rarer air into the denser water at H to reach G along GH. According to the cathetus rule, the image of G must be seen along continuation HD of the visual ray where cathetus GL intersects it. Since K is the point of intersection, that is where the image of G will be seen along AK, just over the vessel's rim.[90]

From this example it is evident that reflection and refraction differ in one key respect: when we drop normal XY to the surface of the water through point of refraction H, we see immediately that angle of incidence AHX and angle of refraction GHY are unequal, whereas in reflection the corresponding angles are always equal. If, however, the similarity between reflection and refraction holds true in all respects, then the relationship between the angles in refraction must somehow be analogous to that between their counterparts in reflection. In order to determine whether they in fact are, Ptolemy suggests a set of three experiments to measure the angles of refraction for given angles of incidence when the visual ray refracts from air into water, from air into glass, and from water into glass. All three experiments are based on the bronze disk used earlier in the experiment to confirm the equal-angles law of reflection.

For the test of refraction from air to water, Ptolemy has us insert the disk upright in a vessel and fill the vessel with water until the water's surface coincides with diameter CAD of the disk and is perpendicular to the disk's surface, as represented in figure 3.12a. After placing a small marker at point A, we are to move the dioptra used for sighting to the top of the disk so that the line of sight through it is along normal BAE. Next, we are to move a small colored peg to point E. Sighting along BA to the marker at A, we see that the peg appears directly behind the marker at A and in line with it along the normal. From this we conclude that no refraction has occurred. Then we are to move the diop-tra to the 10° mark on the disk, sight through it to the marker at point A, and move the peg until it appears to line up with it. With the peg fixed in place, we then measure how many degrees of arc on the disk it lies from E. That will be the measure of the angle of refraction. Repeating the process at intervals of ten degrees all the way up to the 80° mark, we will see that the visual ray has been refracted toward normal EA and that the refraction becomes increasingly severe as the angle of incidence increases. The diagram shows the setup for angle of incidence LAB of 40°, the resulting angle of refraction OAE being a

90. See *Optics*, V, 5–6, in Smith, 230–31, and Lejeune, 225–26.

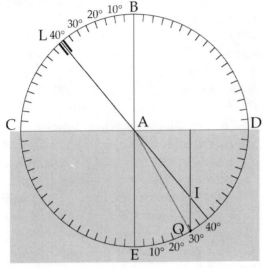

FIGURE 3.12A

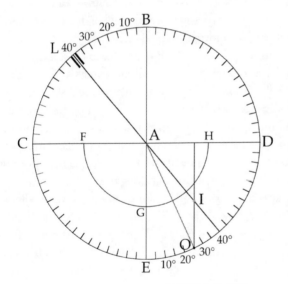

FIGURE 3.12B

bit less than 30°. The image location will be at I, where cathetus OI intersects the continuation of incident ray LA. The results Ptolemy recorded for this experiment are given in table 1 to the right of the diagram. From right to left, it lists the angle of incidence i at ten-degree increments, Ptolemy's measurement for the angle of refraction r at each of those increments, and the modern value of that angle, rounded to the first decimal place. This value is based on the sine law and a value of 1.33 for the index of refraction of water and 1.0 for that of air. It is important to note that, according to modern theory, Ptolemy was testing the refraction of light from the object at O through water into air, that is, from a denser into a rarer medium. Consequently, the angles of refraction are actually angles of incidence, and vice versa.

The setup for determining the angles of refraction when the visual ray passes from air into glass (that is, when light passes in the opposite direction from glass into air) is pictured in figure 3.12b. A glass semicylinder FGH is applied to the face of the bronze disk so that its flat face along diameter FAH is perpendicular to the disk's surface and flush with diameter CAD, its center of curvature coinciding with center point A of the disk. As before, the dioptra is moved from B toward C at ten-degree intervals, and the angle of refraction is measured for each of those intervals. In this case, as in the previous one, we find that the visual ray is refracted toward the normal but that it suffers somewhat greater refraction at each ten-degree interval, which indicates that glass is denser than water. Ptolemy's results, along with the modern values according to the sine law, are given in table 2 to the right of the diagram. The modern values are based on an index of refraction of 1.5 for common crown glass, which is essentially the kind of glass available to Ptolemy at the time.

For the final experiment, which tests refraction from water to glass, we need to take the disk with the glass semicylinder still applied and rotate it 180° so that the semicylinder's convex surface faces upward. Then, as pictured in figure 3.12c, we are to place it upright in the vessel and fill the vessel with water until the water's surface coincides with the flat surface of the glass semicylinder. Then we are to move the colored peg below the water's surface from E toward D at ten-degree intervals, each time moving the dioptra on the opposite arc and sighting along it until the image of the peg comes into view behind the marker at A. As illustrated in the diagram, angle of "incidence" OAE at 40° yields angle of "refraction" LAB of 35°. Ptolemy's results and the corresponding modern values are listed in table 3 to the right of the figure. The modern values listed are based on an index of refraction of 1.33 for water and of 1.5 for glass.

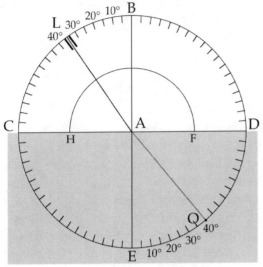

FIGURE 3.12C

TABLE 3
water-glass

i	r (Ptol.)	r (mod.)
10	9.5	8.9
20	18.5	17.7
30	27.0	26.3
40	35.0	34.7
50	42.5	42.8
60	49.5	50.2
70	56.0	56.4
80	62.0	60.8

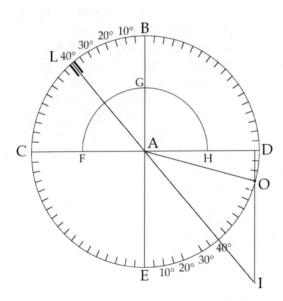

FIGURE 3.12D

The procedure for this experiment is anomalous in two ways. First, unlike the previous two, it actually tests the refraction of light. That is, the presumed line of visual radiation OA is actually a line of light radiation according to modern theory. Second, Ptolemy's procedure requires sighting along the line of refraction rather than along the line of incidence, as was done in the two previous experiments. Physically, of course, this procedure is mandated by the experimental arrangement, which has the water lying below the glass rather than above it. Reversing the placement would have required far more engineering than it merited and, most likely, than was feasible at the time. Moreover, according to Ptolemy's actual setup, carrying out the procedure in the usual way, from the first to the second medium, would have required that the experimenter make all the sightings under water. It is much easier and more convenient, then, to reverse the procedure as Ptolemy did. But that leaves open the question of why he did not simply test the refraction of visual rays from glass to water according to ten-degree increments along BC.

One reason, perhaps, is that Ptolemy meant to be absolutely consistent in testing refraction from rarer into denser media throughout and could therefore brook no deviation from that protocol. Nevertheless, given the physical constraints of this final experiment, it would have made sense for him to reverse the test for the sake of convenience. Had he followed that reverse procedure he would have found that if i in glass = 10°, r in water is around 11°, whereas if i = 20°, r is around 22.5°, and so on, the ray having refracted away from the normal in passing from the denser into the rarer medium.[91] The problem is that this reverse procedure would have come a cropper after i = 60° because in this case the largest possible angle of refraction LAB would be around 62.4° for the largest possible angle of incidence OAE < 90° by some infinitesimal amount.

Hence, if Ptolemy had tested the refraction of visual rays from glass into water according to increments of 10° along arc BC, he would have been unable to see the peg under water, the rarer medium, when he set the line of sight at any point just beyond 62.4° on arc BC and moved the peg toward the normal from the place it had occupied on arc DE when the angle of incidence was 60°. In other words, the experiment would have yielded no results at 70° or 80° on arc BC. By the same token, if Ptolemy had tried to measure the refraction of visual rays from glass into air, as pictured in figure 3.12d, he would have encountered the same problem, only exacerbated by the greater density differential between the two media. Accordingly, when he set the dioptra at any point beyond around 42°, he would have gotten no results. This limitation is

91. The actual values, according to modern theory, would be 11.3° and 22.7°.

air-water

i	r	*first*	*second*
10	8.0		
		7.5	
20	15.5		.5
		7.0	
30	22.5		.5
		6.5	
40	29.0		.5
		6.0	
50	35.0		.5
		5.5	
60	40.5		.5
		5.0	
70	45.5		.5
		4.5	
80	50.0		

FIGURE 3.13

ultimately a function of the critical angle at which light passing from a denser into a rarer medium will no longer penetrate the interface but will instead reflect internally. We will have more to say about this angle in chapter 5, when we deal with Alhacen. Whether Ptolemy discovered this problem empirically or simply skirted it by luck is open to speculation, but the anomalous procedure he followed in the third and final experiment suggests that he had actually encountered it and sought a way around it.

No less anomalous than the procedure Ptolemy followed in that particular experiment are the results he claims to have gotten from all three experiments. Even the most cursory glance at those results shows that they are too clean to be raw data. They had to have been adjusted by rounding, and a closer look reveals how. Take the results for refraction from air to water recapitulated in figure 3.13. Notice that all the values for r are either whole numbers or end in a fraction of one-half, that is, 0.5. The pattern here becomes clear when we subtract successive values for r. Constituting "first differences" these values are listed in the third column. If we subtract these from one another successively, then we arrive at "second differences" of 0.5, as listed in the fourth column. Unlike the values in the previous three columns, these are constant. Consequently, upon reflection, we see that the values for r vary in a regular way, according to first differences that decrease uniformly by decrements of half a degree while i increases uniformly by ten-degree increments. As Otto Neugebauer showed some time ago, this method of regularizing irregularly changing values was used by Babylonian astronomers of the Seleucid period to produce predictive tables (ephemerides) of irregular celestial motions.[92] Ptolemy used the same technique in *Almagest*, III, 1–2, to compute mean solar motion according to hours, days, months, and years.[93]

There is no question, then, that Ptolemy adjusted his tabulations according to a technique borrowed from mathematical astronomy in order to import an underlying regularity to the continually changing values of r, and a look at the tabulations for the other experiments will show that he applied the same technique throughout. Was he simply "doctoring" his results, as one commentator has rather uncharitably claimed?[94] Was Ptolemy, in short, playing fast and loose with his data? Not really. True, Ptolemy conducted his experiments with a heuristic in mind, and, as I have argued elsewhere, this heuristic seems to have been based on an astronomical model of uniform motion about an eccentric circle.[95] Nonetheless, Ptolemy's values for r in tables 1, 2, and 3 are fairly close to the corresponding modern values, generally varying by less than a degree, and sometimes by much less. Even his worst results—for refraction from air to water—are within .6° or less of the modern value, except for the value at 80° of incidence, where the discrepancy between his result and its modern counterpart is slightly over 2°. But given the limitations of his apparatus, which was apparently calibrated to an accuracy of ± 0.5°, that discrepancy is not particularly alarming, especially at an angle of incidence

92. See Neugebauer, *Exact Sciences*, 110–14. As an example, Neugebauer examines an ephemeris from the later second century BC that attempts a reconciliation between solar and lunar motion over a period of 810 years.

93. See Toomer, *Ptolemy's Almagest*, 131–43.

94. Saleh Omar, *Ibn al-Haytham's Optics: A Study in the Origins of Experimental Science* (Minneapolis: Bibliotheca Islamica, 1977), 150. In the same year that Omar published this study, Robert R. Newton published *The Crime of Claudius Ptolemy* (Baltimore: Johns Hopkins Press, 1977), in which he accused Ptolemy of systematically fabricating astronomical data to fit his theory. One might therefore interpret Ptolemy's tabulations for refraction as fabrications of optical data.

95. A. Mark Smith, "Ptolemy's Search for a Law of Refraction: A Case-Study in the Classical Methodology of 'Saving the Appearances' and Its Limitations," *Archive for History of Exact Sciences* 26 (1982): 221–40.

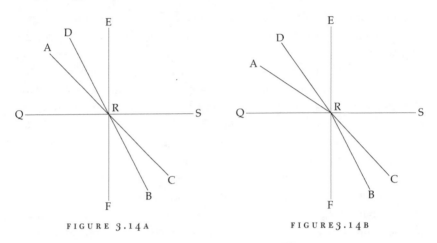

FIGURE 3.14A   FIGURE 3.14B

that is great enough to cause serious refractive distortion of anything seen below the water's surface. The closeness of Ptolemy's values to their modern counterparts therefore suggests not only that he derived his results from actual observation, but also that that he "read" his data in a way that was not inconsonant with the observational limitations of his apparatus and the procedure he followed. Thus Ptolemy was not so much adjusting, or "doctoring," his data as viewing them through a heuristic prism that allowed him to organize and make sense of them according to a rational pattern. He may have been wrong, but he was wrong in a perfectly reasonable way.

On the basis of these experiments, Ptolemy offers two final generalizations about refraction. The first is that "the amount of refraction is the same whichever the direction of passage," by which he means that when a visual ray passes from a given refractive medium into another, it will be broken by the same amount whichever direction it takes.[96] In short, the visual radiation conforms to the principle of reciprocity. Thus in figure 3.14a, if ray AR strikes refractive interface QRS at angle of incidence ARE and is refracted along RB in the denser medium, then if RB were the visual ray, it would be refracted along RA in the rarer medium above. The amount of breaking in both cases will be equal according to angles of deviation ARD and BRC formed by the rays and their alternate rectilinear continuations.[97]

The second generalization, which is considerably weaker than the first, amounts to the claim that if two rays, such as DR and AR in figure 3.14b, re-

96. *Optics*, V, 31, in Smith, 244, and Lejeune, 243.
97. See *Optics*, V, 32, in Smith, 242–43, and Lejeune, 243.

fract into a denser medium along rays RB and RC, respectively, the difference between the angles of incidence will be proportionately greater than the difference between the angles of refraction. Hence, angle ARE:angle DRE > angle CRF:angle BRF, which translates in fractions to ARE/DRE > CRF/BRF.[98] From all this, Ptolemy concludes, a "marvelous fact will be apparent: namely, the course of nature in conserving the exercise of power."[99] Ptolemy never addresses how this fact becomes apparent, but the appeal to conservation of power suggests that Ptolemy had in mind some overarching metaphysical principle, like, or at least analogous to, Hero's least-distance principle, that governs the relationship between angles of incidence and refraction.

The similarity between reflection and refraction extends to the determinacy or indeterminacy of image location. As in reflection from plane mirrors, so in refraction through plane interfaces, the image is invariably determinate because the cathetus and the continuation of the incident ray will always meet in the second medium, whether it is rarer or denser. In refraction through a convex spherical interface, however, such is not the case.

Let the circular segment centered on C in figure 3.15 represent such an interface, with the denser medium below it, and let ray ER from the eye at E refract along RB. In that case, if the object point is located at O_1, its image will appear at I, where the extension of cathetus CO_1 intersects the extension of incident ray ER. On the other hand, if the object point is located at O_2, cathetus CO_2 will be parallel to incident ray ER, so there will be no image location, and if the object point is located at O_3, the intersection will occur behind the eye. Just as in reflection, so in refraction, the visual faculty will transpose the image either to the refracting surface or somewhere between that surface and the eye. The same thing holds for refraction through a concave spherical interface, and it also holds whether the eye is looking from the rarer into the denser medium or vice versa.[100]

Book five ends with a disappointingly incomplete analysis of image formation for object lines seen through plane, spherical convex, and spherical concave refracting interfaces. Ptolemy opens this analysis by advising us to form three containers of clear, thin glass that are open at the top so that they can be filled with water. One of them is to be cubical in shape, another cylindrical, and another cubical, except that one side of it will be concave cylindrical. These containers are thus designed to allow refraction of visual rays into a

98. See *Optics*, V, 34, in Smith, 244, and Lejeune, 245.

99. *Optics*, V, 35, in Smith, 242, and Lejeune, 246.

100. See *Optics*, V, 58–59, in Smith, 252, and Lejeune, 260–61.

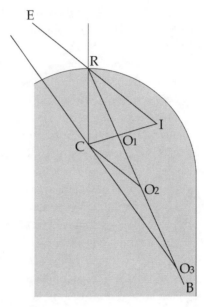

FIGURE 3.15

denser medium to be empirically tested through each type of interface. When they are properly filled, we are to immerse a ruler in each container in order to confirm that the ruler looks magnified when viewed through the plane, convex, or concave face.[101]

Ptolemy's geometrical explanation of this phenomenon is cut short by the abrupt ending of book five, but his account of why things in a denser medium appear magnified when viewed through plane interfaces gives a fairly clear idea of his approach to the problem. Let AB in figure 3.16a be an object line immersed in water, and let E be an eye looking at it. Rays ER and ER′ will be refracted at the water's surface to reach the endpoints of the object along RA and RB. Drop catheti AN and BM from the endpoints to the water's surface, and continue rays ER and ER′ until they intersect their respective catheti at points A′ and B′. A′B′ will thus be the image of AB as seen from E. If object AB were not immersed in water, it would be seen under visual angle AEB, whereas its image is seen under a significantly larger visual angle A′EB′. Thus it appears closer and larger than its object. The opposite will happen when the object is viewed in a rarer medium, as represented in figure 3.16b, where the eye at E is in the denser medium and the object AB in the rarer one. In that case, the

101. See *Optics*, V, 67, in Smith, 254–55, and Lejeune, 256–57.

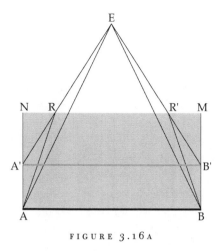

FIGURE 3.16A

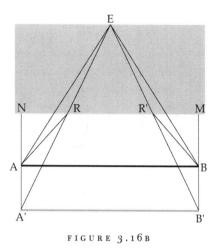

FIGURE 3.16B

image A´B´ will be seen under a smaller visual angle, that is, A´EB´, than angle AEB under which the object would be seen without refraction, so image A´B´ looks smaller and farther away than its object.[102]

After showing that the image of anything seen through a flat refractive interface has the same shape and orientation as its object, no matter whether it appears magnified or diminished in size, Ptolemy barely broaches the subject of spherical convex refractive interfaces before his analysis is interrupted by the premature close of the *Optics*. It seems likely that he was in the midst of following the same analytic steps he took in treating image formation and deformation for objects seen through flat refractive interfaces, but beyond that it is pointless to speculate about how far he might have taken the study of refraction in the unfinished portion of book five.

4. ATMOSPHERIC REFRACTION AND THE MOON ILLUSION

Before leaving Ptolemy, we should briefly discuss two astronomical phenomena addressed in the *Optics*, both apparently involving refraction. The first of these is the upward displacement of celestial objects at the horizon, which makes them appear to lie farther above the horizon than they actually do. Given that this upward displacement affects observational accuracy, we might expect Ptolemy to have at least mentioned it in the *Almagest*, yet there is no clear evi-

102. See *Optics*, V, 70–78, in Smith, 255–58, and Lejeune, 262–65.

dence there that he recognized either the phenomenon or its effect.[103] Not so in the *Optics*, however. "We notice," Ptolemy observes in V, 24, "that [celestial] bodies that rise and set tend to incline toward the north when they are near the horizon and are measured by an instrument for measuring the stars."[104] The instrument in question is probably the armillary sphere described in book 5 of the *Almagest* (see note 64), and the basic technique would be to compare the altitude of a given star when it rises at the horizon and its altitude at the highest point in its nightly transit.

The rationale behind this technique is easily illustrated. Let the circle in figure 3.17a represent the great sphere of the fixed stars with E, the Earth, at its center. Let arc CBA represent the equator, arc GBH the observer's horizon, and N the North Pole, with NE the axis about which the sphere rotates daily in a clockwise direction as seen down through N. Let S be some fixed star on the great celestial sphere carried through arc SS_1S_2 by the daily rotation of the sphere. Consequently, when the star reaches position S_1, where it peeps over the horizon, its arcal distance from the North Pole will be NS_1, as measured by angle NES_1. Then, when the star reaches its highest point above the horizon at S_2, its arcal distance NS_2 from the North Pole, as measured by angle NES_2, should be exactly the same, since the star is supposed to be fixed in place on the sphere. But in fact arc NS_1 will be slightly smaller than arc NS_2, so there is some disparity between the two measurements, the star appearing to lie somewhat closer to N at horizon, when it rises or sets, than it does at its highest point above the observer.

In order to explain this disparity, Ptolemy has recourse to Aristotle's cosmological model of nesting spheres, the centermost of which is the orb of heavy earth surrounded by a shell of less heavy water. This twofold combination of spheres is in turn enveloped by an atmospheric shell consisting of a sphere of lighter air surrounded by a thin sphere of fire, the lightest element of all. Beyond this atmospheric shell lies the celestial realm, which extends to the outermost sphere of fixed stars and is composed of aither, the most sublime and perfect element of all. Since, therefore, aither is less crass or "dense" than the fire and air below it, when a visual ray passes at an angle through the interface between it and the denser atmosphere below, the ray is refracted away from the normal by an amount that depends upon the density differential

103. In *Almagest*, IX, 2, Ptolemy mentions that an interval between two celestial bodies can appear larger at horizon than at zenith, but he gives no explanation of why this might be so; see Toomer, *Ptolemy's Almagest*, 421.

104. Smith, 238, and Lejeune, 238.

between the two media. It bears noting that Ptolemy never mentions fire in his account, referring only to the air in the atmospheric shell. Whether he means "air" to include both air and fire generically, or whether he thinks that fire plays no part in the refraction, is impossible to tell.

Ptolemy explains the effect of atmospheric refraction geometrically according to figure 3.17b, where A represents the observer on Earth, H the Earth's center, AZ the plane of the observer's horizon, E his zenith, GD the interface between the atmosphere below (in gray) and the aither above, and K a star on celestial sphere KZTE. When visual ray AD meets the interface between the atmosphere and the aither at point D to form angle of incidence ADH with normal HDT, it will be refracted away from the normal along DK at angle of refraction TDK > angle of incidence ADH. By the cathetus rule, the image of K will be at X, where the extension of cathetus HK meets the continuation of incident ray AD. Thus the star will be seen along horizon line AX, whose projection is point Z on the celestial sphere, so it will appear shifted upward toward the zenith E according to angle of deviation KDZ. On the other hand, if the star lies well above the horizon, at point K′, the angle of incidence AD′H formed with normal HD′T′ will be far less than angle of incidence ADH at the horizon, so the angle of refraction T′D′K′ will be far less than angle of refraction TDK at the horizon. Hence, the star's image will be shifted far less northward toward the zenith according to a much smaller angle of deviation K′D′Z′. Finally, if the star lies at E, the observer's zenith, there will be no refraction whatever, so its observed location will coincide with its actual location.[105]

Having shown how atmospheric refraction affects celestial observation, Ptolemy suggests the possibility of determining the amount of such refraction for "certain [celestial] bodies whose distance is given—e.g., the Sun and the Moon—and [thus determining] the amount by which the refraction . . . shifts the apparent position upward." But this we could do only "if the distance of the interface between the two media [air and aither] were known." Since that distance is unknown, however, Ptolemy concludes that "it is impossible to provide a method for determining the size of the angles of deviation that occur in this sort of refraction."[106]

Ptolemy's point can be easily grasped according to figure 3.17c, where S represents the Sun at the appropriate position to be seen along ADZ. If the altitude of the atmosphere is relatively high, the angles of incidence and refraction, ADH and TDS, are smaller than corresponding angles AD′H and T′D′S

105. See *Optics*, V, 25–26, in Smith, 239–40, and Lejeune, 238–40.

106. *Optics*, V, 30, in Smith, 241–42, and Lejeune, 241–42.

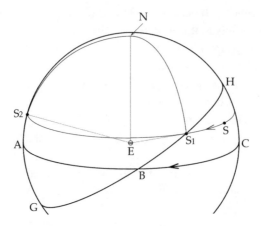

FIGURE 3.17A

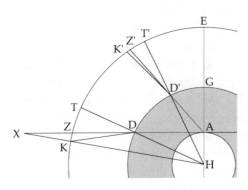

FIGURE 3.17B

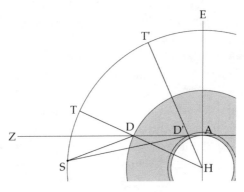

FIGURE 3.17C

if the altitude is relatively low. On the other hand, angle of deviation ZDS in the higher atmosphere is larger than angle of deviation ZD´S in the lower one. This means that the density differential between the atmosphere and the aither must be smaller when the atmospheric shell is thinner. Consequently, without a precise figure for the height of the atmosphere, the actual angle of incidence and, therefore, the angle of deviation cannot be determined. It would also help, of course, to know the density differential between the atmosphere and the aither beyond it. Small wonder, then, that having no means to determine either the height of the atmosphere or the density differential between it and aither, Ptolemy despaired of finding the amount by which visual rays are deviated by atmospheric refraction with any meaningful precision.

The second astronomical phenomenon Ptolemy addressed in the *Optics* is the so-called moon illusion. Apparently related to atmospheric refraction, this phenomenon concerns the noticeable enlargement of celestial bodies, the sun and the moon in particular, when seen at or near the horizon. Adverting to such enlargement in the *Meteorology*, Aristotle attributes it to condensed air or mist through which things appear larger than they are, but he seems to believe that the magnification is due to reflection rather than refraction.[107] Posidonius (fl. ca. 100 BC) also believes that this enlargement is due to mist or vapor rising from the earth's surface, but he imputes it to the refraction of visual rays passing through that vapor rather than to reflection.[108] The resulting magnification

107. *Meteorology*, 3, 4, 373a5–12, in Barnes, *Complete Works*, 602.

108. See Helen E. Ross, "Cleomedes (c. 1st century AD) on the Celestial Illusion, Atmospheric Refraction, and Size-Distance Invariance," *Perception* 29 (2000): 863–71. The dating of Cleomedes, an arch-Stoic and author of a work variously titled *Meteora* ("The Heavens") or *De motu circulari corporum caelestium* ("On the Circular Motion of Heavenly Bodies"), is problematic. He has been located as late as the second half of the fourth century AD and as early as the first century BC. In light of current scholarship it seems doubtful that his floruit should be placed much later than AD 200. He depends heavily upon Posidonius and is thus a key source for our understanding of Posidonius's thought. For a recent evaluation of Cleomedes's life and work, see Alan C. Bowen and Robert B. Todd, trans., *Cleomedes' Lectures on Astronomy: A Translation of* The Heavens (Berkeley: University of California Press, 2004); see especially 100–102 for the relevant passage on the apparent enlargement of the sun when seen through vapor. Toward the end of the *Meteora*, Cleomedes addresses the issue of "paradoxical eclipses" in which both the eclipsed moon and sun appear above the horizon. If this were actually the case, the eclipse could not occur because the earth would not block the sunlight reaching the moon. As a possible explanation of this phenomenon, Cleomedes cites refraction of visual pneuma through "damp and sodden air . . . so that when the ray from the eye is refracted and bends below the horizon, it encounters a sun that has already set"; Bowen and Todd, *Cleomedes' Lectures*, 162–63.

would thus be explained by the increased visual angle under which the image is seen along the incident visual rays refracted through the interface between the rarer air and the denser vapor. This is clearly what Ptolemy has in mind in *Almagest*, I, 3, where he accounts for such magnification on the basis of "exhalations of moisture surrounding the earth" that cause celestial bodies viewed through them to appear larger, "just as objects placed in water appear bigger than they are."[109]

In the *Optics*, however, there is no trace of this explanation. Instead of appealing to refraction, Ptolemy locates the effect in the visual faculty. "Generally speaking," he explains, "when a visual ray falls upon visible objects in a way other than what is inherent to it by nature and custom, it perceives less clearly all the characteristics belonging to them." Accordingly, the visual ray's "perception of the distances it apprehends will be diminished." It is for this reason that "among celestial objects that subtend equal visual angles, those that lie near the zenith appear smaller, whereas those that lie near the horizon are seen in another way that accords with custom. Things that are high up seem smaller than usual and are seen with difficulty."[110] Obscure at best, this explanation is wide open to interpretation.[111] But two things come through clearly. First, things look smaller at zenith than at horizon because they are viewed in an uncustomary or anomalous way. Second, the difference in size at horizon and zenith is only apparent because the body subtends the same visual angle at both positions. In short, the apparent enlargement of objects at the horizon is a visual illusion and therefore psychological rather than optical in origin. Refraction has nothing to do with it.[112]

As already mentioned, it is difficult to determine precisely what Ptolemy had in mind with this explanation of the moon illusion. He may, for instance, have meant to imply that objects look smaller at zenith because they are perceived to be closer at that point than at horizon. In other words, the illusion depends upon a psychological contravention of the size-distance invariance

109. Toomer, *Ptolemy's Almagest*, 39.

110. *Optics*, V, 59, in Smith, 151, and Lejeune, 115–16.

111. See, e.g., Helen E. and George M. Ross, "Did Ptolemy Understand the Moon Illusion?" *Perception* 5 (1976): 377–85. Cf. Abdelhamid I. Sabra, "Psychology versus Mathematics: Ptolemy and Alhazen on the Moon Illusion," in *Mathematics and Its Applications to Science and Natural Philosophy in the Middle Ages*, ed. Edward Grant and John Murdoch (Cambridge: Cambridge University Press, 1987), 217–47, especially 217–27.

112. For an exhaustive, recent study of the moon illusion and efforts to explain it from antiquity to the present, see Helen E. Ross and Cornelis Plug, *The Mystery of the Moon Illusion* (Oxford: Oxford University Press, 2002).

hypothesis. Or he might have meant simply that the object at zenith looks smaller because of the physical difficulty of craning the neck upward. Or, for that matter, he might have meant that the visual flux, with its diminished capacity to perceive upward distance, is unable to grasp the object in its totality, thus giving a diminished impression of its size. Because of its very obscurity, Ptolemy's explanation of the moon illusion is pregnant with possibilities that later commentators explored, Alhacen in particular. Moreover, as we will see in chapter 5, where Alhacen's optical work is addressed in detail, that explanation did not squelch the idea that refraction through thick vapors in the atmosphere also causes magnification of celestial objects at or near the horizon.

5. CONCLUSION

Because of its broad scope and occasional technical complexity, Ptolemy's *Optics* is difficult to summarize and therefore to grasp as a systematic whole. This difficulty stems in great part from the richness of Ptolemy's analysis and his consequent melding of empirical, physical, psychological, and mathematical elements. Nonetheless, despite the overall complexity of that analysis, its primary purpose is clear enough: to explain why things look as they do under widely varying conditions, most of them anomalous. For this reason Ptolemy devotes the vast majority of his account in the *Optics* to visual illusions, because most visual perception is actually misperception.

In order to lay the groundwork for his account of misperception, Ptolemy opens with a sweeping account of vision that serves as a sort of normative platform. The resulting theory is largely, but by no means totally compatible with, those of Aristotle and Galen. All three thinkers cite the inherent color of visible objects as the proper sensible for sight, although Ptolemy adds that the object must be adequately opaque and luminous to be seen. All three think that the spatial characteristics of visible objects—the so-called common sensibles—are inferred from the primary color impression apprehended by the visual faculty. All three stress the need for a continuous intervening medium between eye and object, this medium being luminized air in Aristotle's case, a pneumatic bridge in Galen's, and the cone of visual flux in Ptolemy's. And all three understand vision as a process that unfolds in stages rather than in a single act.

Ptolemy and Galen, meanwhile, part ways with Aristotle in adopting the visual cone as a means of explaining both monocular and binocular vision. Both also ground their accounts in a controlling faculty, the *hēgemonikon* or *virtus regitiva*, that serves as the guiding "self" of visual perception—the "ghost in the machine" Gilbert Ryle derided in his benchmark *Concept of*

Mind of 1949. Unlike Ptolemy, however, Galen places the visual process within a sophisticated anatomical and physiological system rooted in the brain and its pneumatic chambers and pathways. Finally, Ptolemy parts ways with both Galen and Aristotle in two crucial respects. First, he raises light from a mere catalyst to an immediate agent in vision. It may not yet be per se visible, but it is an objective sine qua non for vision. It does not, in short, just activate the medium between eye and object. Second, Ptolemy applies ray geometry systematically throughout his analysis. In this, of course, he is following Euclid's lead, but unlike Euclid, he recognizes the limitations of ray analysis and is quite willing to let psychology trump mathematics when the latter leads to conclusions that conflict with empirical observation.

Where Ptolemy diverges from everyone before him is in his wide-ranging analysis of visual illusions according to a threefold typology. Some illusions are due to anomalous physical circumstances, such as inadequate luminosity or opacity in the object, too much distance between eye and object, and so forth. Some, like the moon illusion or trompe l'oeil representations, are psychological in origin. And some are caused by perturbations in the visual flux. Among these latter illusions, those caused by breaking of the visual flux through reflection and refraction are the most salient and variegated because they involve not only image displacement (that is, the image and object occupy different locations) but also distortions in size, shape, and orientation. They also seem to be peculiarly susceptible to geometrical explanation. Consequently, the lion's share of Ptolemy's *Optics* is devoted to a close, systematic analysis of both phenomena according to their basic principles, mostly shared, and their visual effects, many also shared. It is here that Ptolemy comes into his own, displaying a remarkable creativity in approach. This creativity is clearly manifested in his use of experiment to confirm the equal-angles law in reflection, as well as to determine the relationship between the angles of incidence and refraction in refraction. It is also manifested in his systematic, case-by-case approach to reflection from concave spherical mirrors.

There is no question that in terms of technical complexity, Ptolemy's *Optics* pales in comparison to the *Almagest*. But the difference between the two works at that level reflects not so much a difference in analytic rigor or sophistication as a disparity in the number and complexity of the phenomena to be "saved." Much more needs to be saved in astronomy than in optics because it is subject to many more anomalies. When that point is taken into account, it is no stretch to claim that in its particular domain of study Ptolemy's *Optics* is no less magisterial than the *Almagest*. Both works thus represent the ne plus ultra in their respective disciplines. Nor does the similarity between the two works end there. As will become evident over the next three chapters, although Ptolemy's *Optics* did not have the immediate impact of the *Almagest*, it eventually set the agenda for optics by providing a powerful conceptual and methodological model that determined not only the scope and direction of analysis but also the kinds of problems that would be addressed. As with astronomy, so with optics, that model was modified and brought toward perfection by medieval Arabic thinkers, who then passed it on to the Latin West, where it was assimilated and, in turn, modified (not always for the better) during the later Middle Ages and Renaissance. It is to the beginning of this process of assimilation and modification that we will turn in the following chapter.

CHAPTER 4

Greco-Roman and
Early Arabic Developments

Despite the relative comprehensiveness, coherence, and methodological so-phistication of Ptolemy's *Optics*, it had little or no discernible effect on visual theory in late antiquity. Quite the contrary, it seems to have gone largely un-heeded during that period, as witness its failure to survive in Greek. Of the handful of late antique works that do mention the *Optics*, only one, *The Optical Hypotheses* of Damianos (or Heliodorus) of Larissa, shows more than a passing acquaintance with Ptolemaic optics.[1] Perhaps this failure is symptomatic of what has been traditionally viewed as an intellectual decline in the Greco-Roman world—particularly of scientific thought—during the centuries fol-lowing Ptolemy.[2] As far as the ray analysis of vision is concerned, this view

1. See Lejeune, *L'Optique*, 13*–14*. For a somewhat problematic critical edition of the Greek text of Damianos's short treatise, with facing German translation, see Richard Schöne, *Damia-nos Schrift über Optik* (Berlin: Reichsdrukerei, 1897). Divided into fourteen chapters and mostly Euclidean in tenor, this work summarizes various points about the emission of visual rays, which are taken to consist of ocular light. In chapter 3 there is an explicit reference to Ptolemy's use of an instrument to demonstrate that visual rays move in straight lines. If, indeed, Ptolemy de-scribed such a demonstration, it would have had to be in the missing first book or the missing section(s) after the middle of chapter 5. Chapter 13 of Damianos's treatise, which is by far the longest, deals with the equivalence of visual rays and solar rays. Interestingly enough, chapter 14 refers to Hero of Alexandria's proof of the equal-angles law in reflection. For a useful sum-mary of the work, see Fabio Acerbi, "Damianus of Larissa," *Complete Dictionary of Scientific Biography*, 2008, Encyclopedia.com. Accessed May 30, 2011. http://www.encyclopedia.com /doc/1G2-2830905607.html. Acerbi suggests dating the treatise to the fifth or sixth century.

2. For an extreme articulation of this view, see William Stahl, *Roman Science: Origins, Development, and Influence to the Later Middle Ages* (Madison: University of Wisconsin Press,

seems warranted, at least on the face of it. Very few extant post-Ptolemaic works deal with the geometry of vision, and those that do are unoriginal, if not retrograde.[3] On the other hand, some innovative work was done on the focusing of light rays in concave mirrors. Especially noteworthy in this regard is the analysis of burning mirrors by Anthemius of Tralles (474–534), about whom we will have more to say later.

It is therefore probably best to characterize what occurred in post-Ptolemaic optics, taken in the broadest sense, not as a declension but as a shift of analytic focus toward the physics and psychology of visual perception. Accordingly, visual analysis in late antiquity was primarily "philosophical" in orientation and, as such, found its main arena in commentaries on, and paraphrases of, Aristotle's account of visual perception in the *De anima*.[4] Not surprisingly, certain issues come to the fore in these sources. What is light, and how does it function in vision? Precisely how does vision occur at both the physical and psychic levels? Does the eye, in seeing, act on external objects, or do they act on the eye? How does visual sensation eventually lead to cognition? What are the psychic capacities that make this transition possible? In grappling with these old issues anew, various late antique thinkers elaborated on and transformed Aristotle's perception theory in significant and creative ways.

Key to this transformation was that most late antique Aristotle commentators viewed him through a Neoplatonist (or "Late Platonist") lens crafted by Plotinus (204/5–270) and refined by his disciple and editor, Porphyry (234–305).[5] Aside from compiling Plotinus's *Enneads*, Porphyry was also instrumental in promoting the idea that philosophical differences between Plato and Aristotle are only apparent, that the two should be reconciled when-

---

1962). Stahl's harshly negative view of late antique science has been moderated considerably over the past fifty years.

3. Damianos's treatise, with its emphasis on Euclidean theory, its relatively primitive geometrical descriptions, and its astonishing claim in favor of an equal-angles law in refraction, hardly represents an advance. Nor, for that matter, does the recension of Euclid's *Optics* attributed to Theon of Alexandria (335–405) or the brief series of optical theorems in the sixth book of Pappus of Alexandria's *Mathematical Collection*, which was probably written in the mid-fourth century; see Paul ver Eecke, trans., *Pappus d'Alexandrie* (Paris; Bruges: Desclée de Brouwer, 1933), 445–55.

4. These commentators also drew from the *Parva naturalia*, which includes such important supplementary works as the *De sensu* and the *De memoria et reminiscentia* ("On Memory and Recollection").

5. See Porphyry's *Life of Plotinus*, in A. H. Armstrong, trans., *Plotinus*, vol. 1 (Cambridge, MA: Harvard University Press, 1966).

ever and however possible. The upshot was that subsequent commentators regarded Aristotle as an explicator rather than an opponent of Plato and were therefore convinced that a mastery of Aristotle was prerequisite to a proper understanding of Plato.[6]

As far as visual theory is concerned, this meant that Aristotle's account in the *De anima* had to be coaxed into agreement with Plato's account in the *Timaeus*.[7] In the following section, we will trace this coaxing process in a selective way according to a handful of thinkers that includes, besides Plotinus, Alexander of Aphrodisias (late second/early third century), Iamblichus (245–325), Themistius (317–390), Simplicius (490–560), and John Philoponus (490–570).[8] That this latter figure was Christian brings us to an important point. From the early second century on, various Christian thinkers took a philosophical approach to doctrinal issues, especially those involving Christ's nature and relationship to the godhead. Among these Christian thinkers, Saint Augustine (354–430) is of singular importance for our purposes, and that for three reasons. First, he had much to say about visual perception. Second, he was deeply committed to Platonism as mediated by Plotinus. And third, because of his unparalleled stature as a theological authority, he was widely read in the medieval Latin West.

These thinkers are not randomly chosen. After various works of theirs were translated into Arabic during the ninth century, all of them (except Augustine, of course) informed Arabic thought on visual perception in crucial ways.[9] Among those within the Muslim sphere influenced by them, al-Kindī

6. See, e.g., Henry J. Blumenthal, "Neoplatonic Elements in the *de Anima* Commentaries," in *Aristotle Transformed*, ed. Richard Sorabji (Ithaca, NY: Cornell University Press, 1990), 305–24. For a general account of this development and its implications, see Lloyd Gerson, *Aristotle and Other Platonists* (Ithaca, NY: Cornell University Press, 2005).

7. Plato, of course, discusses the soul and its perceptual capacities, including vision, in other dialogues, most notably the *Republic*, but his treatment of it in the *Timaeus* is arguably the most systematic and comprehensive, particularly with respect to sense perception.

8. For a discussion of the lives, works, and interconnections of these thinkers, see Henry J. Blumenthal, *Aristotle and Neoplatonism in Late Antiquity* (Ithaca, NY: Cornell University Press, 1998), especially part 1. For discussion of specific thinkers, see the relevant articles in Lloyd Gerson, ed., *The Cambridge History of Philosophy in Late Antiquity*, 2 vols. (Cambridge: Cambridge University Press, 2010). Although he predated Plotinus and the beginning of the late Platonist period by almost a century, the Peripatetic commentator Alexander of Aphrodisias was an influential source not only for Plotinus but also for subsequent thinkers throughout late antiquity and the early Arabic Middle Ages.

9. For a brief overview of how late antique ideas were taken up by Arabic thinkers, see Cristina D'Ancona, "Greek into Arabic: Neoplatonism in Translation," in *The Cambridge Com-*

(800–870), Ḥunayn ibn 'Isḥāq (808–873), al-Fārābī (872–950), and Avicenna (or Ibn Sīnā, 980–1037) loom especially large because of the authority they wielded among both Muslim and Latin intellectuals during the Middle Ages. Aside from "philosophical" sources, Greek works on geometrical optics were also rendered into Arabic during the ninth century and were quickly assimilated and critiqued by the likes of al-Kindī and Qusṭā ibn Lūqā (820–912). In addition, under the stewardship of Ḥunayn, Greek medical works, Galen's in particular, were brought into the Arabic ambit during the ninth century. Indeed, it was in large part through Ḥunayn's efforts that Galen gained almost unparalleled status as a medical authority in the Muslim world. Thus by the end of the tenth century, Arabic thinkers had a firm and expansive foundation of Greek optical texts upon which to build their own superstructure during the following centuries.

In the remainder of this chapter we will trace the developments outlined in this background sketch, starting in the next section with a moderately close look at Plotinus's theory of the human soul and its perceptual and cognitive capacities. We will then pass in section 2 to an examination of how subsequent *De anima* commentators interpreted Aristotle's theory of perception and cognition within a largely Plotinian framework. I should note that in order to keep this examination within reasonable bounds, I will ignore, or all but ignore, a host of important related topics, such as language theory, logic, the psychology of emotions, practical reason, and the immortality of the soul. After a fairly close look at Saint Augustine's psychological theory in section 3, we will look in section 4 at how certain analytic strands in late antique *De anima* commentaries were taken up and adapted by key Arabic thinkers during the ninth and tenth centuries. In section 5, finally, we will examine the ways in which Greek geometrical optics was assimilated, modified, and added to within the Muslim sphere between roughly 850 and 1000.

As the foregoing prospectus suggests, the lion's share of this chapter is devoted to theories of the soul and its intellectual capacities, a subject that might seem to lie outside the purview of optics, even taken in its broadest sense. But it is crucial to realize that Greek thinkers, as well as their late antique and medieval followers, tended almost overwhelmingly to analyze mental activity in visual terms; the very vocabulary with which they discussed cognition is deeply rooted in sight words, such as *eidos* ("form" or "appearance"), *eikon* ("image"), and *eidōlon* ("likeness"). In fact, the verb *oida* ("to know") comes

panion to Arabic Philosophy, ed. Peter Adamson and Richard Taylor (Cambridge: Cambridge University Press, 2005), 10–31.

directly from *eidō* ("to see"). Most Greek thinkers also regarded vision as the conduit par excellence for the acquisition of knowledge, closely followed by hearing. Thus viewed as peculiarly symbiotic, vision and cognition were almost invariably explained in reciprocal terms. As a result, the theoretical analysis of each was tightly bound with and tailored to the theoretical analysis of the other.

1. PLOTINUS'S THEORY OF VISUAL PERCEPTION

As we saw with Ptolemy and Galen, both of whom were highly syncretic, the philosophical lines between Aristotelians, Platonists, Stoics, and other philosophical schools had already been blurred by the late second century. Consequently, if we regard philosophical cross-fertilization as reconciliation of a sort, that process was well underway before Porphyry championed the harmonization of Plato and Aristotle.[10] Bearing this point in mind, we should approach Plotinus's theory of visual perception aware that, although largely Platonist in outlook, it incorporates both Aristotelian and Stoic elements. Before analyzing that theory in detail, however, we need to set some background by comparing and contrasting Plato and Aristotle's conceptions of the soul and its perceptual and cognitive functions.

Taking our cue from the late antique commentators, let us start with Aristotle. According to one of his definitions, the soul is "the form (*eidos*) of a physical body having life potentially within it."[11] The potential for life is contingent on the body's organization, and in actualizing that potential, the soul renders the body an animate whole rather than a collection of parts. Like form and matter, soul and body are thus inseparable. Without the body the soul cannot subsist because it has nothing within which to actualize its animating function. Without the soul, the body is nothing but a physical congeries. For Aristotle, therefore, soul is not something inhabiting the body or distributed throughout it; it is, instead, the capacity of certain kinds of bodies to act and react in particular ways.

Those ways fall under three main categories. The first, at the level of the so-called vegetative soul, is defined by the drive in all living things, from plants

10. See, e.g., George Karamanolis, *Plato and Aristotle in Agreement?* (Oxford: Clarendon Press, 2006).

11. *De anima*, 2, 1, 412a21, in Barnes, *Complete Works*, 656. Subsequently, in *De anima*, 2, 1, 412a22 and 412b5–6, Aristotle defines the soul as the actuality (*entelecheia*), then, more specifically, the first actuality (*entelecheia prōtē*) of such a body (in Barnes, *Complete Works*, 656 and 657).

to humans, for self-sustenance, growth, and reproduction. The second category, at the level of the so-called sensitive soul, involves the ability of animals to move about at will in search of what is physically beneficial to them while avoiding what is not. The drive at this level is primarily appetitive. The final category, at the level of the so-called rational soul, is unique to humans and covers their ability to understand and cope with the world according to intellectual judgment.[12] Here, too, the drive is appetitive insofar as humans are naturally inclined to seek knowledge and to act well according to that knowledge. Analytic, not real, this threefold division bespeaks the varying capacities of animate beings according to their organization. Lacking eyes or feet, plants cannot see or move. Lacking reason, lower animals cannot achieve intellectual understanding.

Given that the human soul is fully incorporated in a physical body with both sensitive and intellectual capacities, those capacities must be physically aroused and exerted. Take vision, for example. In discussing Aristotle's account in chapter 2, section 2, we saw that the initial stimulation for vision involves the eye's reception of color transmitted through a continuous medium such as air made actually transparent by the presence of light, which Aristotle characterizes as the "color" of transparency. This reception is couched in terms of assimilation, the eye somehow "becoming" (or at least becoming like) the color that is its special sensible. Aristotle describes this assimilation by analogy to a physical impression stamped on wax by a signet ring.

Aside from color, which constitutes its special sensible, the eye also receives common sensibles, such as shape and size, which are passed with the special sensible to the common sensibility. There, both types of sensibles are combined to form images or phantasms (*phantasmata*) of the physical objects that are being sensed. Comprising the full range of sensibles all five senses provide, these images are stored in the imagination (*phantasia*), where they can be recalled as proxies for the original objects when they are no longer being sensed. In that mode, they represent the objects mnemonically. Subject to recollection (*anamnēsis*), they are the stuff of discursive reason, which scans and judges them according to their logical order and interconnections in order to reach overarching conclusions.[13] When properly drawn, these conclusions are absolutely true and universal. Serving as the objects of intellect (*nous*), they inform it in a way analogous to that in which the special sensibles inform their

12. For a relatively concise outline of this psychological tripartition, see *De anima*, 2, 2, 413a20–414a29, in Barnes, *Complete Works*, 657–59.

13. Aristotle's theory of recollection is most clearly spelled out in *De memoria*.

appropriate organs. At this level, with a full intuitive grasp of its intelligible objects, the intellect is no longer passive; it is truly active and "productive" (*poētikos*).[14]

Among the points that emerge from this brief account, four are particularly salient. First, because the process of perception and cognition Aristotle described is inductive and therefore dependent on sensation, the soul plays a more-or-less passive role. It is to some extent a tabula rasa. Second, although the acts of sensation, perception, and cognition can be interpreted in terms of successive phases, nowhere does Aristotle indicate clearly whether they occur serially or simultaneously. Nor does he indicate whether the faculties responsible for them are actually or only analytically differentiated.[15] Third, while it may be reasonable to characterize special sensibles (for example, color or sound) and common sensibles (for example, shape or size) as "sensible" because they are so clearly interrelated, it is difficult to see how Aristotle's incidental sensibles (for example, "the white thing before me is Diares's son") are so in any meaningful way. And finally, if the soul is thoroughly embodied, then cognition as a psychic act is problematic because its objects, being Universals, are absolutely nondiscursive, incorporeal and incorruptible, whereas the perceptual and cognitive acts leading to their intuitive apprehension are discursive, corporeal, and corruptible. Aristotle in fact acknowledges this problem in *De anima*, 3, 4, 429b5, where he admits that "while the faculty of sensation is dependent upon the body, intellect (*nous*) is separable from it."[16] Having thus actualized its potential to become the Universal, ought not the intellect be separable and immortal in sharing the Universal's perfect incorpo-

14. See, e.g., *De anima*, 3, 5, 430a14–15, in Barnes, *Complete Works*, 684. By "productive intellect" Aristotle means the intellect that produces a full, unitary, and atemporal apprehension from the temporal process of discursive reasoning. This apprehension, which is of the Universal, represents a sort of all-at-once "seeing" of the point of discursive reasoning. In apprehending the Universal, the intellect is brought from potential to actual by becoming that universal. Aristotle's brief discussion of the productive or active intellect in *De anima*, 3, 5 has spawned considerable interpretation and debate over the centuries; for some insight into both, see Michael V. Wedin, *Mind and Imagination in Aristotle* (New Haven, CT: Yale University Press, 1988), especially 160–208; Victor Caston, "Aristotle's Two Intellects: A Modest Proposal," *Phronesis* 44 (1999): 199–227; and Victor Caston, "Aristotle's Psychology," in *A Companion to Ancient Philosophy*, ed. Mary Louise Gill and Pierre Pellegrin (Malden, MA; Oxford: Blackwell, 2006), 316–46.

15. His refusal to isolate common sensibility from special sensibility suggests an analytic rather than physical differentiation.

16. Translation adapted from Barnes, *Complete Works*, 682.

reality and unchangeability?[17] Aristotle leaves this question open with respect to human intellect.

Plato, on the other hand, answers it with an unequivocal "yes" because, unlike Aristotle, he regards the human soul as fundamentally distinct from the body it inhabits—or, rather, the body that inhabits it. A relatively impure derivative of the World Soul that animates and orders the cosmos, the human soul is a mix of almost perfect intelligibility at the top and profoundly imperfect corporeality at the bottom. Tied by its lower nature to the body and thus vitiated, it is nonetheless empowered by its higher nature to regulate the body appropriately. In the *Republic* Plato describes this regulation according to a tripartite psychic scheme not entirely dissimilar to Aristotle's. At the lowest level is the concupiscent part (*epithumetikon*), which, being linked to the physical realm, is entirely subject to bodily appetites. Next comes the spirited part (*thumoeides*), whose proper function is to keep the appetites in check. At the highest level is the rational part (*logistikon*), which is linked to the intelligible realm and directs the spirited part in controlling the appetites.[18] The direction of control in the well-ordered soul is therefore downward from the rational, through the spirited, to the appetitive part. In a poorly ordered soul, the lowest part is allowed to run riot. Letting this happen is irrational and therefore bad because the physical objects to which appetite is tied are almost wholly unreal in comparison to the Forms, of which they are poor and indistinct replicas.

Plato's well-known distrust of sense perception is rooted in the fact that being tightly tethered to the realm of vague, physical replicas, the senses tell us almost nothing useful about reality. Worse, they cozen us into believing that what we perceive by sensation is real.[19] In order to avoid this trap, we must let our reason take charge and direct us toward the perfect, unchangeable Forms that constitute the higher reality exemplified so badly in the lower physical objects. This we do by recollecting those Forms, which are innate to, but lie dormant in our souls. The best that sense perception can do, therefore,

---

17. The Unmoved Mover of Aristotle's *Metaphysics* is just such an intellect: perfectly impassible, perfectly at one with the Universals it knows, and fully actual in thinking its thoughts and, thus, thinking itself. For an argument against conflating this intellect with the actual intellect at the human level, see Deborah K. W. Modrak, "The Nous-Body Problem in Aristotle," *Review of Metaphysics* 44 (1991): 755–74.

18. This tripartite schema is laid out quite clearly in *Republic*, 4, 437d–442d.

19. This, of course, is the point of the Cave Allegory in *Republic*, 7, 514a–521a. Note the difference with Aristotle, for whom sense deception is not rooted in the senses but rather in the judgments carried out in the imagination (see chapter 2, section 2).

is prompt us to recognize the appropriate Forms and thus grasp the Truth. Directed upward toward the Truth, the soul allows its better nature to dominate; indeed, the struggle to attain Truth is a struggle to disembody the soul so as to unmoor it from the trammels of physical distraction.[20] In taking firm control in this way, moreover, the soul is fully active in the cognitive process. Even at the level of vision the soul acts by emitting the gentle fire that passes through the eyes and coalesces with the fire outside to put us in visual touch with the physical world. For Plato, the soul is active throughout the perceptual and cognitive process. In that respect he stands in apparent, sharp contrast to Aristotle, for whom the soul is a mostly passive recipient of perceptual information gained through corporeal assimilation. Or, to put it somewhat starkly and simplistically, in Plato's theory the soul acts through the body, whereas in Aristotle's the body is the conduit through which the soul is acted upon.

One cardinal feature of Plato's thinking about the world is its emphasis on hierarchical order from Good at the top to Evil at the bottom. Truth is good, falsehood evil; Forms good, matter evil; reason good, appetite evil; order good, disorder evil. This zest for hierarchical ordering according to ontological and moral standards is exemplified in Plotinus's theory of the cosmos and the human soul's place within it. At the very top of his hierarchy is the One, which is perfectly good, indivisible, and spatially undetermined. As a wholly transcendent and unitary principle of reality, it overflows with Being to create a succession of entities at increasing levels of divisibility or plurality. Plotinus likens this creative overflow to the emanation of light from a source point, the One being light itself, Intellect the Sun, and Soul the Moon, which is illuminated by the Sun.[21]

The first entity created by this overflow is Intellect with its full complement of Forms. Being closest to the One, it is least divided and most real of all its creations, but its reality is compromised by the plurality of Forms it contains.[22] Next in the succession is World Soul, which is more divided and less real than Intellect, from which it derives. Its job is to replicate the Forms in Intellect within the world of matter and extension. In doing so, it provides the unity of individuation both to the cosmos as a whole and to every object in it. Soul is what binds each thing in the physical universe, from the universe

20. This point emerges clearly from the *Phaedo*.

21. On the One and its creative overflow, see *Ennead* V, 1, in Armstrong, *Plotinus*, vol. 5, 10–53. For the Sun-Moon analogy, see *Ennead* V, 6.4, in ibid., 211. This analogy harks back to Plato's discussion of intellectual illumination in the Cave Allegory in book seven of the *Republic*.

22. Its unity is also compromised by its being the subject of its own thought, so its thinking self is analytically distinct from the self being thought about.

itself to all its constituent objects, into a unified whole within an ontological hierarchy, ranging from the almost perfect unity and reality of incorporeal Intellect and World Soul to the profoundly imperfect disunity and unreality of corporeal beings at the bottom.[23]

Within this schema, human beings have a twofold nature. At the corporeal level, they are animated by an irrational soul through which they have perceptual access to the world of changeable physical particulars. At the incorporeal level they are animated by a rational soul through which they have access to the intelligible world of unchangeable Forms. Sense perception thus involves a passage from the bare apprehension of physical particulars toward an intuitive grasp of their exemplifying Forms. But this passage cannot be upward because, within Plotinus's hierarchical scheme, what is less perfect cannot act upon what is more perfect. Consequently, the irrational soul and its physically based sense perceptions cannot act on the rational soul to bring it from potential to actual knowledge of the Forms. The soul is therefore not perceptually receptive in a physical way, as it seems to be according to Aristotle's theory. "Sense-perceptions," Plotinus insists in *Ennead* IV, 6.1, "are not impressions or seal-stamps on the soul."[24] Perceiving is an inside out rather than an outside in affair, the body serving as a mediating vehicle or tool through which the rational soul exerts perceptual control.

Plotinus's emphasis on the "outward," or ontologically downward, direction of perception underlies his theory of vision. To start with, he repudiates Aristotle's claim that light actualizes the transparency of the intervening medium, be it air or water, in order to make it receptive of color, which it transmits to the eye. For Plotinus, radiated light is perfectly incorporeal and self-subsistent, so it needs nothing corporeal within which to manifest itself. It does so directly and immediately. Otherwise, it would be incorporated in the medium as its "color" in a way analogous to that in which color is incorporated in physical bodies. Colors, which are lights of a sort according to Plotinus, also manifest themselves directly and immediately, and it is through this manifestation that the eye sees both color and the object in which it inheres.[25] This it

---

23. For a detailed but fairly concise discussion of this schema as context for Plotinus's theory of the human soul and its perceptual and cognitive functions, see Eyjólfur Kjalar Emilsson, *Plotinus on Sense-Perception: A Philosophical Study* (Cambridge: Cambridge University Press, 1988), especially 10–22. I have relied heavily on his analysis as a guide in my own brief discussion. For a much earlier, briefer, but still helpful discussion, see Gordon H. Clark, "Plotinus' Theory of Sensation," *Philosophical Review* 51 (1942): 357–82.

24. *Ennead* IV, 6.1, in Armstrong, *Plotinus*, vol. 4, 321.

25. On color as light, see *Ennead* II, 4.5, in Armstrong, *Plotinus*, vol. 2, 115.

does according to its own internal light, which links it sympathetically with the object's color and thereby enables it to grasp the object as a visible whole.[26] The result is a formal impression (*tupos*) of the object, but one with a crucial difference. It is somehow qualitatively like the object but physically unlike it because it is incorporeal and does not replicate the object in a physical way. That, presumably, is what Plotinus means when he says in *Ennead* V, 8.2 that "the [object's] size is drawn in along with [the qualitative sense impression] not large in bulk but large in form."[27]

This qualitative grasp of the object is not perception, though. Properly speaking, perception begins with a sort of cognitive judgment of the object that is formally represented in the sense organ. That the Greek word for organ (*organon*) also means tool underlies the analogy that Plotinus draws in *Ennead* IV, 4.23 between how the soul uses sense organs in perception to how a woodworker uses a ruler in gauging his work. "The ruler," he says, "acts as a link between the straightness in the soul and that in the wood; it has its place between them and enables the craftsman to judge that on which he is working."[28] Like the straightness in the ruler, which imperfectly reflects the absolute straightness conceivable in our minds, the impression of the object in the eye is an imperfect reflection of the perfect Form replicated in that object. As such, it mediates between the sensible particular and its exemplifying Form, so it is through it that we make perceptual judgments about that particular. In order to highlight the active nature of perception, as opposed to mere sensation, Plotinus reminds us that we often sense without being aware that we do so. Perception therefore requires that the soul focus its attention on what is sensed, for "when the soul's activity is directed to [certain] things . . . it will not accept the memory of things [other than] these when they have passed away, since it is not aware of the sense-impression produced by them when they are there."[29]

Because there are five senses, a given object can manifest itself through them in five disparate ways. In order to perceive the object as one and the same, the soul must somehow unify these disparate manifestations. Aristotle remands this function to the common sensibility, but his description of that faculty is

---

26. On the eye's own light, see ibid. On sight as a function of sympathy, see *Ennead* IV 5.1–4, in Armstrong, *Plotinus*, vol. 4, 280–99. Although the notion of sympathy has Stoic overtones, Plotinus is emphatic that it has nothing to do with a pneumatic connection between eye and object.

27. Armstrong, *Plotinus*, vol. 5, 243.

28. *Ennead* IV, 4.23, in Armstrong, *Plotinus*, vol. 4, 201.

29. *Ennead* IV, 4.8, in Armstrong, *Plotinus*, vol. 4, 155.

brief and nebulous. Alexander of Aphrodisias—a key source for Plotinus—attempts to clarify Aristotle's account by locating the common sensibility, as a distinct faculty, in the heart.[30] Furthermore, in order to explain how this faculty operates, he likens it to a circle. By analogy to points on the circumference, the sense organs are linked to the center by radii along which the special and common sensibles converge on the common sensibility. It is at this point, where all the radii meet, that the plurality of sense impressions is brought to unity insofar as the center contains all the radii while yet being perfectly singular and unique.[31] The common sensibility does more than unify sensible impressions, though. Serving as the "ultimate sense organ," it also discriminates among them by judging their differences and similarities.[32]

While Plotinus accepts Alexander's circle analogy, he applies it not to the common sensibility but to the imagination. In fact, he dispenses with the common sensibility altogether, conferring on the imagination the dual function of unifying disparate sensible impressions into an incorporeal impression or "image" (*eidōlon*) of a given physical object as a sensible whole and of retaining that impression mnemonically. The product of judgments (*kriseis*), these perceptual impressions are transformed into quasi-intelligible conceptual impressions.[33] In comparing them to the unchangeable Forms, the reason is able to pass from a qualitative to a quidditative grasp of the sensible particulars from which they are abstracted. At this level we are not just thinking but consciously thinking and remembering our thoughts.[34] Where is this faculty

30. See Alexander of Aphrodisias, *De anima*, 3, 40–61, in Athanasius Fotinis, trans., *The De anima of Alexander of Aphrodisias* (Washington, DC: University Press of America, 1979), 125–36. In locating the common sensibility in the heart, Alexander is presumably following Aristotle's assumption that the heart, or the area around it, is responsible for all vital functions, including perception; see, e.g., *De somno et vigilia* ("On Sleep and Waking"), 456a1–6, in Barnes, *Complete Works*, 724.

31. See Alexander of Aphrodisias, *De anima*, 2, 50, 76. See also Themistius, *De anima*, 86, 18, in Robert B. Todd, trans., *Themistius on Aristotle's On the Soul* (Ithaca, NY: Cornell University Press, 1996), 109; and Simplicius, *De anima*, 200, 14–201, 12, in Henry J. Blumenthal, trans., *"Simplicius": On Aristotle's "On the Soul" 3.1–5* (Ithaca, NY: Cornell University Press, 2000), 56–57.

32. Alexander, *De anima*, 2, 53–55, in Fotinis, *De anima*, 78–79.

33. See Edward W. Warren, "Imagination in Plotinus," *Classical Quarterly* 16 (1966): 277–85. The judgments involved in the perception of sensible particulars can be, and often are, in error, so that, for instance, what we judge as red is actually white seen in red light, and so forth. The result of such judgments is opinion rather than knowledge.

34. See *Ennead* IV, 4.2–3, in Armstrong, *Plotinus*, vol. 4, 141–45; see also Edward W. Warren, "Consciousness in Plotinus," *Phronesis* 9 (1964): 83–97.

located? Following Plato, Plotinus looks to the brain rather than the heart. But the imagination is not *in* the brain because that faculty is incorporeal and dimensionless. Rather, the brain, with the nerves that extend from it to the sense organs, is the vehicle or point of application through which the intellect acts in order to apprehend the supervening intelligibility of the conceptual representations displayed to it in the imagination.[35] Even so, noble as it is, the human intellect needs the light of the highest Intellect to "see" that intelligibility in all its Formal perfection.[36]

This synopsis hardly captures the full complexity, nuance, and frequent obscurity of Plotinus's account, but it suffices to reveal certain crucial features of it that were to influence subsequent thinkers. Most important, perhaps, is his stress on activity rather than passivity in the perceptual process, the body serving as a vehicle rather than as a receptor. Thus while still speaking of "impressions," Plotinus strips them of all materiality or physicality. Rendered totally incorporeal and dimensionless, they are no longer *in* the body (or brain), although they are still *of* it. In addition, because perception involves both the irrational and rational parts of the soul, the two must find a point of juncture, which they do at the level of imagination. Accordingly, for Plotinus, the imagination forms a quasi-intellectual bridge between mere sense perception and the intuitive, intellectual grasp of the Forms that exemplify perceived objects. Perception thus unfolds in an upward ascent at increasing levels of abstraction from the sense organs, through the imagination, to the intellect. The higher the level, the more abstract and unified the impression formed at that level, so in working to grasp the overarching Form in intellect, the perceiver strives to achieve the ultimate good of unity. Plotinus's effort to de-materialize the perceptual process is especially clear in his denial of the need for a physical medium in visual perception. In that regard, he conflicts not only with Aristotle but also with Plato, whose theory of visual fire has clear materialist overtones. So too, he dismisses the Stoic theory of vision based on a pneumatic link between eye and object, although he does explain vision in terms of a sympathetic relationship between the two. Finally, Plotinus's analogy between physical and intellectual light was to have important ramifications among subsequent thinkers, Saint Augustine in particular.

35. See *Ennead* IV, 3.23, in Armstrong, *Plotinus*, vol. 4, 105–7.

36. *Ennead* V, 3.8, in Armstrong, *Plotinus*, vol. 5, 97–98. For a helpful analysis of Plotinus's theory of intellect, see Eyjólfur Kjalar Emilsson, *Plotinus on Intellect* (Oxford: Clarendon Press, 2007).

2. THE LATER *DE ANIMA* COMMENTATORS

By "Aristotelizing" Plato's theory of the soul and its perceptual and cognitive functions, Plotinus bequeathed a somewhat hybrid analytic framework within which a range of subsequent thinkers from Porphyry (late third century) to John Philoponus (mid-sixth century) operated in attempting to harmonize Plato and Aristotle. Not everyone in that range followed Plotinus to the letter, of course. Iamblichus is a prime example. Most likely a student of Plotinus's disciple, Porphyry, he makes no secret of his dissatisfaction with what both his teacher and his teacher's teacher have to say about the soul. The truth, he maintains, is to be found "in those doctrines to which Plato himself and Pythagoras, and Aristotle, and all the ancients who have gained great and honorable names for wisdom, are committed," and Plotinus and Porphyry have strayed from that truth.[37] Nor does Iamblichus restrict his criticism to these two. All philosophical schools after Aristotle, from Platonists and Stoics to the Peripatetics, share the blame in departing from the original intent of Plato and Aristotle. Still, Iamblichus's understanding of the human soul in its perceptual and cognitive capacities is not that different from Plotinus's, at least not as Pseudo-Simplicius and Priscianus report.[38] Like Plotinus, he has the human soul occupy a mean position between pure intelligibility, as manifested in its rational capacities, and pure sensibility, as manifested in its irrational capacities. Like Plotinus, he views the imagination as the mediating faculty between sensation and cognition insofar as it receives impressions from both the senses and the intellect and stores them mnemonically. Like Plotinus, he insists that these impressions are absolutely incorporeal and dimensionless. Like Plotinus, he emphasizes the active nature of perception. And like Plotinus, he regards light as incorporeal and self-manifesting.

Holding all or most of these positions, subsequent commentators faced the task of adapting them to Aristotle's account in the *De anima*. One major problem confronting them was how to incorporate the distinction between irrational and rational soul into Aristotle's analysis of perception and cognition. According to that analysis, almost the entire perceptual and cognitive process, including discursive reasoning, is grounded in the body, which provides a material substrate of sorts. Only at the very end, when the active, or productive,

37. Iamblichus, *De anima*, 7, 365–66, in John A Finamore and John M. Dillon, trans., *Iamblichus* De Anima (Leiden: Brill, 2002), 31.

38. See ibid., 230–78.

intellect comes into play by "becoming" its object—the Universal—does pure reason seem to take center stage. In addition, unlike Platonic Forms, Aristotle's Universals do not constitute the supervening reality of physical objects; they are somehow "in" those objects and must be abstracted from them in order to achieve actualization in the intellect. It is therefore unclear where, if at all, the juncture between irrational and rational souls might be found in the Aristotelian scheme—unless, of course, we restrict the rational soul to the productive intellect alone.

As we have seen, Plotinus locates this juncture in the imagination, which shares in both irrationality and rationality. On the irrational side, it acts like Aristotle's common sensibility by unifying sense impressions of physical particulars and forming a composite "image" that is stored for future reference. Out of such perceptual images, conceptual impressions or images are formed by judgment. These provide the basis for discursive reasoning, which analyzes them according to their intelligibility. Being temporal and aspect-by-aspect rather than intuitive and holistic, this process is not truly intellectual. Yet intellect guides it according to its innate Forms, which provide templates against which the conceptual impressions in the imagination can be formed and apprehended according to their true intelligibility. Standing for the Forms themselves, these conceptual impressions are stored mnemonically in the imagination, whence they are subject to recall. Plotinus thus places an extraordinarily heavy burden on the imagination by making it responsible not only for perceptual unification, judgment, and memory, but also for conceptual judgment and memory. Plotinus lightens that burden somewhat by splitting the imagination into two distinct kinds differentiated by their operations and by what is mnemonically stored in them. The lower form of imagination serves as the depot for sense-based perceptual images; the higher form contains the intellectually validated conceptual images that are the stuff of discursive reasoning. In essence, then, Plotinus posits two kinds of memory: perceptual and conceptual.

Themistius takes a more Aristotelian approach to the irrational-rational divide by lodging it in the distinction between potential and actual or productive intellect.[39] Potential intellect, explains Themistius, contains "enmattered forms, i.e. the universal thoughts assembled from particular objects of perception." In that mode it is unable "to distinguish between [these forms], or make transitions between distinct thoughts, or combine them" discursively. Rather,

---

39. There is divided opinion on whether Themistius was a relatively strict Peripatetic or whether he had strong Platonist leanings; see, e.g., Henry J. Blumenthal, "Themistius: The Last Peripatetic Commentator on Aristotle?" in Sorabji, *Aristotle Transformed*, 113–23.

"like a storehouse of thoughts, or better, matter, it deposits the imprints from perception and imagination through the agency of memory." When, however, "the productive intellect encounters it and takes over this 'matter' of thoughts, the potential intellect becomes one with it, and becomes able to make transitions, and combine and divide thoughts."[40] Like wood to the carpenter who shapes it, the potential intellect provides the material to be informed by the productive intellect. The informing process can be understood in terms of intellectual illumination; by analogy to the sun, the active intellect shines its light on the conceptual forms to reveal their overarching truth to the potential intellect.[41] But the two intellects are not actually separate, as are the wood and the carpenter. Rather, "the productive intellect settles into the whole of the potential intellect, as though the carpenter . . . did not control [the] wood but [was] able to pervade it totally."[42] Analytically but not functionally distinct at the level of discursive reasoning, the two become fully unified when the potential intellect is so perfectly informed as to think itself. In that state, the actual intellect transcends the body and the cognitive functions tied to it, including memory, all of which comprise what he calls the passive (*pathetikos*) intellect. So the productive intellect, being immortal, survives both the body and the passive intellect bound to it, and it does so in an entirely impersonal way because it is not distributed individually. Rather, it is common to and shared by all living humans.[43]

Although subsequent commentators did not share Themistius's view on the impersonal, common immortality of the actual intellect, they were pretty much in agreement with him about the divisions of intellect into passive, potential, and active, the former two associated with the body and the latter somehow apart from it in actualizing the potential intellect's capacity to apprehend the Forms. Particularly salient is Themistius's reliance on the matter-form analogy to explain the soul's perceptual and cognitive capacities. In this vein, he describes "the capacity for perception [as] matter for the imagination . . . the capacity for imagination [as] matter for the potential intellect . . . [and] the potential intellect [as] matter for the productive intellect."[44]

Matter, in this case, is intended metaphorically, yet perception, imagination, and discursive reasoning *are* materialized in varying degrees according to a

---

40. Themistius, *De anima*, 99, 2–10, in Todd, *Themistius*, 123.
41. Ibid., 98, 35–99, 2, 123.
42. Ibid., 99, 16–18, 123.
43. See ibid., 102, 1–104, 13, 126–29.
44. Ibid., 100, 30–33, 125.

hierarchical order determined by how closely bound they are to mere physical sensibility and to each other. Where and how are they actually materialized? Plotinus, following Plato, suggested the brain, with its outlying nerves, as the appropriate location and went even further to cite pneuma as the appropriate vehicle. But he was explicit that it is the vehicle *through* which, not *in* which, perception and cognition operate. Nonetheless, by using the terminology of impression and assimilation, he opened the way for treating pneuma as a material substrate for the images formed in perception and conception. Thus, Simplicius can speak of "the impressing and shaping of the pneuma when the imagination is stimulated" by sense perception, and he can claim that "imagination is similar to sense-perception . . . because it is inseparable from bodies with the pneuma always being impressed by the activity of the imagination."[45]

This tendency to pneumatize perceptual and conceptual functions and, moreover, to locate the responsible faculties in the brain is especially clear in John Philoponus, who was apparently well versed in Galenic physiology.[46] For instance, in his commentary on *De anima*, he gives a relatively detailed description of the eye and its component humors, pointing out that after "first receiving the activity of the outside transparent [air, the transparent cornea and humors of the eye] transmit this through to the optic nerve, in which there is the optic pneuma [that] comes down from the brain and terminates up at the beginning of the lens."[47] The Galenic overtones of this description are impossible to miss. Replete with pneuma, Philoponus explains, the brain is where the perceptual and conceptual faculties are located, the memorative faculty being specifically located in the occipital ventricle.[48]

In transplanting the perceptual and conceptual faculties into the ventricles of the brain and lodging their activity in a pneumatic vehicle, Philoponus is

45. Simplicius, *De anima*, 216, 25–28, in Blumenthal, *"Simplicius,"* 77.

46. See Robert Todd, "Philosophy and Medicine in John Philoponus' Commentary on Aristotle's 'De anima,'" *Dumbarton Oaks Papers* 39 (1984): 103–10.

47. William Charlton, *Philoponus*: On Aristotle's "On the Soul 2.7–12" (Ithaca, NY: Cornell University Press, 2005), 364, line 32–365, line 1, p. 51.

48. *In De anima*, 155, 28–32, in Philip J. van der Eijk, trans., *Philoponus*: On Aristotle's "On the Soul 1.3–5" (Ithaca, NY: Cornell University Press, 2006), 77–78. The association of particular faculties with particular cerebral ventricles and the *pneuma* contained in them goes back at least to Nemesius, bishop of Emesa (late fourth century?), who lodges sense perception and imagination in the twin anterior ventricles of the brain, reasoning in the central ventricle, and memory in the posterior ventricle; see R. W. Sharples and Philip J. van der Eijk, trans., *Nemesius*: On the Nature of Man (Liverpool: Liverpool University Press, 2008), 99–123.

hardly innovative. More innovative is his account of light and transparency.[49] Philoponus starts this account by agreeing with Aristotle that light actualizes the transparency of such media as air and water, but not by transmuting them from potentially to actually transparent. Instead, their transparency is made manifest by the fact that light or illuminated color can be seen through them.[50] Color is thus transformed from potentially to actually visible by illumination, and it manifests itself by radiating through transparent media. Light, for its part, is incorporeal, a point Philoponus defends at length by showing the absurdities involved in assuming that it is material.[51] So far so Plotinian. Philoponus parts ways with Plotinus, however, in his insistence that light is an activity (*energeia*) in the transparent medium and, moreover, that its activity is transmitted through the medium in successive order, starting with the part nearest the light source, which transmits its activity to the neighboring part, and so on. Yet its diffusion is all at once because what is transmitted is incorporeal, so its successiveness is causally but not temporally determined.[52] Philoponus therefore assumes that the medium actually supports the transmission of light and illuminated color. In taking this position, he avoids the problem of action at a distance implicit in Plotinus's explanation of vision on the basis of a physically unmediated sympathetic relationship between eye and visible object.

If light and illuminated color are visible per se, then what role does the eye play in the visual process? Philoponus is emphatic that it does not emit visual rays because to assume that it does leads to physical and empirical absurdities. For instance, if the visual rays consisted of matter, they would have to pass through the whole of the cornea or through pores in it. In the latter case, things would appear spotty rather than continuous, and in both cases the cornea would be ruptured so that the fluid behind it would seep out.[53] To suppose, as

49. For a close study of Philoponus's analysis of light and its theoretical ramifications, see Jean De Groot, *Aristotle and Philoponus on Light* (New York: Garland, 1991). As she points out in the course of that study, there are clear similarities between Philoponus's analysis and that of Alexander of Aphrodisias in his *De anima libri Mantissa*; see ibid., 23–24.

50. *In De anima*, 329, lines 3–35, and 333, lines 13–17, in Eijk, *Philoponus*, 144–45 and 149.

51. Ibid., 324, line 25–331, line 14, pp. 8–14. Philoponus's argument rests in great part on showing that to suppose the eye emits a material flux is as physically problematic as to suppose that light is a material emission from luminous bodies.

52. Ibid., 329, line 35–330, line 19, p. 14.

53. Ibid., 326, lines 15–26, pp. 10–11. Philoponus also argues that if we assume that the visual emission forms a continuous body, it is irrational "to say that from [such] a small [body as the

do the "mathematicians," that visual rays are geometrical lines is no less problematic because those lines would touch outlying objects at dimensionless points, in which case we would see nothing at all.[54] Nor, for various reasons that need not detain us, could the visual emission be incorporeal. Like Aristotle, Philoponus is a fully committed intromissionist, but contrary to Aristotle in *De anima* and *De sensu*, he gives light an immediate rather than mediate role in vision by making it a visible object in its own right rather than a mere catalyst for transparency. Furthermore, he takes pains to show that, just like the visual emission posited by mathematicians, light can be understood to propagate in straight lines, a point that is empirically verifiable.[55] Since light rays and visual rays are geometrically equivalent, then what can be accounted for by the one can also be accounted for by the other. Logic therefore compels us to choose Aristotle's intromissionist theory over the visual ray theory because when geometrized, "it both saves the appearances and avoids the absurdities" that follow from assuming visual rays.[56] All we need do is transform the visual cone into a cone of radiation with its base on the surface of the illuminating object and its vertex in the eye or, more specifically, "the pupil or the centre of the [crystalline] lens," which is where visual discrimination begins.[57]

By analyzing Aristotelian psychology within a Platonic (and Galenic) framework, the thinkers discussed to this point recast Aristotle's account in subtle but significant ways. Having explicitly raised light to the status of per se visible, they rendered the special sensible for sight twofold: light and illuminated color. Also, in emphasizing the active nature of perception, reason, and

---

eye] such large bodies are emitted [to] embrace" the huge portion of the distant heavens seen within the base of the visual cone; *In De anima*, 325, lines 30–32, in ibid., 9–10.

54. Ibid., 326, lines 10–15, p. 10.

55. As an example, Philoponus cites the fact that when light from a lamp shines through a slit to a plank behind the slit, the light on the plank will lie in a straight line with the source; see ibid., 327, lines 17–22, pp. 13–14.

56. Ibid., 331, line 34, p. 16.

57. Ibid., 333, lines 29–33, p. 18. On the basis of this radiative cone, Philoponus goes on to explain binocular vision and diplopia according to the overlapping bases of the cones formed in each eye. The resulting account has a distinctly Ptolemaic ring; see ibid., 339, line 36–340, line 19, pp. 24–25; cf. the description of Ptolemy's account in chapter 3, section 1. It is noteworthy that Philoponus uses the terminology of visual radiation in this explanation, but he has already justified Aristotle's use of such terminology in the *Meteorology* on the grounds that the visual ray hypothesis is "clearer [insofar as] it is not easy to conceive of activities being bent back [i.e., reflected], or of the activities of colours moving through the air at all"; ibid., 333, lines 24–25, p. 18.

cognition, they felt compelled to distinguish the embodied from the disembodied phases of the cognitive process. As a result, they arranged the various faculties engaged in that process in a hierarchical order, with cognition the driving force at the top. Accordingly, they subdivided the cognitive faculty into two basic levels of intellect. At the higher level is the active or productive intellect, which can know all the Forms or Universals in an intuitive and timeless way. At the lower level is the potential intellect, which has the capacity to know Universals but fulfills it through a discursive process grounded in sense perception. Because sense perception is a bodily function, the passage from sensation to cognition must unfold in—or, rather, through—the body and its faculties in a succession, ranging upward from the sense organs, through perception and imagination, to discursive reasoning. Meanwhile, two mnemonic depots are required, one for retention of the perceptual impressions acquired from sensation, the other for retention of the conceptual impressions acquired from discursive reasoning.

All of this is associated with the cerebral ventricles and the pneuma pervading both them and the sensory nerves feeding into the brain. The intellect at this level is therefore "passive" in two ways. On the one hand, it is initially moved to action by a physical effect in the sense organs. On the other hand, and more to the point, it is guided throughout by an overarching agent intellect that reveals the pure intelligibility of the conceptual impressions abstracted through the discursive process. The agent intellect does this by illuminating those impressions so that the potential intellect can "see" their supervening formal reality in the Universal. Intellectual illumination thus makes things intelligible in somewhat the same way as solar illumination makes them visible. The resulting intuitive grasp of the Universal transforms the potential intellect into actual intellect, but in our embodied state we cannot maintain this grasp, so we are usually either oblivious to what we know or engaged in discursively recovering it. Only the agent intellect, in its timelessness, has a continual and perfect grasp of the Universals. It is therefore logical to assume that this intellect lies "outside" or "above" the embodied human soul and the potential and productive intellects associated with it.

Taken as a whole according to logical structure, these ideas translate into a cognitive model based on certain presuppositions about what knowledge is, or is like, and the various psychological stages through which it is realized. Conceptually based on sight and the terminology associated with it, this model reduces the cognitive process to a succession of virtual glimpses of virtual images that ends with a comprehensive gaze at what those images actually represent or mean. As we will see in a moment, the lineaments of this model, with strong Plotinian overtones, are evident in Saint Augustine's ruminations about perception and cognition. But unlike all the thinkers we have examined to this point, except Philoponus, Augustine was Christian rather than pagan. Consequently, his ideas about the soul and its perceptual and cognitive capacities were informed by doctrinal concerns his main philosophical sources did not share.

3. SAINT AUGUSTINE'S PSYCHOLOGICAL MODEL: THE INWARD ASCENT

According to his *Confessions*, begun in the late 390s, Augustine spent much of his early adult life floundering in search of a reason to convert to Christianity.[58] I say "reason" because he was looking for a philosophical system that would demonstratively validate Christianity and thereby bring him to belief through certainty. What he found instead was certainty through belief—hence his famous injunction from Sermon 43, 7: "Believe so that you may understand." Through the mediation of Marius Victorinus (ca. 300–ca. 370), he also found in Plotinus a philosophical system that was more or less compatible with Christianity.[59] Soon after his conversion in 386 and baptism at Milan in 387, Augustine returned to his birthplace in the North African town of Thagaste, where he went into monastic retreat. Called to the priesthood in 391, he became bishop of Hippo Regius four years later, and from that point to the end of his life in 430, he was deeply embroiled in doctrinal disputes. Consequently, all of his extant works, including those written between 386 and his entry into the priesthood in 391, deal explicitly with theological and doctrinal matters and only implicitly with philosophical issues. We are therefore obliged to re-

---

58. For the standard biography of Augustine, see Peter Brown, *Augustine of Hippo: A Biography*, 2nd ed. (Berkeley: University of California Press, 2000); cf., however, James J. O'Donnell, *Augustine: A New Biography* (New York: Ecco, 2005), where Augustine's clay feet come to the fore in fascinating ways. For a somewhat idiosyncratic but useful biographical study focusing on Augustine's religious thought, see Thomas F. Martin and Allan D. Fitzgerald, *Augustine of Hippo: Faithful Servant, Spiritual Leader* (Boston: Prentice Hall, 2011). For a dated but perceptive study of Augustine's theological and philosophical thought, see Étienne Gilson, *The Christian Philosophy of Saint Augustine*, trans. L. E. M. Lynch (New York: Random House, 1960).

59. A prolific author, Marius Victorinus translated Plotinus's *Enneads* into Latin, in which form Augustine claims to have read that work. Victorinus also converted to Christianity in the early 360s and, after conversion, wrote tirelessly on doctrinal matters. For Augustine on Victorinus, see *Confessions*, 8.2.3–5.

construct his theory of perception and cognition from discussions embedded in works focused on other things.[60]

In order to put Augustine's psychological theory into proper context, we must start with his ontological scheme. Generally speaking, Augustine accepts something like Plotinus's ontology, but with major qualifications. For instance, while he implicitly admits the analogy between God and Plotinus's One, Augustine denies God the utter transcendence of the One; He must be immanent, personal, and providential. Second, whereas Plotinus's One is compelled by its very nature to overflow with Being, Augustine stresses the volitional and arbitrary nature of God's creation. Third, God does not create by emanation because He does so ex nihilo without intermediaries. And finally, the ontological passage from God to physical particulars does not represent a descent from Good to Evil. Since the physical world is the creation of a perfectly good God, it must be intrinsically good. Nonetheless, there is a gradation from Truth/Reality/Perfection to Falsehood/Unreality/Imperfection, which means that physical objects have almost no reality in comparison to the intelligible forms they exemplify so vaguely. These forms are contained in Christ, as Intellect, although Christ is not ontologically subordinate to God, as is Intellect to the One. [61]

Because our embodied soul is linked to both the physical and intelligible (or spiritual) realms, we have two basic drives, one in the direction of physical pleasure, the other in the direction of intellectual satisfaction. As with Plato, so with Augustine, we can choose, at least theoretically, which direction to follow according to how our will, which corresponds to Plato's spirited part of the soul, impels us. But in this postlapsarian vale of tears, the human will is so utterly vitiated that it always impels us in the wrong direction, leading us toward physical pleasure rather than intellectual satisfaction. Nor are we able to redirect our will on our own. For that we need Christ's help in the form of grace. Only then can we turn toward the intelligible realm, the realm of God, with the right will. In doing so, we achieve true intellectual satisfaction or happiness (*beatitudo*). Nevertheless—and this is a crucial point—the intellectual satisfaction Augustine has in mind constitutes wisdom (*sapientia*) and lies in knowing and loving God, not in knowing intelligible Forms. This latter knowledge is mere rational apprehension (*scientia*). The role of Christ as mediator is

---

60. See Gerard O'Daly, *Augustine's Philosophy of Mind* (London: Duckworth, 1987), especially 1–6.

61. On Christ as Divine Intellect, see *Against the Academicians*, 3.19.42, in Peter King, trans., *Augustine*: Against the Academicians *and* The Teacher (Indianapolis: Hackett, 1995), 91.

thus of critical importance because it is only through Him that we can acquire the wisdom that yields true beatitude.

Embedded in this theological scheme is a distinction between the true and Truth. God is Truth and, as such, the standard according to which what is true is true. In order to know the Truth, then, we must know what is true, and this knowledge comes from self-reflection. The most obvious thing we know to be true about ourselves, Augustine affirms in *On Free Will*, is that we exist. From this we can deduce that we are alive and, moreover, that we understand that we exist and live. We also know intuitively that to understand (*intelligere*) is superior to simply existing or living.[62] The question, then, is twofold: how do we pass from the perception of physical particulars to an understanding of them, and how can we be sure that this understanding is veridical? In response, Augustine reminds us that we possess two kinds of senses. The inferior, external ones are lodged in the five sense organs, each of which has its special sensible, color for the eye, sound for the ear, and so forth. All of them, moreover, perceive other things, such as size and shape: the Aristotelian common sensibles. Superior to the external senses is an internal sense (*sensus interior*) that can "determine what belongs to just one sense and what to more than one."[63] In short, Augustine's interior sense serves essentially the same purpose as Aristotle's common sensibility or Plotinus's imagination. It unifies the special and common sensibles and, in addition, perceives that it is perceiving, thereby prompting the senses to attend to what they sense or want to sense.[64] The perceptual and conceptual faculties responsible for all this, Augustine observes in his "literal" commentary on Genesis, are associated with the anterior and middle ventricles of the brain.[65]

Above the internal sense, finally, is reason (*ratio*), through which "all of these things [perceived by the external and internal senses], as well as reason itself, become known and are part of knowledge (*scientia*)."[66] The things perceived at all these levels consist of a particular succession of forms (*formae*) or species (*species*), both terms used interchangeably by Augustine. First is the

62. *On Free Will*, 2.3.7, in Thomas Williams, trans., *Augustine*: On Free Choice of the Will (Indianapolis: Hackett, 1993), 33.

63. Ibid., 2.3.8, 34.

64. Ibid., 2.4.10, 37. On Augustine's view of imagination and its specific operations, see O'Daly, *Augustine's Philosophy*, 106–30.

65. *On* Genesis *Interpreted Literally*, 7.18.20–24, in Edmond Hill, trans., *On Genesis* (New York: New City Press, 2002), 332–35.

66. *On Free Will*, 2.4.10, in Williams, *Augustine*: On Will, 37.

species of the sensible body itself, from which comes the species in the sense. Like the impression of a signet ring in wax, this species serves as a sort of image (*imago*) of the object. That image, in turn, yields the species in memory. This species is a likeness (*similitudo*) of the object, but now separated from all corporeal entailments so that it can be summoned up at will by the mind. From this conceptual representation in memory, finally, comes the species "in the gaze of thought" (*species quae fit in acie cogitantis*), which constitutes the Universal exemplified by the concept and seen by the intellect.[67] Consequently, Augustine concludes, "there are two visions, one of perception (*sentientis*), the other of thought (*cogitantis*)."[68]

Augustine insists that we already know the Universals before apprehending them through sense perception. For instance, we all know or can be taught that seven plus three equals ten, and once we know it, we know that it is incontrovertibly true. But this knowledge comes not from sense experience because that is evanescent and deceptive, whereas the fact that seven and three sum up to ten is invariable and necessary. We know this because we have an intuitive understanding of number. The very idea of number must therefore lie outside of sense experience because it is common to all of us, not personal to each of us, as is sense experience.[69] How, then, do we come to know that idea? As expected, Augustine posits an ulterior agent intellect that reveals it to us, but in this case that intellect is not impersonal. It is Christ. He provides the illumination that allows us to see the intelligibility of what we reason about. In doing so, He acts as a teacher, reminding us of what is true about sense experience by helping us recollect the formal reality that renders that experience intelligible.[70] Christ's divine illumination extends to all humans, allowing them to grasp what is true, but it enables Christians to pass beyond mere intellectual apprehension or science, through which the true is known, to wisdom, through which Truth itself is known, albeit only vaguely.[71]

Augustine's most radical departure from the late antique model of cognition lies in his emphasis on will, not intellect, as the superior faculty of the

67. *On the Trinity*, 11.9, in Stephen McKenna, trans., *Augustine*: On the Trinity, *Books 8–15* (Cambridge, New York: Cambridge University Press, 2002), 78.

68. Ibid., 79.

69. *On Free Will*, 2.8, in Williams, *Augustine*: On Will, 44–46.

70. *On the Teacher*, 10.33–14.46, in King, *Augustine*, 135–46. Augustine's discussion in this context is based on words as "signs" in that they signify or point to ulterior things. Likewise, perceptions are signs that point to ulterior things.

71. *On Free Will*, 2.9–13, in Williams, *Augustine*: On Will, 47–57.

soul. It is therefore in the misdirection of will toward physical pleasure (sex in particular) that sin lies, not in the physical objects sought. After all, the search for pleasure in physical things, which are almost wholly unreal, is a quest after virtually nothing. Also, in making will the directive force in the soul, Augustine emphasizes the active nature of all the soul's functions, including perception and cognition. Accordingly, unlike Plotinus, he accepts and indeed promotes the theory of visual radiation.[72] The active and volitional nature of sense perception is manifested in the difference between looking and seeing. With our eyes open, we look continuously, but we rarely see what we are looking at. For that to happen, we must willfully direct our attention to whatever we want to perceive.[73] Here we find a clear echo of Plotinus.

That Augustine's psychological theory falls squarely within the late antique tradition needs no belaboring. If nothing else, his syncretic approach makes this clear; although predominantly Platonic in orientation, Augustine's theory represents the fusion of Platonic, Aristotelian, and Stoic elements so typical of his era.[74] In that regard Augustine manifests the spirit of philosophical reconciliation that prevailed during late antiquity. But of course he added a distinctly Christian twist to the psychological theory he inherited. In that regard, too, he manifests a spirit of reconciliation, this time of theology and philosophy. Augustine achieved this reconciliation not by reducing doctrine to a rationally determined system but by using reason to illuminate doctrine and its scriptural warrant. This gift for bringing philosophy into the arena of faith not as arbiter but as elucidator was one reason that Augustine was so widely read as a theological authority in the medieval Latin West. Another reason is that writing in Latin rather than Greek, he was accessible to a literate, medieval European audience that had become virtually monolingual in Latin. Augustine thus provided a thin but crucial thread of continuity between the ancient Greek East with its rich philosophical tradition and a medieval Latin West that had all but lost direct contact with that tradition. Until the twelfth century, when Europe reestablished this contact through translations of Arabic and, to some extent, Greek texts, Augustine played a critical role as both theologian and schoolmaster.

72. See, e.g., *On the Trinity*, 9.3, in McKenna, *Augustine*: Trinity; and *On the Magnitude of the Soul*, 43, in Joseph Colleran, trans., *Saint Augustine*: The Greatness of the Soul; The Teacher (Westminster, MD: Newman Press, 1950), 66.

73. *On Free Will*, 8.4.10. In *On the Trinity*, 11.2, Augustine refers to this directing of attention as "intention of the soul" (*animi intentio*); McKenna, *Augustine*, 62.

74. On Augustine's reaction to and use of Stoic ideas, see Marsha Colish, *The Stoic Tradition from Antiquity to the Early Middle Ages* (Leiden: Brill, 1985), especially 142–238.

4. THE ARABIC TRANSITION:
THE *DE ANIMA* TRADITION

After Muḥammed's death in 632, Muslim armies swept north from the Arabian Peninsula to gain astonishingly quick control not only of Persia but also of Byzantine territory in an arc from Mesopotamia and the Levant to Egypt. Socially, politically, and religiously tied to Mecca, the Arabs who remained in that arc as occupiers had little or no interest in mingling with the Hellenized and largely Christian local populace or even in converting it. However, the transfer of the caliphal capital to Damascus in 661 by the first 'Umayyad caliph, Mu'āwīyah, marked a crucial shift in the political and cultural center of gravity for Islam. It also marked a shift toward at least partial integration with the local population. As a result, the 'Umayyads relied on Byzantine functionaries for administrative expertise and adopted certain Byzantine cultural and political norms. Conditions would thus seem to have been ripe for the transmission of Greek learning from occupied to occupiers. That this in fact did not happen, Dimitri Gutas observes, is because the Byzantine elite with whom the 'Umayyads dealt "was inimically indifferent to pagan Greek learning" on account of theological developments within the orthodox church. "In this intellectual climate," he continues, "it is impossible to conceive of a translation movement supported by Greek-speaking Christians, of *secular* Greek works into Arabic," at least not without "aggressive promotion by the 'Umayyads, [which] was lacking."[75]

With the accession of the 'Abbāsids in 750 and the eastward shift of the capital to purpose-built Baghdad in 762, conditions became much more favorable for the translation of Greek works into Arabic. For one thing, the educated Syriac Christian population of Mesopotamia remained steeped in Hellenic and Hellenistic learning, and the indigenous Persian and Jewish intelligentsia was sympathetic to that learning. For another thing, starting with the second 'Abbāsid caliph, al-Manṣūr (r. 754–75), there were political and ideological reasons for promoting Arabic translations of philosophical and scientific works not only from Greek but also from Persian. Accordingly, between the time of al-Manṣūr and the early ninth century, there seems to have been caliphal support for translating Greek and Persian works, including a significant portion of Aristotle's *Organon*, as well as Euclid's *Elements*.

The translation movement begun in this early phase became more con-

---

75. Dimitri Gutas, *Greek Thought, Arabic Culture*, (London: Routledge, 1998), 18–19. During the later 'Umayyad period, however, a set of Pseudo-Aristotelian letters to Alexander and the spurious Aristotelian *De mundo* were translated into Arabic.

certed during the second phase, which, legend has it, was set into motion by al-Ma'mūn (r. 813–33), who is said to have sponsored the translation of many Greek philosophical and scientific works, Aristotle's in particular. Just how instrumental he actually was in inspiring this movement, what his motivations for doing so might have been, and what part his famous *Bayt al-Ḥikmah* ("House of Wisdom") might have played in the resulting effort are open to question.[76] Two things are beyond doubt, though. For several decades after the accession of al-Ma'mūn, a vast array of Greek philosophical and scientific works were translated from Syriac and Greek into Arabic, and the resulting translation movement was centered in Baghdad. At the heart of this movement were two circles of translators, one centered on the Muslim polymath Ya'qūb al-Kindī, and the other on the Nestorian Christian savant Ḥunayn ibn 'Isḥāq. In both circles, the primary motivation for translation was research, so the texts selected for translation were chosen not haphazardly but according to specific content.[77]

As far as we know, al-Kindī did not actually engage in translating but relied on others to do it for him, possibly under the patronage of al-Ma'mūn and probably under that of his son, al-Mu'taṣim (r. 833–42). Because al-Kindī's interests were extraordinarily broad, the net he cast for texts was a wide one. Among the works eventually captured in it were a variety of Aristotelian treatises, including at least some of the *Parva naturalia* and a paraphrase of the *De anima*. In addition, al-Kindī had access to parts of Plotinus's *Enneads* under

76. The effort to translate Greek works into Arabic (often through Syriac intermediaries) during the early 'Abbāsid period, particularly under the caliph al-Ma'mūn, has been well documented; for an extensive analysis of this movement and its motivations, see Gutas, *Greek Thought*, especially 28–104; and for a list of Arabic translations carried out between roughly 750 and 1000, see Gutas, "Greek Philosophical Works Translated in Arabic," in Pasnau and Van Dyke, *Cambridge History of Medieval Philosophy*, 802–14. It is fairly certain that Persian astronomical works were translated into Arabic somewhat before the effort to translate Greek works was mounted, perhaps during the late 'Umayyad period. Gutas argues, quite plausibly, that the mastery of paper making during the early 'Abbāsid period facilitated the translation movement; see *Greek Thought*, 13; see also Bloom, *Paper*.

77. See, e.g., Roshdi Rashed, "Problems of the Transmission of Greek Scientific Thought into Arabic: Examples from Mathematics and Optics," *History of Science* 27 (1989): 199–209. On the Kindian circle in particular, see Gerhard Endress, "The Circle of al-Kindī: Early Arabic Translations from the Greek and the Rise of Islamic Philosophy," in *The Ancient Tradition in Christian and Islamic Hellenism*, ed. Gerhard Endress and Remke Kruk (Leiden: Research School CNWS, 1997), 43–76. See also Dimitri Gutas, "Geometry and the Rebirth of Philosophy in Arabic with al-Kindī," in *Words, Texts, and Concepts Cruising the Mediterranean Sea*, ed. Rüdiger Arnzen and Jörn Thielmann (Louvain; Dudley, MA: Peeters, 2004), 195–209.

the title of *The Theology of Aristotle*, as well as Proclus's *Elements of Theology*, also falsely ascribed to Aristotle. Likewise, he had at hand some Platonic works, including part of the *Timaeus* and the *Phaedo*, and as we will see in the next section, he was well acquainted with the Euclidean *Optics*. These are only some of the Greek works available to him during his career, but it should be noted that many of the Arabic versions upon which he relied were either excessively literal or paraphrastic.

The circle surrounding Ḥunayn ibn 'Isḥāq, which included his son 'Isḥāq ibn Ḥunayn and his nephew Ḥubaysh, had a more programmatic approach than the al-Kindī circle. Rather than limit themselves to Greek and Arabic, as did the Kindians, the translators associated with Ḥunayn dealt in three languages, Greek, Syriac, and Arabic. Translating from one to the other with equal facility, they also differed from their Kindian counterparts by being more faithful to the sense than to the letter of the texts to be translated. The programmatic nature of their approach is especially clear in Ḥunayn himself, who is credited with having translated well over one hundred works of Galen into Syriac or Arabic. One of his most important original compositions was an ophthalmological textbook titled *The Ten Treatises on the Eye*. By the time his son Isḥāq ibn Ḥunayn died, in 911, this circle of translators had rendered most of the Aristotelian corpus into Arabic along with a number of late antique commentaries.[78]

A third phase of translation occurred after the Persian Būyids gained control of the 'Abbāsid Empire in 934. The translators involved in this phase, many of them still Christian, were associated with the so-called Baghdad Peripatetics, a group of thinkers that included al-Fārābī. During this phase, several of the previously translated works were retranslated from Greek and, in some cases, from Syriac into Arabic. With this full-scale Arabization of the

78. Most of the information we have about the translation movements between roughly 800 and 980 comes from three biobibliographic sources: Ibn al-Nadīm (936?–995?), Ibn al-Qifṭī (1172–1248), and Ibn Abī Uṣaybiʿa (1203–1270). Especially valuable is al-Nadīm's renowned *Fihrist*, an English version of which is available in Bayard Dodge, ed. and trans., *The* Fihrist *of al-Nadim* (New York: Columbia University Press, 1970). Organized in ten "books" according to specific discipline, the *Fihrist* devotes book seven to philosophical and scientific texts. For a discussion of al-Nadīm's methods, see Devin Stewart, "The Structure of the *Fihrist*: Ibn al-Nadim as Historian of Islamic Legal and Theological Schools," *International Journal of Middle East Studies* 39 (2007): 369–87. On Ḥunayn's translations of Galen, see Max Meyerhof, "New Light on Hunain Ibn Isḥāq and His Period," *Isis* 8 (1926): 685–724. On Arabic translations of Aristotle's *De anima* and *De sensu*, see Francis E. Peters, *Aristoteles Arabus* (Leiden: Brill, 1968), 39–47.

Aristotelian corpus and its constellation of late antique commentaries, Arabic thinkers had an incredibly wide range of Hellenic, Hellenistic, and late antique sources available, and for the most part in a more reliable form than before. In order to see how this evolving corpus influenced Arabic thought on visual perception and its psychological underpinnings, we will briefly examine the theories of four key figures from the period between the mid-ninth and early eleventh centuries: al-Kindī, Ḥunayn, al-Fārābī, and Avicenna.

Let me start with a generalization. All four of these thinkers subscribed to the basic model of cognition discussed in the previous section. Accordingly, they assumed that the passage from sensation to cognition unfolds in stages, starting with sensation, passing through perception and conception, and culminating with intellectual intuition. They were also in general agreement about the basic faculties involved: the sense organs, imagination, memory, discursive reason, and intellect. It would therefore be pointless to describe in detail the psychological theory of each thinker in chronological order. It makes more sense, instead, to examine how they approached certain issues about sight, perception and conception, and intellection and, in grappling with them, modified the basic cognitive model by elaborating on it, rearticulating it, and eventually solidifying it. So let us work upward through the stages of visual perception from sensation to intellectual apprehension with an eye toward three fundamental issues.

The first issue to be considered is whether visual contact between eye and object is established by something emitted from the eye (extramission) or by something emanating from the object (intromission)—or perhaps by both. As we will see in the next section, al-Kindī is unequivocal in supporting extramission. Nonetheless, although he argues that the eye emits a cone of radiation through the air to outlying objects from its vertex at the "pupil," thereby visually illuminating them, he also acknowledges that these objects are seen more clearly when illuminated by strong external light. He thus seems to favor a theory of cooperation between sight and light. This position is consonant with his belief that everything in the universe interacts with everything else through radiation of reciprocal powers or influences.[79]

Ḥunayn takes a somewhat more equivocal stance in favor of extramission than al-Kindī. An ardent disciple and promoter of Galen, Ḥunayn follows his master in supposing that vision results from a pneumatic transformation of

---

79. Al-Kindī proposes this idea in a brief tract titled *De radiis* ("On Rays"), which is currently extant only in a medieval Latin version; for a critical edition and analysis, see Marie-Thérèse d'Alverny and Françoise Hudry, "Al-Kindi: *De radiis*," *Archives d'histoire doctrinale et littéraire du Moyen Âge* 41 (1974): 139–260.

the air between eye and object (see chapter 2, section 3), not from an actual outreach of visual flux. Accordingly, he claims in his *Ten Treatises on the Eye* that, upon receiving the influences "of the visual spirit as well as [the influences] coming from the sunlight, [the intermediate air] has the same relation to vision as the nerve has to the body."[80] This quasi-neurological extension of the eye takes the form of a cone, and its proper sensible is "fire and what is in the nature of fire, *viz*, colour." There are, he continues, "three kinds of fire: flame, red heat, and light," and the interconnection among them is evident from the fact that when light "is concentrated in a glass or in a transparent or shining body, it causes burning."[81]

Ḥunayn was not slavish, though. He was willing to contravene Galen when he saw fit. One such contravention occurs in his account of ocular anatomy at the beginning of the *Ten Treatises*. In the first three treatises of that work, Ḥunayn follows Galen faithfully in describing the seven tunics and three humors of the eye and in pinpointing the crystalline lens as the seat of visual sensation. He also follows Galen faithfully in his detailed description of the functions those tunics and humors serve. Likewise, he follows Galen faithfully in describing how the crystalline lens is rendered sensitive by means of visual pneuma, or spirit, passed to it from the brain through the hollow optic nerves and the retina, "which supplies nourishment to the vitreous humour, and through its nerve the sense of feeling and the luminous spirit . . . to the lens."[82] Where Ḥunayn departs from Galen is in placing the crystalline lens dead center in the eye rather than toward the front (cf. figure 2.3). Why he insists on this modification is unclear, but Bruce Eastwood argues plausibly that he does so for metaphysical reasons: because the lens is functionally central, because the rest of the eye's components subserve this function, and because form should follow function, the lens must be physically central.[83] As we noted earlier, the idea that form follows function is key to Galen's thinking about anatomy and physiology.

Unlike al-Kindī and Ḥunayn, al-Fārābī and Avicenna favor intromission over extramission, perhaps somewhat equivocally on al-Fārābī's part. For in-

80. *The Ten Treatises on the Eye*, trans. Max Meyerhof (Cairo: Government Press, 1928), 25.

81. Ibid., 16; Ḥunayn is apparently referring here to both burning lenses and burning mirrors, which we will discuss in the next section.

82. Ibid., 8; see also 27–31. Ḥunayn explains binocular vision and diplopia according to whether the axes of the cones from both eyes converge on the surface of a visible object. If so, the object will appear single; if not, it will appear double. Image fusion occurs at the optic chiasma, which constitutes the "origin" of sight; ibid., 25–26.

83. Bruce Eastwood, *The Elements of Vision: The Micro-Cosmology of Galenic Visual Theory According to Ḥunayn Ibn 'Isḥāq* (Philadelphia: American Philosophical Society Press, 1982).

stance, in *The Perfect City*, 4.10.1, he refers to sight as "the faculty by which [one] perceives colours and all visible objects like rays of light," which is certainly intromissionist in tenor, yet later, in 4.14.8 of the same work, he refers to "the sight which proceeds from the eye with the ray of vision."[84] Unequivocally intromissionist, Avicenna's theory is of special interest not because of its originality but because of the close attention he pays to vision and its psychological entailments. By far the most comprehensive treatment of the subject is found in book 6 of his vast compendium, *Book of Healing*, but his *Book of Salvation*, an abridgement of the *Healing*, and his *Canon of Medicine* are also important sources.[85] Avicenna mounts his defense of intromission in the *Healing* by attacking extramission on a variety of fronts. If, for instance, the visual cone were composed of discrete rays, then our view of objects would be spotty rather than continuous. If, on the other hand, that cone were continuous throughout, then it would take an unbelievable amount of flux to reach the stars. In either case, if several viewers stare at the same object from roughly the same angle, the combined intensity of their visual illumination should make that object appear commensurately clearer and more distinct than it would look to a single viewer. This applies to both the Galenic and the visual ray theories. In addition, once visual contact is made with external objects, the resulting impressions must be transmitted back to the eye. Why posit both an outward and an inward reach when an inward reach alone will do just as well?[86]

84. Richard Walzer, ed. and trans., *Al-Fārābī on the Perfect State* (Oxford: Clarendon Press, 1985), 165 and 223. In his *Catalogue of the Sciences*, al-Fārābī clearly plumps for extramission; see Elaheh Kheirandish, "The Many Aspects of 'Appearances': Arabic Optics to 950 AD," in *The Enterprise of Science in Islam*, ed. Jan Hogendijk and Abdelhamid Sabra (Cambridge, MA: MIT Press, 2003) for the relevant passage. In *Theories*, p. 43, however, Lindberg construes al-Fārābī as a true intromissionist on the basis of a passage in *The Perfect State*, 13.2; see Walzer, 201, for the relevant passage. Cf. Bruce Eastwood, "Al-Fārābī on Extramission, Intromission, and the Use of Platonic Visual Theory," *Isis* 70 (1979): 423–25, who cautions that this passage can be interpreted as supporting an intromissionist-extramissionist hybrid.

85. Avicenna devotes an entire book of his *Healing* and of his much shorter *Salvation* to sense perception and cognition. For a French translation of the relevant part of the *Healing*, see Ján Bakoš, *Psychologie d'Ibn Sīnā (Avicenne) d'après son oeuvre Aš-šifāʾ* (Prague: Czech Academy of Sciences, 1956), and for an English translation of the relevant part of the *Salvation*, see Fazlur Rahman, *Avicenna's Psychology* (Westport, CT: Hyperion Press, 1952). For a recent biobibliographical sketch of Avicenna, see Jon McGinnis, *Avicenna* (Oxford: Oxford University Press, 2010), 16–26; see also Lindberg, *Theories*, 43–44.

86. See Bakoš, *Psychologie*, 67–106; see also Rahman, *Avicenna's Psychology*, 27–29. For a fairly extensive analysis of Avicenna's attack on extramission, see Lindberg, *Theories*, 43–52; see also McGinnis, *Avicenna*, 107–9.

Most of Avicenna's arguments against extramission echo those leveled earlier by Philoponus in his *De anima* commentary (see section 2 above), and there is good reason to believe that he drew them from that commentary because we know he read it.[87] Like Philoponus, moreover, Avicenna regards transparency not as a potential to be actualized by light but as an inherent property of certain media such as air and water that permits light to radiate through them. Avicenna distinguishes light itself according to two modes. As *ḍaw'*, it is the inherent, formal property of certain bodies such as the sun that renders them luminous. As *nūr*, it is the formal effect of radiated luminosity on opaque objects. Color, for its part, is the inherent, formal property of opaque objects that makes them potentially visible. They become actually visible only when impinging light mingles with their color and empowers it to radiate its form through transparent media. Both light and luminous color are therefore per se visible for Avicenna.[88]

Thoroughly conversant with Galen's account of ocular anatomy and physiology, Avicenna locates the seat of visual sensitivity in the crystalline lens, which is charged with visual spirit passed to it from the brain through the hollow optic nerves. Somewhat flattened at the front and situated at the very center of the eye, as stipulated by Ḥunayn, the lens is disposed to receive the impressions of luminous color at its anterior surface. This it does according to a cone of radiation with its vertex behind the anterior surface of the lens and its base on any given visible object. Accordingly, that object's color radiates its form to the eye through the cone to make an impression on the portion of the lens cut by it. Avicenna uses the well-worn signet-ring-and-wax analogy to describe this impression. Since the amount of the lens's anterior surface the color impression occupies depends on the size of the visual angle formed at the cone's vertex, the apparent size of the object varies with the size of the visual angle; the smaller that angle, the smaller the portion of the lens's surface the color impression occupies, and, therefore, the smaller the object looks. Once the two eyes have received corresponding color impressions from the same object, each impression passes through the visual spirit into the optic nerve of each eye until both impressions are fused at the optic chiasma and then passed to the forefront of the brain for perceptual and conceptual processing.[89]

This brings us to the second issue, which is actually threefold: what are the

87. See Dimitri Gutas, "Avicenna's Marginal Glosses on *De anima* and the Greek Commentatorial Tradition," *Bulletin of the Institute of Classical Studies* 47 (2004): 77–88.

88. See Bakoš, *Psychologie*, 63–66; see also McGinnis, *Avicenna*, 104–7.

89. See Bakoš, *Psychologie*, 106–11; see also McGinnis, *Avicenna*, 109–10.

faculties involved in this processing, where are they located, and what are their proper objects? Avicenna responds to this issue by articulating a model of five faculties or "internal senses." Consisting of common sense, retentive imagination, compositive imagination, estimation, and memory, these faculties are located at specific points in the cerebral ventricles, the first two in the doubled ventricle in front, the second two in the middle ventricle, and the last in the occipital ventricle.[90] The idea of arraying such faculties within the cerebral ventricles in a particular order is hardly new. Ḥunayn is an obvious proximate source for this idea. In his description of the brain and its anatomical and physiological structure in the second book of the *Ten Treatises*, he argues that "in the posterior cavity [of the brain] movement and the act of recollection are accomplished . . . in the anterior part . . . observation and imagination, and . . . in the middle part . . . reflection [that is, thinking and deliberating]."[91] But several things set Avicenna's model apart from that of Ḥunayn and, indeed, that of all the thinkers we have discussed to this point.

For a start, Avicenna distinguishes common sense, as a separate faculty, from imagination, whereas virtually all the post-Plotinian thinkers we have examined so far conflated the two. Avicenna also adds to his array the faculty of estimation, which for the moment we can characterize as "appetitive," and which, according to Harry Wolfson, had already been posited by al-Fārābī.[92] In addition, Avicenna replaces the faculty of discursive reason with compositive imagination, which acts like discursive reason in juggling and associating the images held in the retentive imagination. This faculty is common to both man and lower animals, although in man it takes higher form as the cogitative faculty. Finally, whenever more than one faculty share a given ventricle, Avicenna arrays those faculties from front to back in order of superiority. Thus, for example, the retentive imagination lies beyond the common sense in the anterior ventricle(s), estimation beyond compositive imagination in the middle ventricle.

Like the five external senses, each internal sense has its proper object. For the sense organs themselves these are the sensibles appropriate to them. Thus for sight the proper object is color, for smell odor, for hearing sound, and so on. In its appropriate organ, each sensible produces a formal impression that

90. For the classic account of the internal senses model and its development, see Harry A. Wolfson, "The Internal Senses in Latin, Arabic, and Hebrew Philosophical Texts," *Harvard Theological Review* 28 (1935): 69–133.

91. Meyerhof, *Ten Treatises*, 18. As we have seen, Ḥunayn's model has clear antecedents going back at least as far as Nemesius of Emesa (see note 48 above).

92. See Wolfson, "Internal Senses," 93–95.

constitutes a crude sort of representation or "image" of the sensible that generates it. Along with the common sensibles, the sense representations from all five senses serve as proper objects of the common sense, which combines them into perceptible representations. Remanded to the retentive imagination for mnemonic storage, these formal representations are the proper objects of compositive imagination, which scrutinizes, dissociates, and associates them to produce conceptual representations in the estimative faculty. Conceptual representations, of course, are the proper objects of memory, where they are stored for recollection when needed. In that capacity, they serve as formal representations or "images" of the Universals grasped at the end of the cognitive process. Avicenna thus posits two kinds of memory—perceptual and conceptual—a doubling that can be traced back to Plotinus (see section 1 above).

The mechanism by which these successive formal representations are produced is abstraction (literally, "pulling out from"). For instance, from the sense impressions passed to it through the nerves, the common sense in the first cerebral ventricle forms a perceptible representation of the object. At this level, the formal representation has become more abstract than its sensible antecedent because it has been completely disengaged from its material entailments in the original object. It can therefore subsist even when its generating object is removed from sense. Passed by the common sense to the retentive imagination, these formal representations are grist for the compositive imagination (the cogitative faculty) and the estimative faculty. But beyond representing their objects formally, perceptible representations contain various clues about the ulterior meaning of those objects. Called "intentions" (*ma'ānī*; sing. *ma'nā*) by Avicenna, these clues are somehow "in" the perceptible representations without actually being part of them, much as Venice is "in" a painting of it by Canaletto. Some of these intentions inspire revulsion or attraction. As soon as it spies a wolf, for instance, a sheep recognizes its harmful intentions and flees, whereas the sight of a baby inspires love and the desire to move toward it. Others of these intentions are intelligible and come to light through the judgment of perceptible representations by the cogitative faculty. From all these intentions, finally, the estimative faculty forms conceptual representations that are ensconced in memory.[93]

---

93. See Bakoš, *Psychologie*, 27–33 and 40–46; see also Rahman, *Avicenna's Psychology*, 30–56. For a discussion of Avicenna's notion of intentions, and its ambiguities, see Dag Hasse, *Avicenna's De anima in the Latin West* (London: Warburg Institute, 2000), 127–41. See also Robert E. Hall, "Intellect, Soul, and Body in Ibn Sīnā," in *Interpreting Avicenna: Science and Philosophy in Islam*, ed. Jon McGinnis (Leiden: Brill, 2004), 62–86.

However abstract and general it may be, though, the conceptual representation falls short of the Universal because it is embodied in the spirit pervading the occipital ventricle of the brain and is thus neither immaterial nor incorruptible. How, then, do we come to know the Universal implied, or intended, by this formal representation? This is the third and final issue, and in response, Avicenna appeals to an agent intellect (the "Giver of Forms") that knows all the Universals and thus provides the intellectual illumination that enables the potential intellect to intuitively "see" the true intelligibility of conceptual representations. In short, the agent intellect does for the potential intellect what the sun does for the eye. There is little new here, of course, at least not at the basic conceptual level. Avicenna's appeal to an ulterior intellect is no more innovative than his recourse to the sun-eye analogy in describing its function. Nor does his analysis of the types and modes of intellect break new ground. The distinction he draws between theoretical and practical intellect is as old as Aristotle, and his subdivision of theoretical intellect according to gradations of perfection has clear precedent in al-Kindī's analysis of the soul.[94]

What makes Avicenna's internal senses model significant is therefore not the novelty of its individual components, although some may indeed be new. Rather, it is the way he organizes and refines those components so as to emphasize the inductive nature of perception, reducing it to an almost mechanistic process occurring in clearly defined stages at clearly defined points in the brain within a clearly defined material substrate of cerebral spirit. Avicenna also strikes a delicate balance between passivity and activity throughout the process, some faculties being merely receptive (for example, retentive imagination and memory), others active according to their ability to abstract certain kinds of formal representations. This passive-active dualism is reflected in Avicenna's notion of intentionality. On the one hand, the estimative faculty must be receptive to the implications or intentions borne by perceptible representations; on the other, it must realize or infer those implications intentionally on the basis of its innate capacity to do so. Avicenna thus grants considerable

94. For Avicenna, the gradations run upward from entirely unperfected ("material intellect"), through partially perfected ("dispositional intellect"), to completely perfected ("actual intellect" or "acquired intellect"), the completeness lying in its knowing one or more things perfectly. The state of knowing everything perfectly is reserved for the active or agent intellect. Al-Kindī and al-Fārābī draw essentially the same distinctions; see, e.g., al-Kindī and al-Fārābī's respective treatises, both titled *On the Intellect*, in *Classical Arabic Philosophy*, ed. and trans. Jon McGinnis and David Reisman (Indianapolis: Hackett, 2007), 16–17 and 71–76; see also *The Perfect State*, in Walzer, *Al-Fārābī*, 196–203.

autonomy to the psychological faculties in the formation of concepts according to their innate receptive or productive capacities.

In taking this inductivist tack, Avicenna diverges sharply from Plotinus and his followers, who subordinate the psychological faculties so completely to the agent intellect as to reduce them to mere tools through which it carries out the perceptual process. This top-down approach implies that unaided by the agent intellect, the psychological faculties are unable to achieve anything even approximating knowledge. At its most extreme, in fact, it implies that the perceptual process contributes nothing whatever to the acquisition of knowledge. According to Peter Adamson, al-Kindī can be construed as holding this extreme position, which results in what Adamson characterizes as an "epistemic gap" between sense perception and intellection.[95] But any theory that relies on an ulterior intellect to effect the transition from conception to cognition is subject to such an epistemic gap, and Avicenna's is no exception. Yet although Avicenna fails to close this epistemic gap, he narrows it considerably by granting the psychological faculties the ability to abstract conceptual representations that are somehow knowledge-like.[96] The step from such knowledge-likeness to knowledge itself is thus in principle a fairly short one.

If nothing else, it should be evident by now how deeply rooted Avicenna's internal senses model is in the late antique tradition of cognitive theory, a tradition that harks back to Plato and Aristotle, and indeed beyond. Avicenna's internal senses model is thus the product of a complex process of evolution, which is one reason I have devoted so much space to both the process and the product. Another reason for my focus on Avicenna is that in psychology, as in so many other fields, he was enormously influential during the Middle Ages and Renaissance. In his massively erudite *Muqaddimah*, for instance, Ibn Khaldūn (1332–1406) ranks Avicenna among the four "greatest Muslim philosophers," and Avicenna was so exalted as a physician in the Muslim world that his *Canon of Medicine* remained authoritative for centuries.[97] The same holds for the Latin West, where the *Canon* became a mainstay in the medical curriculum during the thirteenth century and served as an authoritative textbook until well into the seventeenth century. Indeed, Avicenna's cachet in the Latin

95. See Peter Adamson, *Al-Kindī* (Oxford: Oxford University Press, 2007), 121–35.

96. See Jon McGinnis, "Avicenna's Naturalized Epistemology and Scientific Method," in *The Unity of Science in the Arabic Tradition* ed. Shahid Rahman, Tony Street, and Hassan Tahiri (Dordrecht: Springer, 2008), 129–52.

97. Rosenthal, *Muqaddimah*, 374. The other three Muslim greats, according to Ibn Khaldūn, are al-Fārābī, Ibn Bājjah (or Avempace: 1095–1138), and Ibn Rushd (or Averroes: 1126–1198).

West was such that, with a few relatively superficial modifications, his internal senses model predominated from the thirteenth century to the early 1600s.

5. THE ARABIC TRANSITION: GEOMETRICAL OPTICS

The foundational sources for Arabic geometrical optics between roughly 850 and 1000 are essentially those discussed in chapters 2 and 3. The Hellenistic and early Greco-Roman sources consist of the two versions of the Euclidean *Optics*, the Euclidean *Catoptrics*, Diocles's *On Burning Mirrors*, Hero of Alexandria's *Catoptrics*, and Ptolemy's *Optics*, although this latter work appears to have been unknown until the late tenth century. The only source we can confidently date to the post-Ptolemaic period is an early sixth-century study of burning mirrors by Anthemius of Tralles. In addition, Roshdi Rashed has uncovered Arabic versions of two Greek studies of burning mirrors that may postdate Ptolemy. One is attributed to a certain Didymus, the other to "Dtrūms," which may be a mistransliteration of "Didums" = "Didymus."[98] In the remainder of this section, we will discuss four thinkers who dealt specifically with geometrical optics, three from the ninth century—al-Kindī, Qusṭā ibn Lūqā, and Aḥmad ibn ʿĪsā—and one from the very late tenth—Abū Saʿd al-ʿAlāʾ ibn Sahl. In the process, we will try to determine what these figures may or may not have owed to the Greek sources just discussed.

That al-Kindī was thoroughly familiar with the Euclidean *Optics* in some form or another is evident from his *Rectification of the Error and Difficulties due to Euclid in His Book Called* Optics."[99] A critical review of the entire Euclidean text, this work opens with a brief narrative summary of the initial seven postulates of the original. In it, al-Kindī describes the visual cone emanating from the "pupil" of the eye and reaching out to make visual contact with ex-

---

98. See Roshdi Rashed, *Les catoptriciens grecs I: Les mirroirs ardents* (Paris: Les Belles Lettres, 2000), 156–57. Along with critical introductory comments, this work contains critical Arabic texts with facing French translations of the relevant works of Diocles, Didymus, Dtrūms, and the Arabic version of Anthemius.

99. See Rashed, *Oeuvres philosophiques et scientifique d'al-Kindī*, vol. 1 (Leiden: Brill, 1997), 162–335, for the Arabic text and facing French translation. Through a close comparison of al-Kindī's text to both Greek versions of the Euclidean *Optics*, Rashed concludes that no Arabic version of the *Optics* corresponds perfectly with either Greek version, that the version al-Kindī used included elements from both Greek versions, and that both Greek versions derive from the original through various postulated intermediates. For a discussion of the Arabic textual tradition(s) of the Euclidean *Optics*, see Elaheh Kheirandish, *The Arabic Version of Euclid's* Optics (New York: Springer, 1999).

ternal objects within its base.[100] Consisting of a "luminous power," this cone is continuous throughout and therefore not actually composed of discrete, mathematical rays, as Euclid postulated. It is in this postulate, in fact, that the "error" mentioned in al-Kindī's title lies, for if the rays were mathematical lines, al-Kindī argues in a manner reminiscent of Philoponus, they would touch only points, in which case we would see nothing.[101] Al-Kindī suggests that the source of Euclid's error can be traced to proposition one, which claims that since we cannot see any object all at once, we have to scan it with visual rays (see chapter 2, section 5). It is true, al-Kindī affirms, that we need to scan objects, but not because we cannot see them all at once. The real reason is that vision along the axis of the visual cone is clearest. Consequently, if we want to pick out details, we must do so by axial scanning.[102] Having thus invalidated the idea of discrete mathematical rays, al-Kindī corrects the explanation Euclid gives in proposition three for why objects pass out of sight at certain distances. They do so not because they fall between rays but because they subtend such tiny visual angles that no sensible object can be seen under them.[103] According to al-Kindī's analysis, visual acuity is limited by two factors: the position of objects within the base of the visual cone relative to its center, where the axis intersects it, and the size of the visual angle. Al-Kindī adds one more. The closer to the eye an object lies, the denser and stronger the visual luminosity shining on it; just as things are seen more distinctly in strong external light, so they are seen more distinctly in strong visual light.[104]

Al-Kindī explores the similarity between external light and visual radiation at some length in *On the Causes of Differences in Perspective*, which is lost in Arabic but survives in a medieval Latin version under the title *De aspectibus*.[105] In the first six of a total of twenty-four propositions, al-Kindī establishes in a

100. Rashed, *Oeuvres*, 162–63.

101. Ibid., 164–65. Peter Adamson, "Vision, Light, and Color in al-Kindī, Ptolemy, and the Ancient Commentators," *Arabic Sciences and Philosophy* 16 (2006): 207–35, suggests that al-Kindī was influenced by Philoponus indirectly and unwittingly; see especially 217–25.

102. Al-Kindī illustrates the need to scan for detail with a couple of examples drawn from the prologue to the "Theonine" recension of the *Optics*.

103. Rashed, *Oeuvres*, 170–75.

104. Ibid., 168–71.

105. For a critical Latin edition and French translation, see ibid., 438–523. This edition revises the one produced by Axel Björnbo and Sebastian Vogl in *Alkindi, Tideus, und Pseudo-Euclid: Drei optische Werke* (Leipzig: Teubner, 1912), 1–41. For a fairly detailed analysis of this tract, see David Lindberg, "Alkindi's Critique of Euclid's Theory of Vision," *Isis* 62 (1971): 469–89.

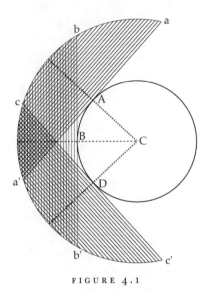

FIGURE 4.1

quasi-empirical way that light radiates in straight lines. He does so by showing that the edges of shadows form straight lines extending from the light source to the boundaries of the bodies casting the shadow and the boundaries of the shadows cast. In proposition six, in fact, he gives a geometrical verification of Philoponus's claim that the rectilinear propagation of light can be empirically demonstrated by projecting light through a slit in a plank (see note 55 above). After arguing in the next five propositions that sight is due to a conical emission from the eye and that the resulting cone is perfectly continuous in form, he turns in proposition twelve to the question of why vision is clearest along the cone's axis. There he undercuts the idea that since the axial line represents the shortest distance between the center of the cone's base and its vertex, the weakening of vision toward the base's edge is due to the increasing length of the rays that illuminate objects placed there. In order to lay the foundations for his counterexplanation, al-Kindī establishes in proposition thirteen not only that light radiates from every spot on the surface of a luminous body, but also that it radiates in all possible directions from each such spot. According to al-Kindī, this radiation occurs instantaneously.[106]

Having proposed omnidirectional radiation from every spot on luminous surfaces, al-Kindī goes on in proposition fourteen to explain why visual clarity varies within the field of view. His account is based on figure 4.1. Let arc ABD

106. Rashed, *Oeuvres*, 486–91.

on the circle be the pupil of the eye, which is centered on C, and assume that all points on that arc radiate visual light. Let abca´b´c´ lie on a circle concentric with circle ABD. Since the ability of point A to radiate its visual light is constrained by the body of the eye, it cannot radiate behind tangent aAa´. The same holds for points B and D. Consequently, point A will illuminate segment abca´ of the outer arc, point B segment bca´b´, and point D segment ca´b´c´. As is obvious from the figure, all three segments overlap at arc ca´, whereas two overlap at arcs bc and a´b´, and none at arcs ab and b´c´. The segment subtending arc ca´ thus represents the area of maximum overlap and intensity. If we were to continue the process indefinitely for other points symmetrically posed between A and D, the area of maximum overlap and intensity would shrink to a spot at the end of "ray" CB, and the amount of overlap would diminish continually, spot by spot, as we move from there toward a and c´—hence the diminishing visual clarity.[107]

Three important points emerge from this analysis. The first and perhaps most obvious of these is that the geometrical ray is an analytic convention, a mathematical abstraction approximating a very thin beam rather than a true line of radiation. Second, al-Kindī's model has light radiate from every spot on the luminous surface in every direction where there is a transparent medium to support it. In theory, any free-floating luminous spot will form a sphere of radiation. Finally, visual radiation originates not in the center of the eye but at the surface of the pupil, which is the seat of visual sensation. Thus, as al-Kindī claims at the beginning of *The Rectification*, the visual "cone will . . . be like the member of a living being by means of which the pupil senses everything on the bodies that [this member] touches."[108] The outward reach of this radiation consists in a transformation of the air along straight lines according to its luminous power: the weaker that power, the weaker the sight.[109] As an exercise of luminous power, therefore, al-Kindī's light has quasi-kinetic and dynamic qualities.

Al-Kindī's younger contemporary and fellow Baghdadi scholar Qusṭā ibn Lūqā proposes a similar theory of vision in his *Book on the Causes of the Difference in Perspectives*, but with some physiological elaboration.[110] Like al-Kindī, he posits a cone of visual radiation with its vertex at the eye and its base defin-

107. Al-Kindī offers a variant of this explanation in *Rectification*, proposition 1, in ibid., 166–69.

108. Ibid., 162.

109. *De aspectibus*, proposition 8, in ibid., 454–55.

110. For the Arabic text and French translation, see ibid., 572–645.

ing the field of view. This radiation, Ibn Lūqā continues, "diffuses from the psychic spirit that emerges from the brain to the pupil through the two hollow nerves that pass from the brain to the pupil." From there "it spreads out . . . in the air to visible objects so as to be like [an] organ of the [viewer]."[111] The Galenic overtones of this account are obvious, and if Ibn Lūqā is following Galen to the letter, the "pupil" mentioned above must be the crystalline lens. Whatever the case, both he and al-Kindī regard the pupil as the seat of visual sensitivity, an idea that harks back to Greek antiquity.

After an aside in which he categorizes three specific kinds of rays—solar rays, fire rays, and visual rays[112]—and after explaining why things appear larger according to larger visual angles, he launches into a surprisingly primitive account of diplopia. This account is of some interest because it is clearly not based on Galen's explanation of binocular vision in the *De usu partium* (see chapter 2, section 3) and indicates that Ibn Lūqā was unexpectedly ignorant of it. It also indicates that he was unaware of Ptolemy's explanation.[113] With these things out of the way, Ibn Lūqā gets to the real point of the treatise, which is to explain image formation and distortion in plane, spherical convex, and spherical concave mirrors.[114] For the most part, the images he analyzes are of the face of someone looking straight into the mirror. Several things are noteworthy about the resulting analysis, which unfolds over twenty-six propositions that include only nine based on geometrical analysis. First, although Ibn Lūqā knows the equal-angles law of reflection and applies it throughout, he makes no effort to prove it mathematically or empirically. Second, in only one case, where he analyzes image formation in plane mirrors, does he locate the image with respect to the reflecting surface—behind it. Third, he never applies the cathetus rule to his analysis, which is doubtless why he deals not with image location but with variations in image size and orientation according to the reflecting surface itself. All of this suggests strongly that he was ignorant of the Euclidean *Catoptrics* or at best knew only parts of it in a vague way. There is no hint whatever that he knew of Ptolemy's *Optics*.

A simple example should help illustrate these points. In proposition twenty-three, Ibn Lūqā undertakes to explain by ray analysis why, when one looks into a convex spherical mirror, his face appears smaller than it should.

111. Ibid., 580–81.

112. Ibid., 582–83.

113. Ibid., 582–87.

114. Ibid., 594–95. Ibn Lūqā does mention cylindrical and conical convex and concave mirrors but restricts his analysis to the spherical form.

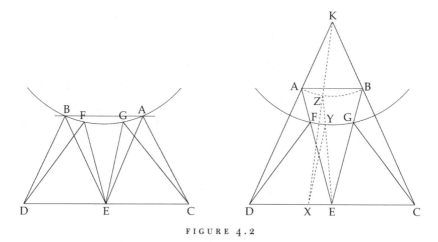

FIGURE 4.2

Let E in the left-hand diagram of figure 4.2 be the eye located on section DC of the face, let straight line BA represent a plane mirror imagined to cut the surface of the convex mirror at points B and A, and let BEA be the visual cone whose limiting rays EB and EA at the edges reflect from plane mirror AB at equal angles along BD and AC, respectively. Earlier, Ibn Lūqā had shown that the image of DC seen under visual angle BEA in the plane mirror would be the same size as its object. Now let rays EF and EG reflect from the convex mirror along FD and GC, respectively. Since visual angle FEG < visual angle BEA, the image of DC will appear smaller than the object would in the plane mirror.

Had he used the cathetus rule, Ibn Lūqā could have proved his point more directly because he could have related the size differential to the actual image rather than to an arc on the reflecting surface. Thus, as illustrated in the right-hand diagram of figure 4.2, image AB on catheti KD and KC is actually located behind the mirror and is clearly shorter than object DC. Furthermore, on the basis of the cathetus rule he could have shown that the image would appear convex, as illustrated by Z, the image of point X, which lies below line AB on the dotted curved line that passes from A to B through Z. Still, Ibn Lūqā's approach is not without ingenuity, especially when applied to the analysis of concave mirrors, where he is able to show how image size and orientation vary according to the placement of the eye with respect to the center of curvature.

Like Ibn Lūqā's analysis of images in the *Book on the Causes*, al-Kindī's brief discussion of the reflection of visual rays in propositions sixteen to twenty-one of the *De aspectibus* shows no definitive trace of either the Euclidean *Catop-*

trics or Ptolemy's *Optics*.[115] At one point, however, al-Kindī's analysis is strikingly similar to one that occurs in the Pseudo-Euclidean *De speculis*, an Arabic compilation currently extant only in a medieval Latin translation.[116] The convergence comes at proposition seventeen of the *De aspectibus*, where al-Kindī offers a maladroit "proof" of the equal-angles law that is essentially parallel to the one given in proposition six of the *De speculis*.[117] All told, as far as the reflection of visual rays and the attendant analysis of images are concerned, the links between these ninth-century theorists and known Greek sources, the Euclidean *Catoptrics* and Ptolemy's *Optics* in particular, are tenuous at best.[118]

When we pass from visual rays to light rays and, more specifically, the analysis of burning mirrors, the picture is far clearer. There is absolutely no question that both al-Kindī and Ibn ʿĪsā were indebted to Anthemius of Tralles. One of the architects responsible for the design and reconstruction of Hagia Sophia after the fire that swept through Constantinople in 532, Anthemius was an exceptionally talented engineer. As such, he took a decidedly pragmatic approach to burning mirrors, focusing more on their construction than on their mathematical aspects. Two of his constructions are of special interest here because they are mathematically sophisticated and because they figure in al-Kindī's and Ibn ʿĪsā's respective studies of burning mirrors in *On Solar Rays* and the *Book of Optics and Burning Mirrors*.[119]

The first construction, which occurs in proposition one of Anthemius's *On Mechanical Wonders*, is illustrated in figure 4.3. Let the thick line containing points A and B be the roof of a building, and let points A and B lie in the plane of the meridian on a line parallel with the horizon. Hence, the entire diagram lies in the plane of the meridian. Open a hole at point B large enough to let a beam of sunlight pass through at midday, when the sun is at meridian. Let BG represent such a beam when the sun is at winter solstice, BL when it is at autumn or spring equinox, and BS when it is at summer solstice. According to the resulting point-by-point construction, whose details need not concern

115. See ibid., 490–513.

116. See Björnbo and Vogel, *Alkindi*, 97–119, for the critical edition of this work. Much of that compilation comes from Hero's *Catoptrics*, although it also contains an analysis of spherical burning mirrors similar to the one provided in the last proposition of the Euclidean *Catoptrics*, both of them based on the incorrect conclusion that parallel radiation into a concave spherical mirror reflects to the center of curvature; see ibid., 105.

117. Cf. ibid., 100–101, and Rashed, *Oeuvres*, 494–97.

118. On possible links between al-Kindī and Ptolemy, see Adamson, "Vision," 207–36.

119. See Rashed, *Oeuvres*, 360–419, for the Arabic text and French translation of al-Kindī's *On Solar Rays* and 652–701 for the text and translation of Ibn ʿĪsā's *Book of Optics*.

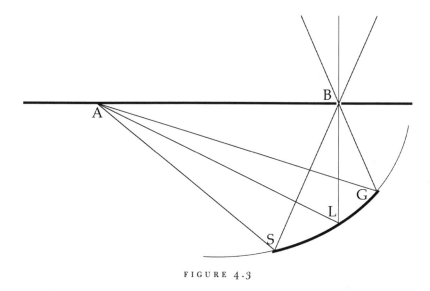

FIGURE 4.3

us, points G, L, and S will lie on a surface that reflects all the incoming beams to point A.[120] As it turns out, this surface is elliptical, line AB being the major axis and points A and B the foci, and the fact that all beams reaching it through B reflect to A follows from a peculiar property of the ellipse—namely, that the sum of the distances between the two foci and every point on the ellipse is equal throughout, that is, BG + AG = BL + AL = BS + AS. As far as we know, Anthemius was the first to recognize this property. Ibn 'Īsā recapitulates this construction in his account of burning mirrors, but for some reason al-Kindī ignores it in his.[121]

The second construction is similar to the first in that it involves producing the figure point by point, the figure in this case being a parabola. Again, the details need not concern us, but the gist of the construction, which occupies proposition five of *On Mechanical Wonders*, can be easily understood by recourse to figure 4.4. Let AB be the diameter of the mirror, and let it be whatever length you please. Bisect it at E and draw line DEC perpendicular to AB. Let D be the focal point, and place it anywhere you like on axis DC. Connect BD. Since B lies on the proposed mirror at one end of its diameter, BD will ipso facto be a reflected ray. Draw BG perpendicular to AB and equal to DB, and continue it to point O. Then draw GC perpendicular to BG. The key to

120. For the full construction, see Rashed, *Catoptriciens*, 285–92.
121. See Rashed, *Oeuvres*, 674–79.

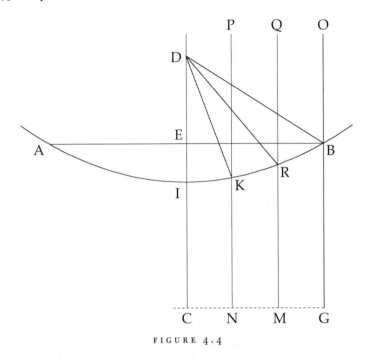

the subsequent construction is the equality between DB and BG, which is perpendicular to line GC. GC constitutes the "directrix," and it is a definitive property of parabolas that any perpendicular dropped from the directrix to any point on a parabola will be equal to the line from that point to the focus. Hence, BG = DB, and CI = DI, which means that I is the parabola's vertex. Anthemius then describes a method for interpolating points between B and I according to the criterion set by the directrix. K and R are two such points, so NK = KD and MR = RD.[122] In this case al-Kindī and Ibn ʿĪsā reverse roles. Al-Kindī includes this construction in his study, but Ibn ʿĪsā leaves it out of his, giving instead a vague account of how to arrange several plane mirrors so that they will focus solar rays upon a body.[123]

In view of their keen, if somewhat untutored, interest in all aspects of optics and their access to the relatively sophisticated treatment of burning mirrors

122. See ibid., 309–12.

123. See ibid., 414–19, for al-Kindī's recapitulation of the construction, and 680–83, for Ibn ʿĪsā's arrangement of plane mirrors, which was apparently based on a more detailed and precise account by Anthemius of how to arrange seven hexagonal plane mirrors to bring light to focus; see Rashed, *Catoptriciens*, 306–9. Al-Kindī follows Anthemius closely in his own discussion of such an arrangement, adding some modifications of his own; see Rashed, *Oeuvres*, 400–407.

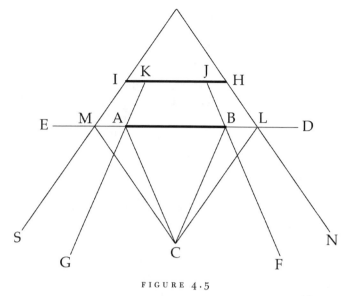

in Anthemius's *On Mechanical Wonders*, it is surprising that neither of these thinkers seems to have known of Ptolemy's *Optics*, at least not directly and certainly not in any detail. Clear testimony to this lack of knowledge can be found in al-Kindī's effort, as reported by Ibn ʿĪsā, to explain why objects appear magnified in water. Suppose, he argues, that AB in figure 4.5 is an object viewed straight on according to visual angle ACB. Now let ED represent the surface of water and sink the object to location IH below it. If ED were a plane mirror, rays CB and CA would reflect at equal angles along BF and AG, respectively, and rays CL and CM would reflect along LN and MS, respectively. By symmetry, rays CB and CA will refract along extensions BJ and AK of the original lines of reflection such that angles JKA and KJB equal angles GAB and FBA, respectively. Hence, the angles of refraction are equal to the angles of reflection. Likewise, rays CL and CM will refract along LH and MI at angles equal to the original angles of reflection. Being touched at the ends by refracted rays MI and LH, the submerged object will be seen under visual angle MCL, which is larger than visual angle ACB, so it will appear magnified.[124] Not only is there no trace of the cathetus rule in this explanation, but it also depends on the patently incorrect assumption that the angles of incidence and refraction are equal, an assumption Damianos made earlier (see note 3 above).

124. See ibid., 424–27, for the Arabic text and French translation.

The earliest uncontested evidence for the influence of Ptolemy comes over a century after al-Kindī with Ibn Sahl, who cites Ptolemy explicitly in a brief treatise on atmospheric refraction titled *Proof That the Celestial Sphere Is Not Perfectly Transparent*.[125] At the very beginning of this treatise, Ibn Sahl mentions his "examination of Ptolemy's book on optics" and adds that he wanted to include the present study "in [a complete] examination of the fifth book of that work," which is devoted to refraction. Of far greater interest than this brief tract, though, is *On Burning Instruments*, which Ibn Sahl composed in the first half of the 980s. As we might expect from the title, he follows in the footsteps of al-Kindī and Ibn ʿĪsā by analyzing parabolic and elliptical burning mirrors and their focusing properties, although his approach is considerably more polished than theirs.[126] Entirely unexpected, however, is the shift he makes from reflection to refraction and the resulting analysis of the focal property of hyperbolic glass lenses.

The point of that analysis is to show that if parallel solar rays pass through a plano-convex hyperbolic lens, they will be brought to perfect convergence at the focus of the opposite branch of the hyperbola. Thus in figure 4.6 SVPO is a planar section through the lens whose convex face is hyperbola SVP. Axis OV of that hyperbola passes through the opposite branch S′V′P′, and F and F′ are the foci for the two branches. AR and BR′ represent parallel rays of sunlight passing through plane face SOP of the lens to strike the hyperbolic face at R and R′. If they were to continue straight through, they would pass toward A′ and B′. XRX′ and YR′Y′ are normal to the hyperbolic face at points R and R′. From Ptolemy we know several things about this situation, assuming, of course, that light rays and visual rays refract equivalently. We know, for instance, that on striking the hyperbolic face inside the glass and refracting into the air, rays AR and BR′ will be diverted away from the normal, so angles of refraction X′RF′ and Y′R′F′ will be larger than the respective angles of incidence ARX and BR′Y. In short, they will not be equal, as al-Kindī assumed. We also know from Ptolemy that as the angle of incidence becomes less oblique, the angle of refraction not only diminishes but approaches equality with the angle of incidence. Accordingly, angle of incidence ARX is smaller with respect to

125. For the Latin text and French translation, see Rashed, *Géométrie*, 53–56.

126. For the Arabic text and French translation of this work, see Rashed, *Géométrie*, 1–52. In addition to analyzing the focal properties of burning instruments, Ibn Sahl also describes how the conic sections out of which these instruments are formed can be drawn in a continuous trace by mechanical means.

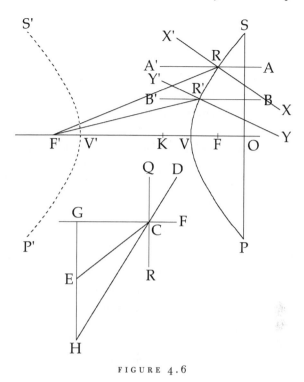

FIGURE 4.6

angle of refraction X′RF′ than angle of incidence BR′Y with respect to angle of refraction Y′R′F′. Granted these conditions, Ibn Sahl undertakes to demonstrate that rays AR and BR′, and indeed all rays parallel to them between S and P, will refract to focus F′ of the opposite branch of the hyperbola.

Yet again, the proof, which is quite elegant, need not concern us here. What does concern us is that it depends upon a lemma of sorts based on the inset diagram, where GF represents the interface between air below and glass (the medium of the lens) above, and DC represents a ray passing through that interface from the denser into the rarer medium to be refracted away from normal QCR along CE. Continue the incident ray along CH, and drop normal GEH through CH and CE to intersect them at H and E. The key here is the ratio between CH and CE. If we mark off point K on the axis of the main diagram such that KV = VF, and if we make the ratio of FF′ to F′K equal to that of CH to CE, two things follow. First, the rays refracted at R and R′ will indeed converge at point F′, and second, the ratio of CH to CE will remain constant throughout, no matter the angle of incidence. That ratio thus measures the index of refraction, and by a simple geometrical conversion it yields the sine law of refraction.[127]

Suffice it to say, this is a remarkable achievement. Not only did Ibn Sahl discover what amounts to the sine law of refraction centuries before it became known in Europe, but he did so through an extraordinarily innovative and sophisticated application of the geometry of conic sections to optics. Ibn Sahl, it would therefore seem, took a major step in the development of modern optics. But a couple of caveats are in order. First, Ibn Sahl's "law" is embedded in the proposition; he does not articulate it explicitly. Second, although his "discovery" of that law is of considerable historical interest, it is of little or no historical importance because, as far as we know, no one picked up on it. It was therefore a theoretical dead end. Still, the very fact that Ibn Sahl had the creativity and skill to accomplish what he did indicates how far both pure and applied mathematics had advanced among Arabic thinkers by the end of the tenth century. It also indicates that by this time the study of refraction had been fully integrated into the optical enterprise through the recovery of Ptolemy's *Optics*. Let me add one more caveat. Although the analysis of burning mirrors and lenses is "optical" in the sense that it deals with light rays, it treats those rays as heat producers, not sight producers. Hence, that study lies outside the mainstream of optics as that science was understood from antiquity onward.[128] The significance of this caveat will become clear in succeeding chapters.

6. CONCLUSION

In this chapter we have followed the development of two modes of visual analysis between the end of the second century and the beginning of the eleventh, one philosophical, the other mathematical. On the philosophical side we have traced the evolution of a hybrid perceptual model based on Aristotle's psychology of faculties and Plato's notion of transcendent intellectual Forms or Universals. According to that model, visual perception unfolds in a succes-

---

127. For the sake of simplicity and brevity, I have taken considerable liberties with Ibn Sahl's proof and procedure. For the actual proof, see Rashed, *Géométrie*, 23–49, and for a fairly detailed synopsis, see Rashed, "A Pioneer in Anaclastics: Ibn Sahl on Burning Mirrors and Lenses," *Isis* 81 (1990): 464–91, especially 479–84. For a briefer synopsis, see Smith, *Alhacen on Refraction*, lxv–lxvi.

128. Al-Fārābī, for example, does not include burning mirrors in the description of optics in his *Catalogue*, where it is clear that for him the sole aim of that science is to explain visual appearances according to unbroken, reflected, and refracted visual radiation; see Abdelhamid Sabra, *Optics* (1989), lvi–lviii.

sion of abstractions, starting with the sensible form in the eye, which yields the perceptible form in the anterior ventricle of the brain. The proper object of discursive reason in the middle ventricle, this form yields the conceptual form, which is remanded to the memorative faculty in the occipital ventricle. The entire process results in a series of virtual representations or "images" that become increasingly abstract and general with each stage. The transition from conceptual to cognitive apprehension is couched in terms of the intellect's intuitively "seeing" what the concept actually represents at the most general level, and it does so by means of the illumination provided by an all-knowing ulterior intellect. Having made this cognitive transition, the potential intellect becomes actual intellect.

On the mathematical side we have seen that despite the promise Ptolemy's *Optics* held out, the ray analysis of vision appears to have languished between roughly 200 and 850. There are occasions, however, when late antique thinkers brought ray theory into the philosophical analysis of vision. In defending intromission, for example, Philoponus adduced a cone of radiation with its vertex in the eye and went on to explain binocular vision on the basis of the cones converging in each eye. This tendency to meld philosophical and mathematical approaches continued with such early Arabic thinkers as al-Kindī, Ibn Lūqā, and Ḥunayn, all of whom combined ray theory with a more or less Galenic model of visual illumination to account for the visual cone.

Meantime, al-Kindī and Ibn Lūqā went beyond this limited use of ray theory to analyze both direct and reflected vision on a Euclidean basis, although al-Kindī repudiated the Euclidean postulate of actual discrete radiation and reduced the Euclidean mathematical rays to mere analytic conventions approximating extremely narrow beams. Al-Kindī also emphasized the similarity between visual radiation and light radiation and stressed the interplay of both in the act of sight. Avicenna went a step further by denying the need for visual illumination altogether, explaining sight on the basis of light and illuminated color radiating in a cone from object to eye. As far as reflection and refraction are concerned, Al-Kindī's and Ibn Lūqā's efforts to account for the visual effects resulting from these two phenomena were apparently hampered by ignorance of both the Euclidean *Catoptrics* and Ptolemy's *Optics*. It is only with Ibn Sahl in the very late tenth century that Ptolemy's *Optics* came to light as an acknowledged source.

Thus when Abū ʿAlī al-Ḥasan ibn al-Ḥasan ibn al-Haytham—or Alhacen in his medieval Latin incarnation—took up the study of optics during the 1020s, he had a rich store of sources and ideas from which to draw in developing the model of vision eventually articulated in his monumental *Book of Optics* (*Kitāb al-Manāẓir*). On the one hand, the geometrical analysis of sight and light had reached a level of sophistication beyond that of the Greek sources, especially Euclid. On the other, the process of visual perception had been structured fairly rigidly according to a succession of psychological faculties lodged in the cerebral ventricles and operating through the pneuma or spirit pervading both them and the sensory nerves feeding into the brain. Furthermore, tentative efforts had been made to incorporate the geometrical analysis of sight into this perceptual model with some limited success. In the next chapter we will see how Ibn al-Haytham contributed to this process of incorporation by recasting and honing Ptolemy's analysis of vision on the basis of light rays rather than visual rays.

CHAPTER 5

Alhacen and the Grand Synthesis

The two main biographical sources for Ibn al-Haytham (965?–1040/41), Ibn al-Qifṭī (ca. 1172–1248) and Ibn Abī Uṣaybīʿa (ca. 1200–1270), are problematic not only because they provide conflicting details about his life and works but also because they may not be talking about the same man. Ibn Abī Uṣaybīʿa refers to his Ibn al-Haytham as "Muḥammed," whereas Ibn al-Qifṭī refers to his as "al-Ḥasan." Sabra is convinced that this discrepancy is nominal in all senses of the term, and if he is right, then Muḥammed/al-Ḥasan Ibn al-Haytham wrote prolifically not just on pure and applied mathematics but also on cosmology, philosophy, and theology.[1] Rashed is equally convinced that the discrepancy is real and therefore that Muḥammed and al-Ḥasan are entirely different people.[2] If he is correct, then al-Ḥasan, not Muḥammed, is the author of the technical, mathematically oriented works attributed to Ibn al-Haytham. Whichever the case, the disparities between the two biographical accounts make it difficult to reconstruct Ibn al-Haytham's life with any confidence. It is no less difficult to establish with assurance the authenticity or chronological order of the works ascribed to him in those accounts. We can be pretty sure, however, that the works on mathematical subjects belong to al-Ḥasan and only a little less certain that the lion's share of them was writ-

1. Abdelhamid I. Sabra, "One Ibn al-Haytham or Two? An Exercise in Reading the Bio-Bibliographic Sources," *Zeitschrift für Geschichte der Arabisch-Islamischen Wissenschaften* 12 (1998): 1–40.

2. Roshdi Rashed, *Les mathématiques infinitésimales du IXᵉ au XIᵉ siècle*, vol. 2 (London: Al-Furqan, 1993), 1–19.

ten between 1027 and 1038 in Cairo, where he lived until his death in very late 1040 or very early 1041.

According to Sabra's cautious reconstruction, Ibn al-Haytham wrote fifteen works devoted specifically to optics, thirteen of them between 1028 and 1038. Among this latter group, the *Book of Optics* is by far the longest, yet it may be one of the earliest.[3] Consisting of seven books, it is organized topically by visual mode. The first three books deal with the analysis of "direct" vision, in which the radial links between object and eye are unbroken. Vision by means of rays broken at reflecting surfaces is the subject of books four to six, and book seven is devoted to vision by means of rays partially broken at refractive interfaces. In terms of subject matter and approach, therefore, the *Book of Optics* conforms almost perfectly to al-Fārābī's definition of optics in the *Catalogue* (see chapter 4, note 128). It also conforms structurally to Ptolemy's *Optics*.

As was pointed out in the introduction, *The Book of Optics* was translated into Latin around 1200 under the title *De aspectibus* ("On Visual Appearances"), the authorial ascription "Alhacen" being a Latin transliteration of the given name "al-Ḥasan." One other optical work by Ibn al-Haytham, the brief *Treatise on Parabolic Burning Mirrors*, was rendered into Latin as the *De speculis comburentibus* at around the same time. As far as we know, these are the only two of Ibn al-Haytham's optical works to have been translated into any European language before modern times and thus the only ones available in the Latin West during the Middle Ages and Renaissance. Also attributed to Ibn al-Haytham is a cosmological treatise whose Latin version is entitled *De configuratione mundi* ("On the Structure of the World").[4] Rashed argues fairly persuasively that this is the work of "Muḥammed" rather than "al-Ḥasan."

Since the *De aspectibus* is far and away the longer and more historically significant of Alhacen's two optical works extant in Latin, I will devote this chapter almost entirely to it. My analysis will be based on the Latin rather than the Arabic version. There are obvious drawbacks to this approach. To start with, the Latin version differs from its Arabic source in significant respects. Not only does it lack the first three chapters of book one, but also a considerable portion of book three in the Latin version is a paraphrase of the original.[5] There are

3. See Sabra, *Optics*, xxiv–liii.

4. For a Latin edition and English translation, see Tzvi Langermann, *Ibn al-Haytham's On the Configuration of the World* (New York: Garland, 1990).

5. For details, see Smith, *Alhacen's Theory*, xxiii–xxiv. For an English translation of the first three chapters of book one, see Sabra, *Optics*, 3–51.

also terminological differences. In the Latin version, for instance, light is distinguished according to *lux* and *lumen*, which reflects Avicenna's bipartition of light into *ḍaw'* and *nūr*, but as far as I can tell, no such differentiation is to be found in the Arabic version of Alhacen's treatise. The *lux/lumen* distinction thus seems to have been imposed by the Latin translator(s).[6] Moreover, as the succinct phrase *traduttore tradittore* puts it, even the most faithful translation is ultimately faithless in one respect or another.

Despite these drawbacks, I think my approach is warranted by historical fact. I went to some length in the introduction to show that it was in the Latin West, not the Arabic "East," that Alhacen's model of visual perception had the greatest impact as well as the most significant ramifications for the development of modern optics. It therefore makes sense to evaluate that model on the basis of the version that enjoyed the vast majority of readership and exerted commensurately greater influence in the marketplace of ideas. Furthermore, most of the divergences between the Latin and Arabic versions make little difference to the overall sense of the text. To take the most glaring example, almost everything Alhacen has to say in the first three chapters of book one missing from the Latin version is either repeated later or can be easily inferred from points made in subsequent chapters and books. There are nonetheless occasions when the divergences really matter, and in those cases I will have recourse to the Arabic version.

Because Alhacen hardly ever cites specific sources by name or work, we are forced to identify them by traces in his analysis.[7] There is no question, for instance, that Ptolemy's *Optics* was a major source for Alhacen. Not only are its traces crystal clear in the *De aspectibus*, but also there is independent evidence that Alhacen was intimately familiar with it.[8] There is also no question that Alhacen knew Ibn Sahl's *Proof That the Celestial Sphere Is Not Perfectly Transparent*. Aside from its traces in Alhacen's conception of transparency, he apparently copied this treatise.[9] Less certain is that Alhacen's model of punctiform light radiation was inspired by al-Kindī's analysis in proposi-

---

6. See *lux* and *lumen* in the Latin-Arabic glossary in Sabra, *Optics*, 188. On the basis of terminological and stylistic differences, I have concluded that there were at least two, and probably three, translators involved in producing the Latin text of the *De aspectibus*; see Smith, *Alhacen on Refraction*, cxxv–cxxvi.

7. Alhacen does occasionally cite Euclid's *Elements* and Apollonius's *Conics*, usually by book and rarely by proposition.

8. Alhacen wrote at least three works dealing explicitly with Ptolemy's *Optics*; see Sabra, *Optics*, xxxii–xxxiii.

9. See Rashed, *Géométrie*, cxli–cxlii.

tion fourteen of his *De aspectibus*, but the similarity between the two models strongly suggests that it was. Similarity can be misleading, though. For instance, Alhacen and Avicenna both support intromission, and Alhacen's understanding of transparency, light, and color is similar to Avicenna's, but there is no evidence to indicate that either thinker was aware of the other. The same holds for the psychological theory undergirding Alhacen's account of visual discrimination and conception. It is like Avicenna's in a general way, but there is no valid reason to suppose that Alhacen got it from Avicenna. All we can safely assume is that since both thinkers were working within the same intellectual context, they shared certain ideas and concepts.

Alhacen was a mathematician of formidable talent and creativity. He was also an extraordinarily systematic and careful thinker. These characteristics are evident in his comprehensive and rigorous approach to visual analysis and the almost seamless way he moves from aspect to aspect of that analysis. Indeed, his zest for system and analytic rigor seems excessive at times, an example being his exhaustive study of various types of light in order to show that they act identically. But bear in mind that at least one of his predecessors, Ibn Lūqā, had distinguished solar rays from fire rays (see chapter 4, section 5), so it was important for Alhacen to demonstrate that such distinctions are operationally irrelevant and that sunlight, moonlight, starlight, firelight, and daylight all radiate, reflect, and refract in exactly the same way. Alhacen's overall approach is therefore predominantly empirical and operational, his focus being on the *how* rather than the *why* of light and sight. Not surprisingly, the resulting theory of visual perception is also predominantly empirical and operational rather than "philosophical." Yet for all his caution in avoiding philosophical issues, Alhacen's account is based on a variety of tacit physical and metaphysical assumptions, some of which have problematic implications. That said, let us begin our examination of Alhacen's theory with a look at its core assumptions.

1. THE ELEMENTS OF ALHACEN'S ANALYSIS

Early in the *De aspectibus*, Alhacen divides optical theorists into two camps. On the one side are the "mathematicians," who posit the extramission of visual rays from the eye to outlying objects. On the other are the "natural philosophers," who posit the intromission of forms from outlying objects to the eye. In determining which alternative to favor, Alhacen offers a familiar clinching argument: since we must suppose that visual rays "take something from the visible object and transmit it [back] to the eye," it is more reasonable to posit a single inward reach than to assume both an outward and an inward reach.

Still, despite the obvious superiority of intromission, the visual ray theorists are not necessarily wrong, at least not "those who suppose that [these] radial lines are imaginary." Hence Alhacen concludes, "both parties have something true to say . . . but neither is wholly satisfactory without the other [to complement it]."[10] For Alhacen, then, the science of optics depends on two core assumptions: first, that vision is due to something formal passing from the object to the eye, and second, that what passes to the eye follows imaginary straight lines to reach it.

Light, according to Alhacen, is just such a thing. It radiates along imaginary lines, and it does so omnidirectionally from every spot on the surface of any luminous object, provided there is a continuous transparent medium to support its radiation. The result is a continuous sphere of propagation each of whose radii constitutes a ray. In short, Alhacen proposes essentially the same model of punctiform radiation that al-Kindī did almost two centuries earlier. As the physical cause of luminosity in the object, light takes the form of *lux*. As its radiative effect in the transparent medium, it takes the form of *lumen*. The two are thus qualitatively different. *Lux* is an inherent, physical property of objects that renders them luminous. *Lumen* has no physical effect whatever on the transparent medium through which it passes. To characterize Alhacen's model of radiation as punctiform is somewhat misleading, though, because he is emphatic that light does not actually radiate from points. Instead, it emanates from tiny areas of "minimal light" (*lux minima*). These are the smallest possible spots of effective light; anything smaller will lack illuminative power. Consequently, we can conceive of physical rays as ultrathin beams of light emanating from spots of *lux minima* and approximating the mathematical rays passing through their axes.[11]

Transparency, Alhacen continues, is a property of certain bodies that allows light to penetrate through them, but it is variable. Air, for example, is more transparent than water, and both are less transparent than aither. The more transparent the body, the less resistance it poses to the passage of light because of its relative lack of density (*densitas* or *soliditas*). In passing obliquely from a rarer, more transparent body into a denser, less transparent body, light is refracted toward the normal because of that resistance. Opacity, which is a natural concomitant of color, is lack of transparency, but it, too, is variable according to the solidity of the body. Certain bodies can be both colored and translucent, and all bodies, no matter how transparent, have some share of

---

10. Quotations from Smith, *Alhacen's Theory*, 373–74.
11. See Smith, *Alhacen on the Principles*, 320.

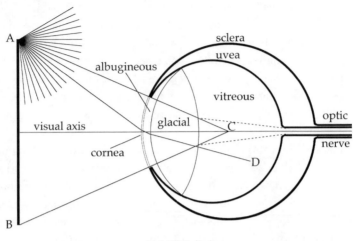

A

sclera

uvea

albugineous

vitreous

visual axis

glacial

C

optic

cornea

nerve

D

B

FIGURE 5.1

opacity as measured by their density. That is why no body, not even aither, is absolutely transparent because if it were, it would have no solidity and therefore nothing within which light could manifest or propagate its radiative effect. Because color has virtually no power to radiate on its own, it cannot be seen unless it exists in a self-luminous body or in a body illuminated by an external source. Once it is illuminated and has trapped the incoming light, color mingles with it and is thereby empowered to radiate from every spot on the surface of its object. Light and luminous color are therefore per se visible, but since all luminous bodies are opaque and thus colored, the light radiating from them is perforce mixed with their color. Consequently, although light and color may be ontologically and analytically distinct, they are physically inseparable. We never see "pure" light.

How the eye is sensibly affected by luminous color is a function of its anatomical and physiological structure. Anatomically, the eye as Alhacen conceived it is illustrated in figure 5.1. The eyeball is a sphere enclosed by the opaque sclera (*consolidativa*) and centered on point C, which lies on the visual axis passing through the center of the hollow optic nerve. The transparent cornea at front forms a perfect continuation of the sclera, so it is centered on C as well. The sclera itself is an offshoot of the outer sheath of the optic nerve, which originates in the dura mater of the brain. From the inner sheath of the optic nerve, which originates in the brain's pia mater, the uvea forms an eccentric sphere inside the eyeball. The lens is nested inside the uvea and covered on the front by a gossamer-thin integument called the spider web (*aranea*) that forms a continuation of what amounts to the retina. Its anterior surface is flattened so as to form a spherical section concentric with the eyeball as a whole and the cornea at front. Its posterior surface is also spherical. At the front of the uvea is an aperture that forms the pupil. The basin between the cornea and the lens is filled with albugineous humor, the lens itself with glacial humor, and the space behind the lens with vitreous humor. These humors are arranged in descending order of transparency, the albugineous humor being less transparent and thus more refractive than the air outside but more transparent than the glacial humor, which is more transparent than the vitreous humor, and so on.[12]

That Alhacen's ocular model is based on Galen's, and to some extent Ḥunayn's, is evident, but it differs from both in significant ways. Unlike Galen, for instance, Alhacen has the cornea concentric with the eyeball rather than bulging outward, but like Galen (and unlike Ḥunayn), he situates the lens toward the front of the eyeball. Like both, he assumes that the lens's anterior surface is flattened, but whereas neither Galen nor Ḥunayn specifies the geometrical form of this flattening, Alhacen does. He makes it perfectly concentric with the spherical eyeball. Furthermore, unlike both Galen and Ḥunayn, Alhacen ignores the retina. He follows both, however, in assuming that the double-sheathed hollow optic nerves originate in the brain, cross at the optic chiasma, and feed into each eye so as to provide conduits for the cerebral spirit that renders the lens visually sensitive (see figure 2.2).[13]

In order to see why Alhacen makes these adjustments to the Galenic model, let us place an illuminated object AB in front of the eye, as in figure 5.1, and let us look at the radiation from point A. According to Alhacen, that radiation takes the form of a continuous sphere within which each of the lines emanating from A represents a ray, understood as the narrowest possible beam of effective luminous color. Let two such rays reach the outer surface of the cornea. The one along AC will strike it orthogonally, so it will pass straight through without refraction, as it will also do after reaching the anterior surface of the lens. If allowed to continue straight through the lens's posterior surface, it would reach center point C of the eyeball. Meantime, the other ray strikes the cornea at an angle, so it is refracted toward the normal in the albugineous humor and

12. Alhacen in fact never makes this order from most transparent/least refractive (albugineous humor) to least transparent/most refractive (vitreous humor) explicit, but his model of optical image selection within the eye requires that order.

13. On the structure of the eye and the function of its components, see Smith, *Alhacen's Theory*, 348–55 and 390–94.

refracted again after striking the anterior surface of the lens at a slant. Consequently, if it were to continue straight through the lens's posterior surface, it would bypass center point C to reach D. The same analysis holds for any other point on AB; of all the rays emanating from it, only the one that strikes the cornea along the perpendicular will pass straight through toward center point C. Consequently, the perpendicular rays from all points on AB form a cone of radiation with its base on the visible object and its vertex at center point C. Constituting what Alhacen calls the "center of sight" (*centrum visus*), this is the viewpoint from which everything encompassed by the cone of radiation is perceptually judged according to its spatial characteristics and location. Before reaching this point, though, the perpendicular rays comprising the cone of radiation are refracted at the interface between glacial and vitreous humors in such a way as to be funneled in proper point-by-point order through the vitreous humor into the hollow optic nerve, as represented by the dotted lines.[14]

The lens is not just optically selective. It is also sensitively selective according to the charge of visual spirit continually infusing it from the brain. It therefore "feels" the impingement of luminous color according to the impression the color makes on it. But the lens is constituted to sense only the impressions made along the perpendicular; the rest it ignores because, being oblique, they are too weak to be felt, just as glancing physical blows make less effective impressions than direct ones. The stronger the impression, the longer it lasts on the lens as an afterimage, and the feeling that accompanies it becomes ever more painful as the impression strengthens. Even the briefest glance at sunlight causes acute pain and a bothersome, long-lasting afterimage.

The sum total of all the perpendicular impressions made on the lens yields a composite color form that serves as a sort of pointillist representation of the object surface that generates it. As such, it corresponds with its generating object surface in perfect point-to-point order from left to right and top to bottom. Constituting what we might (but Alhacen does not) call the visible form, this representation passes through the lens and, after refraction at its posterior surface, is funneled into the hollow optic nerve, maintaining its proper order throughout. Continuing through the optic nerves, the color forms from both eyes are fused at the optic chiasma to yield a single formal representation that is subject to perceptual processing in the brain. This completes the first stage of vision, which Alhacen calls "perception by sense alone" (*comprehensio solo sensu*).

14. Ibid., 355–87 and 417–29.

2. VISUAL DISCRIMINATION, PERCEPTION, AND CONCEPTION

Beyond the mere fact of luminosity and coloring, nothing about the color form has been visually determined to this point. At this stage, the representation can be thought of as a mosaic depiction of the object and the visual field surrounding it. Yet although it consists of nothing more than tiny daubs of colors juxtaposed in particular ways, it implies a host of things that are not actually in it. Alhacen calls these things "visible intentions" (*intentiones visibiles*), and they number twenty-two. An expanded list of Aristotle's common sensibles, these intentions include the two per se visibles, light and color, as well as such things as shape, size, distance, separation, corporeity, transparency, opacity, and even beauty and ugliness.[15] All of these intentions are discerned and judged by the "faculty of discrimination" (*virtus distinctiva*) exerted by the final sensor (*ultimus sentiens/sensator*) at the front of the brain. Much of that discernment involves experience-based deductive inference, which Alhacen characterizes as "syllogistic" (*per sillogismum*). Through repeated experience, for instance, we learn that the color forms we perceive represent three-dimensional bodies, so when presented with a new color form, we deduce the "corporeity" it intends or implies. Likewise, as soon as we close our eyes, we lose sight of what we were seeing, and when we open them, we see it again. Since we know by experience that the eyelids lie between the eye and what we see, we can infer that those things lie some distance beyond our eyelids. From the boundaries of the color form we can infer shape, and by correlating distance with visual angle we can infer size, and so forth.[16]

Alhacen's focus on the experiential nature of perception leads him to suppose that as perceivers, we are all *tabulae rasae* at birth, endowed with the innate ability to reason deductively. Not innate, however, are the actual "first principles" (*propositiones prime*) of reasoning. For instance, we may think we know intuitively that the whole is greater than the part, but in fact we have had to learn it through perceptual judgment. Over time, we learn the difference between wholes and parts through perceptual experience, and this experience also teaches us that parts are smaller than wholes. As a result, we learn that wholes exceed parts and that things that exceed other things are greater.

15. For the complete list, see ibid., 348.

16. On the syllogistic nature of perception, see ibid., 429–38. On the way in which each visible intention is perceived, see ibid., 441–512.

From these premises (*propositiones*) we conclude that wholes are greater than parts.[17] Such deductions have become so automatic that we are no longer conscious of carrying them out, so we take their conclusions to be intuitively obvious when they are not. A considerable amount of perception is like this; although it requires a deductive process, we draw the conclusions so fast and effortlessly that we are not aware of the process.

When we are confronted with a brand-new object and want to determine what it is, we subject it to close visual scrutiny in order to ascertain its every detail in a process of "certification" (*certificatio/verificatio*). This we do by scanning every part of the object's surface with the visual axis. Such scanning is called "intuition" (*intuitio*), and the reason for undertaking it is that visual clarity is greatest along the axis, so passing it over the object gives us the clearest view of its intentional attributes at every spot. For instance, close scrutiny might reveal a thin, jagged black line through the middle of the object. Accordingly, we would perceive the separation intended by that line and conclude that it is a crack. We might also see the difference intended by two subtly contrasting colors, or the sharp difference between the red on one side and the blue on the other.

All of these intentions are apprehended by the final sensor and judged by its faculty of discrimination, and when they are perceptually determined, they are impressed in the imagination for mnemonic recall. One benefit of such a scanning process is that as the intentions of individual spots on the object surface are apprehended axially, the object as a whole is repeatedly seen in a less focused way, so both the object as a whole and its individual intentions are impressed in the imagination. The more often these intentional representations are impressed in the imagination, the more clearly they are remembered, but if they are not reinforced, they will fade from memory. Standing for the object as a discrete individual, the resulting intentional representation in the imagination amounts to perceptible form, which Alhacen designates as a "particular form" or "form of a particular" (*forma particularis*). It is according to this form that we remember friends, favorite locations, and lines in a play, and the process that yields them is what Alhacen calls "perception by syllogism" (*comprehensio per syllogismum*).[18]

Let us assume that we are seeing a blue sphere and a red cube for the first

17. Ibid., 434–35. Cf. Avicenna's notion of induction and "methodic experience" in McGinnis, "Avicenna's Naturalized Epistemology," 141–47.

18. On the process of certification by scanning, perceptual deduction, and memorization of the resulting forms, see Smith, *Alhacen's Theory*, 512–21.

time. Our faculty of discrimination will judge that the color and shape of the one are different from those of the other, but it has no way of differentiating the types of colors or shapes. For that it must be presented with repeated instances of such colors and shapes, and after enough repetitions it begins to develop taxonomies. Having perceived and remembered red enough times, for example, it can discern new instances of red as similar until, finally, it has the taxonomic niche "red" according to which it can recognize all reds by type. Each such niche is filled by a "universal form" (*forma universalis*), which specifies what kind of thing the object is, be it a single visible intention or the full array of visible intentions that define a particular sort of object. Thus when I see a friend, I recognize him as such by his particular form, but I also recognize him as human by his universal form. These latter forms are "quidditative" according to Alhacen, but only in the sense that they represent their objects at the most general physical level. Thus the universal form of a human conveys something like "two-legged moving thing around six feet tall with relatively little hair." Universal forms are equivalent to conceptual forms in that although they are general, they are still tied to the physical properties of what they represent. Consequently, by Alhacen's account, the universal form of human seems not to directly convey "rational, mortal, animal."[19]

Once the particular and universal forms are memoratively "ensconced in the soul" (*quiescere in anima*), they can be recollected as templates against which new perceptions are compared and typed. When I see a specific instance of red, I judge it as such according to the universal form that best represents it, that is, the universal form "red." Likewise, when I see a mule, I judge it as such according to the universal form "mule."[20] This sort of determination constitutes "perception through recognition/precognition" (*comprehensio per scientiam antecedentem*), and it is how we make most of our perceptual judgments. Otherwise, we would be badly hampered if we had to undertake the laborious syllogistic process every time we wanted to determine what we were seeing. Alhacen offers the example of reading. When we are just learning to read and encounter the word DOMINUS on a page, we have to parse it letter by letter in order to grasp what it says, but after enough time, we can bypass that process to grasp it holistically. This is essentially what we do most

19. See ibid., 521–25. Alhacen's universal form is remarkably similar to Avicenna's "vague individual," which constitutes "a wholly indistinct likeness of the individual"; see *Healing*, I, 1, 6–10, in Jon McGinnis, trans., *Avicenna: The Physics of* The Healing (Provo, UT: Brigham Young University, 2009), 7–10, especially 9.

20. According to Alhacen, when we see a familiar object, we perceive it by its universal form before we perceive it by its particular form; see Smith, *Alhacen's Theory*, 523.

of the time, and we even shorten the process by relying on clues or "signs" (*signa*) to make snap judgments about what we are seeing. Thus as soon as I spy something at a distance standing on two legs, I may perceive it as human simply by the "sign" of bipedalism. These snap judgments are often wrong, of course. I might easily read DOMINUS as DOMUS when browsing the text too quickly.[21]

Not all intentions are in the objects themselves; most are in the context within which the objects are seen. Distance is such an intention, and our perception of it is based on continual experience of particular distances, such as a foot, an arm's length, or a pace. Once we have learned to recognize such short distances on the ground, we are able to extend them perceptually, pace by pace, until we can recognize fairly long distances. Longer distances we recognize according to successions of familiar objects of known size. How we come to know a given object's size is by seeing such objects at varying distances time and again and learning to correlate their apparent size, which is a function of visual angle, and their perceived distance. In this way, we learn that certain objects look certain sizes at certain distances, and that in turn teaches us to perceive them as the same actual size at those distances, despite the variation in apparent size. Alhacen therefore subscribes to what psychologists call the size-distance invariance hypothesis. Accordingly, if I have learned that maple trees are generally forty feet tall but look twenty feet tall at a certain distance and ten feet tall at another, then if I see a span of ground on which a succession of maple trees lie at increasing distances of known quantity, I can gauge the overall distance according to the apparent size of those trees as they recede toward the horizon. Such judgments are only approximate, and when the distance becomes too great or when there are no landmarks along the way, I can only estimate the distance. More often than not my estimate will be erroneous, sometimes wildly so, as is the case when I perceive the planets and stars to lie equally, or almost equally, far away.[22]

In fact, errors in visual perception (*deceptiones visus* = "visual illusions") are the rule rather than the exception, and they are caused by a skew in the normative conditions for proper sight. Numbering eight, these conditions are as follows: there must be adequate distance between eye and object, the object must face the eye directly enough to be properly seen, there must be

21. See ibid., 525–29.

22. For Alhacen's full account of distance perception and size perception, see ibid., 448–57 and 475–95.

adequate illumination, the object must be of an adequate size, it must also be adequately opaque, the intervening medium must be adequately transparent, there must be adequate time for proper perception, and the eye must be adequately healthy.[23] These conditions are interrelated. An object may be too distant or too small to be properly perceived in dim light but near enough and large enough to be perceived correctly in stronger light. Or the dim light may be adequate for perception of a larger, nearer object. Or the object may be inadequately opaque to be perceived through haze. Or it may lie at such a slant that its details cannot be perceived, even though it is fairly close and well lit. Or it may lie so far toward the edge of the visual field that both eyes cannot adequately see it, in which case it will appear nebulous or even double. Or the lens of the eye may be inadequately charged with visual spirit for proper perception under otherwise ideal conditions. There is no need to go through every possible permutation, as Alhacen in fact does for all three stages of vision, from vision by sense alone, through perception by syllogism, to perception through recognition.[24] These examples should give a sufficiently clear idea of how he accounts for perceptual errors in direct vision.

Before I turn to the two main sources of visual deception, reflection and refraction, let me briefly review some of the cardinal features of Alhacen's account of direct vision. First, it should be eminently clear that his theory is based on a cone of radiation that is mathematically identical to Ptolemy's visual cone. Moreover, both Ptolemy and Alhacen agree that although the radiation within the cone is perfectly continuous, it can nonetheless be analyzed as if it occurred along imaginary mathematical lines. Both agree that the vertex of the cone is the viewpoint from which all perceptual judgments are made, and both agree that size perception depends on correlating visual angle with distance as measured from that point. Both agree as well that the proper object of sight

---

23. See ibid., 588–89.

24. For Alhacen's entire analysis of visual deceptions, including diplopia (double vision), see ibid., 562–627. Like Ptolemy's, Alhacen's analysis of diplopia is based on a thin plank at one of whose edges the eyes can be placed. Pegs can then be set at various distances on that plank so that the visual axes can converge at certain points. It is worth noting that according to Alhacen's account, all binocular vision is to some extent diplopic insofar as the two images of any point on that surface, except for the one upon which the two visual axes intersect, occupy different spots on the respective lenses, so when they are fused at the optic chiasma, they do not coincide perfectly. The greater the discrepancy in coincidence, the blurrier the fused image until, finally, it is doubled; see ibid., lxviii–lxxvi.

is luminous color, and both cap their accounts with an extensive discussion of visual illusions.

Second, in order to explain vision on the basis of his radiative cone, Alhacen is forced to make certain geometrical adjustments to the Galenic ocular model, reconfiguring the cornea so that it is continuous and concentric with the eyeball and making the anterior surface of the lens concentric with both. He also qualifies the transparency of the vitreous humor in order to ensure that it conveys the color form to the nerve in proper upright and left-to-right order because that is the way we perceive it. Thus, as Alhacen phrases it, "the vitreous [body], along with the receptive capacity that is in [it], is constituted with its sensation of these forms only to maintain their arrangement."[25] In other words, pervaded as it is by visual spirit, the vitreous humor receives and transmits forms in a different way than inanimate transparent bodies do. The same holds for the visual spirit in the optic nerve through which color forms pass to the brain; it too is governed by different rules than those inanimate transparent bodies follow.

Third, the psychological framework within which Alhacen explains visual perception bears some striking similarities to Avicenna's internal senses model. Both thinkers appeal to intentionality, although Alhacen applies it only to the perception of common sensibles. Alhacen's final sensor is at least roughly equivalent to Avicenna's common sense, and for both thinkers the faculty of imagination serves as a mnemonic storehouse for the forms perceptual judgment yields. In addition, both thinkers locate image fusion in the optic chiasma. Like Avicenna, moreover, Alhacen divides the perceptual process into three distinct phases—sensation, perception, and conception—each phase marked by a particular kind of form, the sensible form for sensation, the particular form for perception, and the universal form for conception. In addition, recollection is central to his account of perception, as it is for Avicenna. But Alhacen is extremely vague about details. Despite his emphasis on the syllogistic nature of perceptual judgment, he nowhere mentions a faculty of compositive imagination or discursive reasoning to carry it out. Nor does he have recourse to anything like the estimative faculty. Only once does he advert to intellect, or "mind" (*mens*), and more often than not he remands specific psychological functions to the soul taken generally rather than to any particular faculty in it. The only generalization we can safely make, then, is that the psychological model implicit in Alhacen's account of perception is not incompatible with Avicenna's internal senses model.

25. Ibid., 421.

3. REFLECTION AND ITS VISUAL MANIFESTATIONS

Alhacen devotes the fourth book of the *De aspectibus* to a systematic, empirical verification of the principles of reflection, which Ptolemy has already established. The first principle he addresses is the equal-angles law, and his verification is based on the apparatus illustrated in figure 5.2. Form a thin but sturdy bronze plaque whose arc EAF in the upper left-hand diagram is the section of a semicircle with radius AV of six digits (11.4 cm). Score its top face with lines AV, DV, D´V, and so forth, such that angles DVA, CVA, and BVA = corresponding angles D´VA, C´VA, and B´VA. Then, as represented in the cutaway diagram at the lower left, form a hollow wooden cylinder seven digits (13.3 cm) high and fourteen digits (26.6 cm) in diameter on the outside. Make the inside diameter ten digits (19 cm) so that its wall is two digits (3.8 cm) thick.

Draw a circle on the inner wall of the cylinder at a height of two digits minus half a grain of barley (3.8 cm - 0.43 cm = 3.37 cm) from the bottom, and cut a notch one digit (1.9 cm) deep along it so that the bronze plaque can be inserted snugly into the notch with its upper face perfectly parallel with the cylinder's base and exactly two digits minus half a grain of barley (3.37 cm) above it. Then draw a circle on the outer wall of the cylinder at a height of exactly two digits (3.8 cm), and at points on it that correspond precisely with those on the bronze plaque, drill holes one grain of barley (0.85 cm) in diameter so that the axis of each hole is parallel to the corresponding line scribed on the bronze plaque and half a grain of barley (0.43 cm) above it. Form a wooden block fourteen digits (26.6 cm) square and more than one digit (> 1.9 cm) thick. In the center of the block excise a square hollow four digits (7.6 cm) to a side and exactly one digit (1.9 cm) deep, making sure that its bottom surface is perfectly flat and parallel to the block's top surface. Then attach the cylinder to the block, as pictured in the lower left-hand diagram. When viewed from directly above, as illustrated in the lower right-hand diagram, vertex V of the plaque will lie directly above and in line with the center of the square excised in the base.[26]

Next, Alhacen has us form seven wooden blocks four digits (7.6 cm) wide, six digits (11.4 cm) high, and thick enough to stand perfectly upright on their own. We are then to form segments of seven types of mirrors, that is, plane, convex and concave spherical, convex and concave cylindrical, and convex and concave conical, all made of highly polished iron and all of a size to be inserted into the blocks. Let us restrict ourselves for the moment to the plane

---

26. On the formation of the bronze plaque and the wooden cylinder, see Smith, *Alhacen on the Principles*, 300–304.

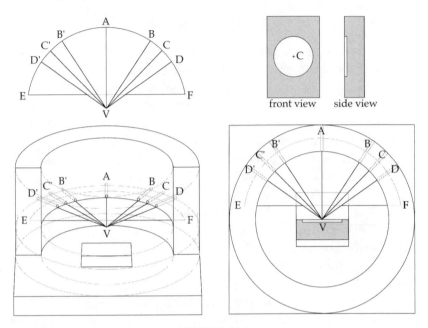

front view   side view

FIGURE 5.2

mirror, which is a thin, flat, iron disk three digits (5.7 cm) in diameter. On the face of one of the blocks we excise a circular hollow three digits (5.7 cm) in diameter and just deep enough to accommodate the disk so that the faces of the mirror and block are perfectly flush and so that center point C of the mirror stands exactly three digits above the base of the block on the block's midline, as illustrated in the upper right-hand diagram. With the mirror properly inserted, we stand the block upright in the hollow at the base of the cylinder and push it forward until the mirror touches vertex V of the bronze plaque, as viewed in the lower right-hand diagram from overhead. V will therefore touch the mirror precisely half a grain of barley (0.43 cm) below its center point C, and the face of the block will be perpendicular to the top face of the plaque.[27]

With everything set up as described, block all of the holes but the one at B′, and pose the apparatus so that sunlight shines through that hole. A thin beam of sunlight should therefore pass through it to the mirror and reflect at an equal angle to B, and the axial lines of both beams should pass through the centers of their respective holes as well as through center point C of the mirror. No matter which hole is tested, the same result should be obtained. We can even approxi-

27. On the formation of the mirrors and blocks, see ibid., 304–8.

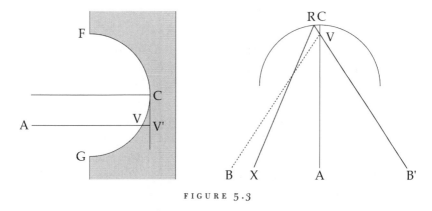

FIGURE 5.3

mate an axial ray by inserting a long copper tube with a narrow opening along
its axis into the given hole, and we should get the same result for all the mirror
segments. In every case, therefore, the equal-angles law is observed, including
the case in which the light enters along perpendicular AV and reflects back
along it. Furthermore, a second principle of reflection is confirmed on the
basis of the experiment—namely, that the axial rays and the associated points
all lie in a single plane of reflection perpendicular to the reflecting surface or
to the plane tangent to that surface at the point of reflection.[28]

The exquisite care with which this experiment is designed can be seen in
the arrangement for the concave spherical mirror. Let FCG in the left-hand
diagram of figure 5.3, which is drawn grossly out of scale, represent the segment of that mirror inserted into its block, as viewed from the side. Let it face
the hole at A such that the hole's axis intersects center point C of its surface
orthogonally (with the axis perpendicular to tangent CV'). When the mirror is
pushed against the plaque, vertex V will touch the mirror ahead of point C by
the amount VV'. As seen from directly above, this situation is illustrated in the
right-hand figure, and we can see from it that the light arriving along B'V will
continue past V to point R on the mirror and will then reflect along RX rather
than VB. In order to rectify this situation, Alhacen instructs us to drill a tiny
hole in the mirror just below C and large enough to allow point V of the plaque
to be pushed through to point V' so as to line up with C on tangent CV'.
Hence, as seen from above, points C and V will coincide, allowing axial rays
B'C and CB to coincide with lines B'V and VB on the plaque. For the convex

28. On the tests and conclusions, see ibid., 308–17.

spherical mirror, we need only pull the mirror back the same amount to keep the axial rays in perfect alignment with the appropriate lines on the plaque.[29]

Some have touted this set of trials as an example of Alhacen's pioneering use of modern experimental methodology, based as it is on controlled testing of the equal-angles hypothesis.[30] Let us consider the various issues wrapped up in this claim, starting with the question of whether the experiment is controlled. Of that there is no doubt. The experimental apparatus is designed with the utmost care to ensure accuracy of results as well as to cover as many cases as feasible according to types of reflecting surfaces. The next issue to consider is whether the experiment is pioneering or innovative. Here the answer is a qualified "no." Alhacen's experiment is no more controlled than Ptolemy's, although Ptolemy's apparatus is less elaborate (see chapter 3, section 2). However, Ptolemy was testing visual rays rather than light rays, and he was doing so indirectly because visual rays are invisible. Alhacen, on the other hand, designed his experiment to test light rays directly and visibly, and in order to do so, he had to reconfigure both the apparatus and the experiment. In that respect his experiment can be legitimately characterized as pioneering.[31] Still, it was Ptolemy who established the basic methodological grounds for it.

The last and most important issue is whether Alhacen's experiment is truly "modern." This issue is complicated. For a start, in order for the experiment to work correctly, the various planes involved must be perfectly aligned. Accordingly, the face and bottom of each block to be inserted into the hollow at the base of the cylinder must be perfectly flat, and both surfaces must be exactly perpendicular to each other so that the block stands orthogonal to the top face of the inserted bronze plaque. This applies to all seven blocks. In addition, the mirrors must be inserted in their blocks so that the points of reflection at C lie directly above and precisely in line with the center of the square excised in the base of the apparatus. This, too, applies to all seven. Meantime, the floor of that excised square must be perfectly flat and parallel to both the top face

29. See ibid., 307–8.

30. On the supposed modernity of this experiment, as well as a subsequent experiment on refraction (which we will discuss later), see Omar, *Ibn al-Haytham's Optics*. For a recent popularization of Omar's thesis, see Bradley Steffens, *Ibn al-Haytham: First Scientist* (Greensboro, NC: Morgan Reynolds, 2007).

31. Although most of the individual tests are based on direct observation of how light reflects inside the apparatus, the equivalent test for "weak" radiation from an illuminated object requires that Ptolemy's method be followed, with the observer looking through the hole to which the radiation should be reflected in order to verify that he sees its reflected image through that hole and not through any other; see Smith, *Alhacen on the Principles*, 323.

of the plaque and the plane passing through the axes of the holes. Not only must these axes all lie in precisely the same horizontal plane, but each axis must be perfectly parallel to the corresponding line on the plaque along the vertical. This applies to all seven holes. A deviation in any one of these conditions for any one block, any one hole, or any one plane will cause some skew in the resulting reflection, and a deviation in more than one will cause an even greater skew. Close is therefore not good enough, even in an apparatus of such relatively small scale.

The inadequacy of "close enough" is clear in the adjustment Alhacen specifies for the concave spherical mirror. As we saw, without that adjustment the incident ray would strike the mirror not at point C but at another point R, so that the reflected ray would go astray. But according to the measurements in Alhacen's construction, the actual discrepancy turns out to be only 0.36 mm, as measured by line VV′ in the left-hand diagram of figure 5.3. So minute is this discrepancy that it makes virtually no perceptible difference in the path of the reflected ray. It is simply too small to matter, so adjusting for it is not worth the effort, especially when all the other parameters will be slightly off. After all, as Alhacen himself acknowledges, when the beams of light pass out of the holes, they disperse over distance to form cones rather than cylinders, so judging where the axes lie within them is a matter of imprecise visual estimation. Clearly, then, Alhacen is imposing mathematical standards of accuracy on physical equipment and processes, which is to say that his experiment has no built-in margin of error. Anything less than perfect equality between angles of incidence and reflection would thus be a disconfirmation of the equal-angle principle. Consequently, the experiment can hardly be called "modern" in any meaningful sense. Indeed, given its obvious unfeasibility as actually described—with all planes perfectly aligned and all measurements perfectly reproduced—the test appears to have been an elaborate thought experiment designed to confirm what Alhacen already took for granted, that is, that light reflects at equal angles.[32] The experiment is therefore intellectually but not physically replicable.

Having supposedly verified that light reflects at equal angles in a single plane perpendicular to the reflecting surface or to a plane tangent to that surface at the point of reflection, Alhacen turns to the physics of reflection. Reflection, according to him, occurs from opaque bodies whose surfaces are

---

32. For a more extensive account of the problems with Alhacen's reflection experiment, see A. Mark Smith, "Le *De aspectibus* d'Alhacen: Révolutionnaire ou réformiste?" *Revue d'histoire des sciences* 60 (2007): 65–81.

extremely smooth, even, and without pores to trap or scatter incident light. Reflection can thus be analogized to physical rebound, incident light acting like tiny, hard, swiftly moving balls striking and rebounding from hard, impenetrable surfaces. Just as we observe in the case of actual physical rebound, reflected light "conserves the force and nature of its previous [incident] motion [and] is reflected back in the direction along which it arrived and along lines that have the same [relative] disposition as the original lines [of incidence]."[33] In other words, the incidence and reflection of light are perfectly symmetrical. As in the case of physical rebound, moreover, light loses some of its force or intensity after reflection. Nevertheless, Alhacen warns us, a body's reflectivity is not due to its physical hardness; water is just as reflective as glass, which is just as reflective as polished iron.

At this point, Alhacen is nearly ready to subject reflection to close mathematical study, but before he can do so, he needs to verify one other principle empirically: that images in reflection lie at the intersection of the reflected ray and the cathetus of incidence, that is, the line dropped perpendicular to the reflecting surface from a given object point. In order to demonstrate this principle, Alhacen uses the mirror segments previously employed in the equal-angles test. Following a suggestion by Ptolemy, he has us hold a short, thin rod perpendicular to the surface of each mirror and view its image from the side. In every case, the image will lie directly beneath and in a straight line with the rod.[34] Why this is so can be explained using the convex spherical mirror as an example. Let ACD in figure 5.4 be the rod posed perpendicular to the mirror's surface, let E be a center of sight, and let the luminous color from A, B, and C reflect to E from the mirror. Since ABCD is perpendicular to the reflecting surface, it lies on the cathetus of incidence, so the images of A, B, and C will appear directly in line with it at C´, B´, and A´, and the same holds for all the remaining points on the rod.

Now that the basic principles of reflection are established empirically, Alhacen is prepared to apply them to the mathematical analysis of reflection. He starts in book five with a set of fifty-four propositions dealing with various aspects of reflection and image location for the seven kinds of mirrors previously tested. The focus throughout is on object points within a single plane of reflection. In some cases the analyses and conclusions are fairly mundane, as for example Alhacen's demonstrations that when a given center of sight

33. Smith, *Alhacen on the Principles*, 321.

34. See ibid., 385–97. The test with the straight rod is one of several that Alhacen describes. For Ptolemy's suggestion of the verification using a rod, see chapter 3, section 2.

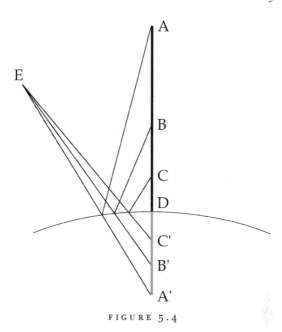

FIGURE 5.4

and a given point of radiation face a plane mirror or the three types of convex mirrors, there can be only one point of reflection and only one image.[35] But the purpose of book five is not to demonstrate individual facts about reflection; it is to lay the foundations for determining precisely where the point or points of reflection will lie on any mirror surface when the center of sight and the radiating point face it at various locations. Trivial for plane mirrors, this problem is far from trivial for the curved mirrors, and its application to convex and concave spherical mirrors has come to be known as "Alhazen's Problem" since the late seventeenth century.[36]

Take the case of convex spherical mirrors. In proposition twenty-five Alhacen poses the problem as follows, according to figure 5.5. If center of sight A and radiating point B face a convex spherical mirror with center of curvature G, and if A and B lie different distances from the reflecting surface—for example, if BG > AG—then find the point of reflection R on the mirror. Simple enough on the face of it, this problem is anything but. Not only is the solution quite complex, but it also hinges on six preliminary lemmas, one of which is

35. See ibid., 399–415.

36. See A. Mark Smith, "Alhacen's Approach to 'Alhazen's Problem,'" *Arabic Sciences and Philosophy* 18 (2008): 143–63.

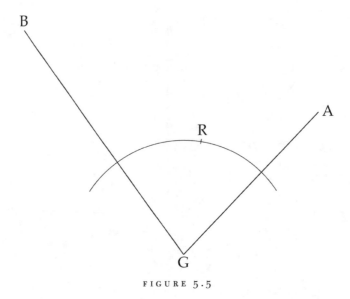

FIGURE 5.5

based on certain properties of hyperbolic sections. Altogether, in fact, the solution occupies more than sixteen pages in the critical Latin text, and its extension to convex cylindrical and conical mirrors occupies another fourteen.[37]

For concave spherical mirrors the problem is even more complex because, depending on where the center of sight and the radiating point are situated with respect to the center of curvature, there can be as many as four reflections and as few as none. The first thing Alhacen must do, therefore, is to determine how many reflections there will be according to particular placements of the two points. As it turns out, he draws directly from Ptolemy for the method, although in typical fashion his approach is more comprehensive and rigorous than Ptolemy's. Specifically, he borrows Ptolemy's technique of forming a circle through the mirror's center of curvature, the center of sight, and the radiating point. Thus, as we saw in our discussion of Ptolemy in chapter 3, if H in figure 3.6a, is the center of sight and Z the radiating point, and if HE = EZ, then reflection will occur at point A, as well as at points T and K where the circle through point H, D, and Z intersects the mirror. If, on the other hand, HE ≠ EZ, as in figure 3.7a, and if circle HDZ cuts the reflecting surface at points K and T flanking AD normal to line HZ, there will be three reflections, two from arc AK and one from arc TL. According to the situation in figure

37. For a detailed summary, without proofs, of the relevant lemmas and the central proposition, see ibid., 146–51; for the actual proofs, see Smith, *Alhacen on the Principles*, 415–46.

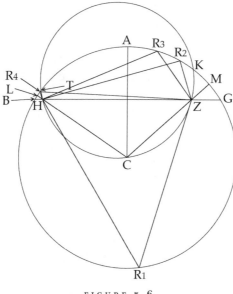

FIGURE 5.6

3.7b, where H lies on the mirror itself, there will be two reflections, both from arc AK, whereas if the circle intersects the mirror at point A on the normal, as in figure 3.7c, there will be only one reflection, and it will occur from arc TL. Finally, if the circle fails to intersect the mirror at all, as in figure 3.7d, there will be only one reflection, and it will occur within arc KL, K being the end point of KD, which bisects circle HDZ. In all these cases, if the circle on the mirror is completed, there will be a reflection from the opposite side, as illustrated in figure 3.8.

Alhacen applies precisely the same technique in a series of theorems, starting with proposition thirty-seven, in order, finally, to show that there can be as many as, but no more than, four reflections.[38] Accordingly, as illustrated in figure 5.6, one reflection will occur at point R_1 on the opposite arc GR_2H, two from points R_2 and R_3 on facing arc AK, and one from point R_4 on facing arc TL. Not every reflection in a concave spherical mirror yields a discernible image, though. Depending upon where the reflected ray and cathetus of incidence intersect, the image may appear behind the mirror or between the mirror and the center of sight. Or it may coincide with the center of sight itself

---

38. See propositions thirty-seven, forty-three to forty-six, and forty-nine, in Smith, *Alhacen on the Principles*, 454–58, 462–70, and 472–74.

or lie behind it. In these latter cases, Alhacen explains, the image will be seen on the reflecting surface and will appear confused with it.[39] Like Ptolemy, in short, Alhacen falls back upon "ghost images" to explain this phenomenon.

Given the way the points are distributed in figure 5.6, Alhacen takes a three-pronged, arc-by-arc approach to determining precisely where they lie, starting with point R_1 on the rear arc, moving to points R_2 and R_3 on facing arc AK, and ending with point R_4 on facing arc TL.[40] Although the procedure differs somewhat for each arc, all three procedures are closely related both to one another and to the procedure used in the case of convex spherical mirrors. Once Alhacen has established that procedure, he extends it to the analysis of concave cylindrical and conical mirrors, demonstrating that as many as four reflections can occur in both, each reflection taking place within a separate plane.[41] Unfortunately, it is impossible to convey the ingenuity, elegance, and conceptual simplicity of Alhacen's solution(s) to his eponymous problem without showing in intricate detail how adeptly he uses both Euclidean and Apollonian geometry.

Alhacen's account of image distortion in book six is somewhat anticlimactic after the virtuosity of his analysis of reflection in book five. Unfolding in a set of thirty-eight propositions, that account adds little beyond mathematical rigor and sophistication to what Ptolemy established in book four of his *Optics*. Like Ptolemy, Alhacen bases his analysis on object lines rather than object points in order to show how things extended in space can appear distorted in size, shape, and orientation in the seven types of mirrors dealt with in books four and five. Accordingly, he shows that images in plane mirrors will appear reversed behind the mirror and the same size as their objects, whereas in the three types of convex mirrors they will appear reversed behind the mirror, but smaller and more curved than their objects.[42]

More extensive and systematic than Ptolemy's, Alhacen's analysis of image distortion in the three types of concave mirrors deals with variations in image size and orientation according to the placement of the object. When it lies close to the mirror, for instance, its image appears behind the reflecting surface, and as it is drawn away from that surface, its image grows increasingly larger until it suddenly becomes incoherent. Remaining incoherent for a while

39. See ibid., 448.

40. For a detailed summary, without proofs, of the relevant lemmas and central propositions, see Smith, "Alhacen's Approach," 151–61; for the actual proofs, see Smith, *Alhacen on the Principles*, 452–72.

41. See Smith, *Alhacen on the Principles*, 475–85.

42. See Smith, *Alhacen on Image-Formation*, 162–204.

as it is drawn even farther from the mirror, the image is suddenly inverted and becomes increasingly diminished in size, although still inverted, as the object is drawn ever farther from the mirror. All of these phenomena are easily explained on the basis of ray geometry and the cathetus rule, yet in none of the resulting demonstrations does Alhacen relate image inversion to the focal point of the mirror.[43] This is an important point because it means that for Alhacen all mirror images, no matter their location vis-à-vis the reflecting surface, are virtual because they are psychological, not physical constructs. There is, in short, no room in his analysis for real images, the kind that can be projected onto a screen.

Although Alhacen's account of image distortion in book six breaks little or no new ground, there is one exception. In proposition three he demonstrates that although objects generally appear diminished in convex spherical mirrors, they can actually appear magnified when seen at the very visible edges of such mirrors. A marvel of ingenuity and mathematical rigor, this proposition is by far the longest in the entire *De aspectibus*, occupying nearly fifteen pages in the Latin edition. Yet Alhacen's inclusion of this proof points to an important feature of his approach to reflection in general and magnification in particular throughout books four to six of the *De aspectibus*. Nowhere does he show an interest in practical application. Never, for instance, does he mention that the magnification of images in reflection could be used to aid readers or craftsmen with weak eyes, and as we will see shortly, the same indifference to practical application extends to his analysis of magnification through refraction. Much of Alhacen's analysis in the *De aspectibus* thus appears to have been driven by pure mathematical considerations: if it can be proven, then it must be proven.

That Alhacen was nonetheless aware of the "practical" aspects of optics is evident from his having written at least four works on burning instruments, one of them devoted to the analysis of the focal property of a glass sphere (what Alhacen called a "burning sphere").[44] As mentioned earlier, only one of these works, a short treatise on parabolic burning mirrors, was translated into Latin under the title *De speculis comburentibus*.[45] That this work falls within the

43. See ibid., 204–32.

44. See Sabra, *Optics*, xxxii–xxxiii. The list includes one work on spherical burning mirrors, one on parabolic burning mirrors, and one on burning mirrors in general. For relevant editions and/or translations, see ibid., notes 50–51, pp. xlii–xliii. For the Arabic text and a French translation of Alhacen's *Treatise on the Burning Sphere*, see Rashed, *Géométrie*, 111–32.

45. For a Latin text and German translation, see I. L. Heiberg and Emil Wiedemann, "Ibn al-Haitams Schrift über parabolische Hohlspiegel," *Bibliotheca Mathematica* 10, series 3 (1909–10): 201–37.

tradition discussed in the previous chapter is evident from Alhacen's explicit citation of Anthemius and Archimedes in the preface, but Alhacen claims that since the proofs offered in previous studies are inadequate, he will rectify that problem. Precisely what studies he has in mind is a matter of conjecture. For instance, although there are certain similarities between Alhacen's approach and that of Diocles, as well as that of Dtrūms, Rashed argues that Ibn Sahl's *On Burning Instruments* is the likeliest proximate source because of several features idiosyncratic to both his and Alhacen's analyses of parabolic mirrors.[46] This is an important point because if Alhacen had access to *On Burning Instruments*, then he had access to Ibn Sahl's analysis of the focal property of hyperbolic lenses and, even more important, the sine law embedded in that analysis. Yet as we will see in the next section, that law never figures into Alhacen's analysis of refraction.[47]

4. REFRACTION AND ITS VISUAL MANIFESTATIONS

Like his analysis of reflection in books four to six of the *De aspectibus*, Alhacen's analysis of refraction in book seven opens with an empirical verification of the underlying principles. For that purpose, he describes in typically punctilious detail the experimental apparatus illustrated in figure 5.7. Its primary component is a bronze pan whose flat bottom is a circle measuring no less than one cubit (50 cm) in diameter and whose rim is no less than two digits (3.8 cm) high, as illustrated to scale in the top diagram. A quarter of the rim is cut out between points A and B. From point F, one digit (1.9 cm) to the right of point A, we are to scribe line FG passing through center point C of the pan's bottom, and we are then to scribe line DCE perpendicular to it. On the inside of the rim we are to scribe a circle at a height of one grain of barley (0.85 cm) above the bottom of the pan, and at point M on that circle, directly above point F, we are to drill a hole one grain of barley in diameter. Its center point will thus lie exactly one grain of barley above point F on the bottom of the pan.

Next we are to form a small, square panel thick enough to stand perfectly

46. See Rashed, *Géométrie*, lxxiii.

47. Rashed argues that Alhacen had read Ibn Sahl's *On Burning Instruments* before writing book seven of the *De aspectibus* and that Alhacen's analysis of spherical refracting interfaces in that chapter was informed by Ibn Sahl's proof of the focal property of hyperbolic lenses. In order to support this argument, however, Rashed is forced to explain away Alhacen's failure to adopt the sine law implicit in that proof. For Rashed's complete argument, see *Géométrie*, lxviii–lxxv, and for a critique of it, see Smith, *Alhacen on Refraction*, lxxxii–lxxxiv.

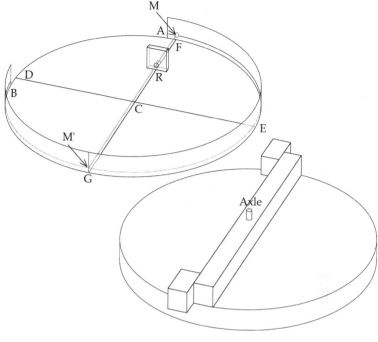

FIGURE 5.7

upright on its own, and we are to drill a hole in it one grain of barley (0.85 cm) in diameter with its center point exactly one grain of barley above its bottom. At midpoint R of radius FC, we are to fix the panel with its hole so that its bottom edge is perpendicular to line FCG and its hole is perfectly in line with the hole in the rim. Accordingly, both holes will share the same axis, MM′, which intersects the rim at point M′ directly above point G on line FCG. Taking the center of the hole in the rim as our zero point, we are to subdivide the circle on the inside of the rim into degrees and then into halves, quarters, or whatever fractions of a degree are feasible. On the back of the pan at its very center we are to attach an axle and fit a bronze strip with two short overhanging segments so that the strip can be rotated, but not too freely, as illustrated in the lower diagram. Finally, we are to form a cylindrical vessel that is just capacious enough to hold the pan when it is placed upright in it. It should therefore be slightly over half a cubit (25 cm) high and slightly over one cubit (50 cm) in diameter at the rim. Accordingly, when the pan is inserted upright into the vessel, it will hang by the strip attached to the back, as illustrated in figure 5.8.

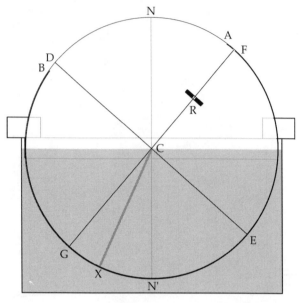

FIGURE 5.8

The pan can thus be rotated about the axle so that line FCG inclines at any angle we wish.[48]

Set up in this way, the entire apparatus can be placed toward the sun to let a beam of sunlight shine directly through the holes at F and R and strike the rim at G, where it will form a circle of light centered on point M´. The axis of that beam will therefore be perfectly parallel to line FG and precisely one grain of barley (0.85 cm) above it. Now we are ready to test refraction from air to water, air to glass, glass to air, and glass to water. In order to test refraction from air to water, for instance, we fill the vessel with water until it reaches point C on the pan, as illustrated in figure 5.8. Then we channel a beam of sunlight through the two holes so as to strike the water obliquely at C, and looking directly through the gap in the rim between A and B, we observe that after refracting there, the beam forms a circle of light at some point X between G and endpoint N´ of normal NCN´. We have thus confirmed that water is less transparent than air and that when it passes into the less transparent medium, the light refracts toward the normal. There is no need to go through every test for air to glass, glass to water, and so on. They all reveal that when light passes from

48. For Alhacen's full description of this apparatus and its construction, see Smith, *Alhacen on Refraction*, 220–24.

one medium to the other along the perpendicular it passes straight through without refraction. They also reveal that when it passes obliquely from a more transparent into a less transparent medium, the light is refracted toward the normal, whereas when it passes in the opposite direction, it is refracted away from the normal. In addition, they reveal that light passes through any given transparent medium in a straight line and, moreover, that the axes of the incident and refracted rays, as well as all the associated points, lie in a single plane of refraction perpendicular to the interface between the two media or to a plane tangent to that interface at the point of refraction.[49]

After these tests, Alhacen attempts to explain the results physically on the basis of the model already used for reflection. Accordingly, he treats light as if it consists of tiny spheres striking resistive bodies at great speeds, except that unlike reflective bodies, refractive bodies are only partially resistant to penetration. As we have seen, this resistance varies with the "density" or "solidity" of the body, so the denser the body the more resistant and less transparent it is. Opaque, polished bodies are therefore so dense as to prevent any penetration, the result being complete rebound. Implicit in this model is that the speed, force, and direction of the light are contingent on the resistance of the medium through which it moves. Thus when light passes from a rarer to a denser medium, its speed, force, and direction are altered, and the more obliquely the light strikes the denser medium, the more radical the alteration. In order to illustrate this point, Alhacen has us consider what happens when we fix a thin wooden plank to an opening and hurl an iron ball at it. When we throw the ball directly toward the plank along the perpendicular, it will break straight through, but as the trajectory becomes more oblique, the blow becomes more glancing, and the resulting path after breakthrough becomes increasingly diverted, until finally the ball bounces off.[50]

With this model in mind, let us assume that light passes along ray AR in figure 5.9 to strike interface ERF along the perpendicular. If there were no density differential at ERF, the light would continue straight to point C at the same speed with which it reached R. Let the medium below ERF be denser in such a way that it poses an upward resistance measured by V´. Slowed by that resistance, the light will reach B instead of C in the same time it took to travel from A to R. Therefore, its "natural" speed in the denser medium is proportional to RB. Now let the light pass obliquely from A´ to R in the same

49. For Alhacen's account of all these tests, see ibid., 224–44.

50. See ibid., 245. As another physical example Alhacen cites the difference in effect between a direct and an oblique sword stroke.

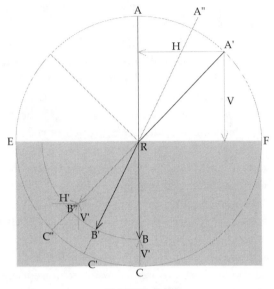

FIGURE 5.9

amount of time. Its motion can be resolved into vertical component V and horizontal component H, and when it strikes interface ERF, it will meet vertical resistance V′ as well as horizontal resistance H′. If the medium is perfectly dense and polished, the light will be resisted along the vertical only, and since the horizontal component will remain unchanged, the light will reflect at equal angles according to the change along the vertical. In the case of refraction, however, the medium is imperfectly dense, so the light will meet both vertical resistance V′ and horizontal resistance H′. Slowed by both, it will reach B″, and its speed as measured by RB″ will be less than its "natural" speed in the medium as measured by RB.

Now according to Alhacen, whenever light passes obliquely into a denser medium, it naturally inclines "toward the direction in which it would pass more easily," and since "motion along the normal is easiest, [the light] must move toward the normal." As a result, "the composite motion that is in it is in no way diminished but only interfered with."[51] In figure 5.9, then, the horizontal resistance along H′ pushes the light into "easier" trajectory RB′, which is the same length as trajectory RB. In that way, the light conserves its "natural" speed in the denser medium. When the light passes in the other direction from B′ to R, the light is impelled rather than impeded by the horizontal and vertical

51. Ibid., 246.

forces, but in order that it not be impelled beyond its "natural" speed in the rarer medium along RA″, which is longer than AR, it is pushed away from the normal to A′, so that the resulting path RA′ is the same length as AR. Note, by the way, that this explanation is based on the principle of reciprocity, which states that in passing along a given path in one direction, the light will follow the same path in the other. Hence, according to Alhacen's analysis, angle of deviation B″RB′ when the light refracts into the denser medium will be equal to angle of deviation A″RA′ when it refracts into the rarer one. Underlying this analysis is the metaphysical principle that natural actions are accomplished as efficiently as possible—or, to put it another way, that nature does nothing in vain.

Ingenious though it is, this vectorial explanation is fraught with difficulties. For one thing, what Alhacen means by "density," as tied to transparency, is unclear. If we confine ourselves to air, water, and glass, then it can be interpreted physically insofar as each medium is obviously denser (and more "solid") than the other in that order. However, while clear glass is physically denser and more refractive than murky water, it is more rather than less transparent. Likewise, while murky water is less transparent than clear water, it is not necessarily denser or more refractive. Furthermore, the dynamics of Alhacen's account are incoherent, because the horizontal and vertical forces act in different ways. When light passes from a rarer into a denser medium, one force impedes the motion, whereas the other effectively enhances it by impelling the light into an "easier" path. Moreover, if we assume that ARC in figure 5.9 is the interface and ERF the normal, then when the light strikes that interface along A′R, it will follow a path between ER and RB″ rather than path RB′. Yet the resistive forces have not changed, only the direction along which they act. What, then, determines whether in one situation the force will act to impede, whereas in another it will act to enhance? Furthermore, the dynamics of this account are peculiar in that everything occurs instantaneously at the refracting interface, so the two forces must act instantaneously as well.[52] Despite these and other problems, though, this model had an important afterlife in the seventeenth century, as we will see in the final chapter.

So far Alhacen has provided a qualitative analysis of refraction, but the experimental apparatus used for that analysis can also be deployed for a quantitative analysis. Let us assume, for instance, that line FCG in figure 5.8 is inclined at an angle of 40° with respect to normal NC and that the beam of sunlight passes along FC and is then refracted along CX. Since the axis of the refracted

---

52. For a more detailed account of these problems, see ibid., lvii–lxiii.

beam lies on the circle scribed on the inner wall of the rim, and since that circle has been subdivided into degrees and fractions of a degree, we can measure the angular distance between points G and X and arrive at a precise measure for angle of deviation GCX, which is how Alhacen measures refraction rather than by angle of refraction XCN´. The resulting angle of deviation should be right around 11.1° (or 28.9° for the angle of refraction). Moreover, since the pan can be rotated within the vessel at any angle we wish, we can determine the angle of deviation for any angle of incidence. Like Ptolemy, Alhacen suggests that we make such determinations at 10° intervals, starting at $i = 10°$ and ending with $i = 80°$, although if we wish, we can use increments of 5° or less. Thus in the case of refraction from air into water, we need only incline line FG at angle FCN of 10°, allow a beam of sunlight to pass straight through the holes to C and refract to X, and then measure the resulting angle of deviation GCX, after which we can repeat the procedure for 20°, and so on up to 80°.[53]

Let us take refraction from glass to air as another example. For this experiment we are to form a quarter-sphere of glass, as illustrated in the top diagram of figure 5.10. Point D is the center of the sphere out of which it is formed, and radius AD of that sphere should be less than distance RC between the front of the small panel fixed to the bottom of the pan and its center point C. Actually, the section is slightly larger than a quarter-sphere by the amount DC´, which is a grain of barley (0.85 cm), so the plane through line HDK parallel to the bottom face of the quarter-sphere cuts it on the equator. When the quarter-sphere is fixed to the bottom of the pan so that point C´ on its bottom edge LC´M coincides with the pan's center point C and is perpendicular to normal NCN´, flat face LAM will be perpendicular to the pan's bottom, and the quarter-sphere's equator will lie in the plane of refraction.[54]

Suppose that the quarter-sphere is fixed such that angle NCF = 40°. If sunlight is allowed to stream through the two holes at F and R, it will pass straight through the convex face of the quarter-sphere without refraction because it will strike it orthogonally. It will then strike flat face LCM at an angle of 40°, and when it passes into the more transparent air, it will be refracted to some point X away from normal NN´. So again, all we need do is remove and refix the quarter-sphere so that angle NCF = 10°, then 20°, and so on, and measure angle of deviation GX for each 10° increment. At least that is what Alhacen instructs us to do. If, however, we follow this procedure to the letter, we find that when we let the light strike the flat face of the quarter-sphere at an angle NCF

53. On this test see ibid., 251–53.

54. On the formation and installation of the glass quarter-sphere, see ibid., 234–35.

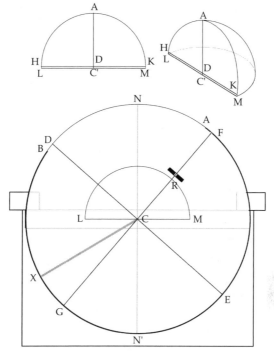

FIGURE 5.10

of 50° or greater, instead of passing into the air, the light reflects internally. This phenomenon is due to the critical angle, which comes into play whenever light passes from a more refractive into a less refractive medium, and in the case of refraction from glass to air the critical angle is just under 42°. Likewise, in the experiment testing refraction from glass to water—for which the vessel is filled with water until it coincides with the bottom face of the glass quarter-sphere— the critical angle is reached at 61°, so the trials at 70° and 80° will not work. Yet Alhacen makes no explicit mention of this problem here or anywhere else in his discussion of refraction.

Even more surprising than his failure to acknowledge the problem of critical angle is Alhacen's failure to provide any values for the angles of deviation. This failure is especially striking in view of the large scale of his apparatus compared to that of Ptolemy's and the potentially greater accuracy of the resulting observations. Why, then, did Alhacen provide no tabulations? Gérard Simon suggests that despite the precision with which the apparatus is designed, various factors, diffraction foremost among them, would have compromised the accuracy of results obtained from it. Those results would therefore have been no better than Ptolemy's.[55] Furthermore, the apparatus is extraordinarily cumbersome to use, particularly when filled with water. It has to be placed just so in order to catch the sunlight at the right angle, and testing the full range of angles from 10° to 80° would require half a day, assuming that the experiment was carried out at an appropriate latitude.[56]

There are also technological issues, especially in regard to the glass quarter-sphere. We can reasonably assume that Alhacen would have used relatively clear, transparent glass more or less equivalent to modern crown glass. We cannot, however, assume that his glass would have been free of bubbles and striations, a problem exacerbated by the relatively large size of the quarter-sphere. It is also doubtful that the quarter-sphere could have been formed to the exacting standards required for the experiment, that is, with the two flat faces perfectly perpendicular to each other and the convex surface perfectly spherical. In addition, there is absolutely no reason to use a spherical section when a cylindrical one will do just as well and is far easier to produce. Like his adjustment for the spherical mirrors in the reflection experiment, Alhacen's specification of a quarter-sphere for testing refraction through glass seems to be driven by the need for physically unachievable mathematical exactitude. We are therefore led to raise the same doubt about feasibility that we did with the reflection experiment, and in this case the doubt is deepened by Alhacen's failure to acknowledge the problem posed by critical angle for the tests for refraction from glass to air and glass to water. In short, there is good reason to believe that he did not carry out the experiment as described, which helps explain his failure to provide any values. That in turn raises serious doubt about the experiment's replicability and, therefore, its "modernity." Furthermore, its originality is questionable in that it is clearly based on Ptolemy's experimental derivation of the angles of refraction.

Alhacen concludes his quantitative analysis with a set of seven governing rules of refraction based on the overarching principle that when light refracts through the same two media, as the angle of incidence decreases, the angle of deviation approaches zero. Or, to put it another way, as the angle of incidence decreases, the angle of refraction approaches equality with it.[57] Of the result-

---

55. See Gérard Simon, "L'Expérimentation sur la réflexion et la réfraction chez Ptolémée et Ibn al-Haytham," in *De Zenon d'Élée à Poincaré*, ed. Régis Morelon and Ahmad Haznawi (Louvain: Peeters, 2004), 355–75.

56. For a more extensive analysis of the problems with Alhacen's experimental procedure, see Smith, *Alhacen on Refraction*, liii–lvii.

57. This is what the second law Ptolemy articulated in *Optics*, V, 34, boils down to; see chapter 3, section 3.

ing seven specific rules, one is of particular concern to us here. It states that "when . . . light passes from a rarer to a denser body, the . . . angle of deviation will always be smaller than . . . the angle [of incidence]." No matter the size of angle A′RA of incidence in figure 5.9, therefore, the resulting angle of deviation B″RB′ will be smaller than it. In its Latin form, this law turns out to be a mistranslation of the original Arabic version, which states that when light is refracted into a denser medium, the resulting angle of deviation is less than *half* the angle of incidence. Thus no matter the size of angle A′RA of incidence in figure 5.9, the resulting angle of deviation B″RB′ will always be smaller than *half* angle A′RA.[58]

To this point Alhacen has established all but one of the basic principles of refraction—namely, that in refraction the image of any object point is located at the intersection of the normal (the cathetus) dropped from that point to the refracting surface and the refracted ray. As expected, he subjects this principle to empirical verification in a variety of ways that need not be rehearsed here.[59] Armed, finally, with the full array of principles as analytic tools, Alhacen turns to the main purpose of book seven, which is to explain why celestial objects look larger at the horizon than at zenith. To that end, he starts with a set of theorems demonstrating that no matter whether the eye faces a plane, spherical convex, or spherical concave interface, and no matter whether it lies in a rarer or denser medium, there will be only one image for any given object point lying in the other medium. However, when the object point is viewed through a convex spherical interface, its image may appear behind the interface, on it, or in front of it, depending on how the center of sight and object point are disposed with respect to the center of curvature, just as is the case with concave spherical mirrors.[60]

Before continuing with his analysis of image formation in refraction, Alhacen makes an abrupt shift to the problem of peripheral vision. As we saw earlier, his account of vision in books one and two of the *De aspectibus* is based on the lens's optical and sensitive selectivity, which allows its anterior surface to be affected only by the radiation that reaches it orthogonally. All such radiation

58. For Alhacen's statement of all seven rules, see Smith, *Alhacen on Refraction*, 259–60; for a discussion of the problematic nature of several of these rules, see ibid., lxiii–lxv.

59. For the full set of verifications, see ibid., 274–80.

60. See ibid., 289–90. Like Ptolemy, Alhacen argues that when the object lies in a denser medium and is viewed through a convex interface, then if the image location lies at the center of sight or when there is no determinate image location, the visual faculty creates a "ghost" image on the surface of the refractive interface, just as it does in the case of concave mirrors; for Ptolemy's explanation, see chapter 3, section 2.

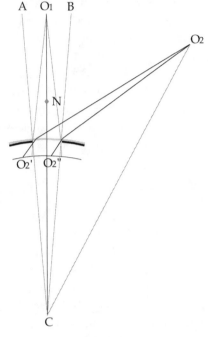

FIGURE 5.11

forms a cone whose base defines the field of view and whose vertex constitutes the center of sight, which is the cardinal reference point for all optical analysis. In the sixth chapter of book seven, however, Alhacen acknowledges that this cone is severely limited by the relatively tiny size of the pupil, so the cone encompasses only a small part of the entire field of view.

Let ACB in figure 5.11 represent that cone as restricted by the pupil. Accordingly, any object between A and B will be seen along a perpendicular ray, so object O_1 will be seen along O_1C. Meantime, object O_2, which lies outside the cone, projects a cone of radiation through the pupil to all points between O_2' and O_2'' on the lens's surface. All these rays strike the lens obliquely, and by Alhacen's earlier account, such rays are selected out according to the lens's optical and sensitive nature. Point O_2 should therefore go unperceived, yet it does not. Why?

Alhacen's response is that the visual faculty interprets all the oblique impressions from O_2 as if they were a single impression arriving along O_2C, which is perpendicular to the eyeball and thus the anterior surface of the lens. Consequently, objects like O_2 lying outside cone ACB are perceived along *virtual* perpendicular rays that are analogous to real ones.[61] The same holds for points like O_1 within cone ACB. They, too, project a cone of radiation through the pupil, and all but one of the rays within that cone strike the lens's surface obliquely. Interpreted by the visual faculty as if they had all arrived along perpendicular O_1C, the resulting impressions from the oblique radiation reinforce the primal impression made along that perpendicular. This extends to all points within cone ACB; their impressions along the perpendicular are reinforced by the oblique radiation projected by them onto the lens. As an empirical verification of this point, Alhacen advises us to hold a needle close to the pupil of one eye on axial ray O_1C. The needle will be perceived as translucent rather than solid because the things it blocks will be seen "through" it according to the oblique radiation that bypasses it to reach the lens's surface. Thus if N in figure 5.11 represents a cross section of the needle on axis O_1C, then point O_1 behind it will be seen according to the oblique radiation that passes by the side of the needle to reach the lens. Both point O_1 and the needle will therefore be seen simultaneously along the same line of sight, giving the impression that the needle is somewhat transparent.[62]

One benefit of this theory is that it explains the continual decrease in visual acuity with distance from axial ray O_1C according to Alhacen's supposition that the intensity with which light strikes the lens varies inversely with the obliquity of the ray along which it arrives. Thus when the object lies toward the very fringe of the field of view, its radiation will reach the lens along extremely oblique lines. The resulting impressions will be commensurately weak, so the image will be extremely nebulous. The main drawback of this theory is that it places an extraordinarily heavy burden on the visual faculty to interpret the impressions correctly. For instance, the radiation within the cones projected from points O_1 and O_2 overlaps at innumerable points on the lens's surface, so the surface at each of those points is affected by an impression from both objects. How, then, does the visual faculty decide which impression belongs to which object? Moreover, the lens must be exquisitely sensitive in order to differentiate among the myriad impressions made on it and then determine which of those impressions should be interpreted as coming along the appropriate virtual ray. Worse, it has to make that determination continually for every object within the field of view. Whatever else it does, Alhacen's account of peripheral vision renders his original visual model incoherent, forcing him to posit virtual rays in order not simply to save the cone of radiation with its

61. See ibid., 110.
62. See ibid., 303–7.

vertex at the center of sight but also to save the mathematical analysis of visual perception in all its modes.

After this relatively brief excursus into the problem of peripheral vision, Alhacen turns his attention in the seventh and final chapter of book seven to the phenomenon of atmospheric refraction in order to investigate its effects on celestial observations. Earlier, in chapter four, he had shown in Ptolemaic fashion that when celestial objects appear at the horizon, they are actually located somewhat below the horizon, which means that their light is refracted slightly toward the normal when it passes through the atmospheric shell. In short, the atmosphere is denser than the aither beyond it.[63] Having therefore demonstrated that the images of celestial objects near the horizon are displaced upward by atmospheric refraction, Alhacen addresses the issue of size distortion. He starts with a general analysis of how objects in a denser medium look magnified when viewed through plane and convex spherical refracting interfaces, pointing out that the apparent enlargement is due to two factors. One is the apparent lifting of the object so that it appears closer, as illustrated in figure 3.16a, where image A′B′ of object AB subtends a greater visual angle according to its apparently shortened distance from center of sight E. But according to the size distance invariance hypothesis, image A′B′ would be *perceived* to be the same size as object AB because it *is* the same size. The second factor is the weakening of the radiation caused by refraction, which makes image A′B′ look dimmer and thus farther away than it actually should. Consequently, its perceived apparent distance is greater than its actual apparent distance, so when the enhanced apparent distance is correlated with visual angle A′EB′, object A′B′ is perceived to be not just closer but also somewhat enlarged. In the case of an object viewed through a convex spherical interface, the magnification is much greater because the increase in the visual angle under which the object is viewed is greater than the decrease in apparent distance.[64]

So far in his analysis of magnification through convex spherical interfaces Alhacen has located the object between the center of curvature and the refracting interface. Now, in proposition seventeen of book seven, he presents a case in which the object lies beyond the center of curvature, as illustrated in the left-hand diagram of figure 5.12. Accordingly, when viewed through a glass sphere by center of sight A, which lies in the air, object line TQ on axis AEGB′ will

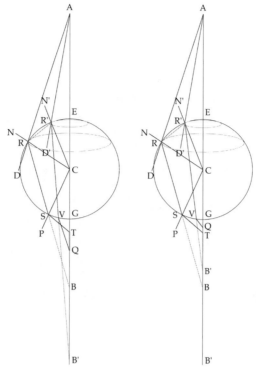

FIGURE 5.12

appear as a ring on the surface of the sphere facing A.[65] Alhacen starts by supposing that the space beyond arc RG is filled with glass and that ray BR from point B on the axis strikes the glass-air interface at angle of incidence BRC so as to refract to A at angle NRA. Choose some point R′ on arc RE, and take A as a point of radiation rather than a center of sight. Let ray AR′ strike the face of the glass at angle of incidence AR′N′ and refract along R′B′ to point B′ on the axis. Likewise, ray AR will strike the glass at angle of incidence ARN and refract at angle CRB. From our discussion of the rules of refraction above, we know that angle of deviation DRB < half angle of incidence ARN, which = angle DRC, and we know that angle of deviation D′R′B′ < half angle of incidence AR′N′, which = angle D′R′C. We also know that angle of deviation

65. See ibid., 318–19. Because Alhacen does not actually enumerate his propositions, I have designated each one by number in my edition, this one being the seventeenth proposition of book seven.

D′R′B′ < angle of deviation DRB. Ray R′B′ will therefore intersect ray RB at some point beyond arc DG before reaching the axis.

Now assume that the glass sphere is surrounded by air, and again take A as a radiating point. As before, ray AR will refract along line RB, but before reaching the axis it will strike arc RG of the sphere at S. Since RC on normal CRN and SC on normal CSP are radii of the sphere, it follows that triangle RCS is isosceles and, therefore, that angles CRS of refraction and CSR of incidence are equal. Consequently, by the principle of reciprocity, angle PST at which the light refracts out of the sphere is equal to angle ARN at which it entered the sphere. So the light will be refracted along ST to point T on the axis. The same analysis holds for ray AR′, which will refract to point V on arc RG and then refract along VQ at the same angle as its original angle of incidence. Hence, rays ST and VQ will intersect beyond arc DG before reaching the axis.

If we reverse the analysis, we see that the light from point T follows path TSRA to center of sight A, and the light from Q follows path QVR′A to center of sight A. Likewise, the light from every point between Q and T refracts at specific points on arcs SV and RR′ to end up at A. The image of each of these points between Q and T lies at the intersection of the refracted ray and the cathetus of incidence, which is axis ACB′, so A is the image location for all of them. In that case, according to Alhacen, the visual faculty will transpose each image to its relevant point of reflection on arc RR′ on the face of the sphere.[66] Consequently, if the entire figure is rotated about axis ACB′, all the transposed image points on arc RR′ will form a ring whose inner and outer boundaries are defined by points R′ and R, respectively. That ring will be the image of line TQ, and according to the intersection of rays, it will be reversed, a point Alhacen neglects to mention. It will also be magnified insofar as chord RR′ of the arc forming the image is longer than object line TQ.

As Rashed points out, this theorem has significant implications for lens theory because it takes spherical aberration into full account according to the radiation from point A. Thus as points of refraction on arc RR′ are chosen at increasing distances from R, the resulting rays refracted at arc SV on the sphere will strike points between T and Q after intersecting beyond arc SG. From this we can infer that just as in spherical concave mirrors, so in glass spheres, there is a focal point on the axis where a certain sheaf of rays refracting out of the sphere will congregate most densely. Alhacen in fact demonstrates this very point in his *Treatise on the Burning Sphere*, which was written after the *De aspectibus* and in which he shows that the focal point will lie a distance

---

66. See note 60 above.

of half the sphere's radius on the axis beyond the sphere's back edge.[67] According to Rashed, then, Alhacen's analysis of radiation through a glass sphere in proposition seventeen represents an important step in the evolution of his analysis of the focal property of glass spheres in the *Treatise on the Burning Sphere*.[68] Proposition seventeen thus has fairly clear implications for the development of a proper theory of lenses, spherical lenses in particular.

Nevertheless, as far as we know, none of Alhacen's pre-Keplerian Latin readers saw these implications. Why not? One reason for this failure is the way the diagram accompanying the Latin version of proposition seventeen is presented. In the Arabic version it takes the form represented in the diagram to the left of figure 5.12. In the Latin version, however, it takes the form represented in the diagram to the right, where none of the relevant rays intersect before reaching the axis.[69] Thus ray R´B´ meets the axis before intersecting RB, as does VQ before intersecting ST, leaving the points of intersection on the axis reversed. In short, the Latin version of the diagram shows no spherical aberration whatever. A likely source for this error is the Latin mistranslation of the rule discussed earlier. Whereas the Arabic version mandates that the angle of deviation be less than *half* the angle of incidence, the Latin version mandates only that it be *less than* the angle of incidence. According to that version of the rule, then, angle of deviation D´R´V in the diagram to the right of figure 5.12 may be larger than half the angle of incidence AR´N´ and in fact large enough to prevent the appropriate ray pairs from intersecting before reaching the axis. This error might have been avoided if Alhacen had pointed out that image RR´ is reversed, but he failed to do so. It might also have been avoided had the translator/illustrator been aware of Alhacen's analysis in the *Treatise on the Burning Sphere*, but that work was unavailable in Latin.

The faulty diagram is not the only thing that masks the lens-theoretical implications of proposition seventeen. The ostensible purpose of that proposition is to explain a particular case of image formation, not to analyze how light rays are affected by passing through glass spheres. True, in tracing the radiation from A through the sphere, Alhacen can be interpreted as undertaking such an analysis, but his real intent in taking this tack is to backtrack

67. Although Alhacen's conclusion that the focal point lies half the radius beyond the sphere's back edge is correct, his method for arriving at that conclusion is problematic; see Smith, *Alhacen on Refraction*, lxxiii–lxxv.

68. See Rashed, *Géométrie*, liii.

69. Five of the eighteen Latin manuscripts lack the diagram entirely; of the remaining thirteen that do have it, all show the diagram as represented in the right-hand diagram of figure 5.12, and this extends to the fourteenth-century Italian translation as well.

to the appropriate points T and Q on the axis from which luminous color will be refracted to a single center of sight on the other side of the sphere. The emphasis is therefore on how all the spots of luminous color on line TQ radiate through the sphere so that the resulting image will appear as a ring on the sphere's surface from the perspective of A. This focus on image formation rather than light radiation is reinforced by Alhacen's empirical verification of proposition seventeen's conclusion. Take a sizeable sphere of clear glass and a ball of black wax the size of a chickpea, he suggests, and look straight through the sphere along an axis. Then, holding the ball on the end of a long needle, move it to and fro along the axis until you see its image as a black circle on the sphere's surface.[70]

Given all these factors, it is easy to understand why Alhacen's Latin readers failed to see the lens-theoretical implications of proposition seventeen. Even Witelo and Friedrich Risner, two of Alhacen's most diligent and perspicacious medieval and Renaissance readers, missed those implications, both of them reproducing the faulty diagram for their reprises of the theorem.[71] Moreover, the context within which proposition seventeen appears further masks its lens-theoretical implications. As mentioned earlier, this proposition is preceded by a set of five theorems explaining how objects viewed in denser media appear magnified. Of this set, the two theorems immediately preceding proposition seventeen deal with the apparent magnification of objects viewed through convex spherical interfaces, so proposition seventeen can easily be construed as a special, albeit somewhat anomalous, case of such magnification. Likewise the very next theorem, which replaces the glass sphere with a glass cylinder, can be construed as a special case of that special case.

This proposition is immediately followed by a concluding set of three theorems in which the center of sight is located in the denser medium and looks at objects in the rarer one. In all three cases, the denser medium is the atmospheric shell enveloped by aither, and in keeping with the analytic theme to this point, the issue is how atmospheric refraction affects the apparent size of celestial objects viewed from a point on earth through the concave interface between the atmosphere and aither.[72] In response, Alhacen shows that just as

70. See Smith, *Alhacen on Refraction*, 319.

71. Witelo's version of this theorem appears as proposition forty-three in book ten of the *Perspectiva*, and of the fifteen manuscripts I have been able to consult, three lack the diagram entirely; the rest have the faulty one. For Risner's use of the faulty diagram in his edition of Alhacen's *De aspectibus* and Witelo's *Perspectiva*, see *Opticae Thesaurus*, 277 and 430.

72. For these three theorems, see Smith, *Alhacen on Refraction,* 321–25.

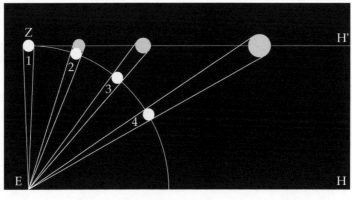

FIGURE 5.13

atmospheric refraction causes celestial objects to appear slightly higher above the horizon than they actually are, it also causes them to appear slightly smaller than they are. He also shows that the closer they are to the horizon, the more their apparent size is affected in the direction of longitude (top to bottom) than in the direction of latitude, a point oddly inconsistent with his subsequent claim that "every star in the sky is perceived as round, so its diameters are perceived [to be] equal."[73]

But if the apparent size of celestial objects is diminished by atmospheric refraction and most radically diminished near horizon, why do the moon and sun appear so much larger at that point than at any other point in the sky? This is the final problem Alhacen addresses in the *De aspectibus*, and he resolves it on the basis of two factors. The principal factor, which is psychological, operates consistently. Since we cannot detect the hemispherical shape of the celestial vault because of its vast distance, we perceive it as flat, lying closest to us at zenith and farthest away at horizon. Likewise, since we cannot detect the circular shape of the solar and lunar orbits, we perceive those bodies as moving in the plane of the flattened celestial vault so that they appear to recede ever farther from us as they move from zenith toward horizon.

Let E in figure 5.13 represent the center of sight on the Earth's surface, EH the line of the horizon, and EZ the zenith line along which the moon appears at position 1 under the visual angle determined by the two tangent lines. As the moon moves on its circular orbit to positions 2, 3, and 4, the visual angle re-

73. Ibid., 325.

mains constant, so the moon ought to appear the same size throughout. How-ever, because of our inability to detect the circularity of that orbit, we perceive the moon as traveling along line ZH′. Consequently, as it reaches successive positions on its orbit, it appears to recede from E along line ZH′, its apparent distance increasing with each new position on its orbit. Since size perception depends on correlating visual angle with perceived distance, the moon appears to get larger as it approaches the horizon despite the unchanging visual angle.

This apparent magnification, Alhacen continues, can be reinforced by a second factor that operates inconsistently. Under certain circumstances, thick vapors rise from the earth to create a wall of vapor denser than the air between it and the center of sight and extending a few degrees above the horizon. Hence, when a celestial object near the horizon is viewed through this wall, it will appear somewhat enlarged by refraction. Added to the perceptual magnifi-cation just discussed, this refractive enlargement causes the object to look even more magnified than usual. As we noted earlier, Ptolemy hinted at Alhacen's psychological explanation by claiming that when celestial bodies are viewed under anomalous conditions, their apparent size will be misperceived, par-ticularly at zenith, where the object presumably looks smaller than it should. He also proposed, but eventually dismissed, the idea that thick vapors cause objects near the horizon to be magnified by refraction. Alhacen thus diverged somewhat from Ptolemy by incorporating both explanations systematically into his analysis of what is now referred to as the moon illusion.

5. CONCLUSION

Over the past few decades Alhacen has become an iconic figure in the history of optics, and justifiably so. Not only was he a brilliant and creative thinker, but also his methodical and comprehensive approach to optics raised that science to a level of analysis far beyond that achieved by any of his predeces-sors, including Ptolemy. Indeed, so far beyond Ptolemy does Alhacen appear to have reached that his optical analysis is now commonly regarded as both revolutionary and modern in approach. After all, by rejecting Ptolemy's visual rays in favor of light rays, Alhacen seems to have reconfigured optics along modern lines, and he achieved much of that reconfiguration on the basis of what looks like modern experimental methodology. Small wonder, then, that he is almost universally credited with having laid the foundations of modern optics on the ruins of its ancient, visual ray forerunner.

I have argued at some length elsewhere that this interpretation is not only simplistic but also ahistorical, so I will not rehearse that argument in detail here.[74] A few points bear repeating, though. For a start, Alhacen did not so much reject Ptolemy's visual rays as replace them with light rays, and he did so with the aim of preserving Ptolemy's radiative cone as the primary analytic vehicle for his account. So intent was he on preserving this cone and the all-important center of sight at its vertex, in fact, that he felt compelled to adduce *virtual* radiation in order to explain peripheral vision within the constraints of this radiative model. Furthermore, if we take him at his word, Alhacen intended neither to undercut nor overturn visual ray theory. He meant instead to reconcile it with the intromissionist alternative by incorporating the best of both into his theory. Nor was Alhacen unique in taking this conciliatory approach. In the previous chapter we saw that both Philoponus and Avicenna transformed the visual cone into a cone of radiation in order to accommodate ray geometry to intromissionist physics and psychology. To characterize Alhacen's theory of light radiation as revolutionary or modern is thus to wrench it from its proper historical context.

The same holds for Alhacen's methodology. It may look modern because of its strong empirical bias and reliance on controlled experiments, but Ptolemy's approach was no less empirical, and it, too, was based on controlled experiments. In addition, Alhacen's two most modern-looking experiments are based on physically unobtainable precision in equipment design and observation, so we are left to doubt that he actually carried them out as described—except, of course, in his mind. And these experiments were not new in conception. They were clearly based on equivalent ones in Ptolemy's *Optics*, although Alhacen had to reformulate them in significant and creative ways to accommodate the testing of light rays rather than visual rays. Then there is the issue of lens theory. We have seen that between them, Ibn Sahl and Alhacen brought the analysis of lenses to a level of sophistication not reached in Europe before the seventeenth century. In both cases, however, the relevant works were never translated, so they could contribute nothing to the development of lens theory in the Latin West. The one proposition in Alhacen's *De aspectibus* that might have influenced that development was diagrammatically misrepresented so as to mask rather than reveal its lens theoretical implications. In his Latin incarnation, therefore, Alhacen contributed little or nothing, beyond perhaps confusion, to the evolution of lens theory in late medieval and early modern Europe.

In light of these and many other points discussed in this chapter, it should be evident by now that Alhacen's theory of light and sight in the *De aspectibus* marks neither a radical turn toward modernity nor a definitive break with its

74. See ibid., xcvii–civ.

Ptolemaic past. That Ptolemy in fact provided the analytic template for Alhacen is abundantly clear from the close parallels between the two accounts and the myriad points at which they converge. Instead of overturning it, then, Alhacen perfected Ptolemaic optics, and he did so by following the basic structure of Ptolemy's analysis while systematically tightening it, building on ideas and approaches Ptolemy suggested, and carrying many of those ideas to completion. Alhacen's method for determining the points of reflection in curved mirrors, for instance, caps Ptolemy's analysis of concave spherical mirrors by resolving a problem left dangling in that analysis. Furthermore, in resolving that problem, Alhacen depended on a particular analytic device Ptolemy developed. This is not to deny that Alhacen's account is substantially different from Ptolemy's, only to say that for the most part the difference is one of degree, not of kind. Alhacen's account may be more systematic, more comprehensive, more detailed, more mathematically rigorous, and more empirically elaborate, but it is still deeply rooted in Ptolemy's account.

To contend that his account is modeled after Ptolemy's is not to denigrate Alhacen's achievement or to accuse him of slavish unoriginality. On the whole, however, his originality lies less in the structure or the constituent elements of his analysis than in the way he reworked them into a remarkably coherent whole. Especially salient in this regard is his implicit use of faculty psychology and the increasingly abstract formal representations associated with those faculties to explain the various levels of perceptual and conceptual processing in the brain. To be sure, there are points at which Alhacen broke new theoretical ground in an effort to accommodate his account of visual perception to the cone of light radiation. An outstanding example is his account of distance perception. Ptolemy relied on the visual faculty's innate sense of the length of the visual rays linking the center of sight to outlying objects to explain our apprehension of moderate distances. Denied this option, Alhacen based his account on inference from bodily measurements correlated to memorized spans of ground, thus shifting the focus of explanation from immediate sense judgment to experientially mediated psychological judgment. At times, though, Alhacen's forays into theoretical innovation lead to serious inconsistencies, as is certainly the case with his account of peripheral vision, which clashes with the account of lenticular image selection laid out with such care in the first two books of the *De aspectibus*.

Though not without problems, Alhacen's theory of visual perception nonetheless represents a masterful reconciliation of ray theory and natural philosophy, which is to say that Alhacen achieved what he set out to do in writing the *De aspectibus*. More important, though, the resulting visual model has considerable explanatory power at both the theoretical and empirical levels. Not only does it explain how vision yields mental depictions of external reality under normative conditions, but it is also designed to insure that these depictions are faithful. More often than not, however, vision does not occur under normative conditions, so the resulting appearances are deceptive. The object may appear displaced from its actual location, it may look magnified, or it may look inverted. The primary goal of Alhacen's visual analysis is thus to rectify these visual errors intellectually in order to restore trust in our ability to "see" the world as it actually is despite misleading appearances. Alhacen, in short, was quite literally saving the appearances. In the next two chapters we will see that a measure of his success in this endeavor is the canonical status enjoyed by the *De aspectibus* as an optical source in the Latin West from the mid-thirteenth to the early seventeenth century.

CHAPTER 6

Developments in the Medieval Latin West

Despite the efforts of numerous scholars since at least the 1920s to rehabilitate it, the medieval Latin West between roughly 500 and 1200 (and even beyond) continues to be viewed at the popular level as culturally and intellectually retrograde.[1] Ripped from its classical roots by the barbarian invasions and the ensuing fall of Rome, so the story goes, the Latin West sank into the cultural and intellectual mire of the Dark Ages. There it languished, in brutish ignorance, until the sudden, fortuitous recovery of classical and Arabic learning through Latin translations of Arabic texts during the twelfth and thirteenth centuries. Only then, with the introduction of these new and newly recovered sources, was the Latin West able to accomplish the intellectual revival of the thirteenth and succeeding centuries. Arab scholars, in short, transformed the Latin West by awakening it from its intellectual stupor.[2] But what exactly did they transform, and how did they transform it? Even more fundamental is the question of just how dark the Dark Ages really were. A bit of historical background will help in sorting out these questions.

What is commonly referred to as the Latin "West," in contradistinction to

---

1. For an early, influential work of rehabilitation, see Haskins, *Renaissance of the Twelfth Century*.

2. See, e.g., Jonathan Lyons, *The House of Wisdom: How the Arabs Transformed Western Civilization* (New York: Bloomsbury, 2009); see also Jim al-Khalili, *The House of Wisdom: How Arabic Science Saved Ancient Knowledge and Gave Us the Renaissance* (London: Penguin, 2011). At the opposite extreme stands Sylvain Gouguenheim, who argues in his highly tendentious and controversial *Aristote au Mont-Saint-Michel* that the Arabs contributed virtually nothing to the intellectual revival of medieval Europe.

the Greek, and eventually Arabic, "East," evolved from emperor Diocletian's late third-century administrative division of the Roman Empire into western and eastern halves along a rough north–south line separating the Balkans from Italy and passing through the eastern reaches of the Gulf of Sidra in Libya. Although politically motivated, this partition reflected a deeper cultural and economic rift between the two halves. Except perhaps for Italy, the Mediterranean stretch of Gaul, and a sliver of North Africa centered on Carthage, the western half was by far the less urbanized, commercially developed, and culturally sophisticated of the two. Predictably enough, the more advanced eastern half soon became the dominant partner. With Constantine I's official establishment in 330 of a "second Rome" at self-named Constantinople, the fix was in. The two halves would drift ever further apart religiously, politically, and culturally, the western half being left increasingly to its own devices under nominal eastern control and protection. As the rift widened, the western half became progressively more Latin in its linguistic and cultural orientation, a tendency accelerated by religious and political tensions between East and West.[3] Even before the barbarian "invasions" of the fifth and sixth centuries, therefore, the political, religious, and cultural ties between East and West were fraying.

By the early eighth century, those ties were frayed to the point of breaking, and despite subsequent efforts at rapprochement, the two halves were destined to go their separate ways. The eastern or "Byzantine" portion had shrunk alarmingly after an Arab siege of Constantinople was barely staved off in 718. Now largely confined to western and central Anatolia and the Aegean region, this segment of the empire maintained precarious and spotty footholds in Italy, Sicily, Corsica, and Sardinia. Meantime, the western half of the empire had become politically fragmented as various Germanic groups established whole or partial control in Britain, Gaul, Germany, Italy, and the Iberian Peninsula during the sixth and seventh centuries. By the 720s, moreover, all but the topmost tier of the Iberian Peninsula had been lost to Muslim armies—as had the whole of Northern Africa—and Arab forces were putting pressure on the southern reaches of Frankish Gaul across the Pyrenees. Arab raiders were also engaged in a process of island hopping in the Mediterranean that would culminate with the conquest of virtually all of Sicily by 902. Consequently, the Latin West's political and cultural center of gravity was inexorably shifting from the Mediterranean toward northwestern Europe.

Geographically isolated, Europe was also intellectually isolated. Once ex-

3. See, e.g., Judith Herrin, *The Formation of Christendom* (Princeton, NJ: Princeton University Press, 1987).

pected of any learned Roman, competency in Greek was all but lost among the educated elite of eighth-century Europe, a tendency already evident in Saint Augustine's failure to master that language.[4] Direct contact with Greek and Greco-Roman philosophical and scientific sources was thus essentially lost, and only a handful of sources were available in Latin translation. Especially important in this regard is Boethius (ca. 475–ca. 525), who not only rendered several of Aristotle's logical works into Latin but also translated Porphyry's *Eisagoge* (or "Introduction" to Aristotle's *Categories*) and went on to compose several commentaries in the late Platonist style of his day. Plato himself was represented in Calcidius's fourth-century translation of most of the first half of the *Timaeus*, and Martianus Capella's early fifth-century encyclopedic treatment of the liberal arts, *On the Marriage of Philology and Mercury*, was a mine of information, and misinformation, about classical mathematics, geography, astronomy, and music theory. Also, as we saw in chapter 4, Saint Augustine provided an important connection to the classical philosophical tradition, albeit shaped to Christian purposes.

Linguistic limitations and lack of sources were not the only constraints on intellectual development in early medieval Europe. The adaptation of the liberal arts curriculum to Christian education during the sixth and seventh centuries and the increasing confinement of that education in monastic (and to some extent cathedral) schools also played a significant part. The goal of monastic education, after all, was to achieve spiritual wisdom grounded in supernature, not demonstrative knowledge grounded in nature, so its focus was on scriptural literacy rather than broad or deep learning in philosophy or science. Furthermore, classical and late antique philosophical learning in its Platonized form was not a mere intellectual pursuit. It had strong religious overtones, and by Christian lights those overtones and all their moral implications were pagan. If that learning was to be put to proper Christian uses, it had to be approached with extreme circumspection, if not avoided entirely.

While such an environment may not have been conducive to freewheeling rationalism or unfettered intellectual curiosity, monasteries were nonetheless crucial to the preservation of classical learning in its etiolated, early medieval form. With the exemplary encouragement of Cassiodorus (ca. 485–585), reproducing and disseminating manuscripts became an integral part of monastic life through the scriptorium. As a result, monastic libraries slowly expanded to include both secular and religious texts. Even if only nominally based on the

---

4. As O'Donnell puts it colorfully: "A word on Augustine's Greek: pathetic," *Augustine*, 126.

classical liberal arts curriculum, moreover, monastic education accorded some value to the study of natural philosophy as a means toward acquiring spiritual wisdom. And as time wore on, increasing emphasis was placed on mastery of logic, especially during the Carolingian Renaissance of the late eighth to mid-ninth century. Consequently, the study of nature was taken more seriously as monastic scholars mined the sources they had at hand with ever-greater intensity and critical acumen. Far from being a cultural and intellectual waste-land, as it is often popularly portrayed, Europe between the early ninth and early twelfth centuries experienced a remarkable revival of learning that was all the more noteworthy for the relatively meager set of classical and late antique sources upon which it was based.

This revival was quickened by the recovery of classical philosophical and scientific sources, along with their late antique and Arabic offshoots, during the twelfth and thirteenth centuries. Made possible at the outset by the recon-quests of the northern half of Spain, of the entire island of Sicily, and of the Levantine littoral toward the very end of the eleventh century, this recovery was based on the resulting availability of Arabic texts in significant numbers and the equal availability of interested European scholars to translate them or have them translated into Latin. In short, Latin scholars were already sophisti-cated enough not only to benefit from Arab learning but also to seek it out. The establishment of the Latin Empire of Constantinople after the Fourth Crusade in 1204 provided the impetus for a second wave of translations, this time directly from Greek rather than indirectly from Arabic. Thus by around 1270, pretty much the entire corpus of Greek and Greco-Roman philosophical and scientific works known to us today, Plato excepted, had been rendered into Latin. And so had a wide variety of Arabic commentaries and indepen-dent scientific works.

In optics, as in virtually every other field of learning, the influx of newly translated sources during the twelfth and thirteenth centuries had a profound impact in the Latin West, and it is the purpose of this chapter to evaluate that impact and its repercussions between roughly 1150 and 1350. Accordingly, in section 1 we will look briefly and selectively at the evolution of visual theory from around 850 to 1150, just before the translation movement began in ear-nest. We will then turn in section 2 to a discussion of that movement, with a particular focus on sources relevant to visual theory. In the third section we will look at how those sources, especially Avicenna, contributed to the develop-ment of a particular model of perception and cognition, and in section 4, we will examine how geometrical optics was fully incorporated into that model on the basis, primarily, of Alhacen's *De aspectibus*.

1. BACKGROUND TO THE TRANSLATION MOVEMENT

The classical and late antique sources for visual theory available in Latin before 1150 are mostly secondary and encyclopedic. As such they deal with light and vision in a nonsystematic way, the result being that the accounts they provide tend to be somewhat vague and suggestive. Lacking entirely, as far as we know, were any texts devoted explicitly to the mathematical analysis of radiation, whether of visual flux or of light. Among the sources that were available, Calcidius's translation of about a third of the *Timaeus* (from 17a to 53c) is especially important because it provided early medieval European thinkers with a rare glimpse at a primary classical source, including the complete account of vision (45b–47c). No less important than the translation itself is the accompanying commentary, in which Calcidius discusses various alternative theories before plumping for Plato's model of cooperation between external fire/light and gentle fire/light emitted by the eye. As an example of intromissionism, he describes the atomist theory of material simulacra passing from objects into the eye. On the other hand, he continues, the Stoics posit an extramission of "innate spirit" (*nativus spiritus*) that takes form as a cone with its vertex inside the eye. The visual ray theory he attributes to the "geometers" (*geometrae*) and, oddly enough, the Peripatetics (Aristotelians).[5] Later on, citing medical authorities (*medici*) and natural philosophers (*physici*), he offers a brief description of the two narrow nerves that link the eyes to the brain, which is where the principal power of the soul resides and from which "natural spirit" (*naturalis spiritus*) passes into the eyes. Although this description has an obvious Galenic ring, Galen does not appear in Calcidius's list of supporting authorities.[6]

Aside from Calcidius, several other Latin sources discuss classical theories of light and sight. For instance, in a wide-ranging discussion of meteorological phenomena in his *Natural Questions*, Seneca (ca. 4 BC–AD 65) offers a fairly clear and detailed description of Aristotle's account of the rainbow in book three of the *Meteorology*. As we saw in chapter 2 (section 1), this account is based on the reflection of visual rays and the resulting perception of solar color

---

5. For Calcidius's translation of *Timaeus* 45b–47c (on vision), see Jan H. Waszink, ed., Timaeus *a Calcidio Translatus* (London: Warburg Institute, 1962), 41–44. For Calcidius's discussion of alternative theories of vision, see ibid., 248–57, especially 248–51.

6. For Calcidius's discussion of the anatomy of the visual system, see ibid., 257. The authorities Calcidius cites in this section include Alcmaeon, Callisthenes, and Herophilus, but not Galen (cf. Lindberg, *Theories*, 89).

images in individual droplets of the rain cloud.[7] Seneca, however, agrees with Posidonius that the cloud actually acts like a concave spherical mirror, a point that "cannot be proved without the help of geometers."[8] Unfortunately, Seneca fails to provide the geometrical proof, but he does advert to image distortion in such mirrors and argues on this basis that "the rainbow is a rough representation (*effigia*) of the sun, not an exact resemblance."[9]

In the seventh book of his *Saturnalia*, Macrobius (fl. ca. 430?) opens his account of vision by describing the atomist theory of simulacra, which he imputes specifically to Epicurus and Democritus.[10] Rejecting this theory on the grounds that if it were so we would see the back of our head, not our face, when looking into a mirror, he offers his own opinion that sight is due to "an innate light [flashing] from the pupil in a straight line." This emission, he continues, "begins from a slender base and ends by becoming broader, [which is] why the eye sees to the depths of the heavens while looking through the teeniest hole [that is, the pupil]."[11] Sight itself, he concludes, is completed "when the light that proceeds from us through clear air strikes an object," but in order for that object to be judged for what it is, "the eyes' perception reports back to our reason the appearance (*visam speciem*) of the thing seen, a memory is summoned up, [and] reason performs the recognition."[12] The visual act thus involves three faculties: perception, reason, and memory. That is why "god the craftsman placed all the senses in the head, which is to say around the seat of reason."[13]

Macrobius's distinction between the physical and psychic stages of visual perception is clearly compatible with Augustine's distinction between exterior and interior senses, and his recourse to memory for the act of visual recognition is reminiscent in a very loose way of Alhacen's account of perception through recognition (*comprehensio per scientiam antecedentem*). Macrobius's mention of Epicurus and Democritus raises an important point about all these classical Latin sources. Not only do they provide the author's own views, but they also provide hints about the views of their predecessors. Subsequent readers were

7. For Seneca on Aristotle's account of the rainbow, see Thomas H. Corcoran, trans., *Seneca VII*: Naturales Quaestiones I (Cambridge, MA: Harvard University Press, 1971), 35–37.

8. Ibid., 53.

9. Ibid.

10. See Robert A. Kaster, trans., *Macrobius*: Saturnalia Books 6–7 (Cambridge, MA: Harvard University Press, 2011), 277.

11. Ibid., 281.

12. Ibid., 283.

13. Ibid., 285.

thus able to piece together a vague understanding of such things as Plato's theory of the soul in the *Republic*, Galenic physiology, and Stoic pneumatic theory without access to the actual sources that spell those theories out. In short, early medieval European thinkers had a piecemeal and sometimes confused grasp of the classical tradition through doxographical allusions.[14]

All these late antique Latin sources, Augustine included, contain the elements of a relatively coherent, though somewhat sketchy, model of visual perception with strong late Platonist leanings. The visual act starts with the emission of luminous flux from the center of the eye along straight lines within a cone of radiation. Originating in the brain's innate spirit and passing to the eye through the hollow optic nerves, this luminous flux meets with external light to put us into visual contact with external objects. The visual information sent back to the eye through the cone of flux provides representations, or species, of those objects, and those representations, in turn, are remanded to the seat of reason in the brain, where they are cognitively judged and stored in memory. Clear traces of this model are evident in the account of visual perception Isidore of Seville (ca. 560–636) gave in the eleventh book of his encyclopedic *Etymologies*, which was considered an authoritative source on "natural questions" throughout the Middle Ages. Vision, he contends, is said to be due either to an aitherial light outside the eye or to a luminous spirit passing from the brain to the eye through narrow channels (presumably the optic nerves). This spiritual substance continues through the tunics of the eye into the air, where it mingles with its like (*similis materia*) to cause vision. The visual faculty (*vis videndi*) is in the pupil (*pupilla*) at the center of the eye, where small images (*imagines*) are seen, presumably at (and perhaps by) the pupil itself. Moreover, the eyes are closest of all the senses to the brain, which is the seat of the soul, so sight is the superior sense. The soul, for its part, is vested with a set of faculties ranging from intellect (*mens*) at the top, through memory and reason, to sense at the bottom, intellect serving as the soul's "head" or "eye."[15]

That these sources, Augustine and Calcidius in particular, were not just mindlessly followed by subsequent European thinkers is evident in the *Periphyseon* (known commonly as *On the Division of Nature*) of the tautologically

---

14. See, e.g., John Marenbon, "Platonism—A Doxographic Approach: The Early Middle Ages," in *The Platonic Tradition in the Middle Ages: A Doxographic Approach*, ed. Stephen Gersh and Maarten Hoenen (Berlin: Walter de Gruyter, 2002), 67–89.

15. See Stephen A. Barney and others, trans., *The* Etymologies *of Isidore of Seville* (Cambridge: Cambridge University Press, 2006), 231–33. For the Latin, see Wallace M. Lindsay, trans., *Isidori Hispalensis Episcopi*: Etymologiarum sive Originum, vol. 2 (Oxford: Clarendon Press, 1911), XI, 10–38.

named John Scottus Eriugena ("John the Irishman, born in Ireland"), whose scholarly career spanned the period from roughly 840 to 870.[16] As a thinker of his time, at the tail end of the Carolingian Renaissance, Eriugena was remarkable if only because he knew Greek and was familiar with the work of several late antique Greek theologians. Particularly influential on Eriugena's thought was his commission in 860 to translate a set of works attributed during the Middle Ages and Renaissance to Dionysius the Areopagite, whose conversion to Christianity by Saint Paul is mentioned in Acts 17:34. Now dated to around 500, the author of these works (restyled Pseudo-Dionysius) was steeped in the late Platonism of his day and, in an effort to reconcile it with Christianity, fell back upon apophatic (or negative) theology as a means of underscoring the utter transcendence of God. In this vein he contrasts the transcendental darkness of God and the immanent brightness of His creative power, which is the metaphysical analogue to the illuminating power of physical light. Accordingly, the created universe is a theophany, a visible and ultimately intelligible manifestation of God as causal, determinative force.[17]

Such ideas inform Eriugena's extensive account of creation in the *Periphyseon*, an account in which mankind figures prominently. Not surprisingly, his discussion, which occupies five long books, is shot through with allusions to light and visual perception. In typically late Platonist fashion, Eriugena distinguishes the irrational from the rational nature of man, the former grounded in his bodily appetites and needs, the latter in his perceptual and cognitive powers, sensation holding an intermediate place between the two. The rational side of man consists of sense—taken as Augustine's interior sense (*sensus interior*)—reason (*ratio*), and intellect (*intellectus*). The external senses, though "called fivefold," are actually simple and unitary and as such serve as "a gatekeeper and messenger" that "announces (*annuntiet*) to the presiding interior sense whatever it lets in from the outside." What is announced consists of qualitative and quantitative likenesses (*similitudines*) of sensible objects that take form as images (*imagines*) in the sense itself and are then transformed into phantasies (*phantasiae*) in the interior sense.[18] Through these phantasies the

16. On Eriugena's life and works, see Deirdre Carabine, *John Scottus Eriugena* (New York: Oxford University Press, 2000).

17. See Paul Rorem, *Pseudo Dionysius* (New York: Oxford University Press, 1993).

18. See *Periphyseon*, 569B20–570A9 and 573A25–573C15, in L. P. Sheldon-Williams, ed. and trans., *Iohannis Scotti Euriugenae* Periphyseon (De Divisione Naturae) *Liber Secundus* (Dublin: Dublin Institute for Advanced Studies, 1972), 98–101 and 106–9. According to Eriugena in *Periyphyseon*, 659B26–659D9, a phantasy (*phantasia*) is the internal representation that comes directly from the senses, whereas its representation in memory is a phantasm (*phan-*

rational faculty "investigates the 'reasons' of all things . . . apprehended by the intelligence or the sense" under the governance of intellect, whose proper goal is to contemplate "God . . . and what lies in Him and subsists about Him, according as it is allowed to ascend."[19] This ascent back toward the font of both creation and understanding is inherent in the human soul but badly hampered by mankind's fall from grace, so following it to completion is "generally in potency only," although a blessed few are capable of attaining it "in act."[20] As one of the gatekeeping senses, "sight is a kind of light which first rises out of the fire in the heart" and passes upward into the brain. From there it flows through "certain channels" to the "pupils of the eye, whence in a very swift rush it leaps forth like the rays of the Sun . . . [and] reaches out to grasp the coloured forms of visible things." In this way the soul acts through the eyes and "in a potential sense . . . is present to receive the phantasies which are everywhere formed in the instruments of its senses."[21]

Eriugena makes explicit reference to Augustine and the *Timaeus* in the course of his ruminations about visual perception and cognition, and the same two sources, Augustine in particular, seem to be at play in a brief account of vision in chapter 6 of Saint Anselm of Canterbury's *De veritate* ("On Truth"), which was written sometime between 1080 and 1086. Anselm's point in this account is that sight is not intrinsically deceptive. As an example he notes that when we look at objects through glass of a given color, they appear to take on that color. What happens in this case, he explains, is that when it "passes through" (*transit per*) the glass, sight (*visus*) takes on its color, adds it to the color of the objects beyond the glass, and then "reports (*renunciat*) the [color] it first receives, either alone or with [the] other." The deception is therefore not due to the exterior sense but to the interior sense that "imputes its mistake

tasma) and thus an "image of an image"; see Sheldon-Williams, ed. and trans., *Iohannis Scotti Euriugenae* Periphyseon (De Divisione Naturae) *Liber Tertius* (Dublin: Dublin Institute for Advanced Studies, 1981), 118–21.

19. *Periphyseon*, 825B19–25, in Edouard Jeauneau, ed., and John J. O'Meara, trans., *Iohannis Scotti Euriugenae* Periphyseon (De Divisione Naturae) *Liber Quartus* (Dublin: Dublin Institute for Advanced Studies, 1995), 196–97.

20. *Periphyseon*, 778C35–5, in ibid., 90–91.

21. *Periphyseon*, 730C10–731B2, in Sheldon-Williams, *Eriugenae Liber Tertius*, 280–83. Eriugena's reference to sight as "a kind of light which first rises out of the fire in the heart" may refer to Galen's theory that the psychic pneuma pervading the brain and passing into the eyes is distilled from the blood-borne vital pneuma flowing from the heart into the rete mirabile (see chapter 2, section 3).

to the exterior sense." The same holds when we are deceived into thinking that a stick is broken when it is partially immersed in water or believing that the image of ourselves we see in a mirror is actually in it. These are all due to misjudgments of the interior sense, not false reporting by the exterior one.[22] Such misjudgments are rectified by the intellect in light of Truth, which is God Himself, and as a source of rectification, reason is a means of bringing us to the recognition of that Truth. This is wisdom.

Although there is nothing in this account to indicate that Anselm drew directly upon Calcidius, he wrote *De veritate* at a time when interest in Calcidius's translation of, and commentary on, the *Timaeus* was surging. As an indicator of this surge and its scope, Anna Somfai points to the proliferation of manuscripts containing the translation and commentary, either together or separately, over the eleventh century—twenty-three versus only five for the preceding two centuries—and the number jumps to fifty-three for the twelfth century.[23] As far as visual theory is concerned, the effects of this growing interest in the *Timaeus* can be discerned in the works of several thinkers active during the first half of the twelfth century. A prime example is Bernard of Chartres, a young contemporary of Anselm's and author of the *Glosae super Platonem* ("Glosses on [the *Timaeus* of] Plato"). Composed some time between 1100 and 1115 and based on Calcidius's translation, this work contains a fairly extensive discussion of vision and its utility to philosophy. Not surprisingly, Bernard imputes sight to the noncaustic "internal fire (*ignis interior*) . . . of the soul [that] passes through the eyes" from the pupil (*pupilla*) at the center. Mingling with its external counterpart, it extends out to a given object, grasps its form (*forma*), and brings it back to the outer surface of the eye, where it "adheres" (*hereo*) before passing into the interior to "announce" (*renuntio*) itself to the soul. All of this occurs in the head, because that is where the highest cognitive faculty of mankind, reason or intellect (*ratio* or *rationalis intellec-*

22. *On Truth*, trans. Ralph McInerny, in *Anselm of Canterbury: The Major Works*, ed. Brian Davies and Gillian Evans (Oxford: Oxford University Press, 1998), 158–59. For the Latin, see Franciscus Schmitt, *Sancti Anselmi, Cantuarensis Archiepiscopi, Opera Omnia*, vol. 1 (Edinburgh: Thomas Nelson and Sons, 1946), 183–85.

23. Anna Somfai, "The Eleventh-Century Shift in the Reception of Plato's 'Timaeus' and Calcidius's 'Commentary,'" *Journal of the Warburg and Courtauld Institutes* 65 (2002): 1–21; see especially the chart on page 8. It should be noted that a Latin translation of *Timaeus*, 27d–47b by Cicero was also available during the Middle Ages, but it is represented by a total of only sixteen manuscripts, as opposed to 156 for Calcidius's translation and/or commentary.

tus), resides.[24] Our dependence on sensation for cognition is due to the soul's being incorporated and thus losing its capacity for pure reason (*pura ratio*).[25]

In his *Cosmographia*, which dates to 1147, Bernard Silvestris follows much the same line of reasoning but with some important specifications. The head, he claims, is the noblest part of the body because it encases the brain and is thus the seat of understanding. That is why the sensory organs are "set . . . close about the palace of the head [so] that judging intellect might maintain close contact with the messenger senses." As one of those messengers, sight is due to the internal light of the eye, which originates in the brain and radiates out to mingle with sunlight. The result is a "beam" (*acumen*) of visual flux that "applies itself to the forms of things and makes a careful record of them." This record is presumably sent back to the eye in the form of a representative image. Polished and round so that such images (*simulacra*) can adhere (*hereo*) more easily to its surface, the eye is protected by a "sevenfold jacket" of tunics and connected to the brain by a nerve that "illumines [it] with its light [drawn] from the brain." The brain itself is soft and fluid in substance so that it can let the "images (*ymagines)* of things impress (*insidere*) themselves more easily upon it." These images are the stuff of the imaginative faculty (*fantasia*), which receives "the forms (*formas*) of things and reports (*renunciet*) all that it beheld to the reason." Equipoised between imagination and memory, reason is so placed in order "to impose its firm judgment" on their operations. This tripartition of the perceptual and rational faculties is reflected anatomically in the ventricular structure of the brain, the anterior cell occupied by imagination, the central one by reason, and the occipital one by memory.[26]

24. *Glosae*, 7.146–430, in *The* Glosae super Platonem *of Bernard of Chartres*, ed. Paul Edward Dutton (Toronto: Pontifical Institute of Mediaeval Studies, 1991), 206–16; see especially 7.140–45, 175–80, 256–57, 341–43, and 368–70.

25. Ibid., 7.422–24, 216.

26. See Winthrop Wetherebee, trans., *The* Cosmographia *of Bernardus Silvestris* (New York: Columbia University Press, 1973), 121–24. For the Latin text, see Peter Dronke, ed., *Bernardus Silvestris* Cosmographia (Leiden: Brill, 1978), 149–51. Bernard's use of the Latin term *insidere*—to impress or engrave—finds an interesting echo in Hugh of Saint Victor's *Didascalicon philosophiae*, which was written some twenty years before the *Cosmographia*: "when a coiner imprints (*imprimit*) a figure upon metal, the metal . . . begins to represent a different thing. . . . It is in this way that the mind, imprinted with the likenesses of all things, is said to *be* like all things and to receive its composition from all things and to contain them not as actual components, or formally, but virtually and potentially"; Jeremy Taylor, trans., *The* Didascalicon *of Hugh of St. Victor* (New York: Columbia University Press, 1961), 47; for the Latin see Charles Buttimer, *Hugonis de Sancto Victore* Didascalicon (Washington, DC: Catholic University Press, 1939), 5–6.

Bernard Silvestris's allusion to the seven-jacketed eye and three-chambered brain is striking because it may reflect an initial wavelet of translations from Arabic to Latin Constantine the African carried out in the late eleventh century. A Muslim of North African origin, Constantine first visited Salerno as a merchant and eventually returned to settle there. Having learned both Latin and "Italian," he converted to Christianity and eventually fetched up at the abbey of Monte Cassino sometime after 1080. Between then and his death, possibly in 1087, he translated a variety of medical texts from Arabic, some of them Galen's and all of them of Galenic inspiration.[27] Constantine was thus instrumental in the dissemination of Galenic medical principles, both theoretical and practical, into medieval Europe.

Among his translated works, two are especially significant for our purposes. The first, variously titled *De oculis* or *Liber oculorum*, is a rendering of Ḥunayn ibn 'Isḥāq's *Ten Treatises on the Eye*. As we saw earlier, Ḥunayn devotes the first three sections of this work to a detailed anatomical and physiological description of the visual system from eye to three-chambered brain (see chapter 4, section 4). We also saw that his anatomical account of the eye, with its seven tunics and three humors, is faithful to Galen in all respects, except that Ḥunayn places the lens at the eye's center rather than toward the front. The second of Constantine's translations at issue here is the *Liber pantegni* or *Liber regalis*, a massive medical compendium based on the *Kitāb al-Malikī* ("Royal Book") of the Persian physician 'Alī ibn al-'Abbās al-Majūsī (d. after 984), known as Haly Abbas in the Latin West. In the course of an extensive account of the human body and its anatomical features in the "theoretical" half of this work, al-Majūsī hews closely to Ḥunayn in his description of the eye and all its components, including the centralized lens and X-formed optic nerves that connect the two eyes with the three-chambered brain.[28] Along with

27. See Herbert Bloch, *Monte Cassino in the Middle Ages* (Cambridge, MA: Harvard University Press, 1986), 98–110. By Bloch's count, Constantine, perhaps with the help of two close colleagues, Afflacius and Atto, was responsible for translating or compiling at least twenty-six works; see ibid., 130–34. In addition, Alfanus, a close associate of Constantine at Monte Cassino before becoming archbishop of Salerno, translated Nemesius of Emesa's *On the Nature of Man* from Greek (see chapter 4, note 48).

28. On al-Majūsī's life and works, see Françoise Micheau, "'Alī ibn al-'Abbās al-Maǧūsī et son milieu," in *Constantine the African and 'Alī ibn al-'Abbās al-Maǧūsī: The* Pantegni *and Related Texts*, ed. Charles Burnett and Danielle Jacquart (Leiden: Brill, 1994), 1–15. On al-Majūsī's description of the eye in the *Pantegni*, see Gül Russell, "The Anatomy of the Eye in 'Alī ibn al-'Abbās al-Maǧūsī: A Textbook Case," ibid., 247–65. It is worth mentioning Constantine's translation of Ḥunayn's *Isagoge*. Attributed to Johannitius and included in the *Articella*, an

subsequent translations of other medical sources, Avicenna's *Canon* in particular, these works formed the basis for ophthalmological theory and practice throughout the remainder of the Middle Ages and well into the Renaissance, a point to which we will return later.

Whether in fact Bernard Silvestris knew, or even knew of, either the *Liber de oculis* or the *Pantegni* is an open question; there is nothing conclusive to show that he did. On the other hand, in the *Dragmaticon philosophiae*, which was written at about the same time as Bernard's *Cosmographia*, William of Conches not only draws upon the *Pantegni* in his account of vision but also cites it explicitly as an authoritative source.[29] Accordingly, he describes the eye as "a spherical disk and clear substance . . . consisting of three humors and seven membranes [that is somewhat] flat in the front surface so that it can better receive the shapes and colors of things." Pervading the brain, meantime, "is a certain airy and subtle substance [that] is termed fire by Plato" and is a distillate of "vapor" (*fumus humidus*) passed from the heart to an "extremely fine net [that is, the rete mirabile] that surrounds the brain." This refined substance passes to each eye through a nerve that originates in the brain and then bifurcates to form an inverted V. Upon reaching the eye and its pupil at the center, the fiery substance conveyed by the optic nerve exits the eye to "join itself to [the] brightness" outside and, in combination with it, to extend out to any given external object in the form of a cone (*conus*). "Taking on the shape and color of that object . . . it then returns . . . through the eye to the cell of the imagination [*phantasticam cellulam*]." There "it represents to the soul the shape and color of the object." Located at the forefront of the brain, this is one of three cerebral cells in succession, the middle one being the seat of discursive reason (*logistica*), the occipital one the seat of memory (*memorialis*).[30]

Culminating with William of Conches's *Dragmaticon*, this admittedly selective overview of visual theory, as it evolved in the Latin West to the mid-twelfth century, allows for a couple of generalizations. First, and most obvious, is that all the thinkers discussed to this point assumed that the eye makes visual

---

extremely popular medical compendium during the Middle Ages and Renaissance, this work contains an account of the eye and brain; see Francis Newton, "Constantine the African and Monte Cassino: New Elements and the Text of the *Isagoge*," ibid., 16–47.

29. On William's reliance on the *Pantegni*, see Italo Ronca, "The Influence of the *Pantegni* on William of Conches's *Dragmaticon*," ibid., 266–85.

30. See Matthew Curr and Italo Ronca, trans., *A Dialogue on Natural Philosophy* (Dragmaticon Philosophiae), *William of Conches* (Notre Dame, IN: University of Notre Dame Press, 1997), 150–51, 154–59. For the Latin text, see Italo Ronca, ed., *Guillelmo de Conchis*, Dragmaticon Philosophiae (Turnhout: Brepols, 1997), 233–34, 239–50.

contact with external objects by emitting internal light or fire toward them. In short, they all subscribed to some variant of Platonic or Galenic extramissionism. Less obvious, perhaps, but more important, is that taken as a whole, these thinkers articulated a fairly clear and coherent model of visual perception based on the following assumptions: That visual contact is initiated at the base of a cone whose vertex lies at the pupil located in the eye's center. That formal representations of the shape and color of visible objects affecting the base of this cone are sent back to the eye, whose visual capacity is due to a spiritual medium flowing to it from the brain through the optic nerves. That once received by the eye, these representations are passed back through the humors and optic nerves to the anterior ventricle of the brain, where they are impressed on the imaginative faculty as images, species, or similitudes of their generating objects. That these similitudes are cognitively processed by the faculty of discursive reason in the brain's middle ventricle. That the results of this cognitive processing are remanded to the occipital ventricle for mnemonic storage. And, finally, that the ultimate cognition of what we see and perceive depends upon a supervening intellect that provides the light of understanding through which we are enabled to "view" the underlying, formal reality of objective appearances.

In its basic structure, this model of visual perception and cognition is remarkably similar to the one that evolved among Arabic thinkers between roughly 850 and 1000 and that came to a head with Avicenna's internal senses account. The main difference is that unlike Avicenna, the Latins clung to extramissionism. That the two models should bear such close resemblance may seem surprising in view of the disparate lines along which they developed, Avicenna's model and its evolution being based on a host of Aristotelian texts and commentaries that were unavailable to Latin thinkers before the second half of the twelfth century. Nonetheless, despite the separate linguistic and textual lines along which they developed, both models were ultimately rooted in the same late antique tradition of Platonized Aristotelian psychology and epistemology. In essence, they evolved in parallel, so it is no coincidence that they came to similar ends. This point is significant because it means that when the relevant texts, including Avicenna's *Healing* and Alhacen's *De aspectibus*, entered the European ambit in Latin translation during the second half of the twelfth century, Latin thinkers were fully prepared to understand and appreciate them because they had the appropriate conceptual and terminological framework within which to do so. Accordingly, the recovery of these texts did not so much redirect or reshape medieval European thought about visual perception as clarify and crystallize it.

2. THE TRANSLATION MOVEMENT AND
THE INROADS OF ARISTOTELIANISM

In the previous section we adverted to a wavelet of Arabo-Latin translations carried out toward the end of the eleventh century under the aegis of Constantine the African. This ripple was followed some fifty years later by a mounting wave that emanated primarily from the city of Toledo, which had been taken from the Muslims in 1085 by Alfonso VI of Castile. A political windfall for the kingdom of Castile, the capture of Toledo was an intellectual windfall for the rest of Europe because it opened direct access to Arabic writings on science and philosophy that were scattered in collections around the city, many of them eventually winding up in the archepiscopal library. Already aware that the Arabs had much to teach them, thinkers from virtually every corner of Europe were attracted to Toledo and its environs by the opportunity to gain firsthand knowledge of thinkers with whose ideas they were only indirectly acquainted. They may also have been attracted by the patronage of Archbishop Raymond (1126–1152), who encouraged systematic translations by teams that typically consisted of Latin scholars and local Jewish or Mozarabic Christian intermediaries who rendered texts from Arabic into Old Castilian or even Hebrew and then into Latin. The standard was thus set for a translation movement that culminated with Gerard of Cremona (1114–1187), under whom, or under whose supervision, over seventy works were rendered into Latin between the 1150s and his death in 1187.[31]

As far as visual theory is concerned, Aristotelian sources, both primary and derivative, entered into the discussion with two key translations dating from around midcentury or somewhat later. The first was a Greek to Latin rendering of Aristotle's *De anima* by James of Venice, who may have been based in Constantinople and may, in fact, have been Greek, as was Eugene of Sicily, the contemporaneous translator of Ptolemy's *Optics* (see chapter 3, introduction). James's Latin version of the *De anima* was supplemented by a translation of the *De sensu* carried out around the same time by someone as yet unidentified.[32] The second key translation—of the sixth book of Avicenna's *Healing*, with its extensive account of psychology—was accomplished by the Christian-

---

31. On the Toletan translation effort, see Charles Burnett, "The Coherence of the Arabic-Latin Translation Program in Toledo in the Twelfth Century," *Science in Context* 14 (2001): 249–88; see especially 276–81 for a list of translations explicitly attributed to Gerard.

32. See Lorenzo Minio-Paluello, "Jacobus Veneticus Grecus, Canonist and Translator of Aristotle," *Traditio* 8 (1952): 265–304.

Jewish team of Dominicus Gundissalinus (Domingo Gundisalvo/Gonzalo) and Ibn Daud (Avendauth in Latin), both of them working in Toledo not long after the death of Archbishop Raymond, probably under the patronage of his successor, Archbishop John (1152–1166).[33] This was one route by which Avicenna's internal senses model was introduced to European thinkers along with his theory of intentionality. Another was the *Tractatus de anima* attributed to Gundissalinus, which gives a relatively faithful and concise overview of Avicenna's internal senses account of perceptual and conceptual abstraction.[34]

Apparently driven by the poverty of Latin sources underlying his own education in the early twelfth century, Gerard of Cremona (ca. 1114–1187) made his way to Toledo sometime around 1150 in order to rectify the situation. Accordingly, he took a programmatic approach with an emphasis on the translation of scientific texts. Among the works emerging from this program, at least six that have a direct bearing on visual theory can be more or less definitively attributed to Gerard. Four of the six—al-Kindī's *De aspectibus*, Tideus's *De speculis*, Ibn Muʿādh's *De crepusculis* ("On Twilights"), and the first three books of Aristotle's *Meteorology*—are expressly devoted to geometrical optics. Gerard may also have been responsible for Arabo-Latin versions of Euclid's *Optics*, Pseudo-Euclid's *De speculis*, and Alhacen's *De speculis comburentibus*. Also, as we mentioned earlier, there is an outside chance that he was involved in the translation of Alhacen's *De aspectibus*. Meantime, on the "philosophical" side, Gerard added Alexander of Aphrodisias's *On the Senses* and, most significant, Avicenna's *Canon* to the list of sources available in Latin at the end of the twelfth century. It was probably in the later twelfth century, as well, that Greco-Latin versions of Euclid's *Optics* and the Theonine "redaction" of his *Catoptrics* were produced (although not by Gerard), and of course Ptolemy's *Optics* had been rendered from Arabic into Latin by the 1160s.[35]

33. For a critical edition of this work, see Simone van Riet, ed., *Avicenna Latinus*: Liber de anima *seu* Sextus de naturalibus, 2 vols. (Leiden: Brill, 1968–72). See also Marie Thérèse d'Alverny, *Avicenne en Occident* (Paris: Vrin, 1993); and Manuel Alonso, "Notas sobre los traductores Toledanos Domingo Gundisalvo y Juan Hispano," in *Dominicus Gundissalinus (12th c.) and the Transmission of Arabic Philosophical Thought to the West*, ed. Fuat Sezgin (Frankfurt: Institute for the History of Arabic-Islamic Science at the Johann Wolfgang Goethe University, 2000), 23–56.

34. A critical Latin edition can be found in J. T. Muckle, ed., "The Treatise De Anima of Dominicus Gundissalinus," *Medieval Studies* 2 (1940): 23–103; see especially 97–100 on the human intellect's ability to see the underlying intelligibility of sensible things through God's illumination.

35. See Lindberg, *Theories*, 209–13, for a list of translated works explicitly devoted to optics and ophthalmology.

Already fairly extensive, this list expanded over the course of the thirteenth century as more sources were translated or retranslated from both Arabic and Greek. Aristotle's *De anima*, for instance, was retranslated from Arabic sometime before the mid-1230s, perhaps by Michael Scot (1175–1236?), who got his start as a translator in Toledo during the second decade of the thirteenth century. Sometime around 1220 he moved to southern Italy, where he found preferment as a court scholar under emperor Frederick II. It was around this time that he may have translated the *Long Commentary* on Aristotle's *De anima* by Ibn Rushd, or Averroes (1126–1198).[36] Not long after Michael's death, however, European scholars became increasingly aware that the Arabo-Latin versions of Aristotle upon which they relied were at best problematic and at worst defective. An obvious solution to this problem was to translate his works afresh from Greek, and the Latin takeover of Constantinople in 1204 paved the way by opening significant portions of the Greek East to settlement by Roman Catholic clerics. One such was the Flemish Dominican William of Moerbeke (1215?–1286). An intellectual with a flair for mathematics, he may have resided at the Dominican convent in Thebes during the 1250s, where he presumably mastered Greek. We know for certain that by 1260 he was at Nicaea and Thebes and that over the next decade or so he produced an enormous body of translations. Along with a host of mathematical works (Archimedes figuring prominently), these included twenty treatises attributed to Aristotle (the *De anima* and *De sensu* among them), as well as Alexander of Aphrodisias's commentaries on Aristotle's *Meteorology* and *De sensu* and Themistius's paraphrase of the *De anima*.[37]

Spanning a period of just over a century, the recovery of these sources affected Latin scholastic thought about visual perception in stages according not only to when they became available but also to when they were regarded as significant enough to be fully assimilated. In this latter respect, for instance, the works dealing specifically with geometrical optics were of at best second-

36. See Richard C. Taylor, trans., *Averroes (Ibn Rushd) of Cordoba: Long Commentary on the* De Anima *of Aristotle* (New Haven, CT: Yale University Press, 2009), cvii–cviii, and Marie Thérèse d'Alverny, "Translations and Translators," in *Renaissance and Renewal in the Twelfth Century*, ed. Robert Benson and Giles Constable (Cambridge, MA: Harvard University Press, 1982), 421–62, especially 455–57.

37. For a list of Moerbeke's translations, see Edward Grant, ed., *A Source Book in Medieval Science* (Cambridge, MA: Harvard University Press, 1974), 39–41. For a brief discussion of his background as a translator, see Marshall Clagett, "William of Moerbeke: Translator of Archimedes," *Proceedings of the American Philosophical Society* 126 (1982): 356–66. Moerbeke also provided the first Latin translation of Hero of Alexandria's *Catoptrics* (see chapter 2, note 92).

ary interest until around 1230, so we will defer discussion of them to section 4. In the following section, therefore, we will confine ourselves to a brief look at how the inroads of Aristotle's *De anima* and associated works influenced scholastic thought about the psychology of visual perception and cognition. As we carry out that examination, we should bear in mind that these works fit within a much broader span of treatises that run the topical gamut from logic to metaphysics.

3. THE SCHOLASTIC ANALYSIS OF PERCEPTION AND COGNITION

With its emphasis on the abstraction of formal representations and intentions, Avicenna's internal senses model was a key—perhaps *the* key—formative source for scholastic perception theory between 1200 and 1350, and indeed well beyond.[38] During the first half of the thirteenth century, in fact, Avicenna was the primary authority around which various other thinkers, Aristotle, Augustine, and Averroes in particular, revolved as corroborating, sometimes contradicting, sources. As might be expected, therefore, the scholastic analysis of perception, conception, and cognition found its theoretical basis in certain Avicennian ideas. Most obvious in this regard was a general shift from extramissionism to intromissionism and a consequent focus on light radiation instead of visual radiation as the physical basis for sight. Already evident in Gundissalinus's *Tractatus de anima*, the Avicennian basis for this shift is clear in John Blund's discussion of *lux, lumen*, color, and sight in his *Tractatus de anima*, which dates to around 1200. "Now sight," he says, quoting almost verbatim from Gundissalinus, who in turn closely paraphrased Avicenna, "is a faculty arranged in the hollow [optic] nerve for apprehending a form of the similitude of colored bodies that informs the crystalline humor by means of rays making an incoming effect at the surfaces of polished bodies."[39] *Lux* and color, he goes on to say, are "the same in subject" (that is, they are qualities inherent in bodies), but *lux* is "a quality that from its essence is the perfection of transparency." Accordingly, "the Commentator [Avicenna] calls *lux* the perfection of transparency; but he calls *lumen* a passion generated in the

38. See Dag Hasse, *Avicenna's* De anima *in the Latin West* (London: Warburg Institute, 2000).

39. See D. A. Callus and R. W. Hunt, eds., *Johannes Blund*, Tractatus de anima (Oxford: Oxford University Press, 1970), 24, lines 5–8; cf. Muckle, "Gundissalinus's De anima," 68, lines 20–23. Cf. also Riet, *Avicenna Latinus*, vol. 1, 83–84, lines 59–62.

transparent [medium], e.g., air."⁴⁰ As "is held by Aristotle," the proper object of vision is color, which is actuated in illuminated objects and causes an alteration in the eye.⁴¹

Blund is clearly at pains here to preserve Aristotle's notion of light as the actualization, or "color," of transparency while yet allowing it to have an immediate radiative effect, in the form of *lumen*, on opaque, colored objects, as well as on the eye. Consequently, his overall analysis of light is somewhat ambiguous because his distinction between *lux* and *lumen* is slightly blurry. Over time, however, various thinkers tried to sharpen and clarify the distinction. For instance, in the anonymous *De anima et de potenciis eius* ("On the Soul and Its Faculties"), written around 1225, the author claims tout court that *"lux* in a transparent medium . . . is *lumen,"* and some forty years later, in his commentary on the *De anima*, Saint Thomas Aquinas (1225–1274) informs us that "the participation or effect of *lux* in the diaphanous medium is called *lumen,"* and "if it takes place in a straight line to an illuminated body [*lucidum*], it is called a ray."⁴² Saint Thomas thus seems to be interpreting *lumen* as a physical effect propagated rectilinearly through transparent media to opaque objects, which it illumines. In an effort to explain how this effect subsists in the transparent medium, Albertus Magnus (1193?–1280) appeals to spiritual existence (*esse spirituale*), which applies also to luminous color. "Hence," he claims in his paraphrase of Aristotle's *De sensu*, "it is necessary that there be a twofold mediation [in the physical act of sight]; one [mediator] is *lumen*, the agent that acts upon colors to make them spiritual; the other [is] the transparent medium that conveys [the resulting effect to the eye]."⁴³ Once rendered spiritual, in fact, light in the medium is an intentional representation of its luminous source, a point, Albert assures us, that Aristotle makes in the *De anima*.⁴⁴

40. Callus and Hunt, *Blund*, 32, line 21, and 32, line 29 to 33, line 1.

41. Ibid., 26.

42. For the Latin text translated from the *De anima et de potenciis eius*, see René Gauthier, ed., "Le Traité *De anima et de potenciis eius* d'un maître ès Arts (vers 1225)," *Revue des sciences philosophiques at théologiques* 66 (1982): 3–55. The translation from Saint Thomas is adapted from Robert Pasnau, trans., *Thomas Aquinas: A Commentary on Aristotle's* De anima (New Haven, CT: Yale University Press, 1999), 219. For the Latin, see René Gauthier, ed., *Sentencia libri de anima* (Paris: Vrin, 1984), 129, lines 318–20.

43. Translation adapted from Cemil Akdogan, ed. and trans., *Albert's Refutation of the Extramission Theory of Vision and His Defence of the Intromission Theory* (Kuala Lumpur: International Institute of Islamic Thought and Civilization, 1998), 101.

44. Ibid., 67.

That Albert is reading a great deal into Aristotle hardly needs saying. In fact, the imputation of spiritual existence to the light effect and color effect in the transparent medium comes from Averroes.[45] The appeal to intentionality, on the other hand, harks back to Avicenna, although by the time Albert wrote his paraphrase (certainly before 1260), the notion of intentionality had been considerably modified under the influence of Averroes, as well perhaps of Alhacen, whom Albert cites as one of his authorities.[46] Just how much Albert was willing to read into Aristotle comes clear in chapter 4 of his paraphrase, when he concludes his refutation of various ancient theories of sight with the claim that

> Aristotle destroyed all these opinions, saying that what is visible [*visibile*], according to its spiritual and intentional existence, is first produced in the medium and then in the eye and that the species of the object seen is moved to the interior of the eye, where the visual power resides in the crystalline humor, and finally proceeds through the hollow of the optic nerves in the spirit [pervading it], the same species being conveyed to the place of the first sensitive [power], which is the common sense, as he showed in the book *De anima*.[47]

Let us pause for a moment to examine some of the theoretical implications of this passage. First, Albert's reference to the common sense and its anatomical and physiological connection to the eye via the optic nerves indicates his adherence to Avicenna's model of internal senses arrayed in the three ventricles of the brain. In this regard he is typical of most scholastic thinkers during the first seven decades of the thirteenth century, including Saint Thomas Aquinas.[48] There are exceptions, though. In book three of his enormously popular and broad-ranging *De proprietatibus rerum* ("On the Properties of Things"), completed before 1260, Bartholomaeus Anglicus specifies only three psychological faculties under the rubric of "the common or interior sense." The first of these, *ymaginativa*, is located in the anterior ventricle of the brain;

---

45. See, e.g., Taylor, *Averroes Long Commentary*, 214: "color has a twofold being, namely, being in the colored body (this is corporeal being) and being in the transparent (this is spiritual being)."

46. See ibid., 37.

47. Translation adapted from ibid., 43.

48. See, e.g., *Summa theologiae*, 1a, quest. 78, article 4, in Robert Pasnau, trans., *Thomas Aquinas: The Treatise on Human Nature*, 73–77; Saint Thomas argues explicitly against the need for Avicenna's compositive imagination. See also Pasnau, *Thomas Aquinas on Human Nature* (Cambridge: Cambridge University Press, 2002), 191.

the second, *logistica*, in the middle ventricle; and the last, *memorativa*, in the occipital ventricle. As his supporting authority he cites Johannitius (Ḥunayn ibn 'Isḥāq), not Avicenna, yet he mentions the estimative faculty (*estimativa*), which is subsumed under *logistica*. Moreover, his description of sight and the eye later in that same book indicates more than passing familiarity with Alhacen's *De aspectibus*, as well as with the *Pantegni* and Aristotle's *De anima*. Bartholomaeus's failure to follow the standard fivefold scheme of psychological faculties was therefore due not to ignorance of the appropriate sources, including Avicenna, but to a desire to make his psychological model conform to Augustine's threefold scheme. This point is borne out by his discussion of the rational soul, where he not only appeals continually to the authority of Augustine but also has recourse to divine illumination to explain the perfecting of human intellect.[49] Bartholomaeus thus falls within a tradition that the great medievalist Étienne Gilson (1884–1978) characterized as "Avicennizing Augustinianism," a tradition marked by the effort to reconcile Avicennian and Augustinian cognitive psychology.[50]

Another aspect of Albert's description of vision that merits discussion is his use of the term "species" (*species*) to describe the formal representation of the object in the eye and common sense. As we saw earlier, Augustine used *species* and *forma* interchangeably to denote a similitude or likeness of the represented object, and by the second half of the thirteenth century, *species* had almost completely supplanted *forma* in the discussion of perception and cognition. The entire perceptual and cognitive process was thus described in terms of successive species, starting with the species in the transparent medium, continuing with the species in the sense, and culminating with a set of species generated in the psychological faculties. Accordingly, the "sensible species" (*species sensibiles*) are formed in the imagination and then projected into the Avicennian compositive imagination or phantasy (*fantasia*) for analytic association and dissociation, out of which the "intelligible species" (*species intelligibiles*) are eventually abstracted by reason for cognitive delectation by the intellect. These in turn are stored as "memorative species" (*species*

---

49. See R. James Long, ed., *Bartholomaeus Anglicus on the Properties of Soul and Body:* De proprietatibus rerum libri III et IV (Toronto: Pontifical Institute of Mediaeval Studies, 1979), 28–30 (on the sensitive soul and its faculties), 31–33 (on intellect and cognition), and 39–45 (on the visual faculty).

50. On Avicennizing Augustinianism, see Gilson, "Les sources gréco-arabes de l'augustinisme avicennisant," *Archives d'histoire doctrinale et littéraire du Moyen-Âge* 4 (1930): 5–149. See also Gilson's introduction to Muckle, "Gundissalinus's De anima," 23–27.

memoriales) in the posterior ventricle of the brain, where they are held for recollection. The mechanism commonly alluded to is "impression," especially within the context of the exterior senses and the imagination, which are viewed as passive recipients of sensible information.

But if species are similitudes or likenesses of the objects they represent, then how are they like those objects? One option is to suppose that they somehow resemble them in a quasi-pictorial way. Accordingly, the visible species in the eye will somehow be "colored," while the sensible species in the imagination will depict the object, as it exists physically, in a way analogous to that in which a painted portrait depicts its subject. Like a portrait, this species will convey such "intended" characteristics as the size, depth, and texture of the object without actually possessing them. By extension, then, we might think of the cognitive likeness of this species, that is, the intelligible species, as a highly generalized, abstract depiction of the object as a *type* of thing, rather than a particular thing, according to its intelligible intentions. The passage from pure sensation to cognition thus unfolds in a replication of increasingly abstract and general depictions of external objects through the psychological faculties, the intelligible species somehow portraying the underlying conceptual reality of which every such object is an instantiation. This, in fact, is "species" in the full, taxonomic sense, devoid of all physical particularity and reduced to the conceptual nub.[51]

Although this pictures-in-the-mind model of perception and cognition accords with Aristotle's assertion that we cannot think without images (see chapter 2, section 2), it is nonetheless problematic. For one thing, if species have real, objective existence, then they ought to be detectable in the transparent medium. Since they are not, they must subsist in that medium without affecting it in a perceptible way. In short, much like the Cheshire cat in *Alice in Wonderland*, the species at every stage must somehow be reduced to its defining "smile." One way of accomplishing this reduction is to posit spiritual rather than crass, material existence, although it must be understood that "spiritual" in this context does not necessarily imply total immateriality or incorporeality. Another is to make the species themselves intentional entities. It is difficult,

---

51. The connection between *species* as a formal representation and "species" as a taxonomical category harks back to Boethius's second commentary on Porphyry's *Eisagoge*. For some discussion of this point, see A. Mark Smith, "Picturing the Mind: The Representation of Thought in the Middle Ages and Renaissance," *Philosophical Topics* 20 (1992): 149–70, especially 152–54.

however, to understand how such things can themselves bear intentions. After all, there must be *something* underlying those intentions, be they Avicenna's maleficent or beneficent intentions or Alhacen's visible intentions, and whatever it may be, it cannot itself be fully intentional and disembodied. Thus no matter how etiolated they may be, species must have at least some physical existence not only in the transparent medium between object and eye but also in the highly refined visual and animal spirit pervading the eye and the cerebral ventricles with their associated psychological faculties. The persistence of afterimages when we look at brightly luminous or illuminated objects is proof that the crystalline lens has been affected by something physical that is somehow "pictorial."

Then there is the problem of what role species actually play in perception and cognition. If, indeed, they are mediating depictions, then does it not follow that they, not the external things they represent, are the objects of perception and cognition? In response to this issue, Saint Thomas insists time and again that species are not what (*id quod*) we apprehend but rather the means by which (*id quo*) we apprehend the actual objects intended by them. Perhaps, then, we should think of the imagination as a mirror of sorts, the sensible species being virtual images viewed mentally through it rather than in it. After all, as Saint Paul says in 1 Corinthians 13:12, "For now we see through a [looking] glass dimly."[52] But even though this analogy helps account for the intentional or purely psychic existence of species, it highlights the problem of mediation because it is the object's mirror image, not the object that is apprehended.

Another issue is the "epistemic gap" between perception and cognition, the corporeal, sensitive soul on one side, the incorporeal intellect on the other. How can this ontological gulf be bridged? As we have seen, the standard response from late antiquity to the mid-thirteenth century was to appeal to an external agent intellect that provides either the intellectual illumination or the universal Forms that enable the potential intellect to actualize its capacity to understand. The virtue of this theory, particularly in its Augustinian form, is that it ensures our knowledge is veridical because it is certified through God's light. The pitfall is that since such externally assisted knowledge does not belong to us individually and personally, it must not subsist in our immortal souls after death. In short, as far as intellect is concerned, there can be no per-

---

52. The notion of the imagination as a mirror occurs within the Platonist tradition, particularly in such late antique figures as Plotinus, Porphyry, and Proclus; see Anne Sheppard, "The Mirror of Imagination: The Influence of *Timaeus* 70e ff.," *Bulletin of the Institute of Classical Studies* 46 (2011): 203–12.

sonal immortality, a point brought to the fore by Averroes under the influence of Themistius.[53]

Saint Thomas is often credited with resolving this problem definitively by integrating the active intellect into the individual soul and vesting it with intrinsic, natural light rather than extrinsic, divine light.[54] But he was by no means the first to suggest this expedient. The anonymous author of the *De anima et de potenciis eius* of ca. 1225 is quite clear on this point: "In this regard Avicenna was wrong because he supposed the agent intellect to be separate from the soul . . . just as the sun is separate from the eye. But there is no doubt that this intellect is an [integral] faculty of the soul."[55] There is also no doubt that the theory of a separate agent intellect, along with that of divine illumination, fell into disfavor during the later thirteenth century, in great part because of its anathematization in the condemnation of 1277.[56] As a result, there was increasing emphasis on the empirical nature of cognition, which was taken to be solidly grounded in physical rather than metaphysical reality—hence the famous Thomistic dictum, "There is nothing in the intellect that is not previously in the sense."[57] The notion that supervening Platonic forms are somehow innately planted or extrinsically projected onto the potential intellect was thus rendered dubious in the extreme.

In addition, as the theory of a separate agent intellect possessing all the

53. See Taylor, *Averroes Long Commentary*, 398–401.

54. On Thomas's integration of agent intellect into the individual soul, see, e.g., *Summa theologiae*, 1a, quest. 79, articles 3–5, in Pasnau, *Aquinas on Human Nature*, 82–89.

55. Translated from Gauthier, "Le traité," 51, lines 453–55.

56. For instance, according to article 32 of the 219 articles listing condemned propositions, it is heretical to believe "that the intellect is one in number for all [men]," and article 123 censures the proposition "that the agent intellect is some separate substance superior to the possible intellect [and] not a form of the human body." Article 115, moreover, prohibits the claim that the universal forms (*species*) are implanted a priori in the human intellect. For the actual texts of these articles, see David Piché, *La condamnation Parisienne de 1277* (Paris: Vrin, 1999), 88, 114, and 116; the English translations are mine. It should be added that some, following Averroes's *Long Commentary* on Aristotle's *De anima*, compounded the problem by assuming that not only the active intellect but also the material or potential intellect is common to, rather than individuated in, humans; see, e.g., Herbert Davidson, "Averroes and Narboni on the Material Intellect," *Association for Jewish Studies Review* (1984): 175–84; and Alfred Ivry, "The Ontological Entailments of Averroes' Understanding of Perception," in *Theories of Perception in Medieval and Early Modern Philosophy*, ed. Simo Knuutila and Pekka Kärkkäinen (Dordrecht: Springer, 2008), 73–87.

57. From Saint Thomas, *Quaestiones disputatae de veritate*, quest. 2, article 3, arg. 19: *Praeterea, nihil est in intellectu quod non sit prius in sensu.*

Universals was on the wane, skepticism about species theory was on the rise in the face of an increasingly trenchant critique. Might it not be the case, suggested Peter John Olivi (1248–1298), that species actually impede rather than facilitate perceptual and intellectual apprehension by forming a "veil" between perceiver and perceived? Perhaps what we take to be mental representation is the act of thought itself rather than an intelligible depiction of what is thought. And even if sensible and intelligible species actually do exist, perhaps they are mere by-products rather than agents of the perceptual and cognitive process. Perhaps, in fact, they can be dispensed with altogether, as William of Ockham (ca. 1288–1349) eventually did, proposing instead that we apprehend external objects directly in a process of intuitive cognition uncluttered by any intermediate entities. Such direct apprehension of physical particulars, he argued, informs us immediately not only that they have specific qualities but also that they are actually present before us. What Ockham calls abstractive cognition, on the other hand, involves reasoning of two kinds. The first is judgment about what is being intuitively apprehended, for instance, that the large, brown object before me is a horse. The second is reasoning based on intuitive cognition of physical particulars no longer present to the sense. Thus, for instance, I might reason abstractively that some, but not all large, brown objects are horses. In neither case do "large," "brown," or "horse," as taxonomic categories, have extramental existence in supervening Universals or in formal representations of them; they are mental states that persist as "habits" or lasting dispositions that can be arranged and rearranged propositionally and syllogistically by the mind. In essence, then, perception and cognition are reduced to action at a distance between mind and object, an idea remarkably (and coincidentally) similar to the unmediated relationship between viewing eye and visible object Plotinus posited (see chapter 4, section 1).[58]

A measure of Ockham's long-term influence, particularly in the realms of logic and theology, is the emergence of an "Ockhamist" school of thought, alongside "Scotist," "Thomist," and "Albertist" counterparts, in the fifteenth century. Yet as Katherine Tachau has shown convincingly, few thinkers in the decades surrounding Ockham, even among his close associates, were willing

58. On the late medieval critique of species theory, see Robert Pasnau, *Theories of Cognition in the Later Middle Ages* (Cambridge: Cambridge University Press, 1997). On Ockham's theory of intuitive and abstractive cognition, see Eleonore Stump, "The Mechanism of Cognition: Ockham on Mediating Species"; and Elizabeth Karger, "Ockham's Misunderstood Theory of Intuitive and Abstractive Cognition," in *The Cambridge Companion to Ockham*, ed. Paul V. Spade (Cambridge: Cambridge University Press, 1999), 168–226.

to jettison species theory in favor of his theory of intuitive cognition.[59] And the same holds for later thinkers, such as Jean Buridan (ca. 1300–ca. 1361) and Nicole Oresme (ca. 1320–1382), who were bent on saving both sensible and intelligible species in the face of problematic implications and entailments.[60] Indeed, the persistence of species theory throughout the later Middle Ages, despite the fourteenth-century critique, is manifest in the account of psychology and cognition Gregor Reisch (ca. 1467–1525) provided in his *Margarita Philosophica* ("Philosophical Pearl") of 1503. A Carthusian monk who served as confessor to Emperor Maximilian I in 1509, Reisch intended this work as "an epitome of the whole of philosophy" and, as such, a survey of the scholastic liberal arts curriculum, including both natural and moral philosophy. Its extraordinary popularity is attested to by its having been reissued and reedited at least twelve times between 1503 and 1600, and it even served as a text for university teaching.[61] Reisch was thus operating at a more popular level than had the scholastic critics of the late thirteenth and early fourteenth century, most of whom dealt with the problem of species from the elevated perspective of academic theology. As a result, Reisch's account of visual perception and cognition is fairly straightforward and uncritical, although by no means philosophically naive.

As expected, the elements of that account include the *lux-lumen* distinction and the standard succession of species propagated, or multiplied (*multiplicare*), through the transparent medium into the eye, then passing through the common sense to the imagination as sensible species, and subsequently projected into the phantasy for analytic scrutiny under the supervision of reason. Thus "cleansed and made abstract" (*depurate et abstracte*), the sensible species, or phantasms, are transformed by the agent intellect into intelligible species that represent the Universals for the potential intellect, which retains them in intellectual memory for recollection. All of this occurs within the standard set of five Avicennian psychological faculties arrayed in the three ventricles of

59. See especially Tachau, *Vision and Certitude*, 180–312.

60. See, e.g., Jack Zupko, *John Buridan: Portrait of a Fourteenth-Century Arts Master* (Notre Dame, IN: University of Notre Dame Press, 2003), 175–82 and 205–26. See also Peter Marshall, "Parisian Psychology in the Mid-Fourteenth Century," *Archives d'histoire doctrinale et littéraire du Moyen Âge* 50 (1983): 101–93.

61. See Andrew Cunningham and Sachiko Kusukawa, trans., *Natural Philosophy Epitomised: Books 8–11 of Gregor Reisch's* Philosophical Pearl (1503) (Farnham, UK: Ashgate, 2010), ix–xiii and xxviii–xxx.

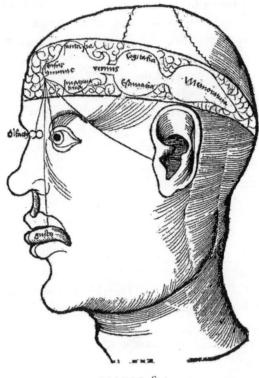

FIGURE 6.1

the brain, which are suffused with animal spirit that passes to the eye via the hollow optic nerves to sensitize the lens.[62]

This array of faculties is illustrated with marvelous but somewhat misleading clarity in the by-now iconic cutaway diagram that accompanies the text and is reproduced in figure 6.1. All the senses, except touch, which is distributed throughout the body, are shown converging linearly at the common sense (*sensus communis*) in the forefront of the first cerebral ventricle. Behind it and at the bottom of the posterior lobe of that same ventricle is the imagination (*imaginativa*), where sensible species are stored as phantasms. Directly above the imagination is the phantasy (*fantasia*), where the sensible

62. On the *lux-lumen* distinction and species multiplication, see ibid., 173–76; on the eye, its connection to the brain, and its visual function, 179–81 and 192–93; on the psychological faculties in the brain, 205; on the operations of common sense, imagination, and phantasy, 205–6; and on the operation of intellect, 222–25; see also 225–26 for Reisch's explanation of intuitive cognition as the apprehension that the object sensed is actually present to the sense.

species are analytically associated and dissociated under the direction of the cognitive faculty (*cogitativa*), which exists only in humans and lies at the top of the middle ventricle. This faculty has access to the phantasy through the "worm-like" opening (*vermis*). Directly below the *cogitativa* is the estimative faculty (*estimativa*), which apprehends nonsensible intentions (Reisch appeals to the wolf-sheep trope), and finally, in the occipital ventricle is the faculty of intellectual memory (*memorativa*). Where this diagram goes awry is in distinguishing *fantasia* from *cogitativa* because they are the same according to Reisch, the latter being the specific form of *fantasia* appropriate to humans, just as Avicenna claimed.[63]

Before turning to the next section, I must acknowledge that for the sake of brevity and simplicity I have run roughshod over a host of significant philosophical differences among late medieval thinkers in regard especially to cognition. Not all of them, for instance, rejected the theory of divine illumination or that of implanted or innate Forms. Many of them also disagreed about whether intelligible species are needed as mediating entities for cognition, and if so, how they serve that function. There were disagreements as well over how those species are formed, some thinkers assuming that they are abstracted from sensible species, others that they are somehow produced by the agent intellect in conjunction with, but apart from, those species. Likewise, there were disagreements about whether intelligible species form the content of potential intellect after its actualization. And there were differences over how species actually represent their objects, whether they do so in a more or less explicit, pictorial way or in an implicit, nonpictorial way. After all, in proper context, the word "lion" somehow represents the large tawny beast without looking like it. All these issues and more were at play throughout the fourteenth and fifteenth centuries, and even beyond. Nevertheless, I think it fair to say that they arose and were discussed within the conceptual framework of Avicennian faculty psychology exemplified in Reisch's model of the sensitive soul and its cerebral organization.[64]

---

63. For a fairly wide variety of medieval and Renaissance diagrams of the brain and its faculties, including Reisch's and several "plagiarized" spinoffs, see Edwin Clarke and Kenneth Dewhurst, *An Illustrated History of Brain Function* (Berkeley: University of California Press, 1972), 10–48.

64. For a somewhat encyclopedic discussion of how these issues were approached, thinker by thinker, over the fourteenth and fifteenth century, see Leen Spruit, Species Intelligibilis: *From Perception to Knowledge*, vol. 1 (Leiden: Brill, 1994), 256–411.

4. GEOMETRICAL OPTICS AND THE
EVOLVING SCIENCE OF *PERSPECTIVA*

We noted in passing that some of the thinkers dealt with in the previous section were acquainted with geometrical optics. Albertus Magnus, for instance, not only cited Alhacen and al-Kindī as authoritative optical sources but also applied ray geometry to the analysis of reflection in his paraphrase of Aristotle's *De sensu*.[65] Yet interestingly enough, despite his vehement defense of intromissionism, Albert predicated that analysis on visual rays rather than light rays. Albert, in short, seems not to have taken geometrical optics seriously enough to incorporate it coherently into his account of light and vision. In this section we will look at a select group of scholastic thinkers, the so-called perspectivists mentioned at the beginning of this book, who did take geometrical optics seriously enough to integrate it fully and coherently not only into their account of light and sight but also into their more general account of visual perception and cognition. As we will see in fairly short order, it was Roger Bacon (ca. 1214–1292) who carried this process to completion by adding Avicennian flesh to the bare bones of Alhacen's empirical and operational account. As far as geometrical optics is concerned, then, Alhacen was central to the perspectivist enterprise Bacon began in the 1260s and Witelo and John Pecham continued in the 1270s. In order to put Bacon's work into proper context, though, we need to look briefly at Robert Grosseteste (ca. 1168–1253).

Grosseteste is not usually counted among the perspectivists because he supposedly knew nothing of Alhacen's *De aspectibus*, although I will argue later that he may have drawn on that work in limited ways. Whatever the case, it is unquestionable that Grosseteste played a critical role in the evolution of perspectivist optics in at least two respects. First, he provided the ideological foundations for that enterprise in his "metaphysics of light," which is exemplified in his brief tract *De luce* ("On Light," ca. 1228).[66] The gist of his argument

65. See Akdogan, *Albert's Refutation*, 81–85 and 89–95.

66. For the Latin text of the *De luce*, see Ludwig Baur, ed., *Die Philosophischen Werke des Robert Grosseteste, Bischofs von Lincoln* (Münster: Aschendorff, 1912), 51–59. For an English translation, see Clare Riedl, *Robert Grosseteste* On Light (Milwaukee: Marquette University Press, 1942). Erwin Panofsky was instrumental in promoting the idea that a Pseudo-Dionysian metaphysics of light was central to high medieval thought and aesthetics as they developed over the twelfth century. Exemplary of such metaphysics, by Panofsky's account, was the birth of gothic architecture at the hands of Abbot Suger of Saint-Denis in the early twelfth century; see Panofsky, ed. and trans., *Abbot Suger on the Abbey Church of St.-Denis and Its Art Treasures* (Princeton, NJ: Princeton University Press, 1946). Cf., however, Peter Kidson, "Panofsky, Suger

there is that light (*lux*), in its purest, simplest, and most spiritual state, consti-tutes the "first corporeal form" (*forma prima corporalis*). Naturally disposed to diffuse, or multiply (*multiplicare*), instantaneously from a point, it creates a sphere of radiation that confers three-dimensionality on the physical universe to the very edge of the firmament of Genesis. Having perfected corporeity at that edge, the light then returns toward its source as *lumen* in "waves" to create the planetary spheres and sublunar spheres of fire, air, water, and earth. Not an act of creation in the strict sense, this diffusion of primordial light is nonethe-less an act of theophany insofar as it reveals the physical world by actualizing its potential corporeity and dimensionality.

As we saw earlier, such a description of theophany by analogy to physical light is found implicitly in a host of late Platonist thinkers, including Augus-tine, and explicitly in Pseudo-Dionysius and John Scottus Eriugena. There is no question that Grosseteste was familiar with the works of Pseudo-Dionysius because he retranslated them from Greek and commented on them toward the end of his life. More questionable is that he read Eriugena's *Periphyseon*; the best we can say so far is that he may have. But establishing a clear line of influ-ence from these thinkers, especially Pseudo-Dionysius, to the *De luce* is prob-lematic because Grosseteste's scholarly work on Pseudo-Dionysius postdates the *De luce* by over a decade.[67] That, of course, does not preclude his having read and absorbed the Pseudo-Dionysian texts much earlier. Still, no matter the source or sources of his discussion of primordial light in the *De luce*, what makes that discussion significant is how closely Grosseteste ties metaphysical and physical light according, for instance, to the *lux-lumen* distinction. This brings us to Grosseteste's second major contribution to perspectivist optics: his conviction that the operations of metaphysical and physical light are so similar that understanding the latter will aid in understanding the former. In short, the study of physical light is theologically warranted in the fullest sense and is thus not a subject of vain curiosity. No less important is his conviction that a proper analysis of physical light must be based on geometry because ge-

---

and St. Denis," *Journal of the Warburg and Courtauld Institutes* 50 (1987): 1–17, who offers a trenchant critique of Panofsky's thesis and, by extension, a critique of the concept of Pseudo-Dionysian light metaphysics as a central organizing principle of high medieval thought.

67. On Grosseteste as translator of and commentator on Pseudo-Dionysius, see James McEvoy, *The Philosophy of Robert Grosseteste* (Oxford: Clarendon Press, 1982), 69–123. On the possible connection between Grosseteste and Eriugena, see McEvoy, "Ioannes Scottus Eri-ugena and Robert Grosseteste: An Ambiguous Influence," in *Eriugena Redivivus*, ed. Werner Beierwaltes (Heidelberg: Carl Winter, 1987), 192–223. On the relative dating of Grosseteste's *De luce* and his works on Pseudo-Dionysius, see McEvoy, *Philosophy*, 69 and 488.

ometry is universally applicable to every aspect of natural philosophy. Thus, as he puts it in the *De lineis, angulis, et figuris* ("On Lines, Angles, and Figures," ca. 1230), "a consideration of lines, angles, and figures is of the greatest utility because it is impossible to gain a knowledge of natural philosophy without them . . . for all causes of natural effects must be expressed by means of lines, angles, and figures."[68]

As this quotation implies, Grosseteste's analysis in the *De lineis* focuses on the cause or agency of natural effects. In general, he maintains, "a natural agent multiplies its power from itself to the recipient [*patiens*], this power [being] sometimes called species, sometimes likeness [*similitudo*]." However, he continues, "the effects differ according to the diversity of the recipient. For when this power is received by the sense, it produces an effect that is somehow spiritual and more noble; [whereas] in matter, it produces a material effect." Every such power, including the visual power that radiates from the eye, is propagated along lines, and the intensity of its effect depends on such things as how short the line between agent and recipient is, how direct or oblique it is, and how straight it is. It also depends on whether the agent acts immediately or mediately.[69]

Reflection and refraction are cases of mediate action because they involve interference with, and breaking of, the radial lines. After rebounding (*redire*) from a perfectly impermeable, reflective surface, for example, light loses intensity, and the more directly it strikes the reflective surface (that is, along the perpendicular), the greater the loss of intensity after reflection, which occurs at an angle equal to the angle of incidence. Alternatively, when the light meets a permeable surface at a slant, it is refracted (*frangere*). In passing from a rarer (*subtilius*) to a denser (*densius*) medium, light verges toward the normal (Grosseteste says "to the right") and away from the normal ("toward the left") in the opposite case. The complete breaking of the ray in reflection weakens light more than its partial breaking in refraction. Refraction in a denser medium, moreover, weakens light least because in that case it is shunted toward the perpendicular, which is the shortest, most direct radial line. Even weaker than reflected and refracted light is the accidental light (*lumen accidentale*) that dimly illuminates every corner of a room when a beam of light enters it directly

68. For the Latin text of the *De lineis*, see Baur, *Philosophischen Werke*, 59–65; and for an English translation, see Edward Grant, ed., *A Source Book in Medieval Science* (Cambridge, MA: Harvard University Press, 1974), 385–87. This and all subsequent quotations from *De lineis* are adapted from that translation.

69. See ibid., 60–61 (Latin) and 385–86 (English).

through a window. Furthermore, although *lux* is naturally apt to propagate its species spherically from a point, when it is distributed throughout a luminous surface, it will exert its power within a cone (*pyramis*) whose base is on that surface and whose vertex lies at a point on the recipient. Consequently, the rays from all the points of *lux* within the cone's base will converge on every exposed point of the recipient, thus guaranteeing that the luminous source will exert maximum effect on the illuminated recipient.[70]

In the course of discussing these points, Grosseteste refers explicitly to Aristotle's *Meteorology* and *Metaphysics*, Boethius's *Arithmetic*, Euclid's *Elements*, and Averroes's *Long Commentary* on the *De anima*. Implicit sources include Pseudo-Euclid's *De speculis*, Euclid's *Optics* (in the version titled *De visu*), and al-Kindī's *De aspectibus*, all of which he cites in other works. There is also a hint of al-Kindī's *De radiis* in Grosseteste's assertion that the analysis applies to *all* natural agents (for example, heat), not just light. Customarily not included among Grosseteste's possible sources are Ptolemy's *Optics* and Alhacen's *De aspectibus*.[71] Yet there are three specific points in his discussion that indicate some knowledge of the latter. The first comes with his claim that refraction in a denser medium mitigates the weakening of light by driving it toward the normal. As we saw in chapter 5, Alhacen makes essentially the same argument. The second point comes with Grosseteste's brief discussion of accidental light, as illustrated by the secondary illumination from light streaming into a room through a window. Alhacen specifically adverts to accidental light (*lux accidentalis*), by which he means the kind of light in illuminated rather than self-luminous objects, but he also refers to this sort of light as "secondary" and offers essentially the same room-with-a-window example to illustrate it.[72] The third point comes with Grosseteste's account of the cones of radiation emanating from luminous surfaces and concentrating their illuminative effect at the vertices. Alhacen offers precisely the same sort of analysis, although he does so in the context of mirrors.[73] Against these possible convergences, however, are two positions that put Grosseteste completely at odds with Alhacen. In the *De iride* ("On the Rainbow," ca. 1230), he insists that the eye emits

70. For Grosseteste on reflection and refraction, see ibid., 61–63 (Latin) and 386–87 (English); on the cones of radiation, 64 (Latin) and 388 (English).

71. See, e.g., Lindberg, *Theories*, 94. On the "De visu" version of Euclid's *Optics*, see Wilfrid Theisen, "*Liber de visu*: The Greco-Latin Translation of Euclid's *Optics*," *Mediaeval Studies* 41 (1979): 44–105.

72. See, e.g., *De aspectibus*, 4, 2.1 and 2.3, in Smith, *Alhacen on the Principles*, 4 (Latin) and 295–96 (English).

73. See ibid., 4, 3.88–95, 30–32 (Latin) and 317–19 (English).

visual rays (*radii visuales*) in order to complete the act of vision. Alhacen, of course, had ruled such radiation out emphatically. In the same work, moreover, Grosseteste asserts that when light enters a denser medium, it is refracted at an angle precisely half the angle of incidence, an assertion that is wholly inconsonant not only with Alhacen's but also with Ptolemy's conclusions about refraction.[74] Hence, if Grosseteste did draw on Alhacen, he did so on the basis of a superficial and selective understanding of the *De aspectibus*.

Although Grosseteste's effort to submit the physics of light to geometrical analysis was only partially successful, he inspired Roger Bacon to bring that effort to fruition. It is perhaps no exaggeration, in fact, to say that in sharing the same Augustinian theological leanings, the same drive for broad learning, and the same enthusiasm for applying mathematics to the analysis of natural philosophy, Bacon was, in a sense, Grosseteste's alter ego. But Bacon had one clear advantage over Grosseteste: he could draw upon a much wider array of sources in carrying out his program. That array included al-Kindī's *De aspectibus* and *De radiis*, Euclid's *Optics* and *Catoptrics*, Tideus's *De speculis*, and Pseudo-Euclid's *De speculis*. More important, it also included Ptolemy's *Optics*, book six of Avicenna's *Healing*, and Alhacen's *De aspectibus*, all of which Bacon not only cited explicitly but also exploited in an intelligent and sophisticated way.

Precisely when Bacon began to study optics seriously on the basis of these sources is open to question. Most likely it was during the long hiatus of intensive research that followed his tenure as master of arts in Paris between roughly 1240 and 1248 and his entry into the Franciscan order in 1256.[75] We are a bit more certain about when he composed his three major optical treatises. The earliest of these, the *De multiplicatione specierum* ("On the Multiplication of Species"), was probably composed in the very early 1260s. An extensive elaboration on Grosseteste's *De lineis* based primarily on Alhacen, this work provides the physical foundations for Bacon's account of visual perception in

74. For the Latin text of *De iride*, see Baur, *Philosophischen Werke*, 72–78 (especially 72–74); and for an English translation, see Grant, *Sourcebook*, 388–91 (especially 389–90). On the half-angle law of refraction, see Bruce Eastwood, "Grosseteste's 'Quantitative' Law of Refraction: A Chapter in the History of Non-Experimental Science," *Journal of the History of Ideas* 28 (1967): 404–14. Grosseteste's insistence on visual radiation may have been a bow to Augustine, whom he regarded as a foremost theological authority. His botched law of refraction indicates ignorance not only of the relevant portion (book seven) of Alhacen's *De aspectibus* but also of the relevant portion (book five) of Ptolemy's *Optics*.

75. See Jeremiah Hackett, "Roger Bacon: His Life, Career, and Works," in *Roger Bacon and the Sciences*, ed. Hackett (Leiden: Brill, 1997), 9–23.

the second and most significant of his optical works, the *Perspectiva*. We know that this work was completed by no later than 1267 because it was included as part five of the *Opus maius* sent to Pope Clement IV in that year. The dating of his third and briefest optical treatise, the *De speculis comburentibus* ("On Burning Mirrors"), is less certain. According to Lindberg, it could have been written any time between 1263 and 1274.[76]

The study of optics is especially important, Bacon tells us at the outset of the *Perspectiva*, because vision is the most informative and engaging of all the senses on account of the variety and beauty of its objects. Hence, "the wisdom [*sapientia*] gained through vision must have a special utility, not found in the other senses."[77] Wisdom, in this context, should be taken in the Augustinian sense, as the spiritual perfection of soul beyond its intellectual perfection in science (*scientia*), so the ultimate value of *perspectiva* is its capacity to lead us beyond intellectual to spiritual fulfillment. Mindful of this higher goal, Bacon opens his analysis with a description of the entire optic complex, starting with its source in the brain and ending with its completion in the eye. Since Bacon follows Alhacen closely in his description of the eye, there is no need to examine it in detail. Consisting of essentially the same nesting tunics arising from the *dura* and *pia mater* of the brain, Bacon's eye is filled with the same three humors and connected at the back to the hollow optic nerve, whose opening is directly in line with the pupil on the visual axis. As the seat of visual sensation, the lens or *glacialis* (the glacial humor) lies toward the front of the eye and is enlivened by visual spirit flowing to it from the brain through the optic nerves and vitreous humor. The resulting structure is represented in figure 5.1, where the eyeball as a whole, the cornea, and the anterior surface of the crystalline lens are all centered on C.[78]

At this point Bacon shifts from Alhacen to Avicenna by adding the three-

76. For critical editions of Bacon's three optical works, see David Lindberg, ed. and trans., *Roger Bacon's Philosophy of Nature: A Critical Edition, with English Translation, Introduction, and Notes, of* De multiplicatione specierum *and* De speculis comburentibus (Oxford: Clarendon Press, 1983); and Lindberg, ed. and trans., *Roger Bacon and the Origins of* Perspectiva *in the Middle Ages: A Critical Edition and English Translation of Bacon's* Perspectiva *with Introduction and Notes* (Oxford: Clarendon Press, 1996). On the dating of those optical works, see Lindberg, *Bacon and the Origins*, xxiii–xxiv. There is also significant optical content in parts four (on mathematics) and six (on experimental science) of the *Opus maius*, part six being where he analyzes the rainbow at length; see Robert Belle Burke, *The* Opus Majus *of Roger Bacon* (Philadelphia: University of Pennsylvania Press, 1928), 131–53 and 587–615.

77. Lindberg, *Bacon and the Origins*, 5.

78. For Bacon's full description of the eye, see ibid., 21–63.

chambered brain explicitly to the optic complex, all three ventricles distinguished according to their pertinent psychological faculties. Thus the anterior ventricle is inhabited by the common sense at front and the imagination behind, the middle ventricle by the estimative and cogitative faculties, and the posterior ventricle by the memorative faculty. The perceptual functions of these faculties follow the standard pattern. The common sense takes in the disparate special and common sensibles and combines them through judgment into sensible species, which are impressed in the imagination for retention and recall in the absence of the physical objects they represent. Sensible species, in turn, give rise to "imperceptible forms" (*insensate forme*) that are apprehended by the estimative faculty. It is through such forms, for instance, that sheep recognize the malignancy of wolves and are impelled to flee. Such forms are thus intentional, in the Avicennian sense, and they are stored as species in the memorative faculty. "Mistress of the sensitive faculties," finally, reason (*logistica*), which is lodged in the cogitative faculty, makes use of both the sensible and insensible species but "possesses [them] in a nobler way" that enables it to participate in the formation of what amount to intelligible species in the rational soul or intellect.[79]

As far as vision is concerned, the origin of species lies in the inherent *lux* and color on the surfaces of visible objects. Like Grosseteste, Bacon assumes that *lux* is naturally inclined to multiply its image or similitude, *lumen*, spherically from points. This it does by a process of self-replication in successive stages through continuous, transparent media. Exerting its power on the razor-thin portion of the transparent medium surrounding it, any given spot of *lux* on a visible surface actuates the potential in that surrounding portion to assimilate its form as *lumen*. Thus actually illumined, each spot within this portion of the adjacent medium shares the same active nature as its cause, so it actualizes the potential of the razor-thin portion of the medium contiguous with it to be illumined, and so forth in a succession of concentric shells as far as the continuous, transparent medium reaches. Each straight line of succession within the resulting, onion-like sphere of propagation constitutes a ray. Because transparent media are disposed to assimilate the species of light and color in an extremely weak, incomplete way, they have an imperceptible effect in such media. That is why we do not see them in the air. On the other hand, they have a much stronger effect on opaque bodies, which assimilate them in a perceptible way according to the light and color they in turn radiate. That is why the full moon appears bright and self-luminous, but more weakly so

79. For Bacon's account of the internal senses, see ibid., 5–21.

than the solar source of its illumination. In addition, since species multiplication occurs in spatial succession, it must take time, albeit an imperceptible amount. As with any other power, the effect of light weakens with distance from its source.[80]

Against Averroes (and evidently Grosseteste), Bacon insists that species in the medium do not have spiritual or intentional existence. Being formal, they must be embodied in order to subsist outside their source. In short, they have real corporeal or material existence. In Aristotelian terms, therefore, species play formal and efficient cause to the medium's material cause, which is why the supporting medium must be both material and continuous. How species manifest themselves is contingent on the disposition of the medium to be informed by them. Transparent and opaque media thus differ radically in that respect, as do crass transparent media, such as air, and the visual and animal spirit pervading eye and brain. The manifestation of species is also contingent on the power of the agent source. Solar *lux*, for instance, is so powerful that when its species is manifested in the eye, it will occlude the weaker species emanating from the stars and moon. Color, for its part, propagates its species so weakly that in order to make a perceptible effect, it needs the added power of illumination.[81]

Thus far Bacon has provided physical specification to Alhacen's account of light radiation and color radiation by basing it on Aristotle's formal, efficient, and material causation, the final cause being the act of sight. Furthermore, by adopting the Alhacenian model of punctiform radiation, Bacon can explain the act of visual sensation in essentially the same way as Alhacen. Each point of *lux* and illuminated color on a visible object's surface multiplies its species radially to every exposed point on the surface of a facing eye, and thence to every exposed point on the lens's anterior surface. Likewise all points of *lux* and color on the visible object's surface radiate their species to each point on the eye's surface and thence to a single exposed point on the lens's anterior surface.

The result is utter confusion on the lens's surface, but because of its sensitivity, the lens makes coherent sense of this confusion by selecting only the species that reach it orthogonally, since they make the strongest impression.

80. On species multiplication and its physical foundations, see *De multiplicatione specierum*, I, 1–3, in Lindberg, *Bacon's Philosophy*, 3–57. Much of what Bacon says about direct, reflected, and refracted vision is dealt with in this work with a view toward explaining its physical foundations; see, e.g., Bacon's discussion of the dynamics of reflection and refraction in ibid., 105–19 and 129–47.

81. On the corporeality of species, their propagation through transparent media, and their various manifestations, see ibid., 187–95.

The rest, having reached it along weaker, oblique paths, are ignored. All the individual impressions selected in this way comprise a pointillist representation of the object's surface that constitutes what we (not Bacon) might call the visible species. Each impressed species at a given point within this composite visible species multiplies straight through the glacial humor toward the center of sight within a cone of radiation based on the object surface. Refracted at the interface between glacial and vitreous humors, the visible species multiplies in proper upright and left-to-right order into the hollow optic nerve and thence to the optic chiasma, or "common nerve" (*nervus communis*), where it is fused with its counterpart from the other eye. It is there, as well, that the final sensor (*ultimum sentiens*) apprehends the species to complete the sensitive phase of vision.[82]

The visible species multiplied from the lens through the optic nerves to the brain conveys the twenty-two visible intentions Alhacen listed, light and color being primary, the rest secondary. Along with the additional seven intentions specific to the other four senses, Bacon considers these to be "inherently sensible" (*sensibilia per se*) insofar as they are perceptible in one way or another to the external senses, the common sense, or the imagination. For their part, the estimative, cogitative, and memorative faculties are naturally disposed to apprehend the "imperceptible sensibles" [*insensate forme*] mentioned earlier. These include the lupine malignancy instinctively inferred by sheep, and in Bacon's terminology they constitute "accidental sensibles" (*sensibilia per accidens*). It is by means of such accidental sensibles, for instance, that I recognize "Peter the Parisian born in the first hour of the night, son of Robert." This, of course, is an Aristotelian "incidental sensible" (see chapter 2, section 2). It is also by means of such accidental sensibles that "when I see a man, I see a substance and an animated thing [because my] sight falls somehow on his substantial nature, and also on his soul, which is a spiritual thing." This grasp of an object's essential core, Bacon concludes, "is [sensible] *per accidens* to the maximum degree," and it is properly suited to the cogitative faculty according to its intelligibility. This faculty, in short, is informed at the most abstract level of all.[83]

Following Alhacen, Bacon analyzes the visual process at three levels, the

82. On the lenticular selection of the visible species and its passage to the optic chiasma, see Lindberg, *Bacon and the Origins*, 69–79; and on the location of the final sensor, 63–69.

83. On the visible intentions, which Bacon does not list in full, on inherent sensibility, and on imperceptible or accidental sensibility, see ibid., 147; on incidental sensibility, as a subset of accidental sensibility, and on accidental sensibility "to the maximum degree," 149.

lowest of which, brute sense perception (*cognitio solo sensu*), involves the mere apprehension of light and color without any grasp of kind. At this level, in short, there is no determination of whether the light is solar or lunar or whether the color is red rather than blue. Determination of this sort occurs at the second level, which corresponds to Alhacen's "perception through recognition" (*cognitio per scientiam* in Bacon's terminology). Such perceptual recognition depends on the taxonomic niches of specific kind, such as "red" or "mule," accruing from repeated perceptions of similar things. Once etched into the imagination, each of these taxonomic niches is represented by a "vague particular" (*particulare vagum*) that is "as common as its universal and is convertible with it." Equivalent to Alhacen's universal forms, and obviously deriving from Avicenna's vague individuals, Bacon's vague particulars provide templates against which we can compare new perceptions in order to fit them into their appropriate taxonomic niches.[84] On this basis we are able "to distinguish universals from one another and from particulars, and particulars from each other by comparison of a thing seen to the same things previously seen." The third and final stage of perception corresponds to Alhacen's "perception by syllogism" (*cognitio per sillogismum*) because, according to Bacon, it "resembles a kind of reasoning" based on deduction and rational judgment. At this level we certify what we see by axial scanning in order to distinguish it as clearly as possible according to its every detail. Since it involves deductive reasoning of a sort, perception at this level is carried out by the power of discrimination (*virtus distinctiva*), which is exerted by the cogitative faculty, mistress of all the others. As perfect as it may be, though, this final stage of perception falls short of true cognition, which requires divine illumination for its actualization in intellect.[85]

Bacon's reliance on divine illumination to bring visual perception to intellectual completion betrays the Augustinian leanings that link him so closely to Grosseteste. So, too, does his insistence that being vested with its own internal light, the eye multiplies its species radially to visible objects in order to cap the visual process. Vision therefore entails both an inward and a complementary

84. On Avicenna's vague individual, see chapter 4, note 19. On its adoption and adaptation by various Latin scholastic thinkers, see Deborah Black, "Avicenna's 'Vague Individual' and Its Impact on Medieval Latin Philosophy," in *Vehicles of Transmission, Translation, and Transformation in Medieval Textual Culture*, ed. Robert Wisnovsky, Faith Wallis, Jamie Furno, and Carlos Fraenkel (Turnhout: Brepols, 2012), 259–92.

85. On the three levels of visual perception, see Lindberg, *Bacon and the Origins*, 155–59, 203–7, and 247–51; on the *particulare vagum*, 157; on certification by axial scanning, 179–81; and on divine illumination, 329.

outward reach of species. But it was not just in blind deference to Augustine, or for that matter Grosseteste, that Bacon posited visual radiation. For one thing, if every natural entity, whether animate or inanimate, exerts its agency on everything else through species multiplication, the eye can be no exception. For another thing, if one reads them correctly (as Bacon considered himself almost uniquely qualified to do), the authorities who appear to reject visual radiation are in fact only denying that it consists of matter pushing through the medium. For yet another thing, in the absence of visual radiation, sight would be reduced to pure passivity. Having the eye cooperate actively in the visual process makes that process intentional—or, as Augustine would have it, willfully "attentional." Finally, and no less significant, most optical authorities Bacon cited, including Augustine, are unequivocal in supporting extramission. Even Alhacen, according to Bacon, "took . . . and expounded" the science of *perspectiva* from that archextramissionist Ptolemy.[86] Consequently, Bacon felt fully justified in using sources that appear contradictory to us because, like Alhacen before him, he was convinced that truth could be winkled out of all of them.

This eclectic mix of apparently contradictory sources comes to the fore in Bacon's account of visual illusions occurring in direct, unimpeded radiation. Such illusions, he argues, are due to some skew in the eight preconditions Alhacen specified for veridical vision. As we saw earlier in chapter 5, these preconditions include adequate illumination, an adequately transparent medium, adequate distance between eye and object, and so forth. Hence, if an object lies moderately far away, we can gauge its distance fairly accurately according to "the continuity of sensible bodies intervening between the eye and [it]." If, however, the distance is immoderate or there are no intervening bodies by which to judge it, then we err, as we do in perceiving the planets and stars to lie equally far from us. Likewise, if the eyes are forced from their accustomed or "moderate" axial convergence, objects will appear double, a fact that Bacon explains experimentally on the basis of the apparatus Alhacen described for that purpose (see chapter 5, note 24).[87]

So far Bacon has been in virtual lockstep with Alhacen. He parts ways with him, however, by including a variety of illusions not mentioned in the *De as-*

86. On the eye's internal light, see ibid., 171; on the need for visual radiation, 101–7; on Alhacen's appropriation of *perspectiva* from Ptolemy, 101.

87. On the preconditions of veridical vision, see ibid., 109–45; on the perception of moderate or properly staked-out distances, 207; on the misperception of planetary and stellar distances, 211; and on diplopia or double vision, 179–87.

pectibus, many of them physiological in origin. For example, Bacon addresses the case Aristotle and Seneca mentioned of drunk or infirm people who are apt to see their own images floating in front of them. This illusion, Bacon explains, is due to the "malign and foul vapours [exuded from such people that] corrupt and thicken the air, so that the part of it near them can behave like a mirror." Bacon also appeals to physiology in explaining the oculogyral illusion Ptolemy described in book two of his *Optics* (see chapter 3, section 1). Citing Avicenna, he argues that this illusion arises when "a circular motion [is imparted] to the spirit in the anterior chamber of the brain [so that] the species [representing any surrounding object] . . . is moved circularly." These are just two of many instances in which Bacon has recourse to sources other than Alhacen in the course of explaining visual illusions, the result being that his account is less systematic and rigorous but more wide-ranging than Alhacen's. Thus unshackled from Alhacen, Bacon addresses a range of phenomena not mentioned in the *De aspectibus*. Why, for instance, does the moon appear to undergo cyclical phases of illumination? Why do the stars appear to twinkle? Why do elderly people tend to be farsighted? Why do some people see better at night than others? All these and more are mingled with such standard Alhacenian questions as why celestial bodies appear larger at the horizon than at zenith or why spherical bodies look like disks when seen at immoderate distances.[88]

Having completed his account of illusions arising in direct, unimpeded radiation, Bacon turns to reflected radiation and the illusions stemming from it. As before, the resulting analysis consists of an Alhacenian core to which elements from other sources are added. For example, in order to establish the equal-angles law, Bacon looks not to Alhacen's or Ptolemy's experimental determinations but to the first three propositions of the Euclidean *Catoptrics* (see chapter 2, section 6). It is no small irony that despite his much-vaunted experimentalism, Bacon fails even to mention Alhacen's or Ptolemy's empirical verifications.[89] The same holds for the cathetus rule of image location. Eschewing Alhacen's empirical verification (see chapter 5, section 3), Bacon attempts to establish it through geometrical assertion based on Euclid's flimsy demonstration of the equal-angles law. Consequently, Bacon's version of the rule depends

88. On seeing one's image in the air when drunk or infirm, see ibid., 281; on the oculogyral illusion, 173; on the lunar phases, 213–23; on the scintillation of the stars, 233–41; on farsightedness, 161–67; and on night vision, 171.

89. For a fairly recent discussion of Bacon's experimentalism, see Jeremiah Hackett, "Roger Bacon on *Scientia Experimentalis*," in Hackett, *Bacon on Sciences*, 277–315.

on two suppositions. The first is that we are naturally disposed to see the species of the object along the visual ray that coincides with the reflected ray. The second is that the image lies on the cathetus because it is formed at the intersection of the cathetus and the extension of the aforementioned visual ray. The circularity of this supposition is obvious, and so is the circularity of the geometrical demonstration of the rule that follows from it.[90]

In the remainder of his discussion of reflection, Bacon applies the cathetus rule to an analysis of the seven types of mirrors Alhacen mandated. A radical distillation of Alhacen's treatment of image formation and distortion in books five and six of the *De aspectibus*, the mathematical portion of this analysis is far outweighed by narrative description. In fact, Bacon provides only two geometrical propositions in his discussion of mirror images. The first is adapted from book five, proposition thirty-four of Alhacen's *De aspectibus*, which shows that in concave spherical mirrors images can be located behind the reflective surface, between that surface and the center of sight, at the center of sight itself, or behind it. The second proposition comes not from the *De aspectibus* but from Euclid's *Catoptrics* and is meant to show why images appear upright or inverted, depending on where the object and the center of sight lie with respect to each other as well as to the reflecting surface. Despite his sparing use of geometrical propositions, though, Bacon reveals a fairly firm grasp of the implications of Alhacen's complex geometrical proofs. He realizes, for instance, that concave spherical mirrors can produce as many as four images and as few as one, and he understands the specific ways in which images can be distorted in size, shape, apparent distance, and orientation in all seven types of mirrors. As before, though, Bacon adds to this Alhacenian core new instances of illusion, such as the iridescence of a dove's neck or the color bands in rainbows, both cases exemplifying how color perception can vary according to the angle at which light reflects from certain surfaces.[91]

Bacon opens his account of refracted radiation with a bare recapitulation of Alhacen's analysis of peripheral vision by means of refracted rays in book

90. On the equal-angles law, see Lindberg, *Bacon and the Origins*, 254–59; and on the cathetus rule, ibid., 262–67.

91. For the first proposition and its accompanying figure, see ibid., 271–73 (cf. Smith, *Alhacen on the Principles*, 450–51, and also figure 5.2.34b, p. 254), and for the second proposition and its accompanying figure, see Lindberg, *Bacon and the Origins*, 275–79 (cf. Euclid, *Catoptrics*, proposition eleven, in Heiberg, *Euclidis Opera*). On the number of possible images in concave spherical mirrors, see Lindberg, *Bacon and the Origins*, 271; on image distortion in curved mirrors, 267–79; and on iridescence and the colors in the rainbow, 279.

seven of the *De aspectibus* (see chapter 5, section 4). He then moves directly to the problem of how image magnification and diminution vary according to the shape of the refractive interface and the relative disposition of eye and object. For the simplest case, with the refracting interface flat, Bacon shows geometrically that when the eye is in the rarer medium, the object will appear magnified because it subtends a larger visual angle and appears closer than the object itself. Evidently unaware of the size-distance invariance hypothesis upon which Alhacen based his explanation, Bacon ignores the apparent distancing of the image because of its refractive dimming (see chapter 5, section 4). Among the remaining ten cases of refractive magnification and diminution Bacon deals with, two are particularly instructive. In the first of these, the object lies in a denser medium, and the eye views it through a concave spherical interface from a point between that interface and its center of curvature. Bacon shows geometrically that under these conditions the object will appear magnified because the angle under which it is viewed will be enlarged. This, he concludes, is why celestial bodies appear magnified near the horizon when seen through the watery vapors that rise from the earth's surface during the summer, just as Alhacen showed at the end of book seven of the *De aspectibus* (see chapter 5, section 4). But Bacon's account assumes that the vapors responsible for magnification extend far beyond the earth's atmospheric shell to envelop the celestial body. Not only is this assumption wholly incompatible with the cosmological theory of his day, but it also flies in the face of Alhacen's stipulation that the vapors form a wall of sorts between the eye and the celestial object. Bacon, in short, badly misunderstood Alhacen's explanation.[92]

In the second of the two instructive cases, the center of sight looks through a convex spherical interface at an object located in a denser medium between that interface and its center of curvature. The resulting magnification can be easily explained on geometrical grounds, and Bacon adverts to such magnification in claiming that spherical chunks of glass less than a hemisphere in size can be placed over "letters and other minute objects" to make them more easily visible. This instrument, Bacon concludes, is of use "to the elderly and those who have weak eyes." In sharp contrast to Alhacen, then, Bacon shows a keen interest in the practical application of optics, and he goes on to discuss several ways in which refractive and reflective magnification can be put to good

---

92. On vision by refracted rays, see Lindberg, *Bacon and the Origins*, 287–93; on magnification through flat interfaces, 295; on magnification through concave interfaces, 299; and on magnification by vapors near the horizon, 313–15.

effect. "Mirrors," for example, "can be so constructed . . . and arranged that . . . one man will appear to be many men, and one army will appear to be many armies." Such expedients, Bacon argues, can be employed to inspire fear in hostile forces so as to keep them at bay by deluding them into believing they are faced with insuperable odds. Likewise, large concave mirrors of the kind supposedly deployed by Julius Caesar to spy on the Britons from the shores of Gaul can be placed so that the doings of a distant enemy can be descried.[93]

The real utility of optics, however, lies in what it can tell us not about physical light but about metaphysical or spiritual light, primarily through parallels. For instance, just as both intromission and extramission of species and power are required for physical vision, so "spiritual vision requires not only that the soul should be the recipient from without of divine grace and powers, but also that it should cooperate by its own power." The analogy between the two types of vision is even more profound than that. "Direct [spiritual] vision," Bacon explains by way of illustration, "we attribute to God; departure from rectitude by refraction, which produces weaker vision, is suited to angelic nature; reflected vision, which is weaker [still], can be assigned to humans." The science of *perspectiva* also helps us grasp the deeper meaning of scripture. What are we to make of the psalmist's appeal, "Preserve us, O Lord, as the pupil ["apple" in the King James version] of [your] eye" (*Custodi nos, Domine, ut pupillam oculi*: Psalm 16:8, Vulgate; 17:8, KJ)? We know from the science of *perspectiva* that the glacial humor is the eye's pupil, and we know that it is encased by two humors and the weblike *aranea*, which are in turn encased by three tunics. We also know that the visual spirit flowing to it from the brain empowers it. Altogether, then, the physical pupil has seven physical guardians. Likewise, the spiritual pupil has seven spiritual guardians in the three theological virtues of faith, hope, and charity and the four cardinal virtues of justice, fortitude, temperance, and prudence. Bacon gives additional parallels, but these should suffice to indicate the genuineness of his conviction that the science of physical vision is a gateway to the wisdom of spiritual vision, both achieved through appropriate illumination. "Just as we see nothing corpore-

---

93. On magnification through convex interfaces, see ibid., 303; on magnification through convex glass lenses, 317; and on the deployment of large concave mirrors, 333. There is some evidence that lenses of the sort Bacon described were used not only in his day but well before, and it is extremely likely that eyeglasses containing such lenses were in use within twenty years of the writing of the *Perspectiva*; see Vincent Ilardi, *Renaissance Vision from Spectacles to Telescopes* (Philadelphia: American Philosophical Society, 2007), 3–49.

ally without corporeal light," he sums it up, "so it is impossible for us to see anything spiritually without the spiritual light of divine grace."[94]

That Bacon did not intend his *Perspectiva* to be a mere epitome of Alhacen's *De aspectibus* is evident from various contrasts between the two works. For a start, Bacon claims to have written the *Perspectiva* "more as a plea than as a comprehensive treatment." Or, to put it in slightly different terms, Bacon wrote the *Perspectiva* as a prospectus. Furthermore, unlike Alhacen (but like Grosseteste), Bacon explicitly justified the study of optics on the basis of its theological utility. Also, in sharp contrast to Alhacen, Bacon was at pains to place the entire visual process—from its inception at visible surfaces to its culmination at the back of the brain—on a firm, fully explicated physical, anatomical, physiological, and psychological footing. In short, Bacon's approach to optics was considerably more (or at least more explicitly) "philosophical" than Alhacen's. In comparison to Alhacen, moreover, Bacon gave short shrift to mathematics, offering only a handful of relatively primitive geometrical demonstrations instead of Alhacen's vast and systematic array of rigorous and sophisticated proofs. The contrast in this respect is startling. Whereas Alhacen's overall analysis of reflection is based on a total of ninety-two geometrical propositions, some of them extraordinarily complex, Bacon's is based on only eight, all of them short and simple. Nor was it for want of at least some facility in Euclidean geometry that Bacon used it so sparingly. Evidence of that facility can be seen in his close, though ultimately confused, analysis of concave spherical mirrors and pinhole images (that is, what amounts to the camera obscura) in the *De speculis comburentibus*. The limitations of his mathematical ability, on the other hand, are abundantly clear from his cursory and uninformed treatment of parabolic burning mirrors at the end of that treatise.[95]

More faithful than Bacon to the spirit of Alhacen's analysis is John Pecham (ca. 1230–1292), who composed his *Perspectiva communis* with a view toward condensing "into concise summaries the teachings of perspective [*perspec-*

---

94. For the analogy between corporeal and spiritual light, see Lindberg, *Bacon and the Origins*, 325; for the three levels of spiritual and physical vision, 329; for the pupil analogy, 323; and for the concluding quotation, 327.

95. For the Latin text of the *De speculis comburentibus*, with English translation, see Lindberg, *Bacon's Philosophy*, 271–341. In his analysis of parabolic mirrors at the end of the treatise (337–41), Bacon refers explicitly to Alhacen's demonstration of the focal property of such mirrors in his *De speculis comburentibus*, and it is clear that he has only the most superficial understanding of that demonstration.

tiva], which [in existing treatises] are presented with great obscurity."[96] Completed toward the very end of the 1270s, this work consists of three fairly short books devoted, in order, to direct, reflected, and refracted vision. Each book in turn is broken into brief, numbered propositions, some of them consisting solely of narrative (for example, book one, proposition one: "Light produces an impression in the eye that is directed toward it") and others of geometrical demonstrations (for example, book two, proposition twenty-three: "Only one image [of a single object] appears in a plane mirror"). In the first and by far the longest of the three books, Pecham follows Alhacen virtually to the letter in his own account of the physical, anatomical, physiological, and psychological foundations of vision. Likewise, in his account of reflection and refraction in books two and three Pecham hews more closely to Alhacen than did Bacon, although, like Bacon, he avoids the mathematical "obscurity" of Alhacen's analysis. A case in point is Pecham's effort in book two, proposition forty-eight, to demonstrate geometrically that there may be as few as one and as many as four reflections/images in a concave spherical mirror according to the relative placement of eye and object with respect to the reflecting surface and its center of curvature. This, in essence, is "Alhazen's Problem," and, as we saw in chapter 5 (section 3), it becomes extraordinarily complex and "obscure" when the two points are placed at unequal distances from the center of curvature. Pecham sidesteps the issue by basing his demonstrations on an equidistant placement of the two points, thus reducing the problem to its simplest form. What he offers, therefore, is a set of easily understood geometrical examples, not conclusive, general proofs, and the same holds for his approach to geometrical demonstration throughout the *Perspectiva communis*.[97]

Firmly grounded though it is in Alhacen's *De aspectibus*, the *Perspectiva communis* bears obvious Baconian traces as well. Perhaps the most salient of these is Pecham's use of the terminology of species multiplication in his account of radiation. In addition, Pecham follows Bacon rather than Alhacen in basing his explanation of image inversion in concave spherical mirrors on proposition eleven of Euclid's *Catoptrics*. Bacon's influence can also be discerned in Pecham's inclusion of such topics as pinhole images, the focusing

96. See Pecham's preface in David Lindberg, *John Pecham and the Science of Optics* (Madison: University of Wisconsin Press, 1970), 61. Like Bacon, Pecham was a Franciscan with strong Augustinian leanings, yet he gives only the barest hint of a theological justification for the study of optics in the preface, when he asserts, in passing, that "the Master—the light of all men—deems the investigation of light worthy of illumination."

97. For book one, proposition one, and book two, proposition twenty-three, see ibid., 63 and 175; on the number of possible reflection points in concave spherical mirrors, 197–205.

of light to a kindling point by concave spherical mirrors, refraction of light through glass spheres, and the rainbow. We will deal with certain of these topics more extensively in subsequent chapters, but for now it is enough to observe that they fall outside the scope of analysis in the *De aspectibus*. On the other hand, even though he agrees with Bacon that "every natural body . . . diffuses its power radiantly into other bodies," Pecham argues against the need for visual radiation on the grounds that it is redundant to posit both an outward and inward transmission of species. He does, however, admit that the natural light of the eye, especially among nocturnal animals, can have an illuminative effect on the colors of external objects. For the most part, then, Pecham's *Perspectiva communis* serves as both a complement to Bacon's *Perspectiva* and an epitome of Alhacen's *De aspectibus*.[98]

Whereas Bacon and Pecham were content to summarize Alhacen's mathematical arguments and thus gloss over the technical details, the third major perspectivist author, the Polish scholar Witelo (ca. 1230–after 1280), took those details with utmost seriousness. Indeed, so closely modeled after the *De aspectibus* is Witelo's *Perspectiva* (completed ca. 1275) that it might be dismissed as a mere recapitulation. There are, however, significant differences between the two works at the level not only of organization but also of topical content. For instance, book one of the *Perspectiva* has no ostensible optical content whatever. Rather, it consists of 137 propositions in which Witelo lays the mathematical foundations for his subsequent analysis of light and vision in the next nine books. Most, but by no means all, of these propositions are drawn from the *De aspectibus*, and in some cases Witelo reveals a somewhat sketchy understanding of Alhacen's mathematical reasoning. Of the remaining nine books, the first three (that is, two through four) deal with direct, unimpeded radiation, the next five (that is, five through nine) with reflection, and the last (that is, ten) with refraction.[99] Aside from expanding the number of books from Alhacen's

98. For Pecham's use of proposition eleven in Euclid's *Catoptrics*, see ibid., 183–85; on pinhole images, 67–83; on the focusing of light in concave spherical mirrors, 209; on refraction of light through glass spheres, 229–31; on the rainbow, 233–35; on radiant diffusion of power, 109; and on the redundancy of positing visual radiation as well as on the eye's natural light, 127–29.

99. Although Witelo's *Perspectiva* first appeared in print in 1535 and then again in a 1551 reprint, the definitive Latin edition is to be found in Friedrich Risner's *Opticae Thesaurus* (Basel, 1572). Critical editions of books one through three and five have been published piecemeal as follows: book one, Sabetai Unguru, ed. and trans., *Witelonis Perspectivae liber primus* (Wroclaw: Polish Academy of Sciences, 1977); books two and three, Unguru, ed. and trans., *Witelonis Perspectivae liber secundus et liber tertius* (Wroclaw: Polish Academy of Sciences, 1991); book five, A. Mark Smith, ed. and trans., *Witelonis Perspectivae liber quintus* (Wroclaw: Polish Acad-

seven to his ten, Witelo imposed a strict Euclidean format on all ten. Accordingly, each book opens with a set of definitions and postulates, when necessary, and then builds in logical order according to distinct propositional chunks, some of them consisting of narrative, the rest of actual geometrical proofs.

Witelo also drew liberally from sources other than Alhacen for supplementary demonstrations. Among these, Euclid's *Catoptrics* and Ptolemy's *Optics* figure most prominently, but Witelo availed himself of others, including al-Kindī's *De aspectibus* and Hero of Alexandria's *Catoptrics*, which William of Moerbeke translated for him in 1269. It was from this latter work that Witelo borrowed the least-lines proof of the equal-angles law of reflection. From Ptolemy he took his tabulations for angles of refraction, although he unaccountably adjusted the value for r from air to water at $i = 10°$ from Ptolemy's 8° to 7.75°. In addition to these structural and propositional elaborations, Witelo extended his analysis to a by-now-familiar range of topics not covered by Alhacen: shadow-casting and pinhole images, the focusing of light in concave spherical and parabolic mirrors, refraction of light through glass spheres, and the rainbow. Like Bacon, Witelo based his analysis of parabolic mirrors on Alhacen's account in the *De speculis comburentibus*, but unlike Bacon, he had a firm grasp of the mathematical principles underlying that account. Witelo also included a determination of the height of the atmosphere (slightly less than fifty-two miles) taken directly from the Latin translation of Ibn Muʿādh's *De crepusculis*, which was often misattributed to Alhacen.[100]

With all these structural and topical elaborations, Witelo's *Perspectiva* is enormous, well over half again as long the *De aspectibus*. It is also less sys-

---

emy of Sciences, 1983). For an unpublished edition of book four, see Carl Kelso, ed. and trans., "Witelonis *Perspectivae* liber quartus" (PhD diss., University of Missouri, Columbia, 1994).

100. For Witelo's use of the least-lines proof (in book five, proposition nineteen), see Smith, *Witelonis liber quintus*, 101–3; on shadow casting and pinhole images (in book two, propositions eight through forty-one), see Unguru, *Witelonis secundus et tertius*, 49–80; on focusing in concave spherical mirrors (in book eight, proposition sixty-eight; also nine, proposition thirty-six), see Risner, *Opticae Thesaurus*, 365–66 and 392–94; on parabolic mirrors (in book nine, propositions thirty-nine through forty-four), ibid., 398–401; on the height of the atmosphere (in book ten, propositions fifty-nine and sixty), ibid., 451–53; on refraction through glass spheres (in book ten, propositions forty-eight and ninety-four), ibid., 443–44 and 456–57; and on the rainbow (in book ten, propositions sixty-five through eighty), ibid., 457–72. On Ibn Muʿādh's *De crepusculis* and its misattribution to Alhacen, see A. Mark Smith, "The Latin Version of Ibn Muʿādh's Treatise 'On Twilight and the Rising of Clouds,'" *Arabic Sciences and Philosophy* 2 (1992): 38–88. For a discussion of the interconnections not only between Bacon and Witelo but also between them and Pecham, see David Lindberg, "Lines of Influence in Thirteenth-Century Optics: Bacon, Witelo, and Pecham," *Speculum* 46 (1971): 66–83.

tematic and elegant in approach and organization. Nonetheless, despite these drawbacks, it not only rivaled but surpassed the *De aspectibus* as an authoritative source during the later Middle Ages and Renaissance. The reasons are manifold. First, by recasting Alhacen's analysis in a strict mathematical format according to discrete theorems, each headed by an enunciation, Witelo made that analysis more accessible by rendering it more topically searchable. Witelo's *Perspectiva* thus served as a comprehensive textbook and as a reference source. Second, in his long prologue to the *Perspectiva*, Witelo offered much the same theological validation for the study of optics as did Grosseteste and Bacon. Witelo, in short, "Christianized" Alhacen's *De aspectibus* for a Latin audience highly attuned to religious imperatives. Third, like Bacon and Pecham, Witelo broadened the scope of his analysis to include a range of phenomena, such as focusing in concave mirrors and refraction through glass spheres, that were ignored in Alhacen's *De aspectibus* because they have little or nothing to do with vision and everything to do with light. Finally, unlike his perspectivist compeers, Witelo insisted on absolute mathematical rigor throughout his analysis, even though that analysis sometimes led to false or misleading conclusions. Consequently, to all appearances Witelo's *Perspectiva* matches Alhacen's *De aspectibus* in analytic precision and cogency and, in addition, surpasses it in topical scope.

5. CONCLUSION

The importance of these three perspectivist thinkers lies not in any technical or theoretical advances they made beyond Alhacen, their primary source. In fact, they often fell short, especially when dealing with such auxiliary topics as the focusing of light in concave spherical mirrors and refraction through convex spherical lenses. Yet as we will see in the next two chapters, the very fact that they integrated these topics into the optical enterprise—and the way they mishandled them—would assume considerable importance during the later Renaissance. Nor was it pathbreaking novelty that made the perspectivists significant. Indeed, it should be abundantly clear by now that they merely tied together certain analytic strands that had been evolving since Greek antiquity. Of overwhelming importance, however, is how they, Bacon above all, tied those strands together by melding Alhacen's model of visual perception with Avicenna's internal senses model of psychology so seamlessly as to render the two mutually reinforcing. On the one hand, Avicenna's psychological model was the natural vehicle within which to embody Alhacen's account of vision in all its technical and operational details. Such a fusion makes perfect sense because, as we saw in chapter 5, both Avicenna and Alhacen shared essentially the same conception of the sensitive soul and its functions. On the other hand, Alhacen's account of vision in all its technical and operational details buttressed Avicenna's internal senses model at every level, from physics and anatomy to physiology and psychology.

The theory resulting from this fusion presupposes that physical reality and our mental "picture" of it are in strict conformance with each other. In short, objective reality actually is as we visualize it in our minds—provided, of course, that our sensitive, perceptual, and cognitive faculties are sound. As we have seen throughout the course of this book, this presupposition was not new with the perspectivists. From at least the time of Plato and Aristotle, every effort, whether philosophical or mathematical, to explain how we apprehend the physical world by means of vision was predicated on it. What sets the perspectivists definitively apart from their predecessors is the exceptionally tight cause-and-effect linkage they forged between objective cause and subjective effect in the visual process. This linkage is manifested in the determinate succession of increasingly abstract and general likenesses or species that mediate between physical reality and its intellectual apprehension. Conveyed by light and color radiating within a cone whose vertex lies at the center of the eye, which constitutes the cardinal reference point for all visual analysis, the physical species in the transparent medium is replicated abstractly in the lens as a punctiform visible species. Along with the common sensibles and the remaining special sensibles, this species is replicated more abstractly in the common sense and imagination as a composite sensible species, from which the reasoning faculty forms the intelligible species that replicates the object most abstractly according to its conceptual core.

These replications, meanwhile, occur within a continuous material substrate that becomes increasingly less crass and more refined or spiritual as the species are multiplied through the external transparent medium and the inspirited eye to the animal spirit pervading the brain. Their abstractness and representative generality is therefore commensurate with the refinement of the material substrate within which they subsist; the more refined the substrate, the more abstract and general the species. Moreover, the stages of replication are tied to specific psychological faculties at specific locations in the brain. The visible species in the eye are transformed into sensible species by the common sense and imagination in the anterior ventricle. These species in turn are transformed into intelligible species—or, as Bacon would have it, vague particulars that are convertible with their Universals—by the reasoning faculty in the middle ventricle and then remanded to the memorative faculty in the occipital ventricle. As long as what is seen is apprehended under appropriate normative conditions, and as long as reason operates according to its proper rules, the mental representations arrived at in this way will be veridical, even if the initial perceptions on which they are based are not. Hence, armed with the proper optical principles, reason tells us that a stick that appears broken when half immersed in water is actually straight or that an image seen behind a mirror is displaced from its generating object in a rationally determinate way. At bottom, then, *perspectiva* is the science not only of vision but also of perceptual rectification.

Earlier, in the introduction to this book, I characterized this visual model as a scientific paradigm in the Kuhnian sense, and I did so not just because of its tight analytic structure and firm theoretical foundations, but also because of its intuitive appeal and empirical scope. Indeed, I would go (and have already gone) so far as to maintain that "what [the perspectivists] offered was less a scientific theory of light or vision than a scientifically justified world view."[101] Perspectivist theory therefore had ramifications that extended well beyond its ostensibly narrow subject matter in light and sight. In the next chapter we will take a brief look at some of these ramifications in such apparently disparate fields as theology, literature, and art. In the process, we will see that between the fourteenth and sixteenth centuries, perspectivist ideas managed to infiltrate a sizeable community of thinkers both inside and outside academia. The result was what, for lack of a better phrase, I will call "optical literacy." Yet we will also see that despite its coherence, sophistication, and explanatory power, the perspectivist visual model did not enjoy an unalloyed triumph during the later Middle Ages and Renaissance. Foremost among those who resisted or at least ignored it were medical thinkers, who generally favored the extramissionist alternative Galen and Ḥunayn/Johannitius offered. We will see as well that during the Renaissance there was a dawning recognition of certain anomalous optical phenomena that could not be adequately explained according to strict perspectivist rules. Accommodating such anomalies to perspectivist theory required that these rules be bent, if not broken. Consequently, by the end of the sixteenth century, just before Kepler undertook his optical research, perspectivist theory had become somewhat ragged around the analytic edges.

101. Smith, "Getting the Big Picture in Perspectivist Optics," 569.

The Assimilation of Perspectivist Optics during the Later Middle Ages and Renaissance

In the first chapter of this book we noted that nearly 190 manuscripts of Alhacen's *De aspectibus* and its perspectivist offshoots are currently known to exist in either complete or fragmentary form. The number rises to just over 210 when Alhacen's and Bacon's *De speculis comburentibus* are taken into account. This is one obvious, albeit crude, gauge of the inroads that perspectivist optics made into the European scholarly community between roughly 1300 and 1500. Another is the sizeable number of works devoted either wholly or in great part to perspectivist optics in the form of paraphrases or scholastic commentaries. Aside from close, topical analyses of the science of *perspectiva* itself, these include commentaries on Aristotle's *De anima* and *De sensu*. Optical matters also figure prominently at points in commentaries on Peter Lombard's *Sentences*. These latter commentaries amounted to doctoral theses for students in theology, the most elite of the advanced disciplines. To such focused, scholarly studies should be added works not explicitly devoted to optics but within which optical allusions crop up to a greater or lesser extent. A prime example is Peter of Limoges's *Tractatus moralis de oculo* (or *De oculo morali*), which was likely written during the last quarter of the thirteenth century and is rife with optical analogies of the Baconian sort based on perspectivist principles.[1]

Textual evidence tells only part of the story, though. Optical lore was also disseminated from the pulpit. Preaching, after all, is teaching, and in order to

---

1. For a recent annotated English translation of this work, see Richard Newhauser, *Peter of Limoges*: The Moral Treatise on the Eye (Toronto: Pontifical Institute of Mediaeval Studies, 2012).

provide spiritual enlightenment, preachers often fall back on examples based on both common experience and scientific principles. It was with just such examples in mind, in fact, that Peter of Limoges wrote his *Tractatus moralis de oculo*, and the alacrity with which it was mined for sermons during the later Middle Ages and Renaissance is evident from the remarkably high number of surviving manuscripts—219 in all. Nor was its appeal limited to a scholastic audience or even one versed in Latin. Various exempla were drawn from it for use in vernacular sermons, and the entire work was translated into Italian toward the end of the fifteenth century.[2]

The vernacularization of medieval optics occurred through other channels as well, some direct, others indirect. An early example of direct vernacularization is the mid-fourteenth-century Italian translation of Alhacen's *De aspectibus*. The very fact that this translation was commissioned, undoubtedly at considerable expense, suggests that there was at least some serious interest in optics outside the academy.[3] Examples of indirect channels other than sermonizing are legion, but two categories stand out. One is vernacular literature, which was disseminated both textually and orally to a fairly wide, mostly aristocratic audience. Another is art, and as we will see later in this chapter, the evolution of naturalistic painting from the early fifteenth century was based to some extent on optical principles and an effort to understand such things as the interplay of light and shadow, aerial perspective, and the visible effects of reflection and refraction. Artists, however, were not free, creative spirits in this period. They were part of the medieval and Renaissance craft tradition insofar as they had not only to master a variety of technical skills from the mixing of pigments to the design of fortifications but also to sell those skills to patrons. Leonardo da Vinci (1452–1519), exceptional though he was, exemplifies the breadth of technical skills that could be commanded by, and demanded of, a Renaissance artist. But of course all medieval and Renaissance artisans had to master various technical skills and pass them on to apprentices and journey-

2. On the number of surviving manuscripts, see ibid., i; to the 219 surviving manuscripts, Newhauser adds another forty-one copies that "can be attested" (ibid., xxix). On the Italian translation, see ibid., xxx. For a useful analysis of Peter's treatise as a preaching manual, see David Clark, "Optics for Preachers: The *De oculo morali* of Peter of Limoges," *Michigan Academician* 9 (1977): 329–43.

3. See Graziella Vescovini, "Alhazen vulgarisé: Le De li aspecti d'un manuscrit du Vatican (moitié du XIVe siècle) et le troisième Commentaire sur l'optique de Lorenzo Ghiberti," *Arabic Sciences and Philosophy* 8 (1998): 67–96; see also A. Mark Smith, "The Latin Source of the Fourteenth-Century Italian Translation of Alhacen's *De aspectibus* (Vat. Lat. 4595)," *Arabic Sciences and Philosophy* 11 (2001): 27–43.

men. The marketplace itself was thus a marketplace of ideas, most but not all of them transmitted orally.

During the Renaissance there were significant changes in this marketplace involving both technical skills and the way they were disseminated. By the fifteenth century, for example, the technology of glassmaking and glassblowing had improved significantly. The resulting availability of better glass in significant amounts and at affordable prices, in turn, enabled craftsmen to manufacture eyeglasses in large quantities. Another outcome of the manufacture of better glass was the production of better mirrors. Consequently, as such optical devices became more commonplace over the fifteenth and sixteenth centuries, an increasingly broad audience became aware of their visual effects. Some among that audience sought to understand how those effects are created; others aimed to put them to practical use in surveying instruments and, eventually, magnifying devices. The advent of print technology was also a crucial factor in the spread of optical lore to a broad, academic and nonacademic audience through both Latin and assorted vernaculars, but we will defer discussion of that until the next chapter.

The purpose of this chapter is to trace some of these developments and their effect on optical thought from the early fourteenth to the mid-sixteenth century across certain disciplinary boundaries. Accordingly, in section 1, we will look at the dissemination of optics as a discipline in the arts curriculum of the late medieval university. We will then pass in the next three sections, 2–4, to a discussion of the popularization of that discipline through theological, literary, and artistic channels. In the process we will see that the entry of optics into the public domain led to a fairly widespread acceptance of the perspectivist paradigm outside the university community. It also led to a close scrutiny of the paradigm from new perspectives that raised some nagging doubts about the universality and coherence of some of its theoretical presuppositions.

1. OPTICS AS A QUADRIVIAL PURSUIT IN THE ARTS CURRICULUM

While the sheer number of extant manuscripts of the *De aspectibus* and its perspectivist derivatives indicates the extent to which the study of optics had infiltrated the arts curriculum during the later Middle Ages and Renaissance, it is only a rough indicator. Not only does that number fail to tell us precisely which texts were actually used for the teaching of optics, but it also tells us next to nothing about how they were taught or studied. We do have scattered clues. Among these are various university statutes that prescribe the use of certain perspectivist texts for the teaching of mathematics. One example is found in a 1390 statute of the University of Prague that specifies Pecham's *Perspectiva communis* for mathematical study at the master's level. Another example is an Oxford statute issued forty-one years later that mandates a choice among Euclid, Alhacen, and Witelo for that same purpose at the bachelor's level. The *Perspectiva communis* is also prescribed for study at the bachelor's level in the statutes of 1389 for the University of Vienna, although there is no indication that it was meant specifically for the teaching of mathematics.[4]

For the period before 1389 we have to fall back on indirect evidence. In the preface to his *Perspectiva communis*, for instance, John Pecham expresses the hope that his study "may be of benefit to young students."[5] The relatively large number of extant manuscripts of his text (no fewer than sixty-four) indicates that his hope was ultimately fulfilled, although how soon after the treatise's completion (around 1280) and at what level is unclear. In addition, several commentaries on the science of *perspectiva* dating from the fourteenth century have survived, indicating that optics was being taught at different universities during that century. Whether these commentaries were delivered to beginning or advanced students is a matter of guesswork.[6] Perspectivist ideas, as well as citations of Alhacen and Witelo by name, also crop up in a wide variety of late medieval scholastic commentaries, as well as literary sources.[7] Then there are the manuscripts themselves. Many of them are interlinearly and marginally annotated, often copiously and in more than one hand. This, of course, suggests

4. For Prague, see Hastings Rashdall, *The Universities of Europe in the Middle Ages*, vol. 1 (Oxford: Clarendon Press, 1895), 442; for Oxford, see David Lindberg's introduction to the reprint edition of Friedrich Risner's *Opticae Thesaurus* (New York: Johnson Reprint, 1972), xxiii; and for Vienna see Rashdall, *Universities of Europe*, vol. 2, part 1, 240. See also Lindberg, *John Pecham and the Science of Optics* (Madison: University of Wisconsin Press, 1970), 30–31.

5. Lindberg, *Pecham*, 61.

6. For some discussion of these commentaries, see Lindberg, *Theories*, 122–32. For a more extensive but somewhat problematic account, see Graziella Vescovini, *Studi sulla prospettiva medievale* (Turin: Giappichelli, 1965), especially 165–267.

7. For some scholastic examples, see Smith, *Alhacen's Theory*, xciv–xcvi; see also Lindberg, *Theories*, 132–46, and *Pecham*, 31–32. Undoubtedly the best-known literary reference is from lines 232–34 of "The Squire's Tale" in Chaucer's *Canterbury Tales*: "They speken of Alocen and Vitulon, / And Aristotle, that writen in hir lyves / Of queynte mirours and of perspectives." The very fact that Chaucer would refer to both Alhacen (Alocen) and Witelo (Vitulon) suggests that he expected his audience to grasp the references. This in turn suggests that his readers and listeners would have at least been aware of perspectivist theory without necessarily understanding it, just as many of us today are aware of quantum theory with no grasp of its technical details.

multiple close readings of the same manuscript, although by and large such annotations consist of corrective textual insertions, indexing remarks (that is, nota bene or the equivalent), or summaries of points made in the body of the work. Nevertheless, despite its overwhelmingly pedestrian nature, such annotation can reveal a good deal about how carefully or critically the text was read, and at times it can even reveal by whom.[8]

How many of these texts were used specifically for classroom instruction and how many for personal study is uncertain at best. Equally uncertain is whether or how often the science of optics was taught as a core subject (in "ordinary" lectures) or peripherally (in "extraordinary" lectures), or at what level in the arts program.[9] We can, however, infer a few things from the clues just discussed. To start with, it appears that Pecham's *Perspectiva communis* was a mainstay in the teaching of optics throughout the Middle Ages and Renaissance. Not only does the relatively large number of extant manuscripts support this conclusion, but also the nature of Pecham's analysis, which is fairly comprehensive but superficial, made it ideal as an introduction to the subject. Indeed, if it were published today, the *Perspectiva communis* would probably be retitled *Perspectiva ad asinos* or *Optics for Dummies*. Somewhat less suitable, but still useful as an introduction, would have been Roger Bacon's *Perspectiva*, by virtue of its broad scope and relative superficiality. Least suitable would have been Alhacen's *De aspectibus* and Witelo's *Perspectiva* because of their technical complexity and inordinate length. If dealt with in their entirety, these two works would most likely have been reserved for intensive, advanced study.[10]

That both Pecham's and Witelo's treatises were prescribed for the study of mathematics at Oxford and Prague has some interesting implications. One such implication centers on what it was that students were expected to learn from those texts. The issue here is focus. Were students meant to treat optics

---

8. For an example of how fruitful such annotation can be (although based on multiple printed copies rather than manuscripts), see Owen Gingerich, *The Book Nobody Read: Chasing the Revolutions of Nicolaus Copernicus*, 2nd ed. (London: Penguin, 2005).

9. Generally speaking, ordinary lectures were delivered in the morning by masters who were fully licensed to teach; extraordinary or "cursory" lectures were delivered in the afternoon by advanced students who had not yet achieved the master's degree.

10. The Latin version of the *De aspectibus* is around 200,000 words long, and Witelo's *Perspectiva* is roughly 70 percent longer. There is a rough parallel here with the teaching of astronomy. Advanced study in that discipline would have been based on Ptolemy's long and technically complex *Almagest* or perhaps Campanus of Novara's somewhat shorter *Theorica Planetarum*, whereas Sacrobosco's relatively concise *De sphera*, like Pecham's *Perspectiva communis*, would have been far preferable for study at the elementary level.

as a mere geometrical exercise with little or no regard to theoretical content? If so, then how much would they have actually learned about optics, even if only by osmosis? This question is especially pertinent to Witelo's *Perspectiva* because its long opening book is devoted solely to geometrical demonstrations abstracted from any real optical content.[11] If instruction were limited to this book alone, then the student would have learned very little, if anything, about optics at either the theoretical or practical level. Pecham's *Perspectiva communis*, on the other hand, integrates geometry and theory so fully that it would be difficult to master the one without the other, even if instruction were focused on ray geometry and its mathematical ramifications. Another implication has to do with the level of instruction and learning in optics. If Witelo's *Perspectiva* was the mandated text, as it could have been in Oxford according the statute of 1431, then that level would have been remarkably high, whether or not instruction was limited to the geometrical content of the first book. The opposite holds for Pecham's *Perspectiva communis* insofar as both its theoretical foundations and geometrical superstructure are fairly rudimentary.

Some things become more confused, others more clear when we take into account the medieval Latin translations of Euclid's *Optics* and *Catoptrics*, Ptolemy's *Optics*, and al-Kindī's *De aspectibus*. Altogether, these works are represented by no fewer than 124 extant manuscripts, most dating from the fourteenth century or later.[12] Such a large number suggests they were used fairly regularly as texts for the teaching of optics. What makes this confusing is that these particular works are based on strict extramissionism, whereas Alhacen and his perspectivist disciples all subscribed to intromissionism with varying degrees of rigor. Contradictory though they may be—and were recognized to be in the later Middle Ages and Renaissance—both theories seem to have been studied in the classroom, and, as we will see in due course, both were seriously entertained in the fifteenth and succeeding centuries.[13] This very fact indicates that while perspectivist theory may have served as an optical paradigm during the later Middle Ages and Renaissance, it had not put paid to the visual ray alternative. Still, the two theories are perfectly equivalent at the mathematical level, so the fact that both seem to have been taught indifferently suggests that the primary stress of that teaching was on ray geometry

11. In the current critical Latin edition, this book alone occupies just over one hundred pages and comprises 137 propositions; see Unguru, *Witelonis liber primus*, 216–316.

12. See Lindberg, *Catalogue*, 21–22, 46–55, and 74–75.

13. On the persistence of extramissionist theories of sight during the Renaissance, see, e.g., Mary Quinlan-McGrath, *Influence: Art, Optics, and Astrology in the Italian Renaissance* (Chicago: University of Chicago Press, 2013).

rather than on visual theory as an integrated whole. Such an approach implies that the analysis of physical cause, that is, light or visual flux and its rectilinear propagation, can be legitimately dissociated from the analysis of perceptual effect and thus studied as an independent subject in its own right. This is important because the dissociation of light and sight implicit in that approach may have helped shape the way optical theorists analyzed reflection and refraction in the sixteenth century.[14] That, for its part, may have influenced Kepler's understanding of how the lens of the eye functions in the visual process.

Given all these points, it seems safe to conclude that in general optics was taught in the university arts program at a fairly low level. Gregor Reisch, for instance, barely touches on ray geometry in his elementary account of direct, refracted, and reflected vision in book ten of the *Margarita Philosophica*, and although he gives lip service to "Witelo, Alhazen [*Alacen* in the original], Bacon, and others" at the end, the only perspectivist source he actually uses is Pecham.[15] Likewise, Reisch's rather murky account of image formation and distortion in reflection, based as it is on how images appear in the handle and bowl of a polished spoon, is descriptive rather than analytic and thus lacking in both theoretical and mathematical sophistication. It also seems safe to conclude that instruction in optics focused more on geometrical superstructure than on theoretical foundations. This conclusion is borne out by Reisch's bare bones ray analysis of refraction, which is heavily dependent on geometrical diagrams but offers virtually nothing in the way of theoretical explanation. We could hardly expect instruction at this level to inspire deep scholarly interest in the subject or to produce critical evaluation of theoretical principles. True, the manuscript traditions of Alhacen's *De aspectibus* and Witelo's *Perspectiva*—by far the most advanced sources of optical knowledge in the later Middle Ages and Renaissance—testify to the fact that some scholars took optics seriously enough to delve into its analytic and theoretical intricacies through private or advanced classroom study. But even at this level we find little in the way of critical or creative thinking, and what we do find of such thinking is often muddled. Thus although various later medieval scholars wrote commentaries

14. On this point see Sven Dupré, "Kepler's Optics without Hypotheses," *Synthese* 185 (2012): 501–25.

15. See Cunningham and Kusukawa, *Natural Philosophy Epitomized*, 183–92. That Reisch did not fully understand the ray analysis of vision is evident from his attempt to explain why we see the sun and moon as disks rather than spheres: "That we see the sun and stars as round comes from the fact that they are very distant, and the cone [of radiation] is no longer taken in the highest point but in a curved line. For this [line] appears larger to a more remote viewer and smaller to a closer viewer" (translation adapted from ibid., 184).

on the science of *perspectiva*, their approach was often piecemeal, according to "questions" that focus narrowly on specific issues.

There are exceptions, of course. Soon after the turn of the fourteenth century, Theodoric of Freiberg, who achieved an advanced degree in theology in the late 1290s, offered an astute explanation of the rainbow in his *De iride* based on refractions into individual raindrops, internal reflections within them, and refractions back out at specific angles tied to specific colors. Not only is his explanation on target conceptually, but it is also remarkably similar to the one proposed concurrently and independently by Kamāl al-Dīn al-Fārisī in his *Tanqīḥ*.[16] Later in that same century, Nicole Oresme proposed a significant modification of Alhacen's account of atmospheric refraction on the basis of logical consistency, insisting that light rays passing down through the increasingly dense layers of the atmosphere should be continually refracted toward the normal to follow a curvilinear rather than a rectilinear trajectory.[17] This modification occurs in the course of a strikingly original analysis designed to show that because of atmospheric refraction and parallax, it is impossible to determine the actual location of any celestial body, the ulterior point perhaps being to demonstrate the futility of judicial astrology.

Such exceptions are rare, though, and if the number of extant manuscripts is anything to go by, Theodoric and Oresme had virtually no influence on the mainstream of optical thought during the later Middle Ages or Renaissance.[18] On the whole, late medieval academics and university-educated scholars, even the most erudite, seem to have been content to accept what they had been taught or had read about optics without questioning it very deeply or considering its possible ramifications at the level of either theoretical principles

16. Still standard is the account Carl B. Boyer gave of each in *The Rainbow: From Myth to Mathematics* (New York: Yoseloff, 1959); see also Raymond Lee and Alistair Fraser, *The Rainbow Bridge: Rainbows in Art, Myth, and Science* (University Park: Pennsylvania State University Press, 2001).

17. According to Alhacen, the atmosphere becomes increasingly dense as it approaches the earth's surface, the fire at the top of the atmospheric shell being least dense and the air at the bottom densest. Yet he insists that when a ray of light is refracted at the interface between the fire at the top of the atmosphere and the less-dense celestial ether beyond, it follows a straight path through the atmosphere; see Smith, *Alhacen on Refraction*, 320–21. Oresme accepts Alhacen's atmospheric model but insists on logical grounds that the ray must be continuously refracted as it passes through the atmosphere to form a curved rather than a straight line; see Dan Burton, ed. and trans., *Nicole Oresme's* De visione stellarum (Leiden: Brill, 2007), 150–61.

18. Lindberg, *Catalogue*, lists four manuscripts for Theodoric's *De iride* (see 75–76), and Burton gives the same number for Oresme's *De visione stellarum* (see 65–67).

or application. As a result, they appear to have lacked both imagination and intellectual curiosity, a point borne out by the failure of any scholar before the mid-sixteenth century (at least as far as we know) to address the problem of exactly how eyeglasses correct poor vision. This failure is ironic because many late medieval scholastic presbyopes and myopes must have worn spectacles while reading the very optical works to which they might have looked, albeit in vain, for a definitive solution.[19]

Far from a vibrant marketplace of new ideas or approaches to optics, the late medieval university thus appears to have been a bastion of conservative group-think. This is hardly surprising. Institutions, by their very nature, tend to foster conformity of thought and outlook and therefore to discourage radical change or innovation in either. Furthermore, the transmission of ideas through manuscripts was limited by the medium itself. The production of codices by hand on parchment was extraordinarily time consuming and expensive, so texts were hard to come by and were hoarded and cherished accordingly. Meantime, the oral transmission of ideas through personal discussion was limited by the relative sparsity and tenuousness of lines of communication and the difficulty of using them. Travel was slow, arduous, and potentially dangerous. Equally slow, personal correspondence was also unreliable and expensive.

Consequently, even with the proliferation of universities and the institutionalization of learning within them, the academic community of the later Middle Ages and Renaissance was dispersed in fairly small, isolated pockets defined by city walls and interconnected in an attenuated network of links that were subject to political rupture at any moment.[20] These pockets, moreover, were recurrently raked by diseases such as bubonic plague, which remained endemic in Europe for centuries after the epidemic of 1347–51. Small wonder that in such an environment the pace of intellectual change was glacial, or that when it happened, such change tended to be local. Nevertheless, despite their deficiencies, late medieval universities were vital to the preservation and transmission of learning throughout Europe, a point attested to in the late fourteenth century by none other than the North African Muslim savant Ibn Khaldūn. "We further hear," he writes in chapter 6 of his *Muqaddimah*, "that the philosophical sciences are greatly cultivated in the land of Rome and along the adjacent shore of the country of the European Christians [and] they are

19. See Ilardi, *Renaissance Vision*, 51–152, where he gives a fascinating and detailed account of how eyeglasses became increasingly common during the fourteenth and fifteenth centuries.

20. As in so many other respects, Italy was something of an exception because of its relatively small size and close-knit urban structure.

said to be . . . taught in numerous classes." Moreover, he continues, "existing systematic expositions of them are said to be comprehensive, the people who know them numerous, and the students of them very many."[21]

2. THEOLOGY AND THE EMERGENCE OF OPTICAL LITERACY

Whatever else it may have been, the late medieval university was not a hermetically sealed community. For one thing, most universities were situated in urban centers, so both masters and students had no choice but to interact at various levels with the broader community within those centers. For another thing, the majority of students who matriculated at medieval universities never achieved a degree. Indeed, as one scholar puts it, "from the perspective of today, the great proportion of medieval university attenders would have to be classified as drop-outs or failures."[22] Once parted from the university, these dropouts carried whatever learning they had acquired with them and doubtless shared it with others who were unable to benefit from higher education. This would certainly have been the case within an urban environment. University dropouts also constituted an audience for ideas disseminated through such popular channels as sermons and literature. As such, they formed a bridge of sorts between academic and nonacademic communities and, in the process, must have raised the level of popular discourse somewhat according to their receptivity to scholastic ideas and allusions. They also entered into that discourse as authors. Just as today, then, so during the later Middle Ages and Renaissance, advanced learning percolated outward and downward.

One salient feature of a university education was its breadth. Every arts student was expected to learn, if not master, a wide variety of subjects from logic to astronomy, and a master of arts degree was prerequisite to entering into the higher faculties of medicine, law, and theology. It was only natural (and indeed expected), therefore, that what was learned in the arts curriculum should be carried over and applied with suitable discretion to the study of these higher disciplines. Accordingly, advanced learning in natural philosophy percolated inward and upward. As regards optics and theology, this upward percolation is manifested in several ways. For instance, in her *Vision and Certitude in the Age of Ockham*, Katherine Tachau has demonstrated a close connection between

21. Rosenthal, *Muqaddimah*, 375.

22. Rainer Schwinges, "Student Education, Student Life," in *A History of the University in Europe*, ed. Walter Rüegg, vol. 1 (Cambridge: Cambridge University Press, 1992), 196.

epistemology and perspectivist visual theory, as mediated by Bacon, in a host of *Sentences* commentaries produced during the late thirteenth and early four-teenth centuries. As we saw in the previous chapter, the focus of her analysis is on whether species are necessary for cognition, and this issue extends beyond human to angelic cognition. How, after all, do angels apprehend the world at large? If they are incorporeal (and thus brainless), they cannot do so by means of the corporeal species or phantasms upon which human perception and cognition seem to depend. What about intelligible species, though? Are they necessary for angelic cognition, and if so, how do angels acquire them without access to the sensible species from which they are supposedly abstracted? Or is angelic cognition immediate and intuitive?[23] This is heady stuff, and so is the attempt by John Wyclif (ca. 1328–1384) to explain Christ's presence in the Eucharist by analogy to reflection. "In the case of the sacrament of the altar," he argues in book four of the *Trialogus*, "even though the bread is broken into three or however many pieces, each of those [pieces] is not really but figura-tively (*habitudinaliter*) the same body [that is, of Christ], just as to the viewer looking into different mirrors, the same face exists intentionally in any one of them."[24] Christ is thus intentionally, or virtually, present everywhere in the consecrated host in somewhat the same way as a single object is intentionally or virtually present "in" one or more mirrors through its representative image.

Wyclif's account implies that the communicant somehow "sees" Christ's presence in the host by means of spiritual vision (Wyclif in fact maintains that Christ has spiritual existence [*esse spirituale*] in the host), and as we have al-ready seen with Bacon, the analysis of physical vision is an obvious organizing principle for the discussion of spiritual vision. But whereas Bacon provided no more than a few examples through analogy, his younger contemporary, Peter of Limoges, approached this analogizing process systematically in his *Tractatus*

---

23. On this question, see, e.g., Tachau, *Vision and Certitude*, 60–61, 249–55, and 263–64. See also Dominik Perler, "Thought Experiments: The Methodological Function of Angels in Late Medieval Epistemology," in *Angels in Medieval Philosophical Inquiry*, ed. Isabel Iribarren and Martin Lenz (Aldershot, UK; Burlington, VT: Ashgate, 2008), 143–53.

24. Gotthard Lechler, ed., *Johannis Wiclif: Trialogus cum supplemento Trialogi* (Oxford: Clarendon Press, 1869), 272; my translation. See also Heather Phillips, "John Wyclif and the Optics of the Eucharist," in *From Ockham to Wyclif*, ed. Anne Hudson and Michael Wilks (Oxford: Blackwell, 1987), 245–58. Interestingly enough, Pope Innocent III (r. 1198–1216) used similar mirror imagery in discussing the Eucharist; see Herbert Kessler, "Speculum," *Speculum* 86 (2011): 1–41, especially 36. Wyclif's focus on reflection is also evident in the analogical asso-ciation he drew between the seven deadly sins and the seven types of plane and curved mirrors canonized by Alhacen.

moralis de oculo. Although Peter never mentions Bacon in that work, the Baconian inspiration for it is manifest. Like Bacon, for instance, Peter distinguishes spiritual vision according to whether it is direct (in the state of postresurrection glory), refracted (in the separated soul after death), or reflected (in this life). Like Bacon, as well, Peter appeals to the sevenfold tunics and humors of the eye as a guide to interpreting the psalmist's plea that God protect us as the pupil of His eye. And in Baconian fashion, Peter likens the need for both intromission and extramission in physical sight to the need for God's infused, grace-giving power and for our own reciprocal effort in spiritual sight.[25]

Peter goes well beyond Bacon, however, in extending the scope of optical analogies, often with considerable ingenuity. "It is proven in the Perspectivist science," for instance, "that a visible object appears to be larger than it is when the eye is in a medium of lesser density [and vice versa] when the eye is located in a medium of greater density." Thus "when a poor person living in the dry land of poverty sees someone overflowing with worldly riches, he will consider the rich man to be great." Conversely, a wealthy person "immersed in the transitory riches of this world . . . considers [a poor person] to be of ordinary size, when in fact he is great in the divine eye." These deceptions can be corrected by spiritual judgment, just the way deceptions in physical vision are corrected by the faculty of judgment in "the common nerve located on the surface of the brain, where the two [optic] nerves come together." The moon illusion affords another example of false magnification. For, as "Alhacen teaches in book 7 of his *Optics* . . . stars located near the horizon . . . appear larger to the eye than when they have risen to the middle of the sky." Just so, "the greater the heights of honors which saintly people appear to ascend, the more they denigrate themselves through humility and the more inferior they desire to appear in the eyes of other people."[26] Proper spiritual judgment will correct this initial misperception of diminished size.

25. On the three types of spiritual vision, see Newhauser, *Peter of Limoges*, 12–13; on the sevenfold protection of the spiritual pupil, 6–7; on spiritual intromission and extramission, 14–15. Newhauser suggests quite plausibly that Peter failed to mention Bacon because at the time Peter wrote the *Tractatus*, Bacon was under a cloud with the Franciscan order and citing him as an authority would have been impolitic. For an analysis of the textual connections between Peter and Bacon, see Newhauser, "*Inter scientiam et populum*: Roger Bacon, Peter of Limoges, and the 'Tractatus moralis de oculo,'" in *Nach der Verurteilung von 1277*, ed. Jan A. Aertsen and others (Berlin: Walter de Gruyter, 2001), 682–703.

26. On the apparent magnification of the wealthy and the corresponding apparent diminution of the poor, see Newhauser, *Peter of Limoges*, 38–41; on spiritual judgment and visual judgment in the "common nerve," 16–17; and on the moon illusion, 41–44.

Peter also likens God to "a mirror without blemish" into which we should look "frequently so that we can discern in ourselves the blemishes of our mind and cleanse them."[27] This is a form of introspection, of course, but it is mediated, quite literally, by reflection, so what we see is not ourselves, to the very depth of our souls, but rather an image or appearance of ourselves. Dallas Denery views Peter's use of this analogy as symptomatic of a growing emphasis during the thirteenth century on self-reflection as the path to true contrition for our blemished nature. The problem, according to him, is that "what [reveals] itself in the soul's forum [is] not the soul itself, but its appearance, and no appearance can guarantee its own truth."[28] But if self-knowledge is uncertain, how much more uncertain must all other knowledge be if it is based on appearance only? Hence, Denery argues, "when he articulated the practice of medieval religious life in the language of perspectivist optics," Peter "simultaneously articulated the problems that those practices had generated for the religious as specifically visual problems . . . arising from a distinction between what appears and [what] really exists." Within this context, Denery concludes, "the epistemological controversies that raged . . . during the first half of the fourteenth century . . . can be read as philosophical elaborations of problems arising from within medieval religious life itself."[29] By framing religious issues in optical terms, Peter put those issues in a particular analytic light that determined how they and their ramifications would be understood.

Peter of Limoges was by no means the only later medieval theologian to draw parallels between spiritual and physical vision on the basis of perspectivist principles. The *Tractatus de luce* of his fellow Franciscan, Bartholomew of Bologna (d. after 1294), exploits such parallels on the basis primarily of the "light metaphysics" of Grosseteste's *De luce*, and Saint Antonino of Florence (Antonio Pierozzi, 1389–1459) links spiritual and physical vision according to a thoroughgoing grasp of perspectivist optics.[30] We will have more to say about Saint Antonino a bit later, but for now we can close this particular discussion with another Florentine example. This one is found in a set of two sermons delivered in 1494 and 1496 by the Dominican firebrand Girolamo Savonarola, the redoubtable and wildly popular preacher who held Florence in spiritual and political thrall for four years until his arrest and execution in 1498. In both

---

27. Ibid., 187.

28. Dallas Denery, *Seeing and Being Seen in the Later Medieval World* (New York: Cambridge University Press, 2005), 72.

29. Ibid., 169.

30. For the text of Bartholomew of Bologna's treatise, see Ireneo Squadrani, "*Tractatus de luce* fr. Bartholomaei de Bononia," *Antonianum* 7 (1932): 139–238, 337–76, 465–94.

sermons he counsels his congregation to don spiritual eyeglasses so as to see the good more clearly. "There are those," he explains, "who see little, and so they wear glasses to see better because through [them] small things appear bigger . . . because the species of the letters that enter the eye by means of the active light strike on the glasses and here they are spread, widen, and appear bigger."[31] In like manner, we need to amplify our spiritual vision for the sake of our souls. Savonarola's homely analogy reveals at least some understanding of perspectivist refraction theory as it applies to magnification, an understanding that he presumably thought his audience would share.

3. OPTICAL MOTIFS IN LITERATURE

If theology provided a natural home for optical ideas in general and perspectivist ideas in particular during the later Middle Ages and Renaissance, no less natural was the home literature provided for such ideas. After all, no form of verbal communication is more self-consciously dependent on imagery than literature, and optical imagery is an obvious structural device. The centrality of light imagery in Dante's *Divine Comedy*, for instance, is impossible to miss, and much has been written about how Dante used optics and what optical sources he drew upon in constructing that imagery. Among the resulting studies, Suzanne Conklin Akbari's recent *Seeing through the Veil: Optical Theory and Medieval Allegory* has two particular virtues. First, it is based on an unusually firm grasp of the philosophical background to medieval optics as it evolved over the thirteenth century.[32] Second, rather than limit herself to a single author, Akbari deals with four, in chronological order, from Guillaume de Lorris (ca. 1200–ca. 1240) and Jean de Meun (ca. 1240–ca. 1305) to Dante (1265–1321) and Chaucer (1343?–1400).[33]

As its title suggests, Akbari's study takes as its theme the "veil" posed by the intermediate entities upon which medieval visual theory is predicated and the problem of how those entities both facilitate and hinder our proper perception of things. Ranging from the transparent, physical medium and the spiritual medium pervading the eyes and sensitive soul, to the species passing through those media and the reflective and refractive interfaces that redirect their pas-

---

31. Quoted from Ilardi, *Renaissance Vision*, 179.

32. See especially Akbari, *Seeing through the Veil* (Toronto: University of Toronto Press, 2004), 21–44.

33. For a fairly typical, recent single-author study, see Peter Brown, *Chaucer and the Making of Optical Space* (Oxford: Peter Lang, 2007).

sage, these intermediate entities can all lead to misperception in one way or another. Yet, ironically, without them perception and cognition would be impossible. The irony here is manifested par excellence in mirrors. On the one hand, they are notorious for distorting what we see in them, giving false appearances that can veil the objects generating them from proper apprehension. On the other, the science of optics allows us to penetrate this veil by rectifying the distortion according to the laws of reflection and image formation. In other words, even mirror imaging, if properly understood, can be a pathway to truth.

When seeing is taken as a metaphor for knowing, as it routinely was during the Middle Ages, all the potential uncertainties of the one extend to the other according to the equivalent veils that need to be "seen" through, including the veil of allegory that is central to so much medieval literature. In this respect, mirrors loom large. We have already seen examples in Peter of Limoges's mirror of introspection, Wyclif's mirror of the host, and the mirror of the imagination into which the mind's eye looks for intellectual enlightenment. Within the context of such image imagery, Akbari shows not only how the four authors selected by her used that imagery in their respective allegories but also how each author relied on particular optical theories in constructing them. She argues, for instance, that Guillaume de Lorris, author of the first segment of the *Roman de la Rose*, which was completed before 1225, based his optical imagery on the extramissionist account of William of Conches, itself based on Plato's *Timaeus*. Key to Guillaume's narrative, Akbari continues, is the mirroring of self-reflection. Accordingly, the pivotal point in that narrative is "located at the fountain of Narcissus and, more specifically, at the moment that the lover looks into the [reflective] surface of that fountain."[34] In doing so, he is deceived into thinking that what he sees there—himself—is the love he seeks. His search is thus grotesquely misdirected toward self-love by his failure to see through the veil of deception the fountain's mirroring surface posed.

Jean de Meun, who completed the much longer second segment of the poem around 1265, takes a different allegorical tack from Guillaume. For a start, Jean's knowledge of optics is far more up to date than Guillaume's, as witness his citation of Alhacen's *De aspectibus*, or "Livre des *regarz*," which is indispensable to anyone "who wants to know about the rainbow." The same holds for mirrors, "for [in Alhacen's book the reader] will be able to discover the causes and the strengths of mirrors that have such marvelous powers that . . . things that are very small . . . are seen as so great and large" that they can be clearly discerned

---

34. Ibid., 47. On Guillaume's reliance on William of Conches, see ibid., 57–63.

at a distance.³⁵ These points, Akbari correctly notes, are falsely attributed to Alhacen. Nowhere does he analyze the rainbow in the *De aspectibus*, nor does he describe how concave mirrors magnify things to make them look near and large. Roger Bacon, on the other hand does both in the *Opus majus*, which leads Akbari to suppose, quite plausibly, that Jean's knowledge of Alhacen was filtered through Bacon, from whom he learned the theory of species multiplication. The impact of this theory is evident, Akbari argues, in Jean's emphasis on the parallel between "the multiplication of visible species [and] the multiplication of the human species . . . in sexual reproduction."³⁶

Written some forty-odd years after Jean de Meun's continuation of the *Roman de la Rose*, Dante's *Divine Comedy*, as well as his *Convivio* and other earlier works, has been closely scrutinized over the past several decades for its optical content. The *Divine Comedy* lends itself especially well to such scrutiny because, as noted earlier, light and vision are so central to its narrative. Indeed, as Akbari interprets it, Dante's imagined ascent from hell, through purgatory, to paradise, is really a passage toward an eventual face-to-face gaze at God that "is mutual, a reflection of the desire of the soul for God and that of God for the soul, expressed as Grace."³⁷ Optical allusions thus crop up regularly throughout Dante's narrative. In order to penetrate the darkness of Hell, for example, Dante and his guide, Virgil, rely upon visual radiation, but "as Dante progresses through purgatory, the language of optics becomes more sophisticated" and, accordingly, more intromissionist in tenor.³⁸ This shift from an extramissionist to an intromissionist visual model is complete by the time Dante enters paradise, where his view of God is mediated by reflection through Beatrice until, finally, his vision becomes perfect enough to face God's ineffable brightness directly. In essence, Dante's ascent from almost complete blindness in hell to perfect vision in paradise is an allegorical reflection of the *Itinerarium mentis in Deum* ("The Journey of the Mind into God") described by the great Franciscan theologian Saint Bonaventure (1221–1274).³⁹ It is also a passage from the relative blindness of seeing dimly through a looking glass to

35. Charles Dahlberg, trans., *The Romance of the Rose* (Hanover, NH: University Press of New England, 1983), 300.

36. Akbari, *Seeing*, 94. On Jean's reliance on Bacon, see ibid., 90–96.

37. Ibid., 170.

38. Ibid., 159.

39. On this "journey" and its ramifications in medieval "popular" culture, see Michelle Karnes, *Imagination, Meditation, and Cognition in the Middle Ages* (Chicago: University of Chicago Press, 2011).

the clarity of the direct, face-to-face vision implicit in 13:12 of Paul's first letter to the Corinthians. Moreover, Dante's use of apparently incompatible visual models (Platonist and Aristotelian) was due not to ignorance or confusion but to his having chosen "among different optical models depending upon the nature of the fiction he wished to construct."[40]

That Dante was not only familiar with scholastic visual theory but also had some technical understanding of it is evident from his mention of the equal-angles law of reflection when describing how he was dazzled by light from an angel.[41] Less evident is precisely where he acquired that understanding. Akbari follows scholarly consensus in tracing much of what Dante has to say about light and vision to Albertus Magnus, but she is on somewhat shakier ground in insisting that over the course of his narrative Dante incorporates ideas from Witelo's *Perspectiva*.[42] Whether Dante was actually familiar with Witelo or any other perspectivist source is somewhat beside the point, though. Much of the optical lore in his narrative, especially during the passage through paradise, is at least consonant with perspectivist theory. To read such theory into that narrative is therefore not entirely implausible, even though it may be a stretch—the sort of stretch Renaissance and early modern readers of the *Divine Comedy* would have been perfectly comfortable making.

Aware of the problems raised by mediation in intellectual vision, Dante nonetheless viewed seeing, with all its problematic entailments, as an apt metaphor for knowing and the achievement of "wisdom" through desire or love. By the time of Chaucer, however, faith in the applicability of that metaphor had faltered as conviction in the universality and certainty of knowledge waned during the fourteenth century. With Chaucer himself, Akbari contends, this loss of faith is manifested over the course of his writing in the way he "abandons vision as a potential mediator between subject and object, and instead

---

40. Akbari, *Seeing*, 115.

41. Ibid., 160: "As when the beam leaps from the water or the mirror to the opposite quarter, rising at the same angle as it descends, and at equal distance departs as much from the line of the falling stone [i.e., the normal], even as experiment and science show; so it seemed to me that I was struck by light reflected there in front of me"; Charles Singleton, trans., *Dante Alighieri: The Divine Comedy: Purgatorio* (Princeton, NJ: Princeton University Press, 1973), 155–56.

42. See Akbari, *Seeing*, 141–42. Witelo was identified as a source for Dante over fifty years ago by Alessandro Parronchi; see "La perspettiva dantesca," in Parronchi, *Studi su la 'dolce' prospettiva* (Milan: Aldo Martello, 1964), 4–90. In *Medieval Optics and Theories of Light in the Works of Dante* (Lewiston, NY: Edward Mellen, 2000), however, Simon Gilson argues that Dante knew nothing of Witelo or the other perspectivists; see also Gilson, "Dante and the Science of 'Perspective': A Reappraisal," *Dante Studies* 115 (1997): 185–219.

turns to the role of hearing." Hence, in his later works Chaucer emphasizes the personal relativity of visual perception and cognition, thereby enabling "the reader to see from the perspective of the teller, and thus to receive a picture of the world that does not pretend to be true or sufficient, but merely an accurate rendering of what it looks like from one point of view."[43] In essence, Chaucer had lost confidence in the Aristotelian visual paradigm scientifically refined and stabilized by the likes of "Alocen and Vitulon . . . that writen . . . of queynte mirours and of perspectives."[44]

As Akbari views it, Chaucer's loss of confidence in that paradigm marks something of a literary turning point, for "in the years after 1400 [the year of Chaucer's death], the use of vision as a metaphor for knowledge continued to decline in the realm of secular writing."[45] Perhaps so, but if we skip ahead some two centuries to the publication in 1599 of Sir John Davies's self-reflective poem, *Nosce Teipsum*, we see that this metaphor was far from dead. Quite the contrary, in fact, it was still vibrantly alive and firmly based on the perspectivist visual paradigm. The following extended excerpt speaks to this point so clearly, completely, and succinctly that any commentary or elucidation would be superfluous:

> This *power* is *Sense*, which from abroad doth bring
> The *colour*, *taste*, and *touch*, and *sent*, and *sound*;
> The *quantitie*, and *shape* of euery thing
> Within th' Earth's center, or Heaven's circle found.

> This power, in parts made fit, fit obiects takes,
> Yet not the things, but forms of things receiues;
> As when a seale in waxe impression makes,
> The print therein, but not it selfe it leaues.

> . . . .

> First, the two *eyes* that haue the *seeing* power,
> Stand as one watchman, spy, or sentinell;
> Being plac'd aloft, within the head's high tower;
> And though both see, yet both but one thing tell.

43. Quotations, in order, from Akbari, *Seeing*, 178 and 233.
44. See note 7, above.
45. Akbari, *Seeing*, 236.

These mirrors take into their little space
The formes of *moone* and *sun*, and euery starre;
Of euery body and of euery place,
Which with the World's wide armes embracèd are:

Yet their best obiect, and their noblest vse,
Hereafter in another World will be;
When God in them shall heauenly light infuse,
That face to face they may their *Maker* see.

Here are they guides, which doe the body lead,
Which else would stumble in eternal night;
Here in this world they do much knowledge *read*,
And are the casements which admit most light:

They are her farthest reaching instrument,
Yet they no beames vnto their obiects send;
But all the rays are from their obiects sent,
And in the *eyes* with pointed angles end:

If th' obiects be farre off, the rayes doe meet
In a sharpe point, and so things seeme but small;
If they be neere, their rayes doe spread and fleet,
And make broad points, that things seeme great withall.

Lastly, nine things to *Sight* requirèd are;
The *power* to see, the l*ight*, the *visible* thing,
Being not too *small*, too *thin*, too *nigh*, too *farre*,
Cleare space, and *time*, the forme distinct to bring.

Thus we see how the *Soule* doth use the eyes,
As instruments of her quicke power of sight;
Hence do th' Arts *opticke* and faire *painting* rise:
Painting, which doth all gentle minds delight.

. . . .

These are the outward instruments of Sense.
These are the guards which every thing must passe

Ere it approch the mind's intelligence,
Or touch the Fantasie, *Wit's looking-glasse.*

. . . .

And yet these porters, which all things admit,
Themselves perceiue not, nor discerne the things;
One *common* power doth in the forehead sit,
Which all their proper formes together brings.

For all those *nerues,* which *spirits of Sence* doe beare,
And to those outward organs spreading goe;
Vnited are, as in a center there,
And there this power those sundry formes doth know.

Those outward organs present things receiue,
this inward *Sense* doth absent things retaine;
Yet straight transmits all formes shee doth perceiue,
Unto a higher region of the *braine.*

. . . .

Where *Fantasie*, neere *hand-maid* to the mind,
Sits and beholds, and doth discerne them all;
Compounds in one, things diuers in their kind;
Compares the black and white, the great and small.

. . . .

Yet alwayes all may not afore her bee;
Successiuely, she this and that intends;
Therefore such formes as she doth cease to see,
To *Memorie's* large volume shee commends.

The *lidger-booke* lies in the braine behinde,
Like *Ianus'* eye, which in his poll was set;
The *lay-man's tables, store-house of the mind,*
Which doth remember much, and much forget.

Heere *Sense's apprehension*, end doth take;
As when a stone is into water cast,
One circle doth another circle make,
Till the last circle touch the banke at last.

. . . .

The Wit. the pupill of the *Soule's* cleare eye,
And in man's world, the onely shining *starre;*
Lookes in the mirror of the Fantasie,
Where all the gatherings of the *Senses* are.

From thence this power the shapes of things abstracts,
And them within her *passiue part* receiues;
Which are enlightned by that part which *acts,*
And so the formes of single things perceiues.

But after, by discoursing to and fro,
Anticipating and comparing things;
She doth all vniversall natures know,
And all *effects* unto their *causes* brings.

. . . .

When she *rates* things and moues from ground to ground,
The name of *Reason* she obtaines by this;
But when by Reason she the truth hath found,
And *standeth fast*, she VNDERSTANDING is.[46]

4. RENAISSANCE ART, NATURALISM, AND OPTICS

Davies's association of "th' Arts *opticke* and faire *painting*" in the tenth stanza of this excerpt provides an appropriate entry point for the discussion of how the two became interconnected during the Renaissance. In one sense, the inter-connection is self-evident because painting is intrinsically visual. By the early

---

46. From Alexander Grosart, ed., *The Complete Poems of Sir John Davies*, vol. 1 (London: Chatto and Windus, 1876), 64–76.

fifteenth century, however, painting and optics became associated at a deeper level with the purported imposition of optical principles on visual space as a means of rendering painted scenes as natural looking as possible. Key to this naturalizing process was the development of linear, or "artificial," perspective, and here, as in so many other contexts, the mirror takes center stage. First, by ordering visual space according to the rules of linear perspective, Renaissance artists aimed to deceive the viewer into seeing what was painted on a flat surface as if it had depth, just as virtual images "behind" a plane mirror appear to exist in three dimensions. By this expedient, Renaissance painters strove to give the illusion of reality in their depictions, and the better the illusion the more naturalistic, or realistic, the depiction, which is to say that "[artificial] perspective rests on the ultimate visual paradox: complete deception in the service of utter veracity."[47]

The link between Renaissance art and mirror imaging is not just figurative. At an operational level, mirroring appears to have been crucial to the development of linear perspective in the early fifteenth century by the Florentine artist/architect, Filippo Brunelleschi (1377–1447). According to several early accounts, Brunelleschi painted two depictions (now lost), one of the octagonal Baptistery in the plaza of the Duomo, the other of the Palazzo Vecchio, both depictions reputed to have been so lifelike that onlookers were lost in amazement. Let us dwell for a moment on the rendering of the Baptistery. Brunelleschi's late fifteenth-century biographer, Antonio Manetti (1423–1497), tells us that he painted it on a panel roughly a foot square (or perhaps two square feet) from a position somewhat less than six feet inside the central portal of the Duomo, directly opposite the east doors of the Baptistery. Then, having drilled a small hole through the middle of the Baptistery's painted door on the panel, Brunelleschi had the viewer hold the back side of the panel up to his eye and look through the hole at a plane mirror held facing the panel's front. The viewer would thus see the Baptistery's depiction reflected in the mirror and,

---

47. Stuart Clark, *Vanities of the Eye: Vision in Early Modern European Culture* (Oxford: Oxford University Press, 2007), 83. It bears noting that Greek and Greco-Roman artists also used various forms of linear perspective as well as modeling techniques, such as highlighting and shading, to give the illusion of depth in paintings and mosaics. Ptolemy, in fact, adverts to two such techniques in his *Optics*, one (aerial perspective) already mentioned in chapter 3, section 1, the other in *Optics*, II, 125 (Smith, *Ptolemy's Theory*, 122). For a useful account of illusionism in Greek and Greco-Roman art, see David Summers, *Vision, Reflection, and Desire in Western Painting* (Chapel Hill: University of North Carolina Press, 2007), especially 16–42.

by standing at the right spot within the Duomo, could presumably compare the mirror image to the reality in order to establish its "utter veracity."[48]

A considerable amount of ink and silicon have been spilled during the past several decades over precisely how Brunelleschi painted this depiction. For the most part, the resulting explanations have been guided by the question of how, or even whether, his method for doing so matched the one Leon Battista Alberti (1404–1472) articulated in his Italian *Della pittura* and Latin *De pictura* of 1435 and 1436, respectively.[49] As Alberti describes it, this method is based on the cone encompassed by the extreme or outer "visual rays" that define the boundary of any plane within the field of view and that form the visual angle. From the vertex of this cone, which lies within the eye and forms what amounts to the center of sight, the "centric ray" extends along the cone's axis to that plane. The "prince of rays," this is "the most vigorous and lively of all" and thus the one along which clearest vision occurs. The remaining "median rays" that fill the space between the extreme and centric rays are "impregnated with the acquired variety of light and colors" on the object surfaces with which they are in contact.[50]

With the visual cone and its constituent rays so defined, Alberti counsels us to take the panel on which we intend to paint our scene, or "historia," and draw a quadrangle, as represented in figure 7.1. This will form a sort of window through which we can imagine viewing that scene. Having arbitrarily established the height AC of a man in the plane of the panel, we subdivide that height into three equal units (each representing one *braccio* = ca. two feet) and then divide bottom edge GF of the quadrangle into as many of these units as feasible. At the height of the man on the face of the panel we pick point C, where the centric ray of an ideal viewer will touch it, and draw lines to that point from all the points of subdivision on GAF. These lines will represent parallels converging at an "almost infinite distance" within the imaginary space beyond the window. In effect, then, the centric point can be taken as the van-

48. See Howard Saalman and Catherine Engass, eds. and trans., *The* Life of Brunelleschi *by Antonio di Tuccio Manetti* (University Park: Pennsylvania State University Press, 1970), 42–44. Manetti's biography most probably dates from the 1480s (see ibid., 10–11).

49. For a useful summary of the various kinds of explanations offered, and their problems, see Martin Kemp, *The Science of Art* (New Haven, CT: Yale University Press, 1990), 9–14 and especially 344–45. Although the question of which version, Italian or Latin, of Alberti's treatise was written first has not been definitively resolved, scholarly consensus now favors the Italian version; see especially Rocco Sinisgalli, ed. and trans., *Leon Battista Alberti*: On Painting (Cambridge: Cambridge University Press, 2011), 3–14.

50. On Alberti's description of the rays according to specific function, see ibid., 26–30.

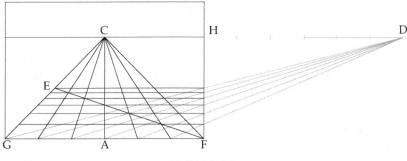

FIGURE 7.1

ishing point of Alberti's projection, so "centric line" CH parallel to GF and passing through centric point C can be taken to represent the horizon. Extending that line to point D so that, for instance, HD = GF, we then drop lines from D to all the intersection points on GF. Finally, from the points where those lines cross HF we draw lines parallel to GF. Each of the trapezoids formed by these transverse parallels with the lines converging on C will represent a square viewed obliquely from the established center of sight. If the procedure has been carried out properly, Alberti assures us, diagonals such as EF extending through a corner-to-corner succession of such squares will pass through all the corners without deviation.[51]

Returning to Brunelleschi's painting of the Baptistery, we can see at least one apparent point of convergence with Alberti's method: the centric ray passing through the panel's eyehole to the center of the east doorway in the mirror image of the Baptistery. Alberti's centric point can thus be imagined to coincide with this point in the virtual "space" of the Baptistery's mirror image. In a recent effort to explain Brunelleschi's method, Samuel Edgerton draws just this imagined connection, arguing that in taking this as the certification point of his painting, Brunelleschi inadvertently discovered Alberti's centric principle and the vanishing point associated with it.[52] How then might Brunelleschi have used this principle to organize his painting? Edgerton suggests that he fell back upon an earlier "bifocal construction" that had been used to produce the equivalent of the trapezoidal checkerboard pattern in figure 7.1. This

51. For Alberti's description of this projection scheme, see ibid., 39–42. Alberti never provides a diagram, so figure 7.1 is a conjectural construction, albeit one that is perfectly consonant with his description.

52. Samuel Y. Edgerton, *The Mirror, the Window, and the Telescope* (Ithaca, NY: Cornell University Press, 2009), especially 48.

construction requires that the bottom edge of a floor or top edge of a ceiling represented in the painting be subdivided into equal units. Two "focal" points are then taken on either side outside the frame of the painting, and lines are drawn from each of them to the points of subdivision. D in figure 7.1 represents one such point, its mate lying equidistant from C on the other side of the painting. The result is a sort of cat's cradle within which certain intersections determine where the lines of the checkerboard should fall. Within this system, the centric point automatically coincides with the midpoint of the line joining the two focal points.[53]

Having produced the grid in this way on the panel, Edgerton continues, Brunelleschi would have reproduced it on a mirror of the same size. Then he would have placed both the panel and the mirror side by side at the same height so that both were exposed to the Baptistery at the appropriate position inside the doorway of the Duomo. Facing both the panel and the mirror with his back to the Baptistery, finally, Brunelleschi would have mapped the mirror image onto the panel, "confident that their collective 'perspective' conformed to the actual catoptrics of the mirror."[54] In short, Brunelleschi's painting was intended as a faithful replication of a mirror image. It is presumably for this reason that one had to view it in a mirror in order to see its "utter veracity," particularly since the mirrored image had to be re-reversed if the Baptistery were to appear in its proper left-to-right orientation. For Brunelleschi, therefore, the Albertian window was actually a plane mirror, yet another "veil" through which the physical world presents itself to us. Nor is this interpretation of the window without some foundation. Alberti himself likens painting to mirroring and on that basis credits Narcissus with its invention, for "to paint, in fact, is what else if not to catch with art that [reflecting] surface of the spring?"[55]

Alberti's appeal to the cone of vision with its three types of rays suggests that his centric projection was based on optical theory and the "natural" perspective embedded in it. Several scholars, following this suggestion, have tried to ferret out the optical sources upon which he depended. Graziella Vescovini, for instance, pinpoints the *Questiones super perspectivam* of Biagio Pelacani di Parma (ca. 1345–1416), in part because it is based on perspectivist theory and in part because a copy of it was available in Florence in 1428, not long before Alberti wrote the two versions of his treatise.[56] Edgerton looks beyond

53. Ibid., 61–64. See also Kemp, *Science of Art*, 9–11.

54. Edgerton, *Mirror*, 66.

55. Sinisgalli, *Alberti*, 64.

56. See, e.g., Vescovini, "A New Origin of Perspective," *Res* 37 (2000): 73–81.

textual sources to the social and intellectual milieu of fifteenth-century Florence, where perspectivist optics "was such a hot topic . . . that artists as well as intellectuals [were] persuaded to learn something about [it], both for improving their pictorial techniques and making decisions about relevant subject matter."[57]

Exemplary of the resulting circle of interest is Lorenzo Ghiberti (1378–1455), whose *Commentario terzo* on art "is a compilation of quotations from various *perspectivist* texts," many of them lifted directly from the fourteenth-century Italian translation of Alhacen's *De aspectibus*.[58] A crucial influence upon, and perhaps within, the circle of interest that included Ghiberti, Edgerton speculates, was Saint Antonino, whom we encountered briefly in the previous section. "Among the most admired and learned intellectuals in Florence during the first half of the fifteenth century," he was expert in perspectivist optics and, like Peter of Limoges, brought that expertise to bear in the pulpit, where he drew the expected analogies between physical and spiritual vision.[59] "Is it possible," muses Edgerton, "that Lorenzo Ghiberti was so moved by [the resulting] optical allegories that he decided to collect and read the optical tracts himself?" Might it not be the case, in fact, that Saint Antonino was directly involved as an advisor in the production of certain paintings?[60]

Whatever the merits or demerits of Edgerton's case, which is circumstantial at best, the central issue is not whether the artists implicated in the development of linear perspective were interested in perspectivist optics but how instrumental perspectivist optics was in that development. In other words, did perspectivist theory play a central, formative role, or was it used as an ex post facto justification for a method developed independently? Not surprisingly, Edgerton favors the stronger position, that is, that perspectivist theory played a formative role, but there are compelling reasons to favor the weaker alternative.[61] For a start, Alberti's method for projecting the transverse paral-

---

57. Edgerton, *Mirror*, 31.

58. Ibid. See also Vescovini, "Alhazen vulgarisé."

59. On Saint Antonino, see Edgerton, *Mirror*, 30–38.

60. Ibid., 36.

61. For examples of the stronger position, see Edgerton, *The Heritage of Giotto's Geometry* (Ithaca, NY: Cornell University Press, 1991); Hans Belting, *Baghdad and Florence: Renaissance Art and Arab Science*, trans. Deborah Lucas Schneider (Cambridge, MA: Harvard University Press, 2011); and Gérard Simon, "Optique et perspective: D'Ibn al-Haytham à Alberti," in *Archéologie*, 167–81. For clear and clearly argued articulation of the weaker position, see Martin Kemp, "Science, Non-Science, and Nonsense: The Interpretation of Brunelleschi's Perspective," *Art History* 1 (1978): 134–61, especially 147.

lels most likely derived not from optics but from a surveying technique for determining ground distances based on sighting from a fixed position through graduated marks on a staff. Edgerton himself suggests that Brunelleschi fell back upon this technique, implying that Alberti did too.[62] Furthermore, in the Latin version of his treatise, Alberti takes an agnostic stance on whether the rays entailed in his projection are extramitted or intromitted. Granted, he writes, "there was no little debate among the ancients concerning these rays as to whether they arise from the surface or from the eyes [but] this is a really difficult controversy that we set aside as completely useless to us." Completely useless, presumably, because all that mattered for Alberti's purposes is that the "rays" be rectilinear and vanishingly thin, like "extremely fine threads."[63]

Then there is the asymmetry between the cone of projection "beyond" Alberti's window and the visual cone in front, from which it supposedly derives. True, they share the same axis, but the vertex of the one lies at an almost infinitely distant point from the plane of the window whereas the vertex of the other lies close to that plane. Since, therefore, the two vertices represent entirely different things, the two cones are not geometrically or optically equivalent. In addition, this "optical" projection is not Alberti's only method for marking out the visual space of his paintings. Later in the *Della pittura/De pictura* he describes the use of a gridded veil through which the painter can view a given scape and map it out point by point within the grid. "Here in a parallel," he explains, you will see "the forehead, in the very next the nose, in the near one the cheeks [so that] you will have placed in the best manner all the things on a panel or on a wall, also subdivided by corresponding parallels."[64] Famously illustrated in the 1538 edition of *Underweysung der Messung* ("Instruction on Measuring") by Albrecht Dürer (1471–1528), this mapping technique is only optical insofar as it assumes that any given line of sight passing through the gridded veil is rectilinear.[65]

Although the extent to which Renaissance naturalistic art in general, and linear perspective in particular, evolved on the basis of optical theory is a matter of dispute, there is no question that most Renaissance artists were persuaded—or at least pretended—that in order to make their paintings look

62. Edgerton, *Mirror*, 64–65.

63. Sinisgalli, *Alberti*, 26.

64. Ibid., 51.

65. See Willi Kurth, ed., *The Complete Woodcuts of Albrecht Dürer* (New York: Crown Publishers, 1946), plate 340. Plates 337 and 338, from the 1525 edition of Dürer's *Underweysung*, show related mapping techniques.

true to life, they needed some understanding of optics and optical effects. They were also aware that the perception of many of these effects has little or nothing to do with the ray geometry of sight. Seeing the apparent convexity or concavity of a painted object, for instance, does not involve sensing actual differences in radial distance between the center of sight and protruding or depressed parts, as Alhacen argues in book two, chapter 3 of the *De aspectibus*, when explaining the perception of physical convexity or concavity.[66] Rooted in the psychology of perception rather than the geometry of radiation, the illusion that objects painted on a flat panel occupy three dimensions is based on the artist's grasp of what Alberti styles the "reception of light." With this in mind, the artist must "remember that a color itself is more vivid and clear in a surface in which bright rays strike and that [as] the force of the light . . . gradually [diminishes], the same color becomes a little darker." Once he has learned this lesson "in an excellent way from Nature," the artist will know to "modify a color by a very light white . . . at the right place [and to] add black in the same way, on the opposite side, in the appropriate place." That way he can render the object in proper relief according to what Leonardo da Vinci later calls *chiaroscuro*. "A mirror will be an excellent judge" of how convincing the portrayal is, Alberti adds, because "every imperfection of a painting appears more deformed in a mirror."[67]

A host of artists after Alberti looked to Nature for instruction on how to mirror such optical effects, and they did so not only through a close examination of the actual phenomena but also through a study of the optical principles underlying them. Among these students of Nature, none was more assiduous than Leonardo, whose copious notes and drawings, which span a period between roughly 1480 and 1520, attest to the remarkable range of his interests as well as to the sharpness of both his tutored eye and his relatively untutored intellect.[68] Not surprisingly, he was keenly interested in light and vision and,

---

66. Smith, *Alhacen's Theory*, 472–74.

67. All quotations from Sinisgalli, *Alberti*, 68–69.

68. For a list of Leonardo's various manuscripts and notebooks, with dating, see Jean Paul Richter, ed., trans., and Carlo Pedretti, commentary, *The Literary Works of Leonardo da Vinci*, vol. 1 (Berkeley: University of California Press, 1977), 92–97. Leonardo himself acknowledges that he lacks "literary learning" but points out that his "concerns are better handled through experience rather than bookishness [and that, moreover] you should praise natural understanding without bookish understanding rather than bookish learning without understanding"; Martin Kemp and Margaret Walker, ed. and trans., *Leonardo on Painting* (New Haven, CT: Yale University Press, 1989), 9.

like Ghiberti before him, looked to one or more perspectivist sources for guidance.[69] How much he may have gleaned directly from those sources and how much indirectly from hearsay we have no way of telling, but his mature ruminations about light and vision are generally consistent with perspectivist theory. Thus after a brief flirtation with extramissionism during the 1480s, Leonardo adopted the intromissionist alternative without reservation, citing afterimage, among other things, as confirmation.[70] On that basis, he assumed that the eye sees by means of radiant cones (*piramide*) that carry "the semblance (*similitudine*) of a body . . . as a whole into all parts of the air" to fill it with an infinite number of images for which the eye is "the target and the magnet."[71] These cones are composed of "infinite separating lines" that form its constituent rays. After passing through the crystalline humor, the images of the light and color lying on the bases of such cones are transmitted through the sphere of the eye along its axis to the common sense (*senso comune*), which is located at the center of the three-chambered brain. This faculty, where all the senses converge, constitutes the seat of judgment.[72]

So far, Leonardo's account of vision accords reasonably well with its perspectivist counterpart. At one point in his ruminations, in fact, the influence of either Alhacen or, more likely, Witelo seems clear. After claiming that "the pupil of the eye has a power of vision all in the whole and all in each of its parts,"

---

69. Leonardo's familiarity with Pecham's *Perspectiva communis* is evident from his extensive quotation from the prologue to that work in Italian translation; see Kemp and Walker, *Leonardo*, 49; cf. Lindberg, *Pecham*, 61. That Leonard at least looked at Witelo's *Perspectiva* is clear from a memorandum of 1499; see Kemp and Walker, *Leonardo*, 265. All of this suggests that despite his lack of formal education, Leonardo was by no means unlearned, even in Latin, which he could apparently read with some competence.

70. See Jean Paul Richter and Irma Richter, ed. and trans., *The Literary Works of Leonardo da Vinci*, vol. 1, 2nd ed. (New York: Oxford University Press, 1939), 132, text 54.

71. Kemp and Walker, *Leonardo*, 50–51. Although Leonardo refers to the images conveyed through the radiant cone as "semblances" or "similitudes" in the quoted passage, elsewhere he refers to them as "species" (*spetie*) in accordance with standard perspectivist usage (Witelo excepted); see, e.g., Richter and Richter, *Literary Works*, 139, text 66, and 140–41, text 70.

72. For Leonardo on rays, see Richter and Pedretti, *Literary Works*, 124; on the three-chambered brain and its connection to the eyes, see Kenneth Clark, *The Drawings of Leonardo da Vinci in the Collection of Her Majesty the Queen at Windsor Castle*, vol. 2, 2nd ed. (London: Phaidon, 1969), plates 12603 recto and 12627 recto; and on the common sense as seat of judgment, see Richter, *Literary Works of Leonardo da Vinci*, vol. 2, 3rd ed. (London: Phaidon, 1970), 100–101, texts 836 and 838. Leonardo relocated the common sense to the brain's middle ventricle because he reserved the frontmost ventricle for the *imprensiva*, which mediates between the senses and the seat of judgment.

Leonardo supports this claim with Alhacen's example of seeing "through" a needle placed directly in front of the pupil. As we saw in chapter 5, Alhacen explained this phenomenon on the basis of oblique radiation reaching the entire exposed surface of the crystalline lens from points behind and blocked by the needle (see section 4).[73] Yet despite his apparent reliance on perspectivist principles, Leonardo either failed to understand them fully or refused to accept all of them. A clear case in point is his analysis of sight in the so-called "Treatise on the Eye," a compilation of ten folios dating from somewhere around 1510, toward the end of his working life.[74]

Much earlier, Leonardo had noted that if a small hole is drilled in the wall of a windowless room facing a sunlit scape, an inverted image of the scape will be projected on a white wall opposite the hole. Moreover, if the "bodies [within the scape] are of various colors and shapes the rays forming the images are of various [corresponding] colors and shapes."[75] What Leonardo describes here, of course, is the projection of real images in a camera obscura ("dark room"), and by the time he wrote his "Treatise," he had concluded that the eye, with its pupil, operates in precisely the same way. That is, the image of what lies outside the eye is projected through the pupil upon the anterior surface of the crystalline lens in inverted and reversed order. The reason, Leonardo explains, is that all the rays entering the pupil intersect there before reaching the lens, so that in effect two radiant cones are formed, one with its base in the field of view, the other with its base on the anterior surface of the

73. For Leonard's use of the needle example, see Edward MacCurdy, ed. and trans., *The Notebooks of Leonardo da Vinci*, vol. 1 (New York: Reynal and Hitchcock, 1938), 238–39. This same example and explanation are given by Witelo in *Perspectiva*, book three, proposition seventeen; see Unguru, *Witelonis secundus et tertius*, 125. In book three, proposition fourteen of the *Perspectiva communis*, Pecham argues that the lens is affected throughout its surface by both perpendicular and oblique radiation from a single point, but he makes no mention of the needle experiment; see Lindberg, *Pecham*, 229. As used by Leonardo, "pupil" (*popilla*) refers to at least three things over the course of his ruminations about sight. One is the aperture itself; another is the lens; and yet another is the cornea. Used by none of the perspectivists, this term does crop up in earlier sources to denominate the seat of visual sensitivity; see, e.g., chapter 3, note 16, on Ptolemy's interchangeable use of "pupil" (*pupilla*) and "viewer" (*aspiciens*).

74. The best and most comprehensive treatment of this treatise and its relationship to Leonardo's earlier ruminations about vision is Donald S. Strong, "Leonardo da Vinci on the Eye: The MS D in the Bibliothèque de l'Institut de France, Paris, Translated into English and Annotated with a Study of Leonardo's Theories of Optics" (PhD diss., University of California at Los Angeles, 1967); published version (New York: Garland, 1979). See also James Ackerman, "Leonardo's Eye," *Journal of the Warburg and Courtauld Institutes* 41 (1978): 108–46.

75. Richter and Richter, *Literary Works*, 141, text 70; see also ibid., 144, text 78a.

lens, and both sharing a vertex at the pupil.[76] Since, however, "everything that the eye sees through the [pupil] is seen . . . upside down but perceived (*conosciute*) upright," the image on the anterior surface of the lens must be reinverted inside the eye before reaching the optic nerve. Otherwise, it would arrive at the seat of judgment upside down and would be perceived that way.[77] Unsure of precisely how or where this second inversion might occur, Leonardo offers one possibility based on assuming that the lens forms a concentric sphere inside the spherical eye. Accordingly, he suggests, the inverted images projected onto the surface of the lens "pass through the center of the crystalline sphere [where] they unite in a point and then spread themselves out upon the opposite surface of this sphere without deviating from their course." From there, they are transmitted through the optic nerve in proper upright order "to the common sense where they are judged." In other words, all the rays reaching the anterior surface of the lens are refracted through its center.[78]

The result of late-life reflection on Leonardo's part, this model of image reception and transmission conflicts with perspectivist principles in several ways, some of them radical. For a start, the perspectivists all located the lens toward the front of the eye, not at its center, where in fact most medieval and Renaissance physicians placed it on the authority, ultimately, of Johannitius (Ḥunayn ibn ʾIsḥāq). Whether Leonardo was appealing to this medical tradition is a matter of conjecture, but it is at least plausible to suppose he was, either because he deemed it superior to the perspectivist alternative or because he did not properly appreciate that alternative and its analytic entailments. In addition to rejecting the perspectivist ocular model, Leonardo also flouted the perspectivist rules of radiation to and through the eye according to a single cone formed by all the rays reaching the cornea and lens orthogonally and converging toward the center of sight. Not only did Leonardo double that cone, but in locating its vertex at the pupil rather than deep inside the eye, he also did away with the center of sight as a systemically defined point. Furthermore, in order to have the rays refract to the center of the spherical lens, Leonardo had to assume they refract not just *toward* the normal but along the normal itself, which is impossible according to perspectivist refraction theory. Even to entertain this impossibility is to misconstrue that theory completely.

76. See ibid., 142, text 71.

77. Ibid., 155, text 77A; English translation adapted from Richter's.

78. See MacCurdy, *Notebooks*, 235. Leonardo provides several diagrams illustrating this intersection at the center of the spherical lens; see, e.g., Clark, *Drawings*, vol. 3, plates 19150v and 19152r.

But it was not ignorance (or at least not *just* ignorance) that drove Leonardo to do such violence to the perspectivist model. Rather, it was the recognition that although "the eye . . . clearly offers proof of its functions," those functions had been misunderstood "by countless writers" up to his own time.[79] Empirical, not theoretical, the "proof" to which Leonardo adverts comes to ground in a simple experiment. With a fairly thick needle punch a hole in a piece of paper and hold the paper not far from the eye so you can stare through the hole at a bright light. Then hold the needle right up to the pupil so that its image is seen clearly in the hole, and move it side to side along the vertical and up and down along the horizontal. In both cases, the needle will appear to move in the opposite direction. This can only happen if the rays by which the needle is seen intersect at or near the hole in the paper.[80] Hence, since the pupil of the eye is just such a hole, the image projected through it to the anterior surface of the lens must be real and inverted. For the perspectivists, on the other hand, the very notion that rays streaming through a small aperture can form actual, physical images on a screen or on the surface of the crystalline lens—a notion central to Leonardo's understanding of the camera obscura— was utterly foreign.[81] They viewed all images as perceptual or psychological constructs, the kinds of imaginary and illusory things we perceive in reflecting and refracting bodies. Without perceivers, in short, images simply cannot exist. Or, as we would put it today, all images are virtual. Less bookish than his academic sources, and therefore less prone to accept, or perhaps even understand, authority than they, Leonardo evidently allowed what he took as empirical fact to override any theoretical imperatives about image formation in the eye. This relative freedom from theoretical constraints enabled him to notice what others had not and to conclude on that basis that the theory needed to be adjusted to empirical fact rather than the reverse.

This same openness to noticing (he was, after all, a gifted artist) is evident in Leonardo's study of what Alberti before him called "the reception of light" and the shadow casting associated with it. Unlike his perspectivist sources, Leonardo recognized that shadows are complex rather than simple because they are not completely defined by the rays extending from the outer edges of the luminous source to the corresponding outer edges of the occluding body.

79. MacCurdy, *Notebooks*, 247.

80. See Strong, "Leonardo on the Eye," 48–49.

81. On medieval explanations of pinhole projections, see David C. Lindberg, "The Theory of Pinhole Images from Antiquity to the Thirteenth Century," *Archive for History of Exact Sciences* 5 (1968): 154–76, and "The Theory of Pinhole Images in the Fourteenth Century, *Archive for History of Exact Sciences* 6 (1970): 299–325.

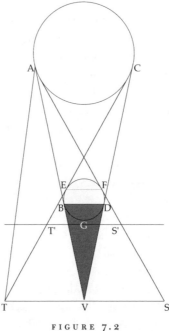

FIGURE 7.2

Take the case of luminous source AC in figure 7.2 illuminating smaller body EBGDF. According to Pecham and Witelo, the two perspectivist sources with whom Leonardo was almost certainly familiar, the shadow will consist of the single, sharply defined cone BDV formed by rays ABV and CDV tangent to both luminous source and illuminated object.[82] Leonardo, however, realized that cones VAS and TCV of radiation form shadows beyond arcs BF and ED, respectively, but that these shadows are lit by the portion of the luminous source to which they are exposed. For instance, the shadow between T and V will be illuminated by an infinite number of rays from arc AC, but it will get darker as it approaches ABV and is exposed to decreasing amounts of light.

Now according to Leonardo, shadows are subdivided into two fundamental kinds. Primitive (*primitiva*) shadows, on the one hand, are attached to the body itself, that is, within arc BGD. Derivative (*dirivativa*) shadows, on the other, are cast by the body outside its boundaries.[83] The derivative shadow

82. See Pecham, *Perspectiva communis*, I, 22–24, in Lindberg, *Pecham*, 99–103, and Witelo, *Perspectiva*, II, 26–28, in Unguru, *Witelonis secundus et tertius*, 65–69; see also Bacon, *De speculis comburentibus*, in Lindberg, *Bacon's Philosophy*, 289–97.

83. See, e.g., Richter and Richter, *Literary Works*, 181, text 159.

in figure 7.2, for instance, consists of a darker core converging at V and a lighter periphery bounded at T and S. The darkest area surrounding the core shadow within this peripheral area constitutes what today we call the penumbral shadow. Clearly, then, the derivative shadow cast on surface TVS will be illuminated almost out of existence. However, as that surface approaches the body—for instance, along line T´S´—the core shadow will not only predominate but also become more intense. Consequently, the closer the composite derivative shadow is to the body itself, the darker and sharper-edged it will be, although it will always be blurred to some extent by the penumbral shadow surrounding it. The same holds for the primitive shadow within arc BGD because rays CET and AFS also form a primitive shadow bounded by line EF. As the rays from points between A and C on the luminous source approach B and D, the primitive shadow formed by them becomes increasingly dark according to diminished illumination. Consequently, like its derivative counterpart, the primitive shadow will be somewhat blurred and indistinct at its edge.[84]

This, I contend, is another case of Leonardo's adjusting theory to observation, although in this case Leonardo did no violence to perspectivist theory, as he did with his biconal account of sight. Instead, he simply followed the logical dictates of that theory to a richer, more nuanced conclusion based on empirical evidence. His artist's eye told him that shade and shadow are neither perfectly sharp nor uniformly dark throughout, so they must be formed in a more complex manner than was commonly supposed by the perspectivists. Simple, but ingenious, his geometrical response to this phenomenon was to include not just the one cone ACV but all the possible cones of radiation extending from A, C, and every other luminous point between them to the illuminated object as base. The result is an infinite number of cones intersecting between that source and the illuminated body. The ensuing analysis thus served both to validate and to explain Leonardo's observation that shade and shadow are blurred and indistinct at the edges. Consequently, the artist must take this blurring into account if he is to render illuminated objects in an appropriately naturalistic

84. For an extended treatment of shade and shadow on Leonardo's part, see ibid., 164–207. On the basis of a close textual analysis of Leonardo's writings on shadows, Dominique Raynaud has recently shown in convincing manner that some of those writings represent direct quotations from or citations of an as-yet-unidentified scholastic Latin source (presentation titled "A Hitherto Unknown Treatise on Shadows Referred to by Leonardo Da Vinci," given at a conference on "Perspective as Practice," held in Paris on September 10, 2013). This raises interesting questions not only about the originality of Leonardo's shadow analysis but also about possible currents of critical analysis in optics during the later Middle Ages.

way according to both the variable brightness of the body's illuminated surface and the variable darkness of its primitive and derived shadows.

Leonardo's commitment to the geometry of vision and its application to painting is encapsulated in his observation that "perspective is to painting a rein and a rudder."[85] By this I take him to mean that linear perspective is a general guide for the painter, not a rigid system to be followed at all costs. After all, even the perspectivists recognized that size perception and distance perception are not necessarily subject to strict mathematical determination. Quite the contrary, in fact, these perceptions often depend on ambient circumstances and an erroneous psychological judgment based on them. That, for instance, is why the moon looks larger at horizon than at zenith or why, when two objects subtending equal visual angles are seen against the same background, the dimmer one will look larger and more distant. Acutely aware of such perceptual aberrations and their basis in ambient circumstances, Leonardo was equally aware that the rigid application of Albertian perspective (*costruzione legittima*) raises serious problems because of a clash of viewpoints. On the one hand, there is the ideal viewpoint that anchors the painting's artificial, or "accidental" (*accidentale*), perspective. On the other, there is the actual viewpoint that provides the natural (*naturale*) perspective according to which the painting is viewed. Since the painting is on a flat surface, equal spaces along the horizontal will not appear equal according to this natural perspective, even though they are portrayed equal according to artificial perspective. For instance, in figure 7.3, object AB in the plane of the painting is the same size as object CD, but from viewpoint E, CD will subtend a smaller visual angle than C′D′, so it will appear foreshortened and smaller than AB. It will also look more distant because it actually lies farther from E. In order, therefore, to look the same size and distance, the two objects must lie on circular arc BC′ such that C′D′ and AB are equal and subtend equal visual angles.[86]

To make matters worse, images are not visually apprehended at a point but, rather, on the exposed surface of the pupil, or crystalline lens, throughout which the visual power is diffused. Consequently, those images will be distorted according to the convexity of that surface. Add to this that vision is normally binocular, not monocular, and the limitations of organizing the visual space of a painting according to strict, geometrical rules become clear. In short, no painting, no matter how well rendered, can faithfully replicate what we

85. Richter and Richter, *Literary Works*, 127, text 40.
86. See ibid., 158–60, texts 107 and 108.

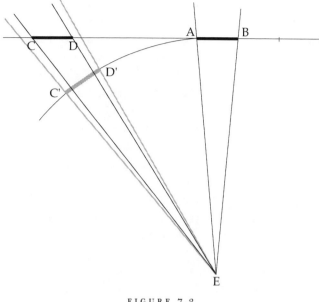

FIGURE 7.3

actually see either directly or in a mirror.[87] The best we can do is make certain adjustments. We can, for example, ensure that the painting is viewed from far enough away that the stereoscopic effect of binocular vision is nullified and the visual angle is small enough to minimize foreshortening toward the painting's edges. Or we can rigorously limit the physical circumstances under which the painting must be viewed. Or we can provide perceptual cues that imply distance and depth in a nongeometrical way according to "aerial perspective" (*prospettiva aerea*) and "color perspective" (*prospettiva de' colori*). Or we can subordinate linear perspective to modeling through chiaroscuro and sfumato (the blurring of primitive shadow) in order to emphasize relief at the expense of strict spatial organization. These adjustments are all tricks of the artist's trade, and the more tricks he knows and can apply effectively, the more deceptively naturalistic the resulting depiction. As for Albert, so for Leonardo, the acid test for the effectiveness of the deception is to view the result in a mirror, which "you should take . . . as your master."[88]

Endowed "with many divine qualities," Giorgio Vasari says of Leonardo

87. See Strong, "Leonardo on the Eye," 385.
88. Kemp and Walker, *Leonardo on Painting*, 202.

in his all-too-brief biography, "he accomplished more by words than by deeds."[89] In written form, those words—many of them dealing with optical matters—occupy thousands of manuscript folios gathered helter-skelter and out of chronological order in various collections scattered throughout Europe and the United States. Leonardo did not intend things to end this way. All indications are that he meant to publish several treatises on the basis of his ruminations in these manuscripts. But Leonardo was, as Vasari implies, long on planning and short on execution, so it was not until the mid-seventeenth century that any of his ruminations saw print in the reconstructed version of his proposed *Treatise on Painting*. As a result, whatever actual influence Leonardo might have exerted on the development of optical theory was limited by the availability of the relevant manuscripts to readers both within and outside the artistic community. That goes a long way toward explaining why his ideas about vision, as innovative and potentially fruitful as they were, lay dormant until fairly recent times.

If Leonardo contributed little or nothing positive to the development of optics, then why deal with him at all, much less at such length in a study that pretends to follow the history of that development? The answer is that in certain respects Leonardo was typical of his day as both a theorist and a keen observer with a strong technical bent. He was, for instance, far from alone in rejecting Albertian perspective as a general system for organizing visual space. In fact, James Elkins claims to have identified more than twenty perspective methods (*modi*) that had been used or proposed between roughly 1430 and 1600. Several of these methods address not the painting's visual space as a whole but specific elements, such as houses or geometric solids, within that space.[90] Even when applying perspective more broadly, painters felt bound to make adjustments to keep it from getting "out of control [so] that nothing seems 'strained, dizzily steep, deformed, or awkward.'"[91] In short, they recognized the need to bend geometry to the dictates of what David Summers calls "the judgement of sense."[92] Otherwise, the painting will look stilted. They also recognized that the eye itself, or rather, the perceptual judgment based

89. Julia Bondanella and Peter Bondanella, trans., *Giorgio Vasari*: The Lives of the Artists (Oxford: Oxford University Press, 1998), 298.

90. James Elkins, *The Poetics of Perspective* (Ithaca, NY: Cornell University Press, 1994), 84–96.

91. Ibid., p. 174; internal quotation from Daniele Barbaro, *M. Vitruvii Pollionis de architectura libri decem* (1567).

92. Summers, *The Judgement of Sense: Renaissance Naturalism and the Rise of Aesthestics* (Cambridge: Cambridge University Press, 1987).

on it, is unruly, that it, too, often flouts geometrical principles in reaching its conclusions about such things as size and distance.

Rather than simply correct, or attempt to correct, for these perceptual shortcomings, many sixteenth-century painters embraced and exploited them in ways that were optically subversive. According to Elkins, for instance, sixteenth-century mannerists, such as Jacopo Pontormo (1494–1557), did not abandon perspective but instead "disassembled, sheared, and disjointed" it in such a way as "to give us what we would see if we had bad seats" in a theatrical production.[93] Not only did sixteenth-century painters play with perspective and viewpoint, but they also played with chiaroscuro, heightening the contrast between light and shade and even using multiple light sources to create complex, often disturbing and "unnatural," patterns of illumination. Caravaggio (1571–1610) epitomizes this tendency. More subtly subversive was the use of perspectival anamorphosis to emphasize the absolute particularity of viewpoint in making sense of what is directly before the eyes. The anamorphic skull in *The Ambassadors* by Hans Holbein (1497–1543) is a stock example; in order to be seen for what it actually is, it must be viewed from an uncomfortably oblique angle. Another popular theme in Renaissance painting, mirror anamorphosis, crops up in such disparate works as *The Arnolfini Wedding* by Jan van Eyck (1390?–1441), with its convex mirror in the background, and the *Self-Portrait in a Convex Mirror* by Parmigianino (Francesco Mazzola, 1503–1540). In addition, several Renaissance artists added trompe l'oeil fillips to their paintings in a playful spirit. However, as Summers points out, subversive though they may be, these optical effects were not applied unreflectively. Like Leonardo, many Renaissance painters were fairly well versed in perspectivist visual theory but not so rigidly committed to it as to let it override their sense of aesthetics. Freed from such constraint, and gifted with tutored eyes, they shared Leonardo's openness to noticing and were willing to bend the rules in order to accommodate them to observable "facts" or aesthetic imperatives. In taking this tack, they followed "natural understanding . . . rather than bookish learning" as their primary guide.

All told, Renaissance art conveys a mixed message about optics in general and visual theory in particular. On the one hand, the "utter veracity" of many Renaissance paintings speaks to the power of optics as a natural and systematic guide to follow in rendering the visible world artificially on a flat surface. In that sense, painting seems to be the very incarnation of optics, an object lesson in its presumed applicability to visual space. On the other hand, the veracity

93. Elkins, *Poetics*, 154.

of any painting is a function of pure deception because it depends on tricking the viewer into seeing the depiction as if it were an image spied through a window or, better yet, in a plane mirror. In that sense, painting is doubly deceptive because it is a virtual reproduction of a virtual image. Worse yet, vision is inherently unstable and prone to misperception, a point that many Renaissance artists highlighted by playing with perspective and lighting or by adding anamorphic elements and the like. The resulting tension between the ideal reliability and real unreliability of vision did not go unnoticed by contemporaries, many of whom deprecated both painting and vision for their inherent deceptiveness. Even the most cursory scan of Stuart Clark's *Vanities of the Eye* leaves no doubt that the two were viewed with deep misgivings, if not outright hostility, by many late Renaissance thinkers.[94]

5. CONCLUSION

In this chapter we have seen how, between roughly 1300 and 1500, knowledge of the perspectivist visual paradigm filtered down from scholastic circles to a broader, less formally educated community of interest through such popular conduits as sermons, literature, and art. At a minimum, what was learned through these conduits formed the basis for what I call optical literacy. It is at this level, for instance, that Chaucer's audience would have gotten his allusion to "Alocen and Vitulon . . . that writen [of] queynte mirours and of perspectives," even though they may have known little or nothing of what those authors actually wrote. Literary sources could be as misinformative as informative, though. A case in point is Jean de Meun's misplaced reference to Alhacen's "Livre des *regarz*" (that is, *De aspectibus*) as an indispensable source for knowledge about rainbows and magnifying devices. Nor can we be certain how well, or even if, Chaucer and Jean de Meun actually knew or understood the sources they cited. Still, misinformed though they may have been, such citations at least served to reinforce the idea that *perspectiva* is not just a science, but a definitive and comprehensive account of visual perception in all its aspects.

The level at which perspectivist theory was understood during this period was not simply a function of formal education. Just as today, so then, there were many intelligent, highly educated people who had but a tenuous grasp of optical theory and its analytic intricacies. On the other hand, there were

94. For the full reference, see note 47, above.

those within the ranks of technically astute but formally unlearned people who had a fairly firm grasp of both. Leonardo belonged to this group. He also belonged to a group, mostly of artists, who sought to use theory as a somewhat loose guide instead of a rigid, absolutely binding system of rules. Far from doctrinaire, therefore, they were quite willing to bend or even break the rules, as they did with Alberti's *costruzione legittima*, in order to make them conform to empirical or aesthetic "fact." This same willingness to bend or break the rules is evident in Leonardo's biconal theory of vision, which was a response to what he saw as incontestable "fact"—namely, that the eye acts like a camera obscura. Not merely conjectural, this "fact" finds confirmation in the apparently backward-moving needle experiment Leonardo described, an experiment that bespeaks his pragmatic and empirical bent.

Annoying though it is, I have been bracketing the term "fact" in quotation marks for a reason. Simply (and tritely) put, facts do not speak for themselves. They tend to conform to theoretical expectations, which is to say that we are prone to see what we expect to see according to various presuppositions about the world around us and how we relate to it. It follows, then, that the fewer or more loosely held such presuppositions, the more open we are to noticing what would otherwise be theoretically unexpected, especially when looking with eyes attuned to subtle distinctions. Of course the belief that we can trust such eyes carries its own theoretical freight. For all we know, they may be nothing but well-tuned liars. Be that as it may, Leonardo and his fellow Renaissance artists were confident enough in their eyes to let observation trump theoretical expectations and were therefore relatively open to noticing. Leonardo's observation that real images can be projected through a small aperture in a dark room is an instance of such noticing insofar as perspectivist theory not only does not predict but actually militates against this sort of projection.[95]

This point bears on an intriguing thesis proposed over a decade ago by the artist David Hockney, who has worked extensively with photography and photo collages. Having noticed with his photographically attuned eyes that certain early fifteenth-century Flemish paintings have a distinctly "optical quality," Hockney wondered whether the artists who produced them used optical aids instead of merely "eyeballing." Abetted by Charles Falco, a professor of optical sciences from the University of Arizona, he concluded that they

95. Alhacen's theory of virtual rays was in fact designed to counter the possibility of inverted image projection through the eye's pupil.

did indeed use such aids, in the form of concave spherical mirrors or convex spherical lenses appropriately posed in a dark room with a small aperture. Under these conditions, Hockney argues, the artists were able, within narrow limits, to project real images of what they wanted to portray on a panel or canvas and used them as the basis for the actual painted depiction.[96]

To say that Hockney and Falco's thesis has not fared well in academic quarters, and some nonacademic ones as well, is to understate the case. Responses to it have ranged from polite but firm skepticism to visceral denunciation, and although some of the harsher criticism has been blunted, the central problem remains: there is simply no clear textual or material evidence, beyond the paintings themselves, to substantiate their claim.[97] Casting about for auxiliary support, Hockney and Falco, especially the latter, have insisted that the artists in question could have learned the method of image projection from perspectivist sources. All they needed to do was put theory into practice. However, as I pointed out earlier, the very concept of real images was foreign to perspectivist opticians because for them all images are psychologically constructed and thus virtual. Nor does the perspectivist analysis of concave spherical mirrors or convex spherical lenses suggest the projection of real images by or through them. Recourse to the perspectivist account of burning mirrors is no help either because, as I mentioned at the end of chapter 4, section 4, these mirrors were analyzed explicitly as heat producers, not image producers.

Hockney and Falco are therefore doubly misguided in supposing that Renaissance artists put perspectivist theory into practice in order to develop the technique of image projection. First, perspectivist theory in no way supports that practice. Second, if the artists in question were as expert in perspectivist theory as Hockney and Falco assume, then that very expertise would have inhibited them from noticing the phenomenon they supposedly put to practical use.[98] Ironically, in attacking that assumption I am actually bolstering their case and granting it a certain plausibility, enough at least to leave me politely but firmly skeptical rather than inimical. In that regard, I stand with Martin

96. This thesis is spelled out in detail in *Secret Knowledge: Rediscovering the Lost Techniques of the Old Masters* (New York: Viking Studio, 2001).

97. For a sample of critical responses to the Hockney-Falco thesis, see the collected articles in the special issue of *Early Science and Medicine* 10, no. 2 (2005), titled *Optics, Instruments, and Painting, 1420–1720: Reflections on the Hockney-Falco Thesis.*

98. For a more detailed articulation of this arguments, see A. Mark Smith, "Reflections on the Hockney-Falco Thesis: Optical Theory and Artistic Practice in the Fifteenth and Sixteenth Centuries," ibid., 163–85.

Kemp. I also stand with him in thinking "that there is more to be gleaned from Hockney's overview of artistic imitation than the windy disputes about the use or non-use of optical devices in particular cases."[99] If nothing else, Hockney and Falco have done a service by foregrounding the undeniable optical quality of Renaissance painting in a provocative way.

As we have seen throughout our discussion not just of art but also of theology and literature, mirror imaging is a recurrent theme. God is the mirror of introspection; phantasy, or compositive imagination, "Wit's looking-glasse"; painting the mirror of nature; and the mirror the artist's master. But every mirror forms a "veil" between object and eye, so what we see through it is not the object itself but, rather, its image or appearance, and then only dimly, according to Saint Paul. The resulting perception is not necessarily false, but neither is it necessarily true, even though it is *of* something true. The problem is to safeguard visual perception from falsity while ensuring that it at least reflects the true. The perspectivists attempted to do this by constructing a systematic model of visual perception rigidly controlled by perceptual judgment under reason's rod, bound by certain optical laws, and subject to normative conditions—all of which are meant to guarantee at least the possibility of veridical perception as a sort of baseline. According to this baseline, visual deceptions occur when one or more normative conditions are transgressed, as happens in reflection and refraction. These deceptions can be rectified by reason according to the laws governing light and sight. Hence, with sufficient care and under appropriate circumstances, visual perception should yield a faithful depiction of physical reality. But no matter how carefully wrought it may be, this depiction is only a representation, not the reality itself, so how can we be certain that it is a *faithful* representation? The answer is that we cannot. We can only assume that it is.

Challenged directly by such scholastic thinkers as Peter John Olivi and William of Ockham, this assumption was addressed somewhat more obliquely by the religious, literary, and artistic figures we have encountered in this chapter. Or perhaps it would be more accurate to say that those figures brought the assumption into the harsh light of public scrutiny and, in the process, highlighted its tenuousness. Even spiritual vision affords not an introspective look

99. Martin Kemp, "Imitations, Optics, and Photography: Some Gross Hypotheses," in *Inside the Camera Obscura—Optics and Art under the Spell of the Projected Image*, preprint 333, ed. Wolfgang Lefèvre (Berlin: Max Planck Institute for the History of Science, 2007), 243–64; quotation from 264.

at the true self but only a glimpse of its appearance, albeit through the perfectly unblemished mirror that is God. The same holds for phantasy, the mirror of mind, but unlike God, it is notoriously unruly and subject to pathological distortion. Indeed, at times it seems to have a mind of its own, and an irrational one at that. Such mirroring metaphors, whether applied to God, painting, or phantasy, bring the problem of imaging into clear focus. After all, images seen in a mirror are virtual or intentional entities, so they exist not in the mirror but in the perceiver who, quite literally, realizes them. The deceptiveness of perception must therefore be due either to the imaging process itself or to the psychological faculties, phantasy above all, within which it occurs.

Explicitly designed to keep this process under tight geometrical, physiological, psychological, and rational rein, the perspectivist paradigm promises a veridical view, but only under optimal conditions. Yet even so this view is mediated by a succession of virtual representations, in the form of intentional species, that are *supposed* in some way or another to be like the objects they represent. Even under optimal conditions this may not be the case. In addition, Renaissance artists could show that the same optimal conditions are quite capable of yielding perfect deception. In just the right light, at just the right distance, from just the right angle, and so forth, a viewer can be deceived into thinking that a faithfully depicted object is the object itself without being aware that he is deceived. It was precisely this sort of veridical deception that brought both vision and painting into disrepute among a host of later Renaissance thinkers. More to the point, the inability to fully account for these sorts of deceptions called attention to the shortcomings of the perspectivist paradigm as a comprehensive and definitive model of vision.

Despite its shortcomings, the paradigm was so well entrenched that most knowledgeable medieval and Renaissance thinkers took it for granted in an uncritical way. Gregor Reisch at the beginning of the sixteenth century and Sir John Davies at the end exemplify this point. There were a few, however, who were less committed than they to the paradigm and thus able to critically appraise its theoretical underpinnings in light of unexpected phenomena, or anomalies. They reacted not by rejecting the paradigm but by tweaking it around the theoretical edges in order to adapt it to those anomalies. This is what Leonardo did in response to image projection in the camera obscura and the unexpected complexity of shadow casting. In the next chapter, we will look at this tweaking process more closely as it evolved in the face of certain anomalies that came to light during the second half of the sixteenth century. As we do, we will see that some of the theoretical adjustments resulting from this process threatened to unravel the foundations of the perspectivist paradigm at key points. Hence, by the time Kepler entered the optical arena at the turn of the seventeenth century, that paradigm had become somewhat frayed at a variety of levels, from its theoretical base to its explanatory superstructure—all within a cultural environment marked by increasing doubt about the reliability of sight.

CHAPTER 8

The Keplerian Turn and Its Technical Background

Kepler entered the optical arena through the back door of mathematical astronomy. What led him to open this door was his brief but stormy apprenticeship with Tycho Brahe (1546–1601), the greatest naked-eye astronomer of the time. Forced to move from his grand Danish observatory of Uraniborg in 1597, Tycho had begun to set up shop outside of Prague in 1599 at the invitation of Emperor Rudolph II, who appointed him imperial mathematician—that is, court astrologer. In early 1600 Tycho invited Kepler to join him as an assistant. After some bickering and dickering, Kepler came on board and, upon Tycho's death in late October 1601, succeeded him as imperial mathematician, a position he held until 1612. Soon after joining Tycho's group, Kepler realized that certain observations (of solar and lunar eclipses in particular) presented problems that could not be resolved without a full understanding of the optical principles behind the instruments used in making the observations. No less critical for Kepler was to understand the optics of atmospheric refraction in order to know precisely how much it skews observations near the horizon. In typically obsessive fashion, he devoted the next few years to achieving that understanding on the basis, primarily, of a close, critical scrutiny of Witelo's *Perspectiva* and Alhacen's *De aspectibus*, both of which had been edited by Friedrich Risner and published in his *Opticae Thesaurus* of 1572. In 1604 Kepler published the results of this study in the *Ad Vitellionem Paralipomena* ("Supplement to Witelo").

Among the observational instruments Kepler examined in the course of his optical research, the camera obscura and the eye are especially significant because of the way the two are related. In fact, as Kepler eventually recognized, the eye is somewhat like a camera obscura with a lens at the pupillary aperture. How he came to that recognition is the subject of this chapter, roughly the first half of which will be devoted to a study of the relatively immediate cultural, technological, and technical background to his optical research. Accordingly, we will start in section 1 by looking at some of the broad technological, social, and cultural changes that occurred between roughly 1450 and 1600, the year that Kepler took up his study of optics. Our primary focus will be on glass technology and printing, both of which led to significant social and cultural changes by the 1500s. In section 2 we will look more closely at the technology of concave mirrors and lenses during the sixteenth century with an eye toward how they were thought to correct visual disorders, focus light, and produce both virtual and real images. Francesco Maurolico (1494–1575) and Giambattista Della Porta (ca. 1535–1615) will emerge as key figures in the effort to account for these phenomena. In the third section we will complete our background sketch by discussing how medieval and Renaissance physicians understood the anatomical structure and function of the eye and how that understanding changed during the sixteenth century, as physicians looked more carefully at ocular anatomy and physiology.

Against this background, we will turn in section 4 to a fairly close examination of Kepler's analysis of retinal imaging. Of particular concern will be the lens theory upon which that analysis was centered. In the fifth section, titled "The Analytic Turn," we will see that contrary to the standard interpretation, Kepler's analysis of lenses and the visual model arising from it owed little or nothing of real import to the perspectivist sources he read. Instead, it was based on a number of conceptual and methodological assumptions that are nowhere to be found in those sources. In short, Kepler's model of retinal imaging was truly innovative and, as such, subversive of its perspectivist counterpart. Just how subversive it was will be the subject of the sixth and concluding section, "The Epistemological Turn," where we will discuss the fundamental incompatibility between Kepler's model of vision, with its basis in retinal images, and the perspectivist model, with its basis in intentional species.

1. TECHNOLOGICAL, SOCIAL, AND CULTURAL CHANGES: 1450–1600

In the previous chapter we noted that the Renaissance marketplace itself was a marketplace of ideas. Every craft, even that of trade, requires specialized knowledge, and some require a considerable amount. Being practical in nature, such knowledge was traditionally viewed as inferior to theoretical knowledge because it deals with imperfect, material objects rather than perfect, mental objects.[1] But this distinction between craft or "technology" (*technē*) and theory (*theōria*) or "science" (*ēpistēmē*) was never as sharp in reality as it was conceived to be in the abstract. After all, an architect may need to know how to manipulate building materials, but he also needs to know a good deal about geometry in order to bend those materials to his design, as well as to conceive the design itself.[2] Even the most theoretical of mathematical astronomers cannot theorize without observational data, and in order to get them, he must have sighting instruments upon whose accuracy he can depend. For that he has to rely on a craftsman who can manufacture such instruments to specification, and the craftsman in turn must understand the purpose for which the instruments are designed in order to manufacture them effectively.[3] In some cases the astronomer himself produced the requisite instruments.

The interplay between theory and technical practice during the later Middle Ages and Renaissance is exemplified to some extent in the development of naturalistic art. In that case theory appears to have redounded on practice, although in Leonardo's case practice also redounded on theory. Glass technology offers another example of how practice can inform theory. Earlier in this book I discussed various empirical studies of refraction through glass described by Ptolemy and Alhacen. These studies were meant to establish or confirm certain theoretical conclusions, and their ability to do so depended on the accuracy of the results they yielded. I also pointed out that the quality of the glass has a direct bearing on those results: the clearer it is and the freer of bubbles and striations, the more accurate the results. Controlling these fac-

1. In the *Republic* Plato draws this negative distinction between craft knowledge (*technē*) and theoretical knowledge (*ēpistēmē*) on the grounds that the former is based on imitation (*mimēsis*) and therefore essentially untrue.

2. Vitruvius opens his *De architectura* in chapter 1 of the first book with a brief discussion of things a proper architect should know in the domain of reason (*ratiocinatio*) as opposed to that of practice (*fabrica*); the former include letters, draftsmanship, geometry, optics, arithmetic, history, moral and natural philosophy, music, medicine, law, and astronomy; see Ingrid D. Rowland, trans., *Vitruvius: Ten Books on Architecture* (New York: Cambridge University Press, 1999). Note that the canonical quadrivial disciplines of arithmetic, geometry, music, and astronomy form part of this list.

3. At the beginning of book five of the *Almagest*, Ptolemy describes the construction of an armillary sphere (*astralabon* in Greek) consisting of six interlocking and pivoting rings; see Toomer, *Ptolemy's Almagest*, 217–19. This was certainly an elaboration on earlier devices of a similar kind, and later Muslim astronomers made improvements while also contributing to the development of astrolabes. By the Muslim Middle Ages, then, astronomical instruments had become extraordinarily complex and technically exacting.

tors is a complex technological problem involving, among other things, the source of silica, the types of fluxes and stabilizers used, and the colorants or decolorants added. In Egypt and Syria this technological problem was more or less resolved before late antiquity, which suggests that Ptolemy's and Alhacen's studies of refraction, if they were actually conducted, were based on reasonably high-quality glass, although poorer than today's window glass.[4]

In medieval Europe that problem remained more or less unresolved until the establishment of the Murano glassworks at Venice toward the end of the thirteenth century. A major step in quality was taken in the mid-fifteenth century, perhaps even earlier, with the production of "crystal" glass, which was especially prized for its clarity and lack of tint, although it, too, was poorer than today's window glass.[5] The ability from that time on to produce blown glass of relatively high quality in significant amounts made possible the mass production of convex and concave lenses for eyeglasses during the fifteenth and sixteenth centuries.[6] The resulting widespread availability of such lenses had both practical and theoretical ramifications in telescopy during the later sixteenth and early seventeenth century because improving such optical devices required some understanding of how they work. This is one reason that so much attention was paid to the analysis of reflection and refraction during that period.

4. From early imperial Roman times Egyptian and Syrian glass set the standard for quality, in part because of the quality of silica from which it was produced and in part because natron was the alkali used in it. In the ninth century, however, there seems to have been a shift from natron to a particular form of plant ash; see Julian Henderson, "Tradition and Experiment in First Millennium A.D. Glass Production—The Emergence of Early Islamic Glass Technology in Late Antiquity," *Accounts of Chemical Research* 35 (2002): 594–602. See also Cesare Fiori and Mariangela Vandini, "Chemical Composition of Glass and Its Raw Materials: Chronological and Geographical Development in the First Millennium A.D.," in *When Glass Matters*, ed. Marco Beretta (Florence: Leo S. Olschki, 2004), 151–94.

5. See W. Patrick McCray, *Glassmaking in Renaissance Venice: The Fragile Craft* (Aldershot, UK; Burlington, VT: Ashgate, 1999), 49–63 and 96–122; see also Rolf Willach, *The Long Route to the Invention of the Telescope* (Philadelphia: American Philosophical Society Press, 2008), 29–38. A crucial source for our understanding of glassmaking around the time of Kepler is Antonio Neri, *L'Arte vetraria* (Florence, 1612), the first book of which discusses the manufacture of glass in general and crystal glass in particular. In the fourth book Neri discusses the addition of lead to crystal glass, but the point of that addition is to color the glass.

6. According to Ilardi, concave lenses for the correction of myopia or nearsightedness were available by no later than the mid-fifteenth century; up to that point only convex lenses for the correction of presbyopia or farsightedness were available; see *Renaissance Vision*, 75–95. See also Willach, *Long Route*, 38–69.

That both theoretical and technical knowledge were exchanged within the Renaissance marketplace of ideas is therefore evident. Less evident is how lively that exchange was, at least up to the end of the fifteenth century. In great part, no doubt, this lack of clarity stems from the reluctance of craftsmen to share the tricks, or "secrets," of their trade, at least the ones that really mattered. Accordingly, craft knowledge was shared in a limited way within a given guild along lines of apprenticeship, and for the most part it was shared orally and through manual example, not textually. Indeed, during the later Middle Ages and early Renaissance, most craftsmen were either illiterate (or "nonliterate") or only functionally (or "pragmatically") literate, so even if they had wanted to share their knowledge in writing, they would have been hard pressed to do so.[7] Leonardo, of course, is one of several exceptions. Also, the lack of copyright safeguards discouraged those who could convey their technical expertise in writing from doing so without couching it in obscure language that only the cognoscenti could decipher.[8]

If guilds and their constituent members were fiercely protective of their monopoly on a given craft and its secrets, municipalities were no less protective of that monopoly. Inventions were increasingly protected by patent over the fifteenth century, the earliest known example being one issued by Florence to Filippo Brunelleschi in 1421.[9] Municipal authorities closely monitored and regulated artisans with particular skills. This was certainly the case in Venice, where emigration of glassmakers was discouraged by the imposition of stiff fines, while immigration was subject to strict control. The flow of raw materials in and out of Venice for glassmaking was also carefully supervised in order to keep those materials out of foreign hands.[10] In the long run such efforts failed.

---

7. Establishing the extent and level of literacy in late medieval and Renaissance Europe is a tricky business, made all the trickier by the coexistence of Latin and local vernaculars as both oral and textual languages. Literacy also varied by time and place, the highest rates and levels probably existing in Italy. Moreover, the ability to read was not necessarily linked with the ability to write. Since the publication of M. T. Clanchy's *From Memory to Written Record: England, 1066–1307* (Cambridge, MA: Harvard University Press, 1979), much of the focus in literacy studies has been on the "literate mentality," which entails a trust in the authority or veracity of the written word. For a good historiographical overview, see Charles F. Briggs, "Literacy, Reading, and Writing in the Medieval West," *Journal of Medieval History* 26 (2000): 397–420.

8. See William Eamon, *Science and the Secrets of Nature* (Princeton, NJ: Princeton University Press, 1994), 87–89.

9. Ibid., 89.

10. See McCray, *Glassmaking*, 43–49. The reason for controlling the trade in raw materials into and out of Venice was to protect the sources of those materials; it was particularly important

Venetian glassmakers eventually set up shop elsewhere in Italy and northern Europe, but even abroad they tended to monopolize the industry, at least for a while. Moreover, by the fifteenth century Florence had taken the lead in the production of optical-quality glass, and unlike their Venetian compeers, Florentine glassmakers were relatively free to move. Thus by the sixteenth century glassworks were established at numerous places north of the Alps, often under Italian auspices. Nuremberg, for instance, became an important center of glass production as a supplier of lens blanks for the Florentine manufacture and export of eyeglasses.[11]

From these few examples we get at least a glimpse of the European marketplace of ideas up to the end of the fifteenth century. That glimpse becomes clearer and more panoramic during the sixteenth century, if only because of the advent of the printing press and the attendant shift from parchment to less-expensive paper as the textual medium. That this technological innovation had a transformative effect on early modern European culture needs no belaboring; the point has been endlessly discussed since the publication of Elizabeth Eisenstein's *The Printing Press as an Agent of Change* in 1979 and its companion abridgement, *The Printing Revolution in Early Modern Europe*, in 1983.[12] I will therefore limit myself to a handful of points, the first of which is that the so-called printing revolution did not take Europe unawares. Print culture, to use Eisenstein's coinage, could not have taken root so quickly had there not already been a readership ready and willing to benefit from printing. The parallel here with the influx of new texts in the twelfth and thirteenth century is inescapable; had print with movable type been developed a century earlier, print culture would probably have taken root more slowly than it did after Gutenberg, and the ensuing "printing revolution" would have probably taken shape differently.

The second point is so obvious it hardly needs making: because printing allowed for the replication of single works in far greater numbers and at considerably lower cost than ever before, it made those works accessible to a

---

to guard the sources of silica and plant ash because the quality of Venetian glass depended on a specific kind and purity of both.

11. See Vincent Ilardi, "Renaissance Florence: The Optical Capital of the World," *Journal of European Economic History* 22, no. 3 (1993): 507–41.

12. Both published by Cambridge University Press. For a fairly recent critical reappraisal of Eisenstein's printing revolution and the early modern "print culture" resulting from it, see Adrian Johns, *The Nature of the Book* (Chicago: University of Chicago Press, 1998). Johns's main target is Eisenstein's insistence on the fixity or standardization of the early modern printed text.

vastly expanded audience. Printing, however, was a business, and in order to turn a profit, printers had to be responsive to demand. Take as an example the printing history of John Pecham's *Perspectiva communis*, which appeared in nine separate issues before the end of the sixteenth century. First published in Milan in 1482/83, it saw print in 1503 at Valencia, in 1504 at Venice and Leipzig, in 1510 at Paris, in 1542 at Nuremberg, in 1556 at Paris, and in 1580 and 1592 at Cologne. It also appeared in an Italian translation published at Venice in 1593.[13] What does this printing history tell us? At a gross level it tells us that there was a fairly strong demand for the book, especially between 1482 and 1510. More than this, though, it tells us that all but the Milanese and Venetian editions of 1482/83 and 1504, respectively, were published at cities with universities. Venice is less exceptional in this regard than might be supposed because it was a major publication center at the time and is a stone's throw from the university town of Padua. Likewise, the university town of Pavia was less than thirty miles from Milan. We can therefore reasonably conclude that these editions were intended for classroom use and, therefore, that Pecham's *Perspectiva communis* was as much a staple for the teaching of optics in the sixteenth century as it was during the later Middle Ages. We can also conclude that by the late sixteenth century there was a sufficiently large community of interest outside the academy—or at least that the printer thought there was—to warrant a vernacular edition.

The printing histories of Witelo's *Perspectiva* and Alhacen's *De aspectibus* tell a different story. Witelo's *Perspectiva* did not appear in print until 1535, at Nuremberg. It was reissued once, in 1551, before it was published along with Alhacen's *De aspectibus* in Friedrich Risner's magisterial compendium, *Opticae Thesaurus* ("Treasury of Optics"), printed at Basel in 1572.[14] Alhacen's *De aspectibus*, on the other hand, never saw print before its inclusion in this compendium. We can reasonably conclude, then, that these two works were used sparingly, and doubtless at an advanced level, for the teaching of optics. Even so, they appear not to have served that purpose until fairly late in the sixteenth century, their use apparently peaking in the second half. It is worth noting that despite its eminent suitability as a text for the teaching of optics, Roger Bacon's *Perspectiva* was not published until 1614.[15] It is also worth noting that Euclid's *Optics* and *Catoptrics* were published in several Latin and

13. For details, see Lindberg, *Pecham*, 71–72.
14. See ibid., 79.
15. See, ibid., 42.

Greek editions during the sixteenth century and that they saw print twice in the vernacular (Italian and Spanish) toward the end of that century.[16]

The *Opticae Thesaurus* of 1572 was not just another edition. Its publisher, Friedrich Risner (d. ca. 1580), was an illustrious scholar with impeccable credentials in both mathematics and optics, and he used his knowledge of both to make his edition as authoritative as possible. To that end, he provided copious annotations to explain various steps of the geometrical demonstrations in both Alhacen's *De aspectibus* and Witelo's *Perspectiva*. In the process he gave specific loci in the relevant mathematical sources upon which those steps are based. This is especially important in Alhacen's case because, as mentioned earlier, he almost never cited sources. Risner also provided a full set of cross-references within each work, as well as between them. In addition, he broke Alhacen's text into theorematic segments and added his own enunciations at the head of each. And, finally, he took pains to reconstruct the diagrams so that unlike those in the manuscripts, they would actually look like what they were meant to represent according to the geometrical descriptions in the text.

Truly monumental in its scholarship, the resulting edition of the two works was not only authoritative; it was carefully designed to facilitate reading and use. Thus Risner avoided all but the most basic abbreviations, and he chose eminently clear roman and italic fonts so that there would be no confusion about textual readings. By segmenting Alhacen's text and adding his enunciations, moreover, Risner made it possible to read the *De aspectibus* selectively according to specific topics. Also, by cross-referencing it with Witelo's *Perspectiva*, he made comparison between the two works easy. More than an authoritative source, therefore, Risner's *Opticae Thesaurus* was a handy reference work. This, in a nutshell, is why it became a canonical optical source from the time of its publication until well into the seventeenth century. That it was through this edition, not through the manuscript tradition, that Kepler learned perspectivist optics is a crucial point to bear in mind because the way he learned it was conditioned by the medium through which it was presented to him.[17]

"The trouble with the publishing business," observed William Targ not so long ago, "is that too many people who have half a mind to write a book do

---

16. See Paul ver Eecke, *Euclide: L'Optique et la* Catoprique, xxxvi–xli.

17. For some insight into how printing fostered significant changes in the way books and information were organized for easier access during the Renaissance, see Ann M. Blair, *Too Much to Know* (New Haven, CT: Yale University Press, 2010).

so." If that is true for recent times, it was no less true for the sixteenth century. The vast majority of texts published over that period ranged from the anodyne to the asinine, at least by modern standards. Specialized treatises of high intellectual quality such as Risner's *Opticae Thesaurus* formed a thin lamina on the blunt cutting edge of publication at the time. Devotional tracts, almanacs, how-to manuals, vicious broadsheets, and brazenly pirated editions, many of them crudely done: these were the staples of the printer's trade. Yet for all their imperfections of reasoning or style, such works were an integral and important part of the marketplace of ideas. Almanacs were packed with medical, astrological, and astronomical lore based on theoretical principles that were accepted at the time. Technical knowledge in the form of "secrets" was openly traded in the piazza and market square by *ciarlatani* (whence "charlatan"), who flogged books of nostrums and recipes to an eager, often gullible, yet sometimes skeptical audience. Most of these "secrets" were old wine in new skins or snake oil, but many were based on tried-and-true practices harking back to such classics as Theophilus Presbyter's *De diversis artibus* (ca. 1120).[18] Meantime, sermon collections, or postils, were published on a massive scale for use either as exemplars or as texts to be read from the pulpit by both Catholic and Protestant clergy.[19] Sermons, as we saw in the previous chapter, were a key source of ideas and knowledge according to the standards of the time. Altogether, then, the sixteenth-century marketplace of ideas was remarkably broad in scope, vibrant, and chaotic—a place where intellectual chaff and wheat were freely exchanged, often without regard to which was which. Adding confusion is the Babel of tongues in which the exchanges occurred.

In retrospect, of course, we might be inclined to judge much or most of the technical and theoretical "knowledge" traded in this marketplace as pseudo-scientific or worse. Part of the problem lies in the obscurantism of so many of the works published during this period. Authors were often deliberately enigmatic either to mask their ignorance or to provide an inkling of some new invention without revealing its details, the point being to advertise the invention and its purported virtues in the hope of attracting a patron. The obscurity

18. See Eamon, *Science and Secrets*, especially 94–193.

19. See John Frymire, *The Primacy of the Postils: Catholics, Protestants, and the Dissemination of Ideas in Early Modern Germany* (Leiden: Brill, 2010). Frymire has identified 535 separate Catholic and Lutheran (plus a few Calvinist) sermon collections for the period between 1520 and 1624. Many of these were republished several times; for instance, according to Frymire's reckoning, the complete collection of Martin Luther's *Hauspostille* saw print eighty-one times between 1544 and 1609 (see 445–65 for the full listing).

of its presentation was thus not necessarily an indication of the worthlessness of the knowledge purveyed or the stupidity of the purveyor. We might also be inclined to make distinctions by language, assuming that works written in the vernacular were intellectually inferior to those written in Latin because they were meant for a less educated and more credulous audience. But there was no clear distinction between "science" and "pseudo-science" at the time. Even the most erudite of thinkers could accept that occult forces were at work in the world and be swayed by arguments based on astrological, magical, or alchemical foundations. So while it may be that works in Latin were intended for the thin, university-trained stratum of the reading public, it does not follow that these works were meant to be studied or taught in universities, or that they ever were.

Giambattista Della Porta's *Magiae Naturalis* is a case in point. First published in four parts or "books" under the title *Magiae Naturalis sive de Miraculis Naturalium Libri IIII* (Naples, 1558), it was republished in a much-expanded form in 1589 as *Magiae Naturalis Libri XX*.[20] As the titles of both editions suggest, Della Porta's purpose was to show how to reproduce marvelous effects by natural means. Thus in the prologue to book seventeen of the 1589 edition, which is devoted to "Strange Glasses" (that is, mirrors and lenses and their optical effects), Della Porta informs us that:

> Now I am come to Mathematical Sciences, and this place requires that I shew some experiments concerning Catoptrick glasses. For these shine amongst geometrical instruments, for Ingenuity, Wonder and Profit: For what could be invented more ingeniously, then that certain experiments should follow the imaginary conceits of the mind, and the truth of Mathematical Demonstrations should be made good by Ocular experiments? What could seem more wonderful, then that by reciprocal strokes of reflexion, Images should appear outwardly, hanging in the Air, and yet neither the visible Object nor the Glass seen? ... We read that Archimedes at Syracuse with burning Glasses defeated the forces of the Romans: and that King Ptolemey built a Tower in Pharos, where he set a Glass, that he could for six hundred miles [see] ... And though venerable Antiquity seems to have invented many and great things, yet I shall

---

20. The immense and long-standing popularity of this work is attested to by the numerous times and places it was republished in both its 1558 and 1589 editions, not only in Latin but also in several vernacular languages, during the remainder of the sixteenth century and much of the seventeenth.

set down greater, more Noble, and more Famous things, and that will not a little help to the Optic Science, that more sublime wits may increase it infinitely.[21]

What follows in the succeeding twenty-three chapters of book seventeen is an odd pastiche of parlor tricks with mirrors and lenses, practical advice on how to manufacture various optical devices, explanations of how they work based on perspectivist ray theory, and displays of classical erudition.

Scion of an aristocratic Neapolitan family, Della Porta never attended university. Privately tutored instead, he was educated in the intellectual and physical skills appropriate to a gentleman, and as part of the arts and letters portion of that education, he was introduced to the quadrivial pursuits.[22] Yet he was obviously interested in more than the theoretical knowledge gained from that introduction. He had an acute interest in the actual phenomena and studied them closely in order to reproduce them physically. In part because of his somewhat naive empirical bent, in part because of his belief in the occult, and in part because of the scope of his interests (he was a renowned comic playwright), Della Porta can easily be dismissed as an impresario, which he was, and a dilettante, which he also was. But he was not a *mere* dilettante, at least not in optics. On the contrary, he was an amateur in the best sense of the term, an empiricist with a solid grounding in perspectivist theory. The firmness of this grounding is evident in his *De Refractione* of 1593, where he undertook a critical, empirically based but theoretically informed analysis of optics in all its aspects from direct radiation to reflection and refraction. Whatever the deficiencies of the resulting account, they were due at least as much to the inadequacy of the perspectivist theoretical principles he followed as to his own inadequacy in applying them.

I will have more to say about Della Porta at the appropriate time, but for now I will hold myself to two observations. First, Della Porta exemplified a growing tendency among sixteenth-century scientific thinkers to meld technical and theoretical knowledge in a nonhierarchical way that favored neither above the other. Craft knowledge was no longer considered contemptible or even necessarily subordinate to theoretical knowledge. Second, Della Porta was not a voice crying in the wilderness; he appealed to like-minded individuals of all stripes, although his efforts were especially oriented toward an

21. English translation from *Natural Magick* (London: Thomas Young and Samuel Speed, 1658), 355. For the Latin, see *Magiae Naturalis Libri XX*, 571–72.

22. For a capsule biography of Della Porta, as well as a discussion of his intellectual leanings, see Eamon, *Science and Secrets*, 194–229.

intellectual elite, some of whom were university educated but many of whom were not. He was thus part of a network of interest that was essentially "public" insofar as it was neither based on nor ensconced in the contemporary university system with its narrowly conceived approach to scholarship.

That network was linked not only through readership but also through personal contact and the sharing of ideas in informal or semiformal gatherings harking back to the Italian "academies" of the fifteenth century.[23] Short-lived, local, and at best loosely structured, these gatherings proliferated throughout the sixteenth century. Della Porta himself is reputed to have organized such a gathering at his own Neapolitan home, the *Accademia dei Secreti*, which seems to have originated sometime in the 1560s and was presumably devoted to the sorts of empirical studies of natural marvels revealed in his *Magiae Naturalis*. Over the course of the sixteenth century, these small "academic" nodes were increasingly linked and widened through personal correspondence, which was facilitated by the availability of relatively cheap paper. As we will see in due course, Kepler's connection to some of these nodes, including the one he joined in Prague, placed him within a matrix of thinkers and ideas that influenced his approach to optics and that ultimately led to his theory of retinal imaging.

2. RETHINKING CONCAVE MIRRORS AND CONVEX LENSES

Toward the end of the sixth chapter I adverted to the perspectivist mishandling of concave spherical mirrors (that is, burning mirrors), as well as spherical lenses, and their focusing properties. In the case of spherical burning mirrors, a critical factor in this mishandling was the lack of appropriate sources. As far as we know, the only relevant source available to the perspectivists was the Euclidean *Catoptrics*, whose final proposition is meant to explain how such mirrors aggregate sunlight into a burning point. As we saw in our discussion of that proposition in chapter 2, Euclid showed two things, both illustrated in figure 2.16a. The first thing is that light from point F on the sun's face will be reflected to some point K on axis FB from a circle perpendicular to that axis and passing through points A and C on the mirror. Whether enough sunlight will be concentrated at K to cause burning is left unstated, but because the light

---

23. Precisely what was meant by "academy" at the time is vague, but it is clear that not every academy was an informal discussion group; see, e.g., James Hankins, "The Myth of the Platonic Academy of Florence," *Renaissance Quarterly* 44 (1991): 429–75.

emanates from only one point on the sun, the implication is that combustion will not occur. The second thing is that all the rays from the sun's face that pass through center of curvature T to reach the mirror, as represented by FB, GD, and HE, will reflect back to T, so enough rays will intersect at that point to cause burning because they emanate from an infinite number of points on the sun's face.

Plausible though it may be, this analysis is incorrect on two counts. First, the focal point of a concave spherical mirror lies not at the center of curvature but at a point exactly midway on the axis between the center and the mirror's surface. Second, as an aggregate, the rays do not intersect at a single point on the axis. They suffer spherical aberration, as illustrated in figure 2.16b, which accompanies propositions two and three of Diocles's *On Burning Mirrors*. Those propositions demonstrate that as parallel rays NL, MG, and KZ approach axis BA of the mirror, they reflect to points E, X, and Y, respectively, on the axis. The closer the incident ray is to the axis, therefore, the higher the point at which the reflected ray will intersect it, the highest point of intersection lying just short of focal point H, halfway between center point B and A. From this, Diocles concludes that the greatest aggregation of radial intersections will occur in a tiny area, not at a point, just below H, so it will be in that area that enough heat will be generated to cause combustion.

Roger Bacon and Witelo both reflect the confusion about spherical burning mirrors that arises from their reliance on Euclid's *Catoptrics*, which was known under the title *De speculis* in Latin. In part 1 of his *De speculis comburentibus*, for instance, Bacon recapitulates Euclid's bipartite argument, as summarized above, showing first how light from point F on the sun's face in figure 2.16a will be reflected from a circle on the mirror to a single point on the axis. Then he shows how a ray from every point on the sun's face can be extended through the mirror's center so as to reflect back to it and produce a confluence of enough rays to cause combustion. Later, in part 7 of the *De speculis comburentibus*, Bacon rejects this second explanation and plumps for the first, arguing that in fact the circles of reflected rays will generate enough heat at the point on the axis where those rays intersect. Since there is an infinitude of such circles, there is presumably an infinitude of combustion points on the axis, each specific to a particular circle of reflection. In other words, there is an unlimited number of "focal" points for any concave spherical mirror, a claim consistent with Bacon's earlier assertion in part 1 that the larger the circle of reflection, the more vigorous the combustion according to the greater number of rays within it. Bacon offers an interesting caveat, though. While it may be true that combustion can occur wherever the reflected rays from a given circle intersect the axis, "the [point of] convergence [after reflection] of rays incident on the least possible circle—that [point] being farther from the mirror's surface than all the rest—cannot lie [at a point on the axis above the mirror that is] more than half the radius of the sphere of which the mirror is a portion." For the modern reader, the implications of this claim are obvious: the reflected rays undergo spherical aberration, and the focal point is equidistant from the mirror and its center of curvature. For Bacon, on the other hand, that point is just a geometrical limit, and spherical aberration is just a geometrical fact without any ulterior import.[24]

Like Bacon, Witelo bases his analysis of spherical burning mirrors on Euclid and, like Bacon, concludes that enough reflection occurs from various circles on the surface of the mirror facing the sun to produce combustion at the appropriate intersection points on the axis.[25] In short, he seems to accept that concave spherical mirrors have an infinite number of "focal" points on the axis. Like Bacon, as well, he adverts to what amounts to spherical aberration in arguing that rays incident on points closer to the axis will reflect to points closer to the center of curvature, but he makes this claim without any recognition of its import for focusing. Unlike Bacon, Witelo specifies no limit point for reflection. Instead, he relates the intersection points to the center of curvature according to whether they are nearer or farther from it. And unlike Bacon, he seems to accept that radiation through the center and reflected back to it will produce combustion, although he never says so explicitly. In this regard, Witelo seems to be more in line with Pecham than with Bacon.[26]

Whereas the phenomenon of spherical aberration is at least implicit in Bacon's, Witelo's, and Pecham's accounts of concave spherical mirrors, it is neither explicit nor implicit in their analyses of radiation through spherical glass lenses. All of them assume that when light from a source point enters such lenses, it is refracted through them and then back out to a single point on the axis. It is at this intersection, of course, that the rays converge with enough intensity to produce combustion. This mistake is glaringly obvious in Pecham's brief account in *Perspectiva communis*, book three, proposition

24. For Bacon's recapitulation of proposition thirty in Euclid's *Catoptrics* in part 1 of the *De speculis comburentibus*, see Lindberg, *Bacon's Philosophy*, 272–77. For Bacon's final conclusion about combustion in concave spherical mirrors in part 7, see ibid., 334–37; quotation adapted from ibid., 335–37.

25. See *Perspectiva*, book eight, proposition sixty-eight, in Risner, *Opticae Thesaurus*, 365–66.

26. Pecham's brief account of spherical burning mirrors, which has no diagram, occurs in book two, proposition fifty-five, of the *Perspectiva communis*; see Lindberg, *Pecham*, 208–9.

sixteen, where the accompanying diagram makes it crystal clear that several rays (which stand for all rays) entering the sphere from a point of radiation meet at a single point on the opposite side.[27] Less obviously mistaken, but no less mistaken, is Bacon's account in *Opus majus*, IV.2.1 of how a hemispherical glass lens with its flat face toward the sun focuses the incoming light to a burning point. Unfortunately, the diagram accompanying the account shows only two outer rays that originate at the sun's center and, after refracting through and out of the lens, intersect at a point. But the implication is clear: all the rays originating at the sun's center will intersect at that point.[28] Witelo takes a slightly different tack in book ten, proposition forty-eight, of the *Perspectiva*. His argument there is that bundles of rays from the sun's center form circles on the glass sphere just as they do in the case of concave spherical mirrors. The rays forming any of these circles will therefore refract to a single point on the other side with enough concentration to cause burning. In this regard, at least, spherical lenses are similar to concave spherical mirrors, and the similarity presumably extends to their both having an infinitude of "focal" points.

Deeply flawed though it is, the perspectivist account of concave spherical mirrors and spherical lenses held sway throughout the later Middle Ages and Renaissance. During the sixteenth century, though, it underwent significant modification in response to new circumstances. Key among these new circumstances was the improvement in glass technology that led to the manufacture of crystal glass by the mid-fifteenth century. Not only was this glass significantly better than any produced in Europe up to that time, but, as it became available in increasingly large amounts over the next century and a half, it became commensurately cheaper. The result was a proliferation of glass artifacts, from drinking vessels and windowpanes to eyeglasses and mirrors. Glass, in short, entered the public domain with a vengeance during the sixteenth century. Eyeglasses were now being mass produced almost to prescription for both presbyopia and myopia, and the increasing quantity of glass mirrors was matched by increased quality, as the lead backing prevalent during the Middle Ages gave way to an amalgam of mercury and tin that made the backing clearer and more secure.

As far as optical theory is concerned, the widespread availability of lenses and concave spherical mirrors of better quality had significant ramifications. For instance, small concave mirrors mounted on stands were used during the

---

27. See Lindberg, *Pecham*, 228–31.

28. See Robert B. Burke, trans., *The* Opus Majus *of Roger Bacon* (Philadelphia: University of Pennsylvania Press, 1928), 132–33.

later Renaissance as magnifying glasses for close reading, but in order to fulfill that function properly, they had to be designed and deployed in just the right way to balance magnification with clarity.[29] Such design and deployment is a technical problem, to be sure, but it has theoretical ramifications because it comes to ground in the mirror's focal length. Likewise, the eyeglass lenses, whether convex or concave, had to be of a particular curvature suited to the particular visual problem. By the mid-fifteenth century, in fact, convex lenses to correct presbyopia were manufactured for specific age groups between thirty and seventy in increments of five years. At the same time, eyeglasses with concave lenses were available to correct two gradations of myopia according to "medium" and "distant" vision.[30] Forming the lenses to such specification has both technical and theoretical ramifications that converge in focal length, just as is the case with concave reading mirrors.

The effective application of these improved optical devices to visual problems led to a renewed fascination with both the devices and their optical effects, particularly the capacity to magnify, which held great promise for telescopy. After all, as Della Porta points out in his summary of "strange glasses" in the prologue to book seventeen of the *Magiae Naturalis*, "We read that Archimedes at Syracuse with burning Glasses defeated the forces of the Romans: and that King Ptolemey built a Tower in Pharos, where he set a Glass, that he could for six hundred miles [see]." There is nothing new in these utilitarian claims. Roger Bacon had already made similar ones, even referring to Julius Caesar's use of a large concave mirror to spy on the Britons from Gaul. Jean de Meun, meantime, had enthused about the marvelous power of concave mirrors to make "things that are very small [appear] great and large" (see chapter 7, section 3). What *is* new is that these claims were being bruited about to a much wider audience than before, and one that was more intimately acquainted with the actual devices in an improved form. That the claims themselves were taken seriously is evident from the fact that during the second half of the sixteenth century, several efforts were made to construct telescopic devices using concave mirrors or mirror and lens combinations. Indeed, if Eileen Reeves is right, even Galileo experimented with concave mirrors before finally battening onto the idea of placing a convex lens at one end of a tube and a concave lens at the other to construct his telescope.[31]

---

29. See Ilardi, *Renaissance Vision*, 43–46.

30. Ibid., 91.

31. See Eileen Reeves, *Galileo's Glassworks* (Cambridge, MA: Harvard University Press, 2008).

A crucial outcome of the renewed interest in these devices and their practical application was the demand to understand how they work. Take eyeglasses as an example. That they somehow "correct" weak sight is problematic because of the way vision was understood to operate not only by the perspectivists but also by medieval and Renaissance physicians. For both groups, the key mechanism for sight was the visual spirit passed to the lens from the brain, so both were in agreement about the physiological basis of sight. Barring physical impediments such as cataracts or corneal lesions, then, loss of visual acuity was thought to be due to an insufficient flow of visual spirit to the lens, either because the optic nerves are constricted or blocked or because the spirit itself is not produced copiously enough. Convex lenses were assumed to compensate for this deficiency refractively by enlarging the visual angle under which the images of things viewed through them are seen. As a result, those things look larger and closer than they actually are. This is presumably what Savonarola had in mind when claiming that "the species of the letters that enter the eye by means of the active light strike on the glasses and here they are spread, widen, and appear bigger" (see chapter 7, section 2). What makes this correction problematic is that it is not a correction at all. It is, rather, a deception because it is based on creating the illusion that the object is larger and closer than it actually is. Moreover, eyeglasses only address the symptom, not the cause, which is why physicians favored remedies designed to improve the flow or manufacture of visual spirit over the use of glasses.[32]

The idea that glasses with convex lenses help the weak-sighted through magnification makes perfect sense. Any sighted person, no matter how compromised his vision, can see that such lenses actually do magnify. But recognition by the mid-fifteenth century that concave lenses correct myopia posed a serious problem because such lenses diminish rather than magnify what is viewed through them. They ought therefore to exacerbate the condition rather than resolve it. The earliest attempt that we know of to explain this anomaly was made in the mid-sixteenth century by Francesco Maurolico (1494–1575). A native of Messina, Maurolico was well versed in mathematical subjects and wrote on a wide range of things from astronomy to music. Among his writings, four on optics that span the period between 1521 and 1567 were published posthumously in a compilation titled *Photismi de Lumine* ("Enlightenment on Light"), whose first part also bears that title.[33] It is in chapter 3 of the third

32. See Ilardi, *Renaissance Vision*, 347–51.

33. *Photismi De Lumine, et umbra ad perspectivam, et radiorum incidentiam facientes* (Naples, 1611)—henceforth *Photismi*. For an English translation of the entire *Photismi*, see

part of this compilation, which dates from 1554 according to the colophon, that Maurolico addresses vision and its improvement by eyeglasses.

On the whole, Maurolico's theory of vision is consistent with that of the perspectivists, two of whom (Bacon and Pecham) he mentions by name. Thus he holds that the crystalline lens is sensitive and lenticular in shape and that it transmits the "species of visible objects" (*visibilium species*) to the opening in the optic nerve at the back of the eye after refraction at the posterior surface of the lens. Where he parts ways with both the perspectivist and medical traditions is in imputing myopia and presbyopia to an improper curvature of the lens's anterior surface. Myopia, he argues, is due to a curvature that is too sharp, presbyopia to one not sharp enough. As a result, the myopic lens brings the incoming rays toward "convergence" (*coincidentia*) too soon, so that the species enters the optic nerve improperly. Conversely, the presbyopic lens brings incident rays toward convergence too late with an equivalent result. Concave lenses thus delay the convergence in myopia while convex lenses hasten it in presbyopia.[34]

Although vague on details, Maurolico's argument seems to be as follows. Let the two circles in figure 8.1 represent the uveal sphere of the eye in cross section, with the lens at front, the opening of the optic nerve at back, and axis AB passing through the centers of both the eye and the optic nerve. For the sake of simplicity, I have dropped the sclera and cornea out of the account and have located the crystalline lens toward the front, whereas Maurolico, following the medical tradition, has it at the center.[35] The diagram to the left represents an unaided myopic eye upon whose lens ray CD is incident at D. The first thing to note is that this ray strikes the lens obliquely rather than along the perpendicular.[36] It is therefore refracted toward the normal at point D and, if allowed to continue straight through, would converge on the axis at F. In reality, though, it is refracted away from the normal at E and enters the optic nerve at point G. The same holds for the symmetrical ray at the right.

Henry Crew, *The* Photismi De Lumine *of Maurolycus: A Chapter in Late Medieval Optics* (New York: Macmillan, 1940)—henceforth, Crew.

34. For Maurolico's full account of vision and the ocular system, see *Photismi*, 69–73; Crew, 105–10. For his account of eyeglasses, see *Photismi*, 73–80; Crew, 110–21.

35. In fact, Maurolico follows Andreas Vesalius explicitly in his description of ocular anatomy. We will have more to say on Vesalius later.

36. Although I have represented the incoming rays as parallel to the axis, Maurolico makes no such specification but is vague about how the incoming rays are actually arranged in space before reaching the eye. He is clear, however, that the only ray that passes into the lens along the orthogonal is the axial one; see *Photismi*, 76; Crew, 115.

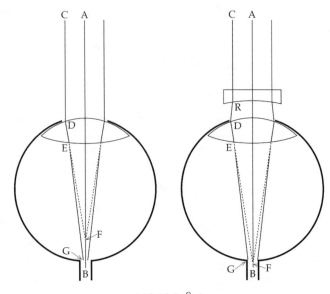

FIGURE 8.1

Consequently, the species encompassed by the two rays enters the optic nerve in too compressed a form.

The same eye is represented in the right-hand figure, but now with a plano-concave lens in front of it. Accordingly, the ray from C passes straight through the lens's flat surface to reach point R on the concave surface at an angle. Emerging into the air at that point, it is refracted away from the normal along RD to strike point D on the crystalline lens at an even more oblique angle than was the case in the unaided eye. Refracted into the lens along DE, the light ray, if allowed to continue straight through, would converge on the axis at F, which now lies below its original intersection in the unaided eye. As before, however, the ray is refracted away from the normal at E along EG, so the species encompassed by it and its mate on the other side enters the optic nerve at the right size. By "spreading" (*dilatare* or *disgregare*) the incoming radiation, then, the concave lens delays the potential convergence at F by just enough to ensure that the species enters the optic nerve in proper form. In a presbyopic eye, on the other hand, a convex lens will ensure that the species, which is originally too spread out, is compressed to the right size when it enters the optic nerve. This it does by "gathering" (*congregare* or *coarctare*) the incident rays.

Three things are especially noteworthy about this account. First, despite operating within the perspectivist framework, Maurolico is forced to break one of its cardinal rules—namely, that the lens is sensitive to perpendicular rays only. Not that this rule was absolutely binding; Alhacen broke it in order to account for peripheral vision according to the lens's ability to detect oblique radiation. Still, he attempted to save the model by positing virtual perpendiculars as the means by which peripheral objects are perceived (see chapter 5, section 4). Maurolico makes no such bow to the model; he simply disregards it. Second, for the perspectivists the incoming rays all converge toward the center of the eye, which is the cardinal reference point for all visual analysis. According to Maurolico's analysis, the point of convergence has no fixed location. Where it lies depends entirely on the type and severity of the visual disorder, so the center of sight is, at best, a floating point. Third, since presbyopia and myopia are due to a deformity in the crystalline lens, artificial lenses actually *correct* those conditions by refraction. For Maurolico, therefore, the two conditions are optical in origin, so their proper remedy is optical, not physiological.

Toward the end of his discussion of eyeglasses, Maurolico adds that in the same way they aid sight by gathering incident rays, convex lenses also bring solar rays to a burning point. This, he reminds us, had already been explained in the first chapter of the third part of the *Photismi*. Before turning to that explanation, let us examine how he deals with such radial gathering in concave spherical mirrors. In proposition thirty-one of the first part of the *Photismi*, which was completed by 1521, Maurolico demonstrates that if luminous point C in figure 8.2 faces concave mirror GD centered on E, then ray CG farther from axis CD will reflect to a point on the axis lower than that to which nearer ray CF will reflect. Both the claim and its proof can be generalized to all rays from C: the farther the incident ray is from the axis, the closer to the mirror's surface the reflected ray will strike it. All the incident rays thus undergo spherical aberration after reflection. With this in mind, Maurolico concludes the first part of the *Photismi* by arguing in proposition thirty-five that solar rays reflected from a concave spherical mirror can be brought to such concentration as to cause burning. But having demonstrated the phenomenon of spherical aberration earlier, he argues correctly that only a small portion of the mirror is involved and that rays from all points on the sun incident upon that portion will reflect not at a point but "within a narrow area" of intense concentration. In other words, the rays will converge in a focal area rather than to a single point. Interestingly enough, Maurolico fails to mention that the focal limit lies halfway between the mirror and its center point.[37]

Dating from 1554, some twenty-three years after his study of reflection from

---

37. Even Leonardo, with far less formal education in mathematics than Maurolico, arrived at that limit in his studies of burning mirrors, as had Bacon some two centuries earlier. See Sven

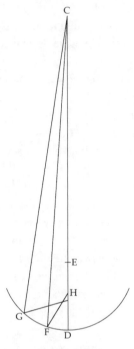

FIGURE 8.2

concave spherical mirrors, Maurolico's account of focusing in spherical lenses follows essentially the same analytic path. For instance, in chapter 1 of the third part of the *Photismi*, he demonstrates in proposition eighteen that "of parallel rays [passing] through a transparent sphere at unequal distances [from the axis], the one farther from the axis with which it is parallel will intersect [the axis] nearer the sphere than the rest." His proof is based on figure 8.3, according to which he demonstrates that ray BG refracted out of the sphere from incident ray AB, which is farther from axis EH, will intersect the axis at some point G higher than H, where ray DH refracted from incident ray CD intersects it. By implication, the farther from the axis the incident ray, the nearer to the sphere its refracted ray will meet the axis, which is to say that parallel rays passing through a transparent sphere undergo aberration.[38]

Dupré, "Optics, Pictures, and Evidence: Leonardo's Drawings of Mirrors and Machinery," *Early Science and Medicine* 10 (2005): 211–36.

38. For proposition eighteen, see *Photismi*, 48; Crew, 65–67—quotation adapted from Crew, 65. In the corollary at the end of this proposition Maurolico generalizes to any and all rays.

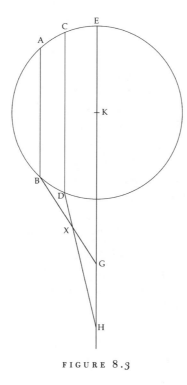

FIGURE 8.3

That, of course, leads to the claim in the twenty-fourth and final theorem that sunlight passing through a glass sphere causes burning by concentrating individual intersections within a tiny area rather than by a convergence of all the rays to a single point. As with his analysis of burning mirrors, so with this one, he specifies no limit point. He does argue that, taken as a whole, all the points of intersection, of which X is one, form a cone whose sides "are not straight, but, because of these sorts of successive radial intersections are curved, and the vertex is the endpoint of the convergences [of rays]."[39] This is known as the caustic curve, about which we will have more to say later, when we deal with Kepler. Altogether, then, Maurolico's respective analyses of burning mirrors and burning lenses show that both devices have certain fundamental things in common—namely, that the rays in both undergo spherical aberration, and both focus light to an area rather than to a point. Moreover, they share one

---

39. For proposition twenty-four, see *Photismi*, 41–42; Crew, 74–75—quotation adapted from Crew, 75.

other significant characteristic: they are both capable of projecting an inverted image onto a screen.[40]

A surprising lapse on Maurolico's part was his failure even to attempt to locate the focal point for either device. As we have seen, the focal point for concave spherical mirrors had already been defined at half the radius of the sphere, but its equivalent for spherical lenses had yet to be determined in such quantitative terms. It was Kepler, in fact, who made that determination accurately—or at least came within a whisker of it—but a significant intermediate step in that direction was taken by Giambattista Della Porta in his analysis of refraction through glass in the *De Refractione* of 1593. Even earlier, in his *Magiae Naturalis* of 1589, he had made some important qualitative observations about both concave spherical mirrors and convex lenses. One involves what he calls the point of inversion (*punctum inversionis*), which is where the image viewed in a concave spherical mirror achieves its greatest magnification before undergoing inversion as the eye is moved a bit farther from the mirror. This is also the focal point of the device, so Della Porta effectively showed that image formation and radial focusing are intimately associated.[41] Della Porta also observed that the clarity of the inverted image in a camera obscura can be improved in two ways. One is by placing a concave spherical mirror toward the aperture and using it to cast the image on a screen between the mirror and the aperture. The other is by placing a convex lens just behind the aperture to focus the image on the back wall. He goes on to suggest that the eye operates somewhat like a camera obscura, the pupil serving as the aperture and the anterior surface of the lens serving as the screen.[42]

Della Porta was not just aware that concave spherical mirrors and spherical lenses are strikingly similar; he exploited that similarity systematically in his analysis of the two devices in book two, propositions one through five, of

40. For Maurolico's discussion of image projection by concave spherical mirrors, see *Photismi*, 28–29; Crew, 43–45. On image projection by spherical lenses, see *Photismi*, 47–48; Crew, 73–74. In neither case does he relate the projection to the device's focal point.

41. Young and Speed, *Natural Magick*, 360; *Magiae Naturalis Libri XX*, 259–60. The Venetian savant Ettore Ausonio (ca. 1520–ca. 1570) actually identified the *punctum inversionis* with the focal point of concave spherical mirrors in his *Theorica speculi concavi sphaerici*, which was composed around 1560, some thirty years before the *Magiae Naturalis* of 1589, but never published; see Dupré, "Ausonio's Mirrors," 160–61. In "Mathematical Instruments and the Theory of the Concave Spherical Mirror: Galileo's Optics beyond Art and Science," *Nuncius* 15 (2000): 551–88, Dupré argues that Della Porta actually knew Ausonio's *Theorica*; see especially 573–75.

42. Young and Speed, *Natural Magick*, 363–65; *Magiae Naturalis*, 587–90.

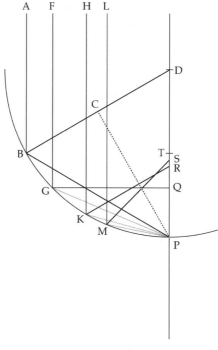

FIGURE 8.4

his *De Refractione*.[43] In the first proposition, he has us imagine that arc BP in figure 8.4 is a segment of a concave spherical mirror whose axis DP passes through center of curvature D. Let chord BP form one side of a hexagon inscribed in the full circle containing the arc, chord GP the side of an inscribed octagon, chord KP the side of an inscribed dodecagon (that is, with twelve sides), and chord MP the side of an inscribed hexadecagon (with sixteen sides). Then let AB, FG, HK, and LM be rays parallel to axis D. In order to determine the angle of reflection, Della Porta instructs us to find the midpoint of the normal to the point of reflection and drop a perpendicular from it to the axis. Where it intersects the axis is where the reflected ray will meet it. Take ray AB as an example. Drop normal BD from center point D to point of reflection B, find midpoint C on it, and drop perpendicular CP to point P on the axis. BP will therefore be the reflected ray. The same method yields reflected ray GQ for incident ray FG, reflected ray KR for incident ray HK, and reflected

43. *De refractione*, 36–47.

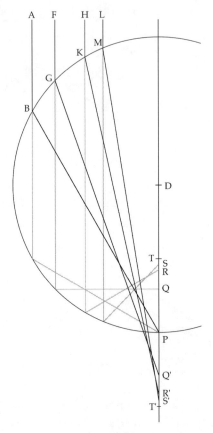

FIGURE 8.5

ray MS for incident ray LM. Accordingly, the angle of reflection will be 60°
at B, 45° at G, 30° at K, and 22.5° at M. It therefore follows that the smaller
the angles of incidence become—in other words, the nearer the incident rays
strike the axis—the closer to midpoint T of radius DP the reflected rays will
meet the axis, so the rays undergo spherical aberration, and T is the mirror's
focal point.[44]

Turning from concave mirrors to spherical lenses in proposition two, Della
Porta uses essentially the same figure as he did for proposition one. This time,
however, the circular segment in figure 8.5 forms part of the cross section of
a glass sphere with the rays striking its front surface at points B, G, K, and M.
Mark off points Q′, R′, S′, and T′ below P on the axis such that Q′P = QP,

44. See ibid., 36–41.

R´P = RP, S´P = SP, and T´P = TP.[45] The refraction for each incident ray will therefore be determined by the point corresponding to the previous point on the axis according to reflection. Ray AB incident at 60° will therefore refract along BP, ray FG incident at 45° along GQ´, ray HK incident at 30° along KR´, and ray LM incident at 22.5° along MS´. By analogy with the concave mirror, therefore, as the incident rays draw closer to the axis, the refracted rays intersect it at points increasingly closer to T´.

The procedure Della Porta outlines here constitutes theoretical ray tracing insofar as it is based on actual quantification. On the one hand, the geometrical model according to which the angles of incidence are produced allows them to be readily translated into numerical values (that is, 60°, 45°, 30°, and 22.5°). On the other hand, the angles of refraction yielded by that model can be easily calculated by trigonometry. The angle of refraction at B, for instance, is precisely 30°, at G just over 25.5°, at K just under 17.7°, and at M just under 13.4°.[46] Moreover, the analysis entails a full recognition that the refracted rays undergo spherical aberration and that T´ is the limit point beyond which the refracted rays will not intersect the axis—in short, the focal point.[47] Consequently, as in the case of spherical concave mirrors, so in this one, only a small portion of the sphere surrounding the axis is implicated in focusing.

Quantitatively determined though it may be, this model is only hypothetical. In order to test it physically, Della Porta turns to the process of empirical ray tracing. One such test is described in proposition four, where he has us form a plate of very clear glass or crystal around 2 cm high and large enough to accommodate a semicylindrical hollow around 30 cm in diameter and 2 cm high. Illustrated in figure 8.6, this plate is to be attached to a table on whose surface all the relevant lines and points in figure 8.5 are incised. ABP, FGQ, HKR, and LMS in figure 8.6 will thus correspond precisely and proportionately with ABP, FGQ´, HKR´, and LMS´ in figure 8.5, and point D will be the center of curvature for the semicylindrical hollow. Extend AB to X, and draw line BB´ to form angle B´BX = angle PBX. Draw lines GG´, KK´, and MM´ the same way.

45. See ibid., 41–43.

46. These values are rather badly off for crown glass, whose index of refraction is ~1.5. In that case, the values for r should be 35.3°, 28.1°, 19.5°, and 14.8°, respectively. On the other hand, for lead glass, whose index of refraction is ~1.7, the values for r would be 30.5°, 24.6°, 17.1°, and 13.0°, respectively, which are much closer to Della Porta's predicted values of 30°, 25.5°, 17.7°, and 13.4°, respectively.

47. By sheer coincidence, T´ would be the actual focal point if the sphere were formed from crown glass and the rays underwent double refraction into and out of the sphere.

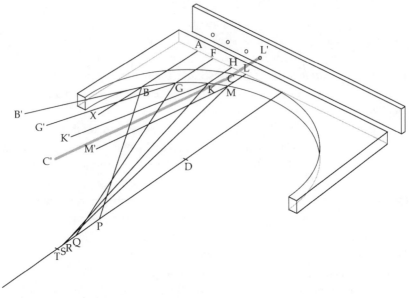

FIGURE 8.6

In a thin strip of wood or metal drill four small holes that will line up perfectly with AB, FG, HK, and LM when the strip is placed upright behind the glass plate, and make sure the holes lie well below the level of the top of the glass plate, while the strip itself stands well over 2 cm high. The entire apparatus should then be positioned so that direct sunlight streams through all the holes into the glass plate. When all but the hole at L′ are blocked, a narrow beam will be projected through it to the rear of the glass plate and, after passing straight through that surface, will strike the circular interface between the glass and the air at point C, directly above point M. There it will form an angle of incidence of 22.5°. It will then refract away from the normal on entering the air and, according to the theory, follow course CC′ directly above and parallel to line MM′. This course is dictated by the principle of reciprocity because the angle of refraction formed by MM′ is equal to the angle of refraction formed by MS, which is the path the light would follow were it refracted from air into glass at the same angle of incidence.[48]

Theoretically, then, when all the holes and their corresponding angles of incidence have been tested, the beams of light passing through the holes in the

48. See *De refractione*, 45–46.

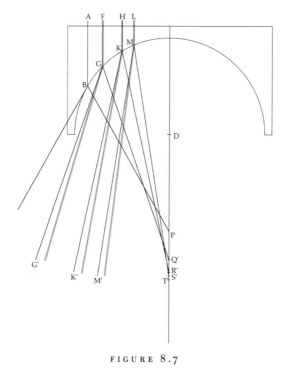

FIGURE 8.7

strip and refracted at the concave surface of the glass plate should follow their respective lines MM', KK', and GG' exactly (line of refraction BB' can be discounted because it enters the glass without emerging into the air). What actually happens depends on the glass's index of refraction, and unfortunately we have no way of determining precisely what kind of glass Della Porta used. Most likely it was equivalent to crown glass, whose index of refraction is around 1.5, so there would be a significant disparity between the predicted and actual angles of refraction, as illustrated in figure 8.7, where the light gray lines represent the actual refracted rays. On the other hand, it is possible, though highly unlikely, that he used lead glass with an index of around 1.7. In that case there would have been a much closer fit between predicted and actual angles. But figure 8.7 does not take into account such factors as diffraction and dispersion that would have seriously compromised the accuracy of this experiment, so the results would have been ambiguous enough that Della Porta could easily have interpreted them as fulfilling his theoretical expectations no matter what kind of glass he used.

3. RETHINKING THE EYE

With Della Porta we have come a long way from the confused and crude per-spectivist account of spherical lenses. We have also come a long way toward Kepler's account of spherical lenses, which was central to his broader analysis of retinal imaging. But before dealing with that account in the next section, we need to take a quick look at how medical authorities during the Middle Ages and Renaissance understood the anatomical structure and function of the eye and how that understanding gradually changed over the sixteenth century. Until around 1500, physicians generally followed the Galenic visual model as Johannitius modified it. They therefore assumed that the lens forms a sphere dead center in the ocular globe and that vision is accomplished by means of cerebral spirit fed to the lens from the brain and then extramitted radially through the cornea.[49] This, for instance, is the anatomical model Leonardo adopted in his "Treatise on the Eye" (see chapter 7, section 4), although he adapted it to an intromissionist theory based on light rays. Generally speaking, though, medieval and Renaissance physicians favored extramissionism, even though some, like Jacques Despars (ca. 1380–1458), were aware that it conflicts with the perspectivist model. Despars in fact acknowledged the conflict in his commentary on Avicenna's *Canon*, which was composed between 1432 and 1453 and published posthumously in 1498, but he favored the extramission-ist alternative on the grounds that although not necessarily true, it is more appropriate and convenient for medical purposes. Like Alberti, in short, he

49. Typical of the medieval understanding of the eye is the *Tractatus de oculis* of Petrus Hispanus, or Pedro Julião (ca. 1215–1277). No less typical, and even more popular than Petrus's treatise, was the *De oculis eorumque egritudinibus et curis* ("On the Eyes and Their Maladies and Cures") of Benvenutus Grassus (d. late thirteenth century?). In both cases, the treatment of ocular anatomy and physiology is extremely perfunctory because the primary focus of both treatises is on the diagnosis and cure of maladies that affect the white and cornea of the eye, not its internal components. On Petrus Hispanus, see A. Mark Smith, "Petrus Hispanus' 'Treatise on the Eyes,'" in *The Treatise on the Eyes by Pedro Hispano*, by A. Mark Smith and Arnaldo Pinto Cardoso (Lisbon: Alêtheia Editores, 2008), 8–56. For a Latin edition of his treatise, see Albert Maria Berger, ed. and trans., *Die Ophthalmologie* (Liber de oculo) *des Petrus Hispanus* (Munich: J. F. Lehmann, 1899), and for a recent English translation, see Walter J. Daley and Robert D. Yee, trans., "The Eye Book of Master Peter of Spain—a Glimpse of Diagnosis and Treatment of Eye Disease in the Middle Ages," *Documenta Ophthalmologica* 103 (2001): 119–53. For an English translation of Grassus's treatise, based on the first printed edition of 1474, see Casey Wood, *De oculis eorumque egritudinibus et curis* (Stanford, CA: Stanford University Press, 1929).

was somewhat agnostic about the direction of radiation and thus the ultimate physical cause of vision.[50]

Even with the renewed interest in, and enthusiasm for, anatomizing during the first half of the sixteenth century, physicians were slow to revise the Johannitian ocular model, and when they did, they did so incrementally. For instance, by the time Andreas Vesalius (1514–1564) published his anatomical magnum opus, *De humani corporis fabrica libri septem* ("Seven Books on the Structure of the Human Body," Basel, 1543), it was commonly accepted that the crystalline lens is somewhat flattened at front rather than perfectly spherical. Vesalius went even further by having the lens radically compressed both front and back into a distinctly lenticular shape. Yet even he, despite his consummate skill as an anatomist, kept it at the center of the eye rather than shifting it toward the front. Thus despite his highly critical attitude toward the Galenic/Johannitian model, Vesalius was still committed enough to it that he failed to "see" where the lens actually belongs in the eye.[51] Somewhat less committed to that model, perhaps, Realdo Colombo (1516–1559) took Vesalius to task in his *De re anatomica libri XV* (Venice, 1559), claiming that the lens is slightly displaced from the center toward the front of the eye. He also noted, as did Vesalius before him, that the crystalline lens is optically equivalent to a reading or magnifying glass (*specillum*), for if you remove it from the eye and hold it close to letters, "they appear larger and are made out more easily." As to whether vision is due to extramission (*emissio*) or intromission (*immissio*), Colombo remained agnostic, arguing that this issue "is not easily explained and is a controversial matter yet to be resolved."[52]

As Lindberg observes, the structural modifications to ocular anatomy these thinkers imported were relatively minor and did little to alter the traditional Galenic/Johannitian understanding of ocular anatomy or of the crystalline lens as the primary seat of visual sensation.[53] That, however, changed dramatically with the publication of *De corporis humani structura et usu* ("On the Structure

50. See, e.g., Danielle Jacquart, *La médicine dans le cadre parisien, XIVe–XVe siècle* (Paris: Fayard, 1998), especially 412–13.

51. Nonetheless, in book four of *De Fabrica*, Vesalius correctly observes that the optic nerves are displaced from the axis and that they are not hollow; see William Frank Richardson and John Burd Carman, trans., *Vesalius* On the Fabric of the Human Body, *Books III and IV* (Novato, CA: Norman Publishing, 2002), 183–85.

52. See *De re anatomica*, 216–19. For a more detailed account of sixteenth-century approaches to ocular anatomy and function, see Lindberg, *Theories*, 168–77.

53. See Lindberg, *Theories*, 173–75.

and Utility of the Human Body") in 1583 by the Swiss physician Felix Platter (1536–1614). Platter altered both the Galenic and the perspectivist models in two fundamental and radical ways. First, he was unequivocal in taking the retina, not the lens, as the "primary component of vision" (*pars primaria visionis*). Second, locating it well toward the front of the eye, Platter reduced the lens to a mere "looking glass" (*perspicillum*) through which the species of light and color are projected onto the retina. Precisely how the retina functions as the seat of vision Platter left unexplained, and he had little to say about the optics of the lens beyond adverting to the way it spreads the incoming species out over the retina's surface. In other words, just as it does artificially, outside the eye, the lens in its natural context, inside the eye, does nothing more than magnify what is seen through it so as to give a clearer view.[54]

How indebted Platter may have been to earlier sources for his theory of retinal sensitivity is unclear. There was precedent for the idea that the retina is somehow implicated in vision, at least insofar as the *aranea* covering the sensitive, anterior surface of the lens was thought to be an extension of the retina.[55] But lodging visual sensitivity entirely in the retina and thus denying any share of it to the lens flew in the face of both perspectivist and medical theory up to his time. It did, however, resolve the argument over whether vision is due to extramitted or intromitted radiation. Leaving absolutely no room for extramission, Platter's visual model requires that vision be due solely to incoming rays that are transmitted optically, not sensitively, through the lens. Aside from the retina itself, then, Platter's eye is just an optical instrument, no different in kind from a pair of eyeglasses or a fluid-filled globe with a convex lens between the pupillary aperture at front and the retinal screen at back. Neither the radical nature nor the historical importance of this idea can be overstated because it was Platter to whom Kepler ultimately looked for his understanding of both the structure and function of the eye as a whole and the crystalline lens as an image projector within it. Once he accepted that idea, all that remained for Kepler was to demonstrate precisely how the lens fulfills this function according to strict optical principles. This he did in chapter 5 of the *Ad Vitellionem Paralipomena*, or "Supplement to Witelo," to which we will now turn.

---

54. Platter's full description of the eye, in tabular form, is to be found in *De corporis humani structura*, book two, 186–87; see p. 187 for his account of the retina and the lens, and see book three, plate 49, for the full set of figure illustrating the components of the eye.

55. See Lindberg, *Theories*, 171–72.

4. KEPLER'S ANALYSIS OF RETINAL IMAGING

The very title of Kepler's optical masterpiece, "*Supplement* to Witelo," is misleading. Far from making a few additions or amendments to Witelo's—and thus the perspectivist—theory of light and sight, Kepler engaged in a wide-ranging and detailed critique that prompted a drastic overhaul of the visual model at its center. Indeed, as he puts it in his preface to the *Paralipomena*:

> For because many things, not only about the direct ray, but also about the reflected and refracted ray, were overlooked by Witelo, and many things that should have been explained *a priori* were only brought in extraneously from experience and set up in place of axioms, I thought it a good idea to look a little more deeply into the nature of light.[56]

That deeper look yielded some significant results. One was Kepler's account in chapter 2 of why the image of a luminous object, such as the sun, projected through a small, irregularly shaped aperture onto the back wall of a camera obscura, takes the shape of the object rather than of the aperture. Kepler's explanation is logical and straightforward: when it is far enough from the aperture, the projected image consists of an infinite number of overlapping images of the aperture that combine to take the shape of the luminous object. In other words, the resulting image is not homogeneous or holistic, as Pecham and others supposed, but rather a congeries of individual, aperture-shaped daubs of light. According to this analysis, Kepler also determined that such images have a penumbra that augments their apparent size, a point crucial to measuring the actual diameter of the image as a means of establishing the relative size of the luminous object (whether sun or moon) as precisely as possible. This, then, is a case in which understanding the optics of an observational instrument makes a critical difference in the accuracy of the results obtained through it.[57]

---

56. William Donahue, trans., *Johannes Kepler, Optics: Paralipomena to Witelo and Optical Part of Astronomy* (Santa Fe, NM: Green Lion Press, 2000), 16—henceforth, Donahue. For the Latin see *Ad Vitellionem Paralipomena, Quibus Astronomiae Pars Optica Traditur* (Frankfurt, 1604), 4—henceforth, *Ad Vitellionem*.

57. For Kepler's full account of the camera obscura, see Donahue, 55–71; in *Ad Vitellionem*, 37–56. On the camera obscura as an observational instrument in astronomy, see Giora Hon and Yaakov Zik, "Kepler's *Optical Part of Astronomy* (1604): Introducing the Ecliptic Instrument," *Perspectives on Science* 17 (2009): 307–45. Kepler's references to Pecham ("Ioannes Pisanus") in chapter 2 of the *Ad Vitellionem* ("Several years ago, some light shone forth upon me out of

More significant and innovative than Kepler's account of the camera obscura as an observational instrument, however, was his account of the eye as a somewhat equivalent instrument according to Platter's description. Central to that account is the conception of the crystalline lens as a purely optical device rather than a partly optical and partly sensitive selector of visual information. The idea that the lens is sensitive and, moreover, sensitive only to perpendicular rays, was especially troubling to Kepler, and that for two reasons. First, because "there is hardly any difference in illuminating between the perpendiculars and those very close to them," it is difficult to understand how the lens makes such fine distinctions that it can isolate the perpendiculars from all the others. And if it does not isolate them, then the impression at any given spot on the lens ought to be confused, not clear. Second, the perspectivist account of peripheral vision is incoherent because it "implies that oblique and direct radiation is received by the same point" on the lens.[58] The result, then, ought not to be mere confusion but sheer visible noise. It was presumably in the face of these and other lesser inconsistencies in the perspectivist account of vision that Kepler was driven to conclude that the visual image is not formed *on* but, rather, *by* the lens, which somehow projects that image onto the retina at the back of the eye. Key, therefore, to defending this conclusion was to demonstrate how the lens actually accomplishes that projection, so let us turn to this demonstration and the lens theory embedded in it.

Kepler starts by assuming that the eye's lens is roughly equivalent to a sphere of water and on that basis examines how light will be refracted through it. Let ADG in figure 8.8 represent such a sphere suspended in air. Let AZE be its axis, and let rays MB, NC, and OD parallel to it strike the front surface of the sphere to form angles of incidence MBV = 30°, NCX = 60°, and ODY < 90° by just enough that ray OD will enter the sphere rather than skim by along the tangent. For all practical purposes, ODY can be taken as exactly 90°. Kepler then invites us to make two more assumptions: first, that the resulting angles of refraction are constantly proportional to the angles of incidence (ZCE = 2ZBE, and ZDE = 3ZBE) and second, that all the rays refracted in the water will meet at point E if allowed to pass straight through surface DG. Granted these assumptions, Kepler demonstrates by reductio ad absurdum that in fact

---

the darkness of Pisanus," Donahue, 56; *Ad Vitellionem*, 39) suggests that he had studied optics sometime during his stint at the University of Tübingen (from 1589 to 1594) on the basis of the *Perspectiva communis*.

58. *Paralipomena*, 5.4, in Donahue, 220; in *Ad Vitellionem*, 204–5.

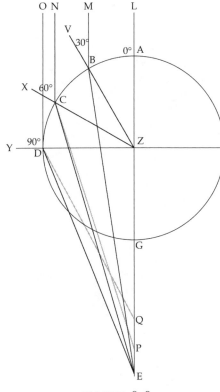

FIGURE 8.8

ZCE > 2ZBE, so refracted ray CE must be lifted to some point P above E on the axis to lessen the angle enough that ZCP = 2ZBE. By the same token, ZDE > 3ZBE by a proportionally greater amount, such that refracted ray DE will have to be lifted to point Q above P on the axis to make angle ZDQ = 3ZBE.[59]

So far, Kepler has shown that all rays of light entering the globe of water along parallel lines will undergo spherical aberration and that the greater the angle of incidence (that is, the farther from the axis the ray strikes the sphere's front surface), the closer to point G the refracted ray will intersect the axis. In addition, since angle ODY is the largest possible angle of incidence at which refraction can occur, point Q is the "nearer limit" (*quod terminet citime*) to which parallel radiation through the sphere can reach the axis. In short, no

59. *Paralipomena*, 5, proposition 9 and lemma, in Donahue, 196–97; in *Ad Vitellionem*, 184–86.

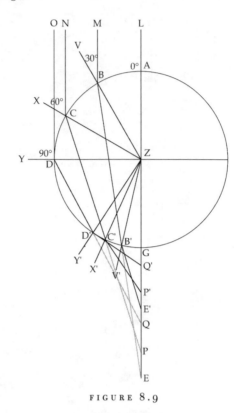

FIGURE 8.9

parallel ray can be refracted through the sphere to any point between Q and G.[60] From all this it follows that if the initially refracted rays undergo a second refraction into the air at surface DG, they will suffer spherical aberration after leaving the sphere, which they do at an angle of refraction equal to the original angle of incidence according to the principle of reciprocity. Consequently, ray DQ in figure 8.9 will refract at point D′ along D′Q′ at angle Y′D′Q′ = angle ODX = 90°, CP will refract at C′ along C′P′ at angle X′C′P′ = angle NCX = 60°, and BE will refract at B′ along B′E′ at angle V′B′E′ = angle MBV = 30°. Point Q′ will therefore mark the nearer limit of double refraction through the sphere, just as Q marked that limit for single refraction into it.[61]

Having established these points under the false assumption that the angles

60. *Paralipomena*, 5, proposition 13, in Donahue, 201–2; in *Ad Vitellionem*, 188–90. Strictly speaking, the nearer limit for Kepler is the point nearest G to which no ray will reach the axis.

61. See Donahue, 199; *Ad Vitellionem*, 186.

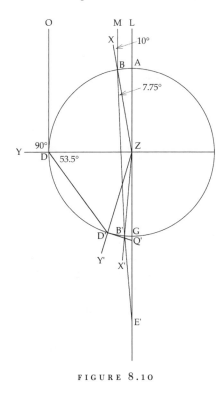

FIGURE 8.10

of refraction are constantly proportional to the angles of incidence, Kepler is ready to undertake the process of theoretical ray tracing in order to establish precisely where the refracted rays will intersect the axis. Take ray OD in figure 8.10. For all practical purposes, as established earlier, this ray strikes the sphere's surface at angle ODY = 90°. According to Kepler's method for calculating refraction, such an angle of incidence will yield angle of refraction ZDD´ = 53.5° (which is nearly 5° too high), and by the principle of reciprocity, ray DD´ will refract out of the sphere along D´Q´ at angle Y´D´Q´ = angle ODY = 90°. According to trigonometric analysis, therefore, GQ´ will be slightly less than one-twentieth the sphere's radius (~0.044 the radius, or ~0.046 the radius by Kepler's reckoning). Now let ray MB strike the sphere's surface at angle MBX = 10°. Drawing on Witelo's value for refraction from air into water at *i* = 10°, Kepler assigns a value of 7.75° to angle of refraction ZBB´ (see chapter 6, section 4). Accordingly, refracted ray BB´ will emerge from the sphere along B´E´ at angle X´B´E´ = angle MBX = 10°. On this basis, Kepler calculates that the distance between point E´, where refracted ray B´E´ strikes the axis, is barely longer than the radius of the sphere.[62] He then concludes that all the radiation striking within arc BA on the sphere will meet the axis near point E´, the last possible intersection being just beyond it. This point, which Kepler designates the "last limit" (*finis ultimus*), constitutes the focal point of the sphere, so it follows that the densest aggregation of intersections will occur just short of that point and will be produced by the radiation falling on a circle within 10° of the axis.[63] As with spherical concave mirrors, so with spherical lenses, only a small portion around the axis is implicated in focusing.

Thus far Kepler has been engaged in theoretical ray tracing. In order to test the conclusions yielded by that process, Kepler turns to empirical ray tracing. For that purpose he uses a urinary flask, which consists of a spherical base of blown glass connected to a neck. Filling the base with water and placing it so that a shaft of sunlight shines through it, he holds a sheet of paper against the side directly opposite that at which the sunlight enters the sphere and examines the light cast on it.

With the paper held right up against the sphere, as represented at position 1 in figure 8.11, Kepler notices that a bright circle of light is projected onto it. This circle, he concludes, is produced by the intersection of outer rays, which form an envelope of bright light. Then, removing the paper along the axis to position 2, he sees a spot of light emerging in the middle of the bright, outer circle at nearer limit Q of refraction, which lies around one-twentieth of the radius (Kepler mistakenly says "diameter") beyond the sphere's edge. It follows, therefore, that the central spot of light is formed by radial intersection on the axis itself. As he moves the paper ever farther from the sphere's edge along the axis, Kepler observes that the bright, outer circle diminishes at a decelerating rate. Thus the circle at position 1 exceeds the circle at position 2 by the same amount as the circle at position 2 exceeds that at position 3, and the same holds for the circle at position 4, even though the axial distances become increasingly longer. This leads Kepler to conclude that the envelope of radiation forms a curved cone, whose curvature he later concludes is hyperbolic. Finally, when the paper is brought to position 5, at the vertex of the cone, just shy of focal point F, the bright, outer circle and the inner spot of light coalesce to form an aggregate illumination so intense "that gunpowder in cold water is ignited" by it. When the paper is removed

62. *Paralipomena*, 5, proposition 14, in Donahue, 204–5; in *Ad Vitellionem*, 190. According to modern theory, that distance should be precisely one radius.

63. *Paralipomena*, 5, proposition 15, in Donahue, 205–6; in *Ad Vitellionem*, 190–91.

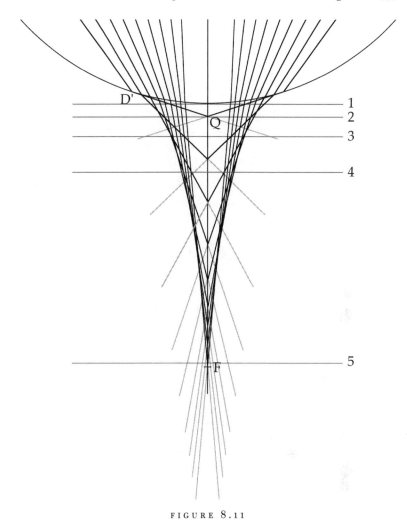

FIGURE 8.11

much beyond that point, Kepler adds, "the figure quite vanishes, all the rays having intersected."[64]

Just before detailing this experiment with the water-filled flask, Kepler makes an important specification: "While up to this point an image (*imago*)

---

64. *Paralipomena*, 5, proposition 19, in Donahue, 210–11; in *Ad Vitellionem*, 193–94. For Kepler's argument that the caustic curve formed by the cone, already described in a general way by Maurolico, is hyperbolic, see *Paralipomena*, proposition 24, in Donahue, 214; in *Ad Vitellionem*, 198.

has been [taken as] a mental entity (*ens rationale*), now let the figures (*figurae*) of objects that really exist on paper or upon another screen (*pariete*) be called paintings (*picturae*)."[65] By modern lights, the distinction Kepler draws here is between virtual and real images, the former being psychological, viewer-dependent constructs, the latter being physical, viewer-independent constructs. Within the context of perspectivist optics, however, no such distinction exists because the perspectivists viewed *all* images as virtual. It is presumably in recognition of this point that Kepler designates what is physically projected on a screen as a "painting" (*pictura*), since it consists of individual daubs of color on a material surface, rather than as an "image," which is an intentional entity existing in a spiritual medium. Hence, in figure 8.11, the "figure" of the sun is painted at focal point F.

With that in mind, Kepler turns from the projection of parallel rays through the sphere to the projection of rays from a single luminous point at a finite distance from it. In this case, just as in that of parallel rays, there is a nearer as well as a last limit. However, as the luminous source point approaches the sphere, both limit points migrate outward, and the closer the source point gets to the sphere, the farther outward the two limit points will be driven.

In figure 8.12, for example, where the source point is I, nearer limit Q and last limit L have both been forced outward in comparison to their respective locations Q′ and E′ in figure 8.10. Accordingly, the figure of point I will be "painted" with maximum clarity on screen XY placed at, or just in front of, point L. The figures of points I′ and I″ flanking I will also be painted on that screen at points L′ and L″, respectively, but with decreasing clarity according to their increasing distance from I. The clarity will be further compromised by the extraneous radiation from each of the points that impinges on the primary painted spots at the foci. Also, the resulting composite picture these painted spots form will be inverted, since L′ to the left of L is formed by I′ to the right of I.[66]

Now let us pose a diaphragm with a small opening in front of the sphere, as represented by DG in figure 8.13, EF being the opening. Under these circumstances, of all the light emanating from I, only a thin sheaf of rays, represented in solid black, will be allowed to enter the sphere, the rest (in gray) being oc-

65. *Paralipomena*, 5, proposition 18; translation adapted from Donahue, 210; in *Ad Vitellionem*, 193. Elsewhere, in *Paralipomena*, 5, proposition 4, Kepler maintains that "an image is in part an intentional entity, a product of vision. Thus, when the act of seeing ceases, the image ceases to exist"; translation adapted from Donahue, 193; in *Ad Vitellionem*, 180.

66. *Paralipomena*, 5, proposition 20, in Donahue, 211–12; in *Ad Vitellionem*, 194–95.

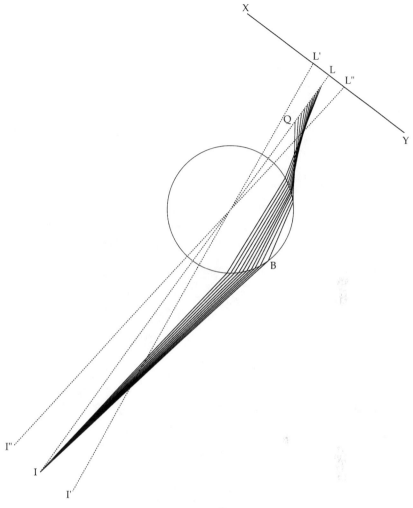

FIGURE 8.12

cluded by the diaphragm. The rays within the entering sheaf will intersect at
various points after refracting through the sphere, but there will be a relative
concentration of such intersections near point M. It is therefore at this point
that the clearest and brightest depiction of I will be produced on a screen. Let
I′ and H be luminous points in the field of view, I′ and I lying equidistant from
H. Point H will be depicted most clearly at or near focal point K, whereas I and
I′ will be depicted most clearly at M and M′, since the radiation that would
intersect near the focal point has been cut off. By blocking that radiation, more-

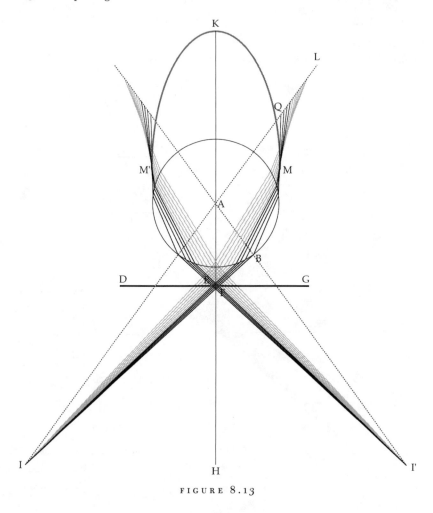

FIGURE 8.13

over, the diaphragm reduces the amount of extraneous light that would diminish the clarity of the depictions at M, K, and M′. In fact, none of the radiation
from I or I′ will reach K, so the depiction there will be as clear as possible. At
M and M′, on the other hand, the depictions will be less clear not because of
extraneous radiation but because the intersections are less concentrated and
involve fewer rays than at the focal point.

Applied to the eye, this analysis accounts for both its structure and its functioning. The diaphragm is formed by the uvea (the iris) with its small pupillary
opening, and the retina, represented by curved line MKM′, forms the screen
on which each point in the field of view is depicted with maximum clarity in

inverted order, all within a completely enclosed, dark space. Thus as Kepler puts it, "the use of the opening of the uvea . . . is in part apparent; likewise, the purpose for which the sides of the retina are moved closer to the crystalline [lens] than in the fundus."[67] This latter point is illustrated in figure 8.13, where points M and M′ are considerably closer to the water-filled sphere than point K, so the retina has to curve toward them in order to capture their depictions as clearly as possible.[68] Even so, as we have already noted, those depictions are less clear than the one at point L, which explains why peripheral vision is less distinct than axial vision. One clear benefit of Kepler's visual model is that it takes the mystery out of eyeglasses. Since the eye is nothing more than an optical system, eyeglasses must provide optical correction to optically deformed eyes. In the case of presbyopia, close vision is poor because the lens focuses light from near objects at points beyond the retina. Convex lenses correct this problem refractively by bringing the intersections forward to the retina. This correction is therefore due to refocusing rather than to magnification, as was commonly supposed in Kepler's day. In the case of myopia, or nearsightedness, on the other hand, distant vision is compromised because the eye is elongated along the axis, which causes the appropriate radial intersections to occur too soon. Concave lenses therefore drive them back to the retina according to the necessary change in refraction.[69]

5. THE ANALYTIC TURN

Superficially at least, the theory of retinal imaging just described marks a radical departure from the perspectivist theory of lenticular imaging. Not only is Kepler's image formed optically at the back of the eye rather than sensitively toward the front, but it is real and inverted as well. Furthermore, Kepler dispenses with the visual cone so central to the perspectivist account, and with it the center of sight. Yet according to Lindberg, these apparent divergences mask fundamental links at a deeper analytic level that reduce the apparent dis-

67. *Paralipomena*, 5, proposition 23, in Donahue, 212–14; in *Ad Vitellionem*, 196–97.

68. Figure 8.12 misrepresents both the lens and the retina as they actually exist. In the real eye, according to Kepler, the lens lies well toward the front and consists of a small spherical segment in the front and an equally small hyperboloidal section in the back, so it is actually lenticular rather than spherical in shape. Likewise, the retina is more or less spherical in shape, but since its center lies well behind the lens, the distance between it and the lens grows increasingly small as it approaches the lens. See *Paralipomena*, 5.1, in Donahue, 179–80, 184, and the diagram on p. 188; in *Ad Vitellionem*, 167–68, 173, and 176d.

69. *Paralipomena*, 5, proposition 28, in Donahue, 216–18; in *Ad Vitellionem*, 200–203.

parities between the two theories to differences of degree rather than of kind. After all, Lindberg maintains, Kepler labored "within a conceptual framework given him by the medieval perspectivists," and this conceptual framework "furnished him with a variety of theoretical commitments . . . factual information, unresolved problems, methodological rules, criteria of success, and the like." Committed to the perspectivist paradigm, in other words, Kepler was guided throughout by its theoretical and analytical precepts. As Lindberg sums it up, "by taking the medieval tradition seriously, by accepting its most basic assumptions but insisting on more rigor and consistency than the medieval perspectivists themselves had been able to achieve, [Kepler] was able to perfect [that tradition]." Aside from logical rigor and consistency, in short, Kepler added little or nothing of substance to what his perspectivist sources furnished.[70]

As far as basic refraction theory is concerned, there is no question that Kepler followed the perspectivist rules; for instance, that light rays are rectilinear, that light refracts toward the normal on entering a denser medium obliquely and away from the normal on entering a rarer one, that the severity of refraction is a function of the angle of incidence and the density differential between the two media, that the plane of refraction is perpendicular to the refracting interface, that refraction is governed by the principle of reciprocity, and so on. These, I have no doubt, Kepler learned from the perspectivists. Yet as I was at pains to show in chapters 3 and 5, these rules all trace back directly from Alhacen to Ptolemy. Granted, Alhacen and his perspectivist followers transformed Ptolemy's visual rays into light rays and thus reversed the direction of radiation. And granted, this reversal of direction was crucial to Kepler's account because it hinges on the radiation of light into the eye and would make no sense if based on visual rays. Nevertheless, the rules themselves are indifferent to the direction of radiation. No matter whether it is a visual ray passing from a denser into a rarer medium, or a light ray passing in the opposite direction, the paths will coincide according to the principle of reciprocity. Hence, while Kepler's basic refraction theory came from the perspectivists, theirs ultimately came from Ptolemy. Even the value of $7.75°$ for the angle of refraction from air to water at $i = 10°$, which Kepler took from Witelo, is a permutation based on Ptolemy's tabulations in book five of the *Optics*. The lack of originality in Kepler's basic refraction theory is thus fully matched by that of Alhacen and the perspectivists.

So much for theoretical commitment. What about Kepler's methodological

70. See Lindberg, *Theories*, 206–8.

commitment? Let us start with his analysis of spherical aberration. If Lindberg is correct, then this analysis must have been based on perspectivist sources. As we saw in section 2 of this chapter, however, the perspectivists assumed that radiation from a luminous source point through a glass sphere is refracted to a single point, not a focal area, on the opposite side. Nowhere does spherical aberration figure in such a model. We also saw in chapter 5 that the one place in Alhacen's *De aspectibus* where Kepler might have gotten wind of spherical aberration—book seven, proposition seventeen—would have been no help whatever. In fact, it would have been a hindrance because of the way the figure for the proposition misrepresents the analysis in that theorem (see chapter 5, section 4). A look at the diagram to the right of figure 5.12 makes this point clear. This, in essence, is the diagram as it appears in all the manuscripts of Alhacen's *De aspectibus* and Witelo's *Perspectiva*. It is also the way it appears in both texts, as edited by Risner, Kepler's immediate source.

Note, first, that rays ST and VQ fail to intersect before reaching the axis, so they show no spherical aberration whatever. Note, as well, that after double refraction, the outermost ray AN from luminous source point A strikes the axis at T. According to the analysis represented in the diagram, this defines the last limit of radiation through the sphere and thus its focal point. For Kepler, on the other hand, it defines the nearer limit, the focal point lying at a distance of slightly more than a radius of the sphere from point G. Furthermore, as I pointed out in the discussion of this theorem, its purpose is not to analyze how light refracts from luminous point A through the sphere but to demonstrate how an annular image of object line QT is formed on the front of the sphere as seen from center of sight A. It is for these reasons, I argued, that the perspectivists failed to grasp the implications of the theorem and were thus convinced that light radiating through a sphere intersects at a single point rather than suffering spherical aberration. Clearly, Kepler could not have learned about that phenomenon from his perspectivist sources, Alhacen included. Equally clear, he could not have gained a proper understanding of the focal point from those sources. Nor could he have learned the twofold method of theoretical and empirical ray tracing from them because none of them goes beyond a qualitative to a quantitative analysis of refraction. Not even Alhacen, for all his elaborate experimental apparatus, provided tabulations for refraction, and although Witelo did, he cribbed them from Ptolemy rather than producing his own.

These four things—the recognition and use of spherical aberration, a proper understanding of the focal point, theoretical ray tracing, and empirical ray tracing—are absolutely central to Kepler's analysis. And they are nowhere to be found in his perspectivist sources. Where, then, did Kepler get them?

Della Porta is the obvious candidate because all four elements, factual and methodological, figure in his analysis of radiation though glass spheres in the *De Refractione*. Not only does Kepler cite Della Porta several times in chapter 5 of the *Paralipomena*, but also he is lavish in praising him. "As far as I am concerned," Kepler gushes, "you [Della Porta] have [in me] a grateful admirer and publicist of your name" because, among other things, "thou hast blessed us, O excellent initiate of nature," by definitively resolving the issue of "whether vision occurs by reception or emission."[71] He also praises Della Porta for having rightly conceived of the eye as a sort of camera obscura, although he chides him gently for assuming that the anterior surface of the lens, rather than the retina, is the screen upon which the images are cast. But all of these references are to the *Magiae Naturalis* of 1589, not to the *De Refractione* of 1593, which Kepler claims to have sought diligently without success.[72] Therefore, if we take him at his word, Kepler could not have gotten the four key elements of his analysis directly from Della Porta, despite certain striking similarities between their respective approaches. That, of course, does not preclude his having learned about them indirectly, through spoken or written report.

This last point bears on the ways in which information and ideas were disseminated during the sixteenth century. Discussed at some length in the first section of this chapter, the vehicles for such dissemination included not only books but also manuscripts, which were sometimes widely circulated. For example, sometime between 1592 and 1601, Galileo, a Florentine, managed to acquire a manuscript copy of the unpublished *De speculi concavi sphaerici* composed by Ettore Ausonio, a Venetian, some forty years earlier.[73] But knowledge was also transmitted through personal correspondence as well as discussion, either individual or in semiformal "academic" gatherings. At a personal level, for instance, Kepler was probably steered to Platter's *De corporis humani structura* by the Slovak physician Johannes Jessenius, or Ján Jesenský (1566–1621), whom he befriended at Prague. Kepler also had close connections to the electoral court of Dresden and participated in at least one public "experimentum" dealing with the camera obscura in the Kunstkammer there, an experiment described briefly in the *Paralipomena*.[74]

71. *Paralipomena*, 5, 4, in Donahue, 224; in *Ad Vitellionem*, 209.

72. *Paralipomena*, 5, 4, in Donahue, 225; in *Ad Vitellionem*, 210.

73. See Dupré, "Ausonio's Mirrors." Galileo had access to Ausonio's manuscript through the library of Gian Vincenzo Pinelli (1535–1601), an avid collector of printed books and manuscripts who befriended Galileo while he was at the University of Padua.

74. See Sven Dupré, "Inside the *Camera Obscura*: Kepler's Experiment and Theory of Optical Imagery," *Early Science and Medicine* 13 (2009): 219–44, especially 220–22. For Kep-

Kepler, however, was not limited to the two nodes at Prague and Dresden. Like every other active scholar of his day, he formed part of a broad, international network through which knowledge filtered in complex ways, node by node, both chronologically and geographically, private libraries, such as Pinelli's in Padua, serving as important clearinghouses. Within such a web, Kepler may well have been influenced, however distantly and indirectly, by a host of previous and contemporary thinkers, including Maurolico, who himself was part of a southern Italian network of scholars sharing the same interests. Whatever the actual lines of influence, though, it is clear that Kepler's understanding of, and approach to, spherical lenses falls squarely within the sixteenth-century tradition of analysis, not within the perspectivist tradition exemplified by Witelo or by Alhacen in his Latin incarnation. Kepler, in short, was following a markedly different analytic path from his perspectivist sources—a path already blazed to some extent by the likes of Maurolico, Ausonio, and Della Porta.[75]

6. THE EPISTEMOLOGICAL TURN

Kepler was well aware that his account of retinal imaging poses certain perceptual and epistemological problems. One such problem is image inversion. Indeed, "for my part," Kepler admitted, "[I] tied myself in knots for the longest time, trying to show that . . . another inversion occurs [so that] the parts . . . to the left again became right before they reach the retina." Like Leonardo almost a century earlier, Kepler felt bound to make the "picture" reach the seat of sensation upright so that it would be perceived as such. Ultimately, he persuaded himself that the original inversion could stand because the retinal reception of the picture "is an effect contrary to [the] action" of the object's illumination. Hence, because "recipients of action must be directly opposite the acting things," the receptive points on the retina must be directly opposite

---

ler's description of the "experiment," see *Paralipomena*, 5, proposition 5, in Donahue, 194; in *Ad Vitellionem*, 181.

75. Cf. David Lindberg, "Optics in Sixteenth-Century Italy," *Annali dell'Istituto e museo di storia della scienza* 2 (1983): 131–48, for a rather negative assessment of Maurolico's and Della Porta's significance in the history of optics. See, however, Robert Goulding, "Thomas Harriot's Optics, between Experiment and Imagination: The Case of Mr. Bulkeley's Glass," *Archive for History of Exact Sciences* 67 (2013): DOI: 10.1007/s00407-013-0125-1. As Goulding shows, in analyzing the focal property of Mr. Bulkeley's lens, Harriot did much the same sort of theoretical ray tracing as Della Porta and, more to the point, Kepler, but he did so on the basis of the sine law of refraction.

the visible points acting on them. Such direct opposition is determined by an imaginary double cone through whose common vertex all the point-to-point lines of action pass between object and retina, so the image inversion is both necessary and natural. Kepler could, of course, have appealed to psychological intervention, as did the perspectivists, in accounting for such things as peripheral vision. Instead of taking that route, though, he preferred to fall back on a physico-mathematical argument.[76]

Perhaps less obvious but considerably more profound than the problem of image inversion is that of how the faculties lodged in the cerebral ventricles make perceptual and cognitive sense of retinal images. The perspectivists addressed this issue by assuming that the entire visual process is based on the passage of intentional species through the air to the lens, and thence into the cerebral ventricles via the hollow, spirit-infused optic nerves. But Kepler's retinal image is real, not intentional, so it is not suited to the cerebral spirits within which sensation, perception, and cognition supposedly occur. Kepler was thus at a loss to explain precisely "how this image (*idolum*) or picture (*pictura*) is joined together with the visual spirits that reside in the retina and in the nerve" or how, after that juncture, it enters into "the caverns of the brain" to be judged. Perhaps, he ventured, something is sent out by the soul through the optic nerves, like an emissary, to "meet this image" and presumably to report back to the judging power in the brain. But rather than take a definitive stand, Kepler was content to "leave [it] to the natural philosophers (*physici*) to argue about [these issues], for the arsenal of opticians (*opticorum*) does not extend beyond [the] opaque wall [of the retina]."[77]

With this demurral, Kepler made it eminently clear that as far as he was concerned, the science of optics deals only with the physical passage of light through the transparent components of the eye to the retinal screen, where it forms an inverted, spot-to-spot picture of everything within the field of view. What happens in the nerves and brain after that is wholly beyond its scope. In other words, optics deals with the physical cause of vision only, not with its physiological or psychological effects. This is so because the laws of optics strictly govern the physics of vision, whereas the physiology and psychology of perception are subject to entirely different rules. Accordingly, in response to the assumption (which he attributed to Witelo) that the act of sight is completed when visual images reach the optic chiasma from both eyes, Kepler demanded to know

76. See *Paralipomena*, 5, 4, in Donahue, 221–22; in *Ad Vitellionem*, 205–7.
77. *Paralipomena*, 5, 2, in Donahue, 180; in *Ad Vitellionem*, 168.

what can be pronounced by optical laws about this hidden confluence [of images at the optic chiasma], which, since it goes through opaque, and therefore dark, parts, and is administered by spirits, which differ entirely in kind from humors and other transparent objects, has already completely removed itself from optical laws.[78]

Kepler's point is obvious: however it might be accomplished, image formation in the nerves and brain is not determined by optical principles because the media through which it occurs cannot transmit light. And this includes the spirits that are assumed to support image transmission to and through the brain. For Kepler, therefore, the retina is not even a veil through which the perceptual and cognitive faculties somehow apprehend external objects. It is an impermeable wall that separates the eye from the brain and, thus, the objective cause from the subjective effects of vision.

As we saw in chapter 6, the perspectivist account was designed to ensure that the objective cause of vision, in the form of luminous color, corresponds to its subjective effects according to a rigidly interconnected succession of intentional species. The interconnection is determined by similitude, each species in the succession forming an increasingly abstract likeness of the predecessor that gives rise to it. From beginning to end, therefore, the visual process is supposed to yield a veridical, albeit virtual, depiction of outlying objects and, in the process, an understanding of what they really are as individuals and types. In Kepler's account, on the other hand, the linkage between objective and subjective manifestations of light and color, so painstakingly wrought by the perspectivists, is ruptured by the retinal wall. So how does the retinal image enter the brain to be perceptually and cognitively judged when it is both too large to fit into the optic nerve and not optically transmissible by the spirits pervading that nerve and the brain? To suggest, as did Kepler in passing, that the faculty of judgment in the brain might send an emissary to look at that image and report back begs the question. For one thing, whatever the brain dispatches constitutes a second viewer whose mode of seeing has to be explained. For another, the retinal image, not the object itself, is what is apprehended by that second viewer, so the faculty of judgment has to make an interpretive leap from the emissary's reported representation to the object it represents at two removes. The same sort of interpretive leap would have to be made if the retinal image were to produce an immediate, intentional image of itself in the spirits pervading the retina and optic nerve because it, not the

78. *Paralipomena*, 5, 2, in Donahue, 180; in *Ad Vitellionem*, 168.

object itself, would be represented by that intentional image.[79] Much the same objections had been raised by fourteenth-century thinkers in regard to intentional species as veils between perceiver and object, but in the case of retinal images the veiling problem is even more obvious and more acute.

By disjoining the physics of light from the psychology of sight, both domains subject to wholly different laws, Kepler brought the problem of correspondence into stark relief. How can we be sure that our internal impressions of external objects match them in a meaningful way? Kepler's model of retinal imaging destroys all hope of establishing such certainty because physical cause and perceptual effect are nothing like each other. The best we can do is to posit a covariance between the two, such that, for instance, the internal impression of "red" always follows from a specific but dissimilar physical stimulus. Nor can the gap between physical cause and perceptual effect be bridged by intentional species because they are psychic entities and thus wholly unlike the physical pictures from which they supposedly derive. In effect, then, Kepler not only drastically widened the "epistemic gap" lurking in the wings of medieval cognition theory, but he also transposed it from the end to the very beginning of the perceptual process. In view of all these points, it should be evident that instead of perfecting perspectivist visual theory, Kepler struck at its epistemological heart by repudiating not only intentional species but also the assumption of similitude that gives them explanatory force.

7. CONCLUSION

Kepler may not have fully grasped the radical implications of his theory of retinal imaging, but he certainly understood that it jeopardized the perspectivist paradigm by reversing the order of analytic priority. For the perspectivists, that priority was given to the act of visual perception in all its ramifications, so the physical analysis of light had to be accommodated to the greater goal of explaining that act. In other words, the perspectivists subordinated light theory to sight theory. It is for this reason that Alhacen, and the perspectivists following him, did everything in their power to preserve the visual cone, in

79. Oddly enough, Kepler proposed this expedient in his *Dioptrice* of 1611, where he refers in proposition sixty-one to the species "received by the retina [that] passes through the continuity of the spirits to the brain [where it] is . . . delivered to the threshold of the faculty of the soul" (Alistair Crombie, trans., "The Mechanistic Hypothesis and the Study of Vision," in Crombie, *Science, Optics, and Music in Medieval and Early Modern Thought* [London: Hambledon Press, 1990]); for the Latin, see *Dioptrice* (Augsburg, 1611), 23. Kepler himself therefore seems not to have been fully alive to the perceptual consequences of his theory of retinal imaging.

the form of a cone of radiation with the center of sight at the vertex. This was meant to ensure a smooth passage of similitudes from the object to the lens and then through the optic complex into the brain while providing a determinate location from which those similitudes could be visually judged. For Kepler, on the other hand, physics took priority over psychology, so any explanation of the visual act had to be accommodated to his physical analysis of light, which nullified the cone of radiation and the center of sight associated with it. In other words, Kepler subordinated sight theory to light theory. The upshot, as we have seen, was a poor fit between the physics-mandated retinal image and its perceptual translation in the optic nerves and brain. Kepler attempted to improve this fit by reinverting the image before it reached the retina, but he eventually realized the futility of this effort. All he could do, finally, was throw his hands up and fob the problem of perceptual translation off onto natural philosophers. In doing so, he created an implicit but clear distinction between physical and psychological optics and went even further to limit the scope of optics itself to physical optics only.

Kepler thus recognized that his theory of retinal imaging was not just an incremental adjustment to the perspectivist paradigm but a sharp departure from it. He also recognized that to subscribe to this theory meant that the paradigm had to be overhauled in fundamental ways. In the long run, as we will see in the next chapter, it was overhauled into oblivion during the seventeenth century, partly in reaction to, and partly in the service of, a new physics of light and color. Consequently, Kepler's theory of retinal imaging represents a pivot-point in the history of optics, the point at which that science took a definitive turn away from classical Greek visual theory, as perfected by Alhacen and the perspectivists, toward modern light theory, as it emerged in the seventeenth century.

This turn was not entirely unheralded. Well before Kepler came onto the scene, theologians, *littérateurs*, artists, and scholastic thinkers of all stripes had poked and prodded the perspectivist paradigm. Found wanting at various points, it was tweaked from time to time, as it was by Maurolico, who bent the rules, by Leonardo, who broke them, and by Della Porta, who quantified them. Meantime, as the reliability of vision was called increasingly into question, faith in the perspectivist paradigm as a definitive account of sight was shaken. In addition, the technological and cultural changes that occurred over the fifteenth and sixteenth centuries led to the recognition of optical anomalies and the production of improved optical devices that stimulated new analytic approaches. The perspectivist paradigm that came down to Kepler was thus sapped of much of its explanatory power, although it was still almost universally accepted, if only faute de mieux. Its hold over Kepler is evident in both his initial effort to reinvert the retinal image and his inability to think beyond the internal senses model of perception and cognition at its heart. The weakness of that hold, on the other hand, is evident in his adamant support of a physical account of vision that conflicts with the internal senses model at the deepest level. Thus because he was only loosely committed to the perspectivist paradigm, Kepler had no qualms about reworking it as he saw fit, despite the epistemological inconveniences that follow from that reworking.

In the next, and concluding, chapter of this book we will examine some of the interrelated ways in which the Keplerian turn and the resulting divorce of light theory from sight theory played out during the seventeenth century. One of those ways, we will see, led to the rapid improvement and deployment of telescopes and microscopes and an attendant dispute over the epistemic validity of such instruments. Another led to the development of new approaches to the physics of light and color. And yet another led to a radical rethinking of thinking according to both the objects of perception and thought and their ultimate physical and metaphysical grounds. Key among the theorists who will figure in our examination are Galileo Galilei (1564–1642), René Descartes (1596–1650), Christiaan Huygens (1620–1695), Robert Hooke (1635–1703), and Isaac Newton (1643–1727).

CHAPTER 9

The Seventeenth-Century Response

In the preceding chapter I presented the Keplerian turn in such stark terms that it may look as though I am claiming two things that I am not. The first thing I am not claiming is that Kepler's analysis of retinal imaging was directly and causally responsible for the scientific developments I will address in this chapter. That analysis was simply one of several factors that conduced to or were absorbed into those developments and, as such, served more as reinforcement than as underlying cause. Kepler's analysis, to put it in slightly different terms, was published at just the right time to enter effectively into a broader discourse about the constitution of the physical world and how we, as perceivers and knowers, have access to it. Within this broader discourse, the effort to define that world and its constituent objects according to absolute, perceiver-independent characteristics loomed large. This effort gave rise to the sharp distinction between "primary" and "secondary" qualities Galileo famously described in his *Assayer* of 1623:

> Upon conceiving of a material or corporeal substance, I immediately feel the need to conceive simultaneously that it is bounded and has this or that shape; that it is in this place or that at any given time; that it moves or stays still; that it does or does not touch another body; and that it is one, few, or many. . . . But that it must be white or red, bitter or sweet, noisy or silent, of sweet or foul odor, my mind feels no compulsion to understand as necessary accompaniments. Indeed, without the senses to guide us, reason or imagination alone would perhaps never arrive at such qualities. For that reason I think that tastes, odors, colors, and so forth are no more than mere names so far as pertains to

the subject within which they reside, and that they have their habitation only in the sensorium (*nel corpo sensitivo*).[1]

To be sure, the distinction Galileo draws here between real, physical characteristics and their intentional, sensible corollaries is perfectly consistent with the decoupling of physical cause from perceptual effect in Kepler's account of retinal imaging. But just because both thinkers followed a similar track does not mean that Kepler actually laid the rails for Galileo. Far more likely, as we will see in a moment, is that the two arrived at comparable positions because, like Avicenna and Alhacen before them, they were operating within effectively the same conceptual framework.

The second point I may seem to be making is that acceptance of Kepler's theory of retinal imaging automatically entailed rejection of the perspectivist paradigm at every level. That this was not the case is evident from the analysis of sight the Jesuit savant Christoph Scheiner (1573–1650) offered. Best known for his debate with Galileo over sunspots, Scheiner was unusually erudite, discerning, and progressive in his scientific outlook.[2] All three characteristics are evident in his detailed study of the eye and vision in *Oculus*, which he published at Innsbruck in 1619. The erudition and discernment come through in his description of ocular anatomy in the first book of the treatise. Far more accurate than Platter's account, Scheiner's acknowledges that the lens lies at the very front of the eye, that the cornea bulges outward, and that the point where the optic nerve enters at the back is significantly offset from the visual axis passing through the centers of the lens and eyeball. Scheiner also adopted Kepler's model of retinal imaging and, with it, the consequence of image inversion. In the *Rosa Ursina* ("Orsini's Rose") of 1630, in fact, Scheiner claims to have seen or demonstrated both the formation and inversion of such images at the back of the eye, once in a human eye freshly removed from a cadaver and several times in animal eyes taken from carcasses. Yet for all his willingness to adopt Kepler's analysis of retinal imaging, Scheiner still looked to visible species (*species visibiles*) as the mediating entities in sight. In addition, despite insisting that vision is due to intromission only, he consistently referred to visual

1. Stillman Drake and C. D. O'Malley, trans., *The Controversy on the Comets of 1618* (Philadelphia: University of Pennsylvania Press, 1960), 309. It is perhaps worth noting that Galileo's distinction harks back to the atomic theory of Democritus.

2. For a thorough analysis of the sunspot debate between Scheiner and Galileo, along with the relevant texts in English translation, see Eileen Reeves and Albert Van Helden, *Galileo Galilei and Christoph Scheiner on Sunspots* (Chicago: University of Chicago Press, 2010).

rays (*radii visorii*) rather than light rays in order, presumably, to emphasize their perceptual function rather than their physical operation.[3]

Scheiner was not alone in refusing to part ways entirely with the perspectivist paradigm, despite fairly strong Keplerian leanings. Some who adopted Kepler's model during the early decades of the seventeenth century (including Kepler himself in 1611) glossed over the fundamental incompatibility between that model and its perspectivist alternative. Others temporized. In response to the problem of image inversion, for example, Pierre Gassendi (1592–1655) and Nicolas-Claude Fabri de Peiresc (1580–1637) floated the idea that the retina is not just opaque but also reflective. Forming a concave mirror, so their reasoning went, it would reflect the lens-projected image back to the lens in reinverted order to be seen in its proper, upright orientation.[4] This idea eventually sank of its own weight, but the key point is that from quite early on several thinkers felt the need at least to cope with Kepler's theory, if not to accept it unconditionally.

Like any new theory, therefore, this one required time to sweep the field. Perspectivist sources, especially Alhacen and Witelo, as published by Risner, continued to be read and cited during much of the seventeenth century. As late as 1667, in fact, Newton's mentor, Isaac Barrow (1630–1677), grudgingly included in Lecture IX of his *Lectiones Opticae XVIII* a version of Alhacen's method for determining the reflection point in convex spherical mirrors, "stripped of that horrible combination of prolixity and obscurity, and of his uncouth barbarity of speech." Two years later, in 1669, Christiaan Huygens offered an exquisitely concise resolution of "Alhazen's Problem" to determine all possible reflection points in any spherical mirror, whether convex or concave, when the luminous point, the center of sight, and the cross section of the mirror lie randomly within a given plane of reflection.[5] By that time, however,

3. For Scheiner's description of ocular anatomy according to dissection, see *Oculus*, 1–29; see especially the diagram on p. 17. For his use of both visible species and visual rays, see the title page of *Oculus*. For his claims about having observed retinal imaging in dissected eyes, see *Rosa Ursina*, II, 23, p. 110.

4. See, for example, Saul Fisher, *Pierre Gassendi's Philosophy and Science: Atomism for Empiricists* (Leiden: Brill, 2005), 33–37. For a more detailed and nuanced account, see Robert A. Hatch, "Coherence, Correspondence, and Choice: Gassendi and Boulliau on Light and Vision," in *Actes du Colloque International Pierre Gassendi* (Digne-les-Bains: Société Scientifique et Littéraire des Alpes de Haute-Provence, 1994), 365–85.

5. For a discussion of how seventeenth-century authors used and cited Alhacen and Witelo, see Lindberg, introduction to reprint edition of *Opticae Thesaurus*, xxiv–xxv. For Barrow's version of Alhacen's method for determining reflection points in convex spherical mirrors, as

Alhacen's *De aspectibus* and the optical theory flowing from it had been reduced to little more than historical artifacts. As of the late seventeenth century, in short, the Keplerian turn and its consequences had become so firmly rooted as to all but obliterate the perspectivist past.

How did this happen, and in such a relatively short time? That is the formative question for this chapter, and I will address it in five sections. In the first section, I will provide a brief sketch of the conceptual and cultural context within which the Keplerian turn manifested itself. Then I will show in a highly selective way how that turn played out within three lines of development. The first line, which I will touch on in section 2, involves the use and improvement of telescopes and microscopes during the seventeenth century, some of that improvement being due to, and some of it conducing to, new theoretical insights. In sections 3 and 4, I will show how those theoretical insights were reflected in new conceptions of light and color, as well as of the laws governing their physical manifestations. Finally, in the fifth section, I will show how these two lines of instrumental and theoretical development not only fed into but also were nourished by efforts to explain or explain away the "epistemic gap" implicit in Kepler's visual model.[6] As we will see in the course of this chapter, all three lines of development were intertwined in such a way that many of the thinkers implicated in one line were also implicated in the others. René Descartes is a cardinal example.

1. THE CONCEPTUAL AND CULTURAL CONTEXT
FOR THE KEPLERIAN TURN

That the scientific enterprise—or, more properly, the pursuit of natural philosophy—underwent significant, even explosive change over the seventeenth century has never been in question. Far more controversial than the fact of scientific change during this period is how to characterize and explain it. Was it revolutionary? Was it somehow unitary in its wellsprings? Did it grow

---

provided in Lecture IX of his *Lectiones Opticae* (which were given in 1667 and published in 1669), see H. C. Fay, trans., *Isaac Barrow's* Optical Lectures (London: Worshipful Company of Spectacle Makers, 1987), 117–19. On Huygens's general solution of "Alhazen's Problem," see Smith, "Alhacen's Approach," 162–63.

6. Another line of development that merits at least mention is that of seventeenth-century northern European art, Netherlandish art in particular. For a classic study of this line of development and its bearing on the Keplerian turn, see Svetlana Alpers, *The Art of Describing* (Chicago: University of Chicago Press, 1983), especially chapter 2. See also Ofer Gal and Raz Chen-Morris, *Baroque Science* (Chicago: University of Chicago Press, 2013).

out of methodological concerns and a concomitant emphasis on experimentalism and hypothetico-deductive reasoning? Was it essentially reactionary, a conscious repudiation of an Aristotelian system increasingly perceived as bankrupt? Is it best explained on social grounds, in terms of culturally determined knowledge production? Was it therefore rooted in social change? Did it emerge on the basis of new canons of certitude that may or may not have been socially determined? Were those canons developed in response to skeptical currents that emerged in the later sixteenth and early seventeenth centuries? That the answer to each of these questions is a qualified "yes" and an equally qualified "no" bespeaks the complexity of the scientific change that occurred during that century. It also bespeaks the difficulty of accounting for it in terms of the "Scientific Revolution," an umbrella concept that has fallen out of favor over the past few decades.

I have no intention of entering into the debate over the definition and cause, or causes, of scientific change during the seventeenth century, much less the dispute over whether it was revolutionary. Instead, I will limit my discussion to three complementary trends in seventeenth-century scientific thought that provide a context within which the optical developments to be analyzed later can be more fully appreciated.[7] For the most part, what is new about these trends is a matter of emphasis rather than of conceptual novelty. And not every scientific thinker of the seventeenth century followed them because to do so meant rejecting or drastically revamping a complex system of "Aristotelian" natural philosophy that still made sense in many respects. Nonetheless, resistance to these trends dwindled over time as more and more thinkers were persuaded that they offered better, more fruitful ways to make sense of nature and her workings.

The first of the three trends at issue evolved from the idea that the physical universe is atomic, or at least particulate, in structure. Hardly new to the seventeenth century, this idea had been around since Greek antiquity and was entertained from time to time over the Middle Ages. Nevertheless, the way it was refined and exploited during the seventeenth century recast it in fundamental ways that converged in the "Mechanical Philosophy" espoused variously and in varying degrees by the likes of Pierre Gassendi, Thomas Hobbes (1588–

7. The literature on seventeenth-century science and the "Scientific Revolution" is so voluminous and the interpretive slants so variegated that I cannot begin to cite an adequately representative sample. For a mere taste, see the textbook accounts and accompanying bibliographies in Shapin, *The Scientific Revolution*; Peter Dear, *Revolutionizing the Sciences* (Princeton, NJ: Princeton University Press, 2001); and John Henry, *The Scientific Revolution and the Origins of Modern Science* (New York: Palgrave Macmillan, 2008).

1679), Robert Boyle (1627–1691), and of course René Descartes. For a start, the constituent elements of such a materialist universe can be strictly defined according to the limited set of primary qualities Galileo described in the passage quoted earlier: being "bounded and [having] this or that shape; [being] in this place or that at any given time"; moving or at rest, and so on. Since these characteristics pertain only to bodies and their spatial relations, it follows that all physical change must be due to spatially determined interactions among bodies, whether through collision or agglomeration at the macroscopic or microscopic level. All physical change, in short, distills to matter in contact or in motion. To assume this is to reduce the universe to a vast machine whose internal workings are governed by mechanical principles alone. It is also to deny or reconceptualize the occult forces and influences upon which such disciplines as astrology and alchemy depend.

The second trend, which is closely related to the first, is encapsulated in another well-worn quotation from Galileo's *Assayer*: "[Natural] philosophy is written in this grand book, the universe . . . in the language of mathematics, and its characters are triangles, circles, and other geometric figures without which it is humanly impossible to understand a single word of it."[8] At first blush, one might interpret Galileo here as affirming what had already been assumed for nearly two millennia: that mathematics is a crucial analytic tool for the study of natural philosophy or science. With its heavy reliance on ray geometry, the science of optics exemplifies this point. But in this quotation Galileo is not just touting mathematics as a means, albeit a critically important one, of making sense of nature. He is singling it out as the *sole* means to that end because the physical universe is mathematical to its structural core. Nature is a "book" written in mathematical sentences that can only be parsed in geometrical and arithmetical terms. Any other reading is false or misleading. By implication, moreover, the perfect certitude of "pure" mathematics should carry over to its application in natural philosophy. This assumption, which both Galileo and Kepler shared to a great extent, dovetails with the conception of a mechanical universe whose constituent elements, as well as the interactions among them, are defined by spatial qualities such as size, shape, distance, speed, and so forth that lend themselves naturally to mathematical description and analysis.

The third trend took form in the effort to establish firm, clear criteria of scientific certainty. As we saw in chapter 7, the reliability of visual perception specifically, and sense perception in general, had been called into question on a variety of fronts well before the seventeenth century. An obvious corollary

8. Drake and O'Malley, *Controversy of Comets*, 183–84.

to such distrust of sense perception is doubt about the truth or certainty of its cognitive results. This doubt comes to the fore in the dissociation of objective cause from subjective effect in Kepler's account of vision, and the resulting lack of correspondence between the two is enshrined in the distinction between primary and secondary qualities, the former being absolute, real, and spatially determined, the latter relative, virtual, and psychically constructed. Utterly different in kind, the two cannot be linked through similitude. That leaves two options. Either empirical observation must be abandoned as a legitimate ground of scientific inquiry, or a strict and meaningful correlation must be established between physical cause and perceptual effect according to something other than representational likeness.

For obvious reasons the option of abandoning empiricism was rejected out of hand, although some, like Descartes, kept it to a controlled minimum. On the other hand, and perhaps somewhat paradoxically, a host of seventeenth-century scientific thinkers embraced empiricism warmly and, with it, the "Experimental Philosophy" godfathered by Francis Bacon (1561–1626) and avidly promoted by Robert Boyle, who was also an advocate of the Mechanical Philosophy. "The truth is," says Robert Hooke, another champion of the two philosophies, "the science of Nature has been already too long made only a work of the Brain and the Fancy: It is now high time that it should return to the plainness and soundness of Observations on material and obvious things."[9] Again, it was emphasis more than sheer novelty that set this movement apart from the experimentalism of, say, Ptolemy or Alhacen. One such emphasis was its increasing reliance on sophisticated instrumentation that provided not just controls, but artificial, or "unnatural," controls for observation. The air pump, with its ability to create near-vacuum conditions for experimentation, has become a stock example, but telescopes and microscopes also provided such artificial controls by extending vision beyond its natural limits.

Furthermore, the empiricism upon which seventeenth-century experimentalism was based had a decidedly positivist slant insofar as its practitioners claimed to be (and perhaps actually thought they were) observing the "facts" objectively without allowing preconceived notions to color their perceptions of them. Naive in the extreme, if not simply disingenuous, this claim carried with it the obligation to explain precisely how such factual observations can be trusted. Suffice it for now to say that the resulting efforts to resolve this problem were based on assumptions about appropriate experimental controls, the probability of being right, the credibility of the reporting observer(s), or

9. *Micrographia* (London, 1665), "Preface," v (unnumbered).

the sense in which the observational facts should be construed. Taken to its extreme, the idea that observations must be construed implies not only that observational facts do not speak for themselves but also that what they appear to say is entirely different from what they mean. Yet another quotation from Galileo, an experimentalist of the first water, captures this point perfectly. "There is no limit to my astonishment," he marvels in the Third Day of his *Dialogue* of 1632, "when I reflect that . . . Copernicus [was] able to make reason so conquer sense that, in defiance of the latter, the former became mistress of [his] belief."[10] The irony is inescapable. Even though firmly based on observational data, Copernicus's heliocentric hypothesis flies in the face of what our senses tell us in no uncertain terms: that the earth is perfectly stable and motionless.

No less important, finally, than these scientific trends themselves was the social matrix, or marketplace of ideas, within which they evolved. During the seventeenth century that marketplace expanded dramatically as the network of nodes and links established during the late Renaissance was enlarged and crystallized. Informal gatherings gave rise in some cases to formal societies, starting with Prince Federico Cesi's Roman *Accademia dei Lincei* in 1603 (to which Della Porta was granted membership in 1610 and Galileo a year later) and culminating with the founding of the English Royal Society in 1660/62 and the French Académie Royale des Sciences in 1666. All three societies, and others as well, were established with a strong experimental or pragmatic bias and an attendant stress on both luciferousness and fructiferousness, to use Francis Bacon's coinage. Networks of correspondence also expanded considerably, in some cases coming to focus in a single archcorrespondent, such as Peiresc, Marin Mersenne (1588–1648), Ismaël Boulliau (1605–1694), Samuel Hartlib (1600–1662), and Henry Oldenburg (1619–1677).[11] In tandem with the ex-

10. Stillman Drake, trans. *Galileo*: Dialogue concerning the Two Chief World Systems (Berkeley: University of California Press, 1967), 328.

11. For a comparative look at the correspondence networks of Peiresc, Mersenne, Boulliau, and Oldenburg, see Robert A. Hatch, "Between Erudition and Science: The Archive and Correspondence Network of Ismaël Boulliau," in *Archives of the Scientific Revolution: The Formation and Exchange of Ideas in Seventeenth-Century Europe*, ed. Michael Hunter (Woodbridge, UK: Boydell Press, 1998), 49–71. Hatch has argued elsewhere that communities formed through correspondence signaled a "shift from relatively polite 'pamphlet wars' to larger groups involving more bare-knuckled exchanges" and that while "new freedoms encouraged excess, the speed and freedom of letters also invited a larger community to anticipate events and work collectively [while promoting] competition and new ways to establish priority without public

panding networks of correspondence, the first learned journals were established for the purpose of disseminating scientific knowledge to large numbers of subscribers. Both forms of learned communication, public and private, responded to a growing need for scientific awareness as well as fast, cheap, and timely exchanges of ideas. As those exchanges became easier and quicker, scientific change accelerated commensurately.

By the end of the seventeenth century, therefore, the marketplace of scientific ideas had taken on a decidedly modern, international cast in certain key respects. In others, however, it was still rooted in the Renaissance, particularly during the first half of the century. Consensus on proper scientific norms had yet to be reached. Naive empiricism was still the order of the day in many cases. Occultism was by no means dead, although it was on the wane or at least taking new forms. And private patronage was still critical to the pursuit of science. Johannes Kepler, Galileo Galilei, René Descartes, and Christiaan Huygens were all supported either by outside funding from ducal, royal, or imperial patrons or, in Descartes's case, by drawing on his own financial resources.[12] Nonetheless, we can discern a clear tendency by the end of the century toward scientific professionalization and a resulting emphasis on science as a cooperative, even communal and self-regulating affair. In terms of methodological commitments, conceptual foundations, shared ideals, disciplinary boundaries, and social structure, late seventeenth-century science was still a long way from "modern," but it was certainly evolving in that direction.

2. EXTENDING VISION IN BOTH DIRECTIONS

Precisely when the first practical telescopic device was constructed, or even conceived, has yet to be satisfactorily determined, and it seems unlikely that it ever will be. Priority is conventionally given to Hans Lipperhey, a spectacle maker from Middelburg in Zeeland, on the basis of a document dating from the beginning of October 1608 that indicates he was the first to apply for a

---

commitment"; see Hatch, "The Republic of Letters: Boulliau, Leopoldo, and the Accademia del Cimento," in *The Accademia del Cimento and Its European Context*, ed. Marco Beretta, Antonio Clericuzio, and Lawrence M. Principe (Sagamore Beach, MA: Science History Publications, 2009), 165–80, especially 177–78.

12. Even Descartes ended up accepting the patronage of Queen Christina of Sweden in 1649, although he enjoyed it only briefly, dying in 1650 only a few months after having arrived in Stockholm.

patent.[13] Not by much, though. Within a few weeks, two other Netherland-
ers sought patents for such an instrument from the same granting authority.
Admittedly slim and somewhat slippery, all evidence points to the conclusion
that the three devices, which amounted to mere spyglasses, were "Galilean"
insofar as they consisted of a convex objective lens and a concave eyepiece
inserted at opposite ends of a sighting tube. If, as we may reasonably suppose,
Lipperhey chose the lenses for his instrument from the selection of eyeglass
lenses at hand in his shop, the resulting device, perhaps a foot or so in length,
would have been fairly unprepossessing. First, because of the convex objective's short focal length and the comparatively long focal length of the concave
eyepiece—20 inches at best for the former and 8 (or, more properly, -8) for
the latter—its magnifying power would have been no more than 2.5x.[14] Nor
would lack of magnification have been its only drawback. The lenses themselves would have been problematic because, being hand ground only to rough
specification, they would have been fairly astigmatic, or aspherical, in curvature. Moreover, the glass itself would have been of relatively poor quality in
terms of both clarity and internal consistency. Good enough for eyeglasses, in
short, the two lenses combined in the spyglass would most likely have yielded
poor resolution and bothersome image distortion.

Whether Lipperhey was the very first to envision this particular arrangement of lenses is an intriguing question whose answer is "perhaps not." As
we saw in the previous chapter, the hope of creating a telescopic device was
very much alive in the second half of the sixteenth century, and several people
attempted to fulfill that hope using concave mirrors or mirror-and-lens combinations. It is not implausible, therefore, to suppose that in the process someone stumbled upon Lipperhey's arrangement, although the lack of documen-
tary evidence to that effect makes it seem unlikely. This was, after all, an age

13. See Albert Van Helden, *The Invention of the Telescope* (Philadelphia: American Philosophical Society Press, 1977), for what is still the canonical account. For useful supplementary
studies, see Reeves, *Galileo's Glassworks*, especially 81–166; and Willach, *Long Route*, especially
85–99. For more detailed, focused, and up-to-date studies, see the articles in Albert Van Helden
and others, eds., *The Origins of the Telescope* (Amsterdam: KNAW Press, 2010)—especially
Sven Dupré, "William Bournes' Invention: Projecting a Telescope and Optical Speculation in
Elizabethan England," 129–45; Albert Van Helden, "Galileo's Telescope," 183–201; and Giuseppe Molesini, "Testing Telescope Optics of Seventeenth-Century Italy," 271–80.

14. Simply put, the magnifying power of a telescope = focal length of objective divided by
focal length of eyepiece (in this case 20 ÷ 8 = 2.5). A magnification of 2.5 means that the image
will appear two-and-a-half times closer than the object when seen with the naked eye and will
thus subtend an appropriately larger visual angle.

of exaggerated, published claims of priority and ingenuity. What we can be certain of, though, is that quite soon after Lipperhey applied for his patent, which he failed to get, news of the device began to spread. At the same time, spyglasses of Lipperhey's sort began to pop up for sale, one perhaps as early as October 1608 at the Frankfurt Fair, and by 1609 such instruments were on offer at several places in Europe. In late May of that year, in fact, the English polymath Thomas Harriot (1560–1621) pointed such a low-power device at the moon in order to map its details. This, as far as we know, is the first "scientific" use of the instrument.

Lipperhey and Harriot notwithstanding, it was Galileo who actually "invented" the telescope, as opposed to the spyglass, by giving it specific form and function. True, as he himself acknowledges in his *Sidereus Nuncius* ("Starry Messenger/Message") of 1610, he got the idea for his instrument from a report in May or June of 1609 that credited it to "a certain Dutchman," presumably Lipperhey. But if the idea was not original to Galileo, what he did with it transformed both the idea and the device in such a way as to make them uniquely his. For it was with Galileo that Lipperhey's spyglass became not only a "telescope," in the true sense, but also a "Galilean" telescope at that. For a start, Galileo vastly improved both the instrument and its design, and he did so in remarkably short order. The telescope he trained on the moon in December 1609, had a magnification of around 20x, and soon after that he increased it to around 30x.

In order to achieve such results Galileo had to make significant modifications to the original design. First and foremost, he had to increase the focal length of the objective and decrease that of the eyepiece beyond the limits found in contemporary eyeglass lenses. That is, he had to make the objective "weaker" and the eyepiece "stronger." He therefore undertook the painstaking and laborious task of grinding his own lenses, the convex ones from relatively large blanks. As the magnification increased, however, he found that placing a diaphragm, or stop, on both the objective and eyepiece sharpened the image, in part by reducing peripheral radiation and in part by isolating the "sweet spot" of the lenses. But adding the diaphragms shrank the field of view and, in the process, diminished the image's brightness. In response to these problems, Galileo mounted his telescope on a stand so as to keep it as steady as possible for close, continuous observation.

Because of its limited field of view the Galilean telescope is extremely difficult to use, even under the best of circumstances. Yet Galileo quickly mastered it and by winter of 1609, while still on the faculty at the University of Padua, managed to discern the surface structure of the moon in such detail that he was able to calculate the height of a lunar mountain (four "Italian miles").[15] By January of the next year he plotted the four largest moons of Jupiter and soon after that began mapping certain constellations using a micrometer of some sort. In a rush to reap the benefits of these observations, Galileo published them in his *Sidereus Nuncius*, which appeared in March 1610, with a dedication to Cosimo II de' Medici, Grand Duke of Tuscany. In a not-so-subtle bid to curry additional favor with Cosimo, Galileo named the four newly discovered Jovian satellites "Medicean Stars" (*Medicea Sidera*). The gambit paid off. Cosimo offered Galileo the post of court philosopher and mathematician at a handsome salary. After settling in Florence in September 1610, to take up that post, Galileo availed himself of his newfound freedom from teaching to focus on his program of celestial observation while improving the optical quality of his telescopes. Within a few months he had observed the phases of Venus, and even earlier he had noted the "ears" of Saturn formed by the bright edges of its ring, which he took to be moons.

Something of a blockbuster because of its startling revelations about the moon, the satellites of Jupiter, and new stars invisible to the naked eye, *Sidereus Nuncius* thrust both Galileo and his telescope into the limelight. Within a few months of its publication, telescopes were being deployed throughout Europe to confirm or disconfirm Galileo's observations and to make new ones—and this despite some initial misgivings about the reliability of both the instrument and what Galileo claimed to have seen through it. It was with a telescope, for instance, that Christoph Scheiner made his observations of sunspots in 1611, and the publication of those observations eventually led to the debate with Galileo over whether the spots are planets in close orbit about the sun (Scheiner) or blotches on the solar surface carried around by the sun's rotation (Galileo). Hence, almost as soon as Galileo publicly demonstrated its utility, the telescope was on the way to becoming not only an astronomical instrument but also a scientific one insofar as it provided the observational basis for reconceptualizing the universe. If nothing else, Galileo's observations constituted definitive evidence against not only Aristotelian cosmology but also Ptolemaic astronomy and, as such, challenged the core assumption of absolute geocentricity. By default, Copernican heliocentricity and Tychonic geoheliocentricity

15. For this calculation, see Albert Van Helden, trans., *Sidereus Nuncius* (Chicago: University of Chicago Press, 1989), 51–52. For arguments that Galileo's ability to recognize the moon's rugged surface structure stemmed from his training as an artist, see Samuel Edgerton, *The Heritage of Giotto's Geometry* (Ithaca, NY: Cornell University Press, 1991); and Mary G. Winkler and Albert Van Helden, "Representing the Heavens," *Isis* 83 (1992): 195–217.

emerged as the only reasonable alternatives, and the telescope held promise as a means of deciding between them.

Whether Galileo designed and modified the telescope according to theoretical principles is a vexed question because, despite vague claims to having arrived at such principles by reasoning, Galileo never actually articulated them.[16] It is therefore possible that he proceeded by trial and error without any theoretical guidance. As Giora Hon and Yaakov Zik have recently argued, though, it is equally possible, and (they argue) more likely, that Galileo understood the optics of his instrument well enough to form lenses of just the right curvature (and thus focal length) to effectively increase the magnification of his telescopes to such high levels.[17] Whichever the case, it was Kepler who offered the first explicit theoretical analysis of lenses and, on that basis, of the telescope in his *Dioptrice* of 1611. One important result of his lens analysis was the description of a telescope whose eyepiece is convex rather than concave but still relatively strong. Telescopes with this configuration have since become known as "astronomical" or "Keplerian." Even earlier, in his *Dissertatio cum Nuncio Sidereo* ("Conversation with the Sidereal Messenger," published at Prague in early May of 1610), Kepler made another important contribution by suggesting that if the curved surface of a plano-convex lens is hyperbolic rather than spherical, parallel rays passing through its plane surface will be brought to perfect point focus, thus obviating the problem of spherical aberration and the blurring of images that results from it.[18] As we saw earlier, Ibn

16. In *Sidereus Nuncius*, Galileo claims to have constructed his telescope "on the basis of the science of refraction." Then, after a superficial and qualitative account of how the telescope magnifies, he promises that "on some other occasion we shall explain the entire theory of this instrument." This he never does. Later, in the *Assayer*, he claims to have arrived at the telescope through reasoning, although again the reasoning he gives is not based on any overt theoretical principles. See Van Helden, *Sidereus Nuncius*, 36–39, and Drake and O'Malley, *Controversy*, 212–13; see also A. Mark Smith, "Practice vs. Theory: The Background to Galileo's Telescopic Work," *Atti di Giorgio Ronchi* 54 (2001): 149–62.

17. Hon and Zik, "Magnification: How to Turn a Spyglass into an Astronomical Telescope," *Archive for History of Exact Sciences* 66 (2012): 439–64; see also Hon and Zik, "Galileo's Knowledge of Optics and the Functioning of the Telescope," at http://arxiv.org/abs/1307.4963. The theory behind magnification is simplicity itself: the longer the focal length of the objective in relation to that of the eyepiece, the greater the magnification.

18. For Kepler's account of the "Keplerian" telescope, see *Dioptrice*, 42–45; for his brief discussion in the *Dissertatio* of the focus point in hyperboloidal lenses, see Edward Rosen, trans., *Kepler's Conversation with Galileo's Sidereal Messenger* (New York: Johnson Reprint, 1965), 20.

Sahl demonstrated this very point toward the end of the tenth century, but without any reference to images or image resolution (see chapter 4, section 5).

Both suggestions were taken up in relatively short order. Perhaps as early as 1613, and certainly by 1617, Scheiner adopted a telescope of Keplerian design, and by the 1640s the Keplerian telescope began to supersede its Galilean counterpart. Despite one signal drawback, the Keplerian telescope has two significant advantages over its Galilean rival. The drawback is that it inverts the image. The advantages, as Scheiner points out in his *Rosa Ursina* (folio 130r), are that it gives a much wider field of view and therefore allows for commensurately greater magnification. Increasing magnification, in turn, carries with it the benefit of decreasing both spherical and chromatic aberration, the latter due to the prismatic effect of the lens. Thus Keplerian telescopes with objectives of ever-longer focal lengths were produced over the seventeenth century until, finally, the telescopes became too long and cumbersome to be practical, some of them reaching well over 100 feet in length by the later seventeenth century. Meantime, the disadvantage of image inversion was resolved by the placement of one or more erector lenses between the objective and eyepiece.[19]

The addition of lenses, however, compounded the problems of spherical and chromatic aberration. Starting with Descartes in the late 1620s, several thinkers, including Huygens and Newton, undertook to correct spherical aberration by designing machines to form the hyperbolic lenses Kepler suggested, but these efforts ground to a halt in the later seventeenth century because of insuperable technical difficulties.[20] No less intractable was the problem of chromatic aberration. One of the earliest stabs at resolving it was made in 1668 by Isaac Newton, who recognized that such aberration does not occur in reflection and therefore designed a telescope with a concave mirror replacing the objective lens. In 1672 he presented such a reflecting telescope to the Royal

19. For a still-useful discussion of the evolution of the telescope during the seventeenth century, see Albert Van Helden, "The Telescope in the Seventeenth Century," *Isis* 65 (1974): 38–58. For technical details about early telescopes, see Henry C. King, *The History of the Telescope* (Cambridge, MA: Sky Publications, 1955), 34–119. Since the image in a Keplerian telescope is formed between the objective and eyepiece, an appropriately calibrated micrometer can be placed in the plane of the image for measuring, which gives that type of telescope yet another advantage.

20. See D. Graham Burnett, *Descartes and the Hyperbolic Quest* (Philadelphia: American Philosophical Society Press, 2005). This is a useful source for understanding the techniques, both machine based and manual, for lens grinding during the seventeenth century.

Society, but the design was impractical because of the difficulty of grinding the mirror accurately and because the alloy of tin and copper out of which he formed it was inadequately reflective.[21]

For the most part, the improvement of telescopes over the seventeenth century depended on resolving technical problems rather than on developing new theoretical insights. One exception was the "discovery" of the sine law of refraction sometime before the 1630s and the ability on that basis to calculate focal lengths accurately according to the refractive index of the glass, the thickness of the lens, and the degree of curvature. Lens grinding was still an art, not a science, although machines were used for some of the steps in the process, and certain craftsmen became so adept that they were able to grind and polish fairly large lenses of nearly perfect sphericity and clarity. Eustachio Divini (1610–1685) and Giuseppe Campani (1635–1715) were especially noteworthy for the perfection of their lenses, and their telescopes were prized accordingly during the second half of the seventeenth century. Also, adjusting the length of the telescope according to its single or multiple drawtubes could compensate for lack of exactitude in measuring the focal lengths of the lenses.

Essentially a telescope in reverse, with a strong objective and a relatively weak eyepiece, the compound microscope probably evolved at around the same time. By no later than the early 1620s, for example, Galileo acknowledged explicitly that he could transform his telescope into a microscope by reversing it and letting the relatively strong concave eyepiece serve as the objective.[22] At around the same time, microscopes of the Keplerian type, with a strong convex objective, were produced and modified in various ways over the century to improve magnification, scope, and image clarity. As with the telescope, so with the microscope, these modifications were mostly based on technical adjustments, not theoretical insights, and involved such things as improving the illumination of specimens and adding lenses to enhance resolution. Thus by the time Hooke undertook his microscopic work in the early 1660s, the compound device he crafted for that purpose used a sophisticated lighting system and had an additional field lens to give a wider, more uniform view.[23] Even with these

21. On the problem of chromatic aberration, see M. Eugene Rudd, "Chromatic Aberration of Eyepieces in Early Telescopes," *Annals of Science* 64 (2007): 2–18.

22. Toward the bottom of page 105 of the *Assayer*, Galileo refers to *un Telescopio accommodato per veder glio oggeti vicinissimi* ("a telescope modified for seeing very close objects").

23. For Hooke's description of his microscope, see *Micrographia*, "Preface," xx–xxi (unnumbered). The lettering in his illustration of the device in Scheme 1, figure 6, does not accord with that in his description.

modifications, though, the compound microscope was surpassed for many purposes, especially microdissection, by the "simple" version. Consisting of a single convex lens, the most basic of these was the low-power "flea-glass," but types containing a high-power biconvex or spherical lens, some as small as 1 mm in diameter, could raise the magnifying power to over 300x with good resolution within an extremely narrow field of view. Antonie van Leeuwenhoek (1632–1723) brought this simple device to such perfection that he was able to observe a host of microorganism (including human spermatozoa) that were either undetectable or only barely detectable in even the best compound microscopes of his day.[24]

If Galileo's *Sidereus Nuncius* put the telescope on the map, so to speak, Robert Hooke's *Micrographia* of 1665 did the same for the microscope, not so much by demonstrating its scientific utility as by broadcasting the wonders that could be seen through it. Adding to the wonder of these wonders is that they spring from the most mundane things. Under the microscope, for instance, the sharpest, smoothest of needles turns out to be "broad, blunt, and very irregular," while the common flea appears "all over adorn'd with a curiously polish'd suit of sable Armour, neatly jointed, and beset with multitudes of sharp pinns, shap'd almost like Porcupine's Quills, or bright conical Steel-Bodkins." And lest anyone doubt the veracity of his descriptions, Hooke included marvelously intricate illustrations, the one of the flea occupying a four-page foldout in all its glory.[25]

But Hooke had a broader purpose than simply to titillate his readers. He was appealing to those who "still acknowledge their most useful informations [*sic*] to arise from common things, and from diversifying their most ordinary operations on them." For it is through such common observations and operations that we may eventually find "that those effects of Bodies, which have been commonly attributed to Qualities, and those confess'd to be occult, are perform'd by the small Machines of Nature, which are not to be discern'd without these [optical] helps." Hooke's *Micrographia* was thus an earnest of the utility of instrumental aids in the pursuit of scientific truth. It was also an incitement to continue improving such aids, for,

> 'Tis not unlikely, but that there may be yet invented several other helps for the eye, as much exceeding those already found, as those do the bare eye, such

24. For a dated but still useful discussion of the technical details of early microscopy, see Savile Bradbury, *The Evolution of the Microscope* (Oxford: Pergamon Press, 1967), 6–87.

25. On the needle, see *Micrographia*, 1–3; on the flea, ibid., 210–11.

as by which we may perhaps be able to discover . . . the figures of the compounding Particles of matter, and the particular Schematisms and Textures of bodies.[26]

All we need do, then, is perfect observational instruments to the point that we can behold the entire machinery of nature from the tiniest cogs to the celestial wheels they drive. And what could be more appropriate than that the grand instrumentality of nature be apprehended through the instrumentality of telescopes and microscopes?[27]

As curator of experiments for the Royal Society, Hooke had reason to be both enthusiastic and optimistic, as were many others at the time. Microscopy and telescopy had yielded and would continue to yield significant results over the century. The microscope-aided observation of pulmonary capillaries in 1661 by Marcello Malpighi (1628–1694) provided the missing link in William Harvey's theory of circulation. Jan Swammerdam (1638–1680) used the microscope to uncover the developmental and anatomical details of insects during the 1670s, and he went on to make significant contributions to the study of animal anatomy and physiology on the basis of microscopic dissection. What Swammerdam did for insects, moreover, Nehemiah Grew (1641–1712) did for plant anatomy. As a result, Catherine Wilson observes, "in revealing layer after layer of articulated structure, the microscope gave solidity and accessibility to what theory delivered up as an atomized or mathematized world."[28]

Meantime, using a telescope of his own manufacture in the mid-1650s, Huygens was able to determine that the "ears" Galileo observed on Saturn were actually the edges of a ring surrounding the planet. He also saw Titan, its largest moon and later, in 1659, not only detected the rotation of Mars but also estimated (correctly) its period at twenty-four hours. Some twenty years later, the director of the Royal Observatory at Paris, Giovanni Cassini (1625–1712), was able to find three more Saturnian moons and to make out the division in the ring that has since taken his name. It was through the telescope, as well, that planetary sizes and distances were measured with increasing, albeit varying, accuracy over the century, the inner planets (Mercury and Venus)

26. *Micrographia*, viii (unnumbered); the remaining quotations in the paragraph are from ibid., xxv (unnumbered). It is of course well known that Hooke was the first to observe "cells" in cork and other plant substances; see *Micrographia*, 112–16.

27. On the resulting focus on "instrumentalism," see Gal and Chen-Morris, *Baroque Science*, 79–113.

28. Catherine Wilson, *The Invisible World: Early Modern Philosophy and the Invention of the Microscope* (Princeton, NJ: Princeton University Press, 1995), 69.

according to solar transits and the outer ones (Mars, Jupiter, and Saturn) by parallax. By the very early 1700s, in fact, William Whiston, who succeeded Isaac Newton as Lucasian Professor of Mathematics at Cambridge, gave values for the sun-planet distances of earth and the five other known planets that are on average only about 11 percent lower than the modern ones.[29]

Yet despite these and many other significant observational breakthroughs, enthusiasm for telescopy and microscopy had waned considerably by the end of the century, and that for a variety of reasons. For a start, Hooke's confident expectation in 1665 that these instrumental aids would reveal "the small Machines of Nature" had not been met. On the contrary, the ever-widening and deepening view of nature at both macroscopic and microscopic levels showed her structure to be far more complex and multilayered, and thus far less easily resolvable into simple substructures or mechanisms, than anticipated. Furthermore, as time wore on, it became increasingly evident that instead of sharply curbing the "Fancy," as Hooke had hoped, telescopic and microscopic observation often yielded imagined or at least contestable results.[30] Thus while extending sight enormously in both directions, such artificial eyes did not necessarily improve it or make it more certain. To make matters worse, the artificial eyes themselves proved to be as fallible as the natural eyes whose deficiencies they were meant to rectify, and adding technical or theoretically based modifications to them seemed to bring them no closer to true perfection. Far from it, such modifications often brought to light new problems that needed resolution.[31]

Not only was the reliability of these instruments at issue, but so also was their utility. However much the microscope may reveal to the physician of the fine structure of things, that revelation tells us nothing about their medicinal properties. "The nettle," Girolamo Sbaraglia (1641–1710) observed toward the end of the century, "does not owe to the microscope its faculty of restraining the force of the blood, though the instrument reveals its prickly surface . . .

---

29. For Whiston's values, see the frontispiece to his *Astronomical Lectures* (London, 1715). For a detailed account of telescopy and its application to measuring planetary sizes and distances, see Albert Van Helden, *Measuring the Universe* (Chicago: University of Chicago Press, 1985), especially 105–59. See also Van Helden, "Telescopes and Authority from Galileo to Cassini," *Osiris* 9 (1994): 8–29; "The Importance of the Transit of Mercury of 1631," *Journal of the History of Astronomy* 7 (1976): 1–10; "Saturn and His Anses," *Journal of the History of Astronomy* 5 (1974): 105–21; and "Christopher Wren's 'De Corpore Saturni,'" *Notes and Records of the Royal Society of London* 23 (1968): 213–29.

30. See Wilson, *Invisible World*, 215–50.

31. See, e.g., ibid.; see also Burnett, *Hyperbolic Quest*, 123–27.

[indeed] everyone knows that no solid advantage to medicine has proceeded from such [microscopic] studies."[32] Even worse than their possible inutility, perhaps, was the sense of dislocation microscopes and telescopes created. As Wilson puts it, "the suggestion that what appears with more magnification might explain what happens at a level closer to the surface is troublesome, for to get one level in view is to lose the other."[33] No less troubling to some, Margaret Cavendish (1623–1673) serving as a fascinating example, was the very idea that the "prosthetic use" of such artificial devices to augment natural vision is appropriate because it contravenes the naturalness of vision. Is it not presumptuous, perhaps dangerously so, to meddle with nature in such a way? And is it not equally presumptuous to think that we can get the true measure of the universe with such unnatural devices? Perhaps, then, the scientific enterprise espoused by the likes of Hooke, with its emphasis on instrumentality, mechanism, and controlled experiment, is not simply misguided but perverted.[34]

3. NEW THEORIES OF LIGHT

The scientific outlook Cavendish and her ilk decried is epitomized by Descartes's approach to optics in *La Dioptrique* and *Les Météores*, which were appended (along with *La Géométrie*) as "essays" to the *Discours de la Méthode* of 1637. Descartes's interest in optics predates this treatise by nearly two decades, and at the time he published it he had certainly read Witelo and Alhacen and was probably familiar with other perspectivist sources as well. Of all the optical authorities known to Descartes, however, Kepler was his lodestar. Indeed, he admitted as much ("Kepler a été mon premier maître en optique"), and a considerable share of his *Dioptrique* is devoted to confirming and elaborating on points Kepler made in both the *Paralipomena* and *Dioptrice*.[35] In the fifth Discourse of the *Dioptrique*, for instance, Descartes provides a detailed analysis of retinal imaging and even recapitulates Scheiner's experiments with dissected eyes in order to confirm such imaging empirically. In the seventh and eighth Discourses, he explains how vision can be "perfected" by lenses and caps his discussion by demonstrating that not only hyperboloidal but also

32. Quoted from Wilson, *Invisible World*, 234.

33. Ibid., 231.

34. See Frédérique Aït-Touati, *Fictions of the Cosmos* (Chicago: University of Chicago Press, 2011), 133–90, especially 174–90; I owe the phrase "prosthetic use" to her.

35. On Descartes's optical sources and the problem of identifying them, see A. Mark Smith, *Descartes's Theory of Light and Refraction* (Philadelphia: American Philosophical Society Press, 1987), 8–12.

ellipsoidal lenses will bring parallel light rays to point focus. On that basis, he describes in Discourse Nine how hyperboloidal lenses, which are preferable to ellipsoidal ones, can be deployed for both close and near magnification and concludes in Discourse Ten with an account of his method for machine grinding such lenses.[36]

In all these cases, where the focus is on technical ray analysis, Descartes was following Kepler's lead. When it came to the physics of light, though, he forged off in an entirely different direction from Kepler. Like Grosseteste before him, Kepler understood physical light by analogy to the creative impulse emanating from God and multiplying outward to form the universe. "What wonder, then," he proclaimed early in the *Paralipomena*, "if that principle of all adornment in the world [light], which the divine Moses introduced immediately on the first day into barely created matter, [is] the most excellent thing in the whole corporeal world . . . and the chain linking the corporeal and spiritual world." And what wonder if it should "follow the same laws by which the world was to be furnished." Light is therefore to be understood as an incorporeal "outflowing" that takes form as a spherical surface, its propagation being radial because it is only along such lines that it can maintain equality.[37]

Descartes would have none of this metaphysical reasoning, and he was no less inimical to the perspectivist conception of radiated light (*lumen*) as an intentional species of the objective quality of luminosity (*lux*). Impatient with such qualitative accounts, Descartes offered an alternative based on his firm conviction that the physical universe operates in mechanically determinate ways that are explicable in terms of primary qualities only. Such a universe has no place for incorporeal effluences or species.[38] Accordingly, as he conceived it in *Le monde*—which was in the works by 1629 although not published until 1664—the universe comprises three microscopic "elements" differentiated by size and shape. The grossest of the three, Earth, consists of relatively large and

36. For Descartes on retinal imaging, see Charles Adam and Paul Tannery, eds., *Oeuvres de Descartes*, vol. 6 (Paris: Léopold Cerf, 1902), 114–39 (henceforth referred to as AT, followed by appropriate volume number); on perfecting sight with lenses and the analysis of hyperboloidal and ellipsoidal lenses, 146–96; on deploying hyperboloidal lenses, 196–211; and on machine grinding such lenses, 211–27.

37. See *Paralipomena*, 1, 5–7, in Donahue, 17–20; in *Ad Vitellionem*, 5–7. See also David Lindberg, "The Genesis of Kepler's Theory of Light: Light Metaphysics from Plotinus to Kepler," *Osiris* 2 (1986): 5–42.

38. On Descartes's commitment to mechanism and its methodological and ideological wellsprings, see Stephen Gaukroger, *Descartes: An Intellectual Biography* (Oxford: Clarendon Press, 1995).

irregularly shaped particles that bunch together to form crass, physical objects such as eyes and planets. At the other end of the material spectrum is Fire, whose particles are infinitesimally small and combine to form such luminous sources as candle flames and the sun. Ether (or Air) holds an intermediate position between these extremes, its particles being smaller than those of Earth and larger than those of Fire. They are also perfectly inelastic and spherical.

As the sole components of the physical universe, these elements form a plenum in which all the space between crass, Earthy objects and fine, Fiery light-sources is filled with Ether, the result being a perfect "contiguum" rather than a true continuum. Descartes's universe is thus corpuscularist but not atomist in structure because it has no void spaces. Within this system, the constituent Fire-particles in any light source are assumed to rotate continuously and at great speed, thus forming subvortices within a universe resolved into an arrangement of larger, interlocking vortices. Consequently, the particles on the surface of the light source, which rotate most swiftly of all, have the same tendency to fly off centrifugally as a stone whirled in a sling. Being hemmed in by the tight press of Ether surrounding them, though, these surface particles cannot actually fly off. Instead, they exert a constant outward pressure on the rigid Ethereal envelope. This pressure, in turn, translates into a rectilinear impulse at every point of contact between that envelope and the revolving Fire particles. Transmitted instantaneously throughout the surrounding array of inelastic Ether particles, this outward pressure is what we call "light," the rectilinear impulse a "ray," and the capacity of the Ether to propagate it "transparency." Hence, light radiation involves not motion but, rather, the *tendency* toward motion (*inclination à se mouvoir*). It is, in short, matter in *virtual* motion. Therefore, Descartes concludes, radiating light must be subject to the same laws as real, projectile motion.[39]

According to Descartes, then, since light acts like a projectile—a hard-hit tennis ball being his example—its interaction with gross bodies can be analyzed in terms of collisions and their consequences. This is the basis for his notorious derivation of the sine law of refraction, and although the actual "discovery" of this law, like that of the telescope, has several claimants, Descartes was the first to publish it and to justify it on theoretical grounds.[40] That justification can be

39. For a more detailed account, with appropriate references, of Descartes's light theory and its grounding in the cosmology set out in *Le Monde*, see Smith, *Descartes's Theory*, 13–19.

40. Aside from Ibn Sahl, who never explicitly articulated the sine law, Thomas Harriot and Willibrord Snel (1580–1626) are credited with having arrived at it independently in 1601 and 1621, respectively; see Smith, *Descartes's Theory*, 81. In fact, as Goulding shows in "Thomas Harriot's Optics," it is highly likely that Harriot knew the sine law in the 1590s. See also A. I.

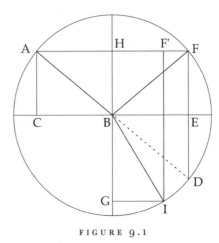

FIGURE 9.1

summarized as follows. Suppose, first, that CE in figure 9.1 is the surface of a plane mirror and that the light incident along ray AB strikes it at angle ABH. The "force of its movement" (*la force de son mouvement*) along AB can be resolved into two "determinations" (*déterminations*), one downward along vertical AC, the other along horizontal AH. Upon rebounding at equal angles along BF, which is equal to AB, the light will maintain the same horizontal determination, so HF = AH. On the other hand, the vertical determination along EF will be in the opposite direction from AC but will remain otherwise unchanged. The light's force of movement will therefore remain constant throughout.

Now suppose that CE is a refractive interface and that the medium below it is 1.5 times as dense as the one above. Suppose, further, that the light's force of movement in the denser medium increases in proportion to its density. In this case, then, the light will go half again as fast in the denser medium, so in the time it takes the light to travel distance AB in the rarer medium, it will go 1.5 times as far in the denser one. Or, to put it another way, in the denser medium the light will travel a distance equal to AB in two-thirds the time. Finally, let us stipulate that just as in reflection, so in refraction, the horizontal determination remains constant and is thus conserved throughout. Accordingly, let the light passing along ray AB in figure 9.1 refract at B along some ray BI equal to AB. It will traverse that distance in two-thirds the time it traversed AB, and according to the principle of horizontal conservation, the resulting determi-

---

Sabra, *Theories of Light from Descartes to Newton* (Cambridge: Cambridge University Press, 1981), 99–105. Although Descartes was accused of stealing the law from Snel, that accusation is most likely a canard.

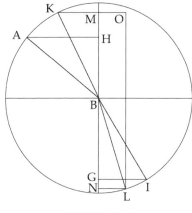

FIGURE 9.2

nation HF′ will be two-thirds AH. Drop perpendicular F′I from F, draw ray BI, and drop GI perpendicular to HG. GI will thus be two-thirds AH. Then, in figure 9.2, take another incident ray KB, whose horizontal determination before refraction is KM, and MO after refraction along ray BL. MO will thus be two-thirds KM. Since perpendicular LN dropped from refracted ray BL is equal to MO, it will also be equal to two-thirds KM. Therefore, AH:GI = KM:LN. But each of these lines is proportional to the sine of its subtended angle, so the ratio translates to sine ABH:sine IBG = sine KBM:sine LBN. It therefore follows that sine i:sine r is constant. Taken as a quotient, this proportion yields the index of refraction, which in this case is 1.5.[41]

How Descartes arrived at this derivation is something of a mystery, but three things about his overall argument suggest that he constructed it on perspectivist foundations. First, there is an obvious correspondence between his composition of horizontal and vertical determinations and the composition of motions in the perspectivist analysis of refraction. Second, the analogy (*comparaison*) he draws between light and a tennis ball is strongly reminiscent of the perspectivist analogy between light and a small, iron ball.[42] And finally, like the perspectivists (except Bacon), he appeals to virtual motion in order to account for the dynamic effects of light without having it actually move through space and time. But whatever its actual inspiration, Descartes's "demonstration" of the sine law is problematic at several levels. For one thing, the principle of

41. For Descartes's complete account of refraction and the resulting derivation of the sine law, see *Dioptrique*, 1 and 2, in AT, vol. 6, 81–108.

42. See, e.g., Sabra, *Theories*, 69–85 and 93–99, and Smith, *Descartes's Theory*, 19–66.

horizontal conservation is arbitrary. Why should determination be conserved at all, much less in that particular direction?[43] No less arbitrary is the supposition, which the perspectivists shared, that virtual motion is governed by the same laws as actual motion. Why is that necessarily the case? Most glaring, however, is the problem posed by Descartes's assumption that the speed, or force of movement, of light is directly proportional to the density of the medium through which it moves. In his physical account of radiation, Descartes argued that light is a mechanical impulse transmitted *instantaneously* through a perfectly rigid medium. How, then, can it go faster or slower when it has no speed? And how does it make sense to have it accelerate in a denser medium, which ought to pose greater resistance to it?

Oddly enough, these inconsistencies in Descartes's reasoning did not cause his demonstration to be rejected out of hand. As late as 1668, in fact, Newton praised Descartes for having demonstrated "the Truth" of the sine law "not inelegantly," no small concession from someone who spent most of his career attacking the program of Cartesian physics.[44] Not everyone was as forgiving as Newton, though. Descartes's younger contemporary, Pierre de Fermat (1601–1665), was suspicious of his demonstration from the outset, and he was no less suspicious of the "law" that followed from it. Perhaps, he surmised, that law is only an approximation. Fermat thus set out to find a more rigorous way to determine the true law, and he looked to the principle of natural economy as his guiding criterion. This, of course, is the principle underlying Hero of Alexandria's least-distance demonstration of the equal-angles law of reflection, but refraction cannot be subject to this version of the principle because the shortest distance through any refractive interface is along a straight line. Fermat therefore concluded that in refraction the path followed by light must be dictated not by spatial brevity but by temporal brevity and on that basis managed to construct a satisfactory proof by 1662.

Conceptually, if not mathematically, simple, Fermat's reasoning comes to ground in the assumption that instead of increasing, as Descartes would have it, the speed of light decreases in a denser medium because of its resistance. The denser and more resistant the medium, the more slowly light passes through it. Accordingly, let CB in figure 9.3 be a refractive interface dividing a

43. Actually, Descartes had attempted to justify this principle in his discussion of rotary motion in *Le Monde*, but fearing repercussions after Galileo's condemnation in 1633 because the underlying cosmology of this treatise was essentially Copernican, he abandoned his plan of publishing it; see *Le Monde; ou, Traité de la lumière* (1664), in AT, vol. 10.

44. Sabra, *Theories*, 300.

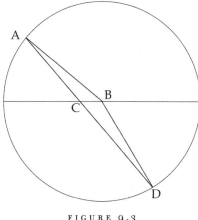

FIGURE 9.3

rarer medium above from a denser one below, and let ACD and ABD be two possible paths for light to follow through it. Representing the shortest possible distance between A and D, rectilinear path ACD would therefore be the mandated Heronian one. However, the total time it takes light to travel from A to D along distance ACD is the composite of the time it takes to traverse AC at a higher speed and CD at a lower one. On the other hand, in longer path ABD, distance AB along which the light moves faster is longer than AC, whereas distance BD along which the light moves more slowly is shorter than CD. Hence, the total time it takes light to traverse both arms of longer path ABD may be less than that needed to traverse both arms of shorter path ACD. It was after a comparative analysis of this sort, based on his method of maxima and minima, that Fermat was drawn to conclude (to his surprise) that the shortest possible temporal path is in fact the one dictated by the sine law.[45]

Fermat's least-time proof merely confirmed a law that had long been established and accepted on empirical grounds, as well as on the theoretical grounds Descartes laid. In his *Micrographia*, for instance, Hooke describes an apparatus "by which the refraction of all kinds of Liquors may be most exactly measur'd" so that "by tryals of that kind [we can] see that the laws of Refraction are not only notional."[46] Nonetheless, by transforming Descartes's virtual motion into

45. For a good account of Fermat's procedure and its evolution, see ibid., 116–58. See also Michael Mahoney, *The Mathematical Career of Pierre de Fermat* (Princeton, NJ: Princeton University Press, 1994), 170–95 and 387–402.

46. *Micrographia*, "Preface," xix–xxi. Suffice it to say, such empirical "tryals" only corroborated what was already assumed—namely, that the sine law governs refraction, which Hooke evidently learned from "the most incomparable Des Cartes."

actual motion and thereby converting light from a virtual to a real projectile, Fermat lent credence to the atomist conception of light the likes of Pierre Gassendi and Robert Boyle favored. He also reinforced the idea that light takes a finite amount of time to travel and, therefore, that it propagates neither instantaneously nor at such unimaginable speed that its passage is virtually instantaneous, no matter the distance. In fact, Ole Roemer (1644–1710) confirmed this notion of finite speed in 1676 on the basis of discrepancies between the time at which Jupiter's moon Io was predicted to appear and when it actually did appear after emerging from behind the planet. Recognizing that those discrepancies vary according to how close or distant the Earth was in its orbit from Jupiter, Roemer estimated that the light from Io took around eleven minutes to pass a distance equal to the radius of the Earth's orbit.[47]

Despite these developments, which militated against Descartes's corpuscularist theory and gave support to the atomist alternative, there was no scramble to jump the Cartesian ship. The preferred strategy, instead, was to fix the leaks, and no one was better at it than Christiaan Huygens. As we saw earlier, Huygens was deeply interested in telescopy and was thus drawn to the study of dioptrics by the very early 1650s. This marked the beginning of a lifelong, and ultimately inconclusive, effort to produce a complete theoretical analysis of refraction with particular application to lenses and their use in telescopy. Not until the early 1670s, however, did Huygens apply himself to the physics of light with an eye toward developing his own theory "in accordance with the principles accepted in the Philosophy of the present day." Those present-day principles, of course, were the ones specific to the Mechanical Philosophy. By 1678, so he tells us, he was confident enough in his new theory to present it to the Académie Royale des Sciences. He then sat on it for more than a decade before finally publishing it in the *Traité de la lumière* of 1690.[48]

Although sympathetic to Descartes's corpuscularist physics, Huygens was well aware of its shortcomings. "For it has always seemed to me," he admits early in the *Traité*, "that even Mr. Des Cartes, whose aim has been to treat all the subjects of natural philosophy (*Physique*) intelligibly, and who assuredly

47. For a facsimile of the report of Roemer's findings, given to the *Académie Royale de Sciences* in 1676, see René Taton, ed., *Roemer et la vitesse de la lumière* (Paris: Vrin, 1978), 151–54. See Darrigol, *History*, 65, for a simplified explanation of Roemer's method.

48. For the English quotation, see Sylvanus S. Thompson, trans., *Treatise on Light* (repr.; New York: Dover, 1962), 2; for the French original see *Traité*, 2; see *Treatise*, "Preface," v (*Traité*, *2), for Huygens's claim to have perfected his theory by 1678. For an extensive treatment of Huygens's research on dioptrics, see Fokko Jan Dijksterhuis, *Lenses and Waves* (Dordrecht: Kluwer, 2004).

has succeeded in this much better than anyone before him, has said nothing that is not full of difficulties, or even inconceivable, in dealing with Light and its properties."[49] Particularly vexing to Huygens was Descartes's assumption that light is transmitted instantaneously. Convinced that it takes time, Huygens welcomed both Fermat's least-time proof and Roemer's empirical determination of the finite speed of light as confirmation of his "hypothesis." Yet he was unwilling to abandon Descartes's impulse theory in favor of the atomist alternative, according to which light consists of actual projectiles zipping through space. How, then, to reconcile Cartesian impulse theory with the temporal passage of light? The answer, Huygens realized, was to treat light as analogous to sound, which propagates through the air at a finite speed according to a succession of longitudinal waves emanating spherically from percussive sources. Like sound, therefore, light must propagate through ether in the form of successive wave fronts centered on luminous sources. And like sound-bearing air, light-bearing ether must be highly elastic rather than perfectly inelastic. Unlike air, however, ether must be "of a substance as nearly approaching to perfect hardness and of a springiness as prompt as we choose" that it can transmit light waves at the speed Roemer's tabulations indicated.[50]

Following Descartes's lead, Huygens traced the origin of these waves to the tiny particles composing the light source, each of which exerts percussive impulses on the particles of ether in contact with it. Those impulses are passed in turn through the neighboring ether particles and thence through all the rest in a succession of individual compressions and rebounds that leave the particles virtually unmoved in place. Each fire particle thus creates its own spherical wave through the ether, and what results is an infinitude of such waves emanating from every particle in the light source. As these individual waves get farther from the source, they "unite together in such a way that they sensibly compose one single wave only" (the so-called Huygens-Fresnel principle). Forming a sphere centered on the light source, this aggregate wave front constitutes optically effective light, and it is on this basis that Huygens not only explained reflection and refraction but also derived their governing laws.[51] The pièce de résistance for Huygens, though, was his explanation of the opti-

49. Translation adapted from Thompson, *Treatise*, 7 (*Traité*, 6–7).

50. For Huygens's discussion of Roemer's determination, see ibid., 7–9 (*Traité*, 7–8); on Fermat's proof, ibid., 42–45 (*Traité*, 39–41); and on the analogy between sound and light, ibid., 4 and 12 (*Traité*, 3–4 and 11).

51. Quotation from ibid., 18 (*Traité*, 17). For Huygens's complete account of spherical radiation, see ibid., 12–22 (*Traité*, 11–20); for his treatment of reflection, ibid., 22–28 (*Traité*, 21–26); and for his treatment of refraction, ibid., 28–45 (*Traité*, 26–41).

cal anomaly presented by Iceland spar, a transparent form of calcite. Brought to public notice in 1669 by Erasmus Bartholin (1625–1698), what is anomalous about this crystal is that it produces double images by refracting light in two different ways. In order to explain this phenomenon, which today we attribute to polarization, Huygens showed how the initial wave front can split into two separate wave fronts on entering the crystal. The light refracted normally forms "spherical" wave fronts in the ether permeating the crystal, whereas the light refracted abnormally forms "spheroidal" (in fact, ellipsoidal) wave fronts in the particles of the crystal itself. The genius of Huygens's account lies in how he tied the formation of these abnormal wave fronts to the geometrical structure of the crystal with such precision as to yield the deviant angle of refraction as determined by careful empirical measure.[52]

4. RECASTING COLOR

Whatever their differences in analytic details, all the theories examined to this point have one thing in common: by reducing light to the mechanical inter-action of bodies distributed in space, they also reduced it to a secondary quality. No longer could it be imagined to exist objectively, as an inherent property of luminous bodies or as a formal effect in transparent media. Indeed, within the context of this mechanistic approach, it makes no sense whatever to conceive of light as objectively "in" anything. And the same holds for color, light's ontological doppelgänger. It too must be reducible to a mechanical effect and thus to a secondary quality lacking objective existence. To deny color objective existence was of course to repudiate the long-standing idea that specific colors are a blend of black and white in particular proportions, but this idea had already lost much of its luster by the late Renaissance.[53]

As with light, so with color, it was Descartes who made the first meaningful attempt to explain it on mechanistic grounds. Articulated in the midst of a close, quantitative analysis of the rainbow in the eighth Discourse of the *Météores*, Descartes's account links color to the speed with which the Ether particles rotate within a given line of light impulse.[54] In its natural state, Descartes

52. For Huygens's ingenious and mathematically dense analysis of double refraction in Iceland spar, see ibid., 52–105 (*Traité*, 48–101).

53. See, e.g., Alan E. Shapiro, "Artists' Colors and Newton's Colors," *Isis* 85 (1994): 600–630, especially 600–610. See also Eric Kirchner and Mohammed Bagheri, "Color Theory in Medieval Islamic Lapidaries: Nīshābūrī, Tūsī and Kāshānī," *Centaurus* 55 (2012): 1–19.

54. Descartes also adverts to the rotational speed of ether particles and color in *Dioptrique*, 2, in AT, vol. 6, 92.

claims, light is white because the rotational speed of the spherical Ether particles transmitting it is commensurate with the light's radial "action or movement" (*l'action ou le mouvement*). Any other rate of spin will cause the light to appear colored. The faster the spin, the more the color tends toward red, whereas the slower the spin, the more it tends toward violet. The spectrum of colors—from red, through yellow, green, and blue, to violet—therefore unfolds according to the relative swiftness of rotation, red being due to the fastest spin, violet to the slowest. What determines the rate at which these Ether particles spin is their interaction both with gross matter and with each other. Descartes looks to the prismatic effect of refraction as his analytic example. The resulting explanation rests on two factors. First, at the instant it strikes the surface of a denser medium obliquely, a light-impelled particle changes its rate of spin. But that rate is also affected by the rotation or counterrotation of the surrounding Ether particles, which can either accelerate or retard it.[55] This complex of external factors will therefore dictate the resulting color.

All of this is supposedly in aid of explaining the formation of the spectral color bands in the primary and the secondary rainbows. Whether coincidentally or not, Descartes's account of this formation is similar in concept to that of Theodoric of Freiberg, who, as mentioned earlier in chapter 7 (section 1), based his on refractions and internal reflections in individual drops within the raincloud.[56] Recourse to figure 9.4, which is grossly out of scale, should make Descartes's version of the account fairly easy to grasp. Let the lower circle represent a drop within a raincloud, and let solar ray SB refract into it along BC according to the sine law. Then let it reflect internally along CD and refract back out along DE. Let E represent the viewpoint of someone observing the raindrop, and draw line EX parallel to SB so as to form angle DEX. If this angle is 42°, point D will appear red to the viewer at E. If, on the other hand, it is 41°, that point will appear violet, so all the other colors will appear at angles in between these limits. Descartes bases this claim on experiments with light passing into a water-filled glass sphere.

Let the upper circle represent another raindrop in which solar ray SF enters the drop and, after *double* reflection at G and H, emerges along KE. If angle KEX is 51°, point K will appear red to the viewer at E, but more faintly

55. See *Météores*, 8, in AT, vol. 6, 331–34.

56. The only source Descartes mentions is Maurolico, who offered an account of the rainbow in the second part of the *Photismi* (49–68; Crew, 79–108), but only to chide him for getting the basic angles wrong. In addition, Maurolico's explanation does not take refraction into or out of the raindrops into account, so it differs markedly from both Theodoric's and Descartes's explanations.

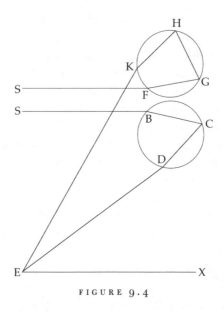

FIGURE 9.4

so than in the lower drop because the second reflection weakens the light. If it is 52°, on the other hand, K will appear violet, so the original spectrum in the lower drop will appear reversed between the limits of 51° and 52° in the upper drop. Imagining these exaggeratedly large drops reduced to the size of actual raindrops, we can see that the individual raindrops ranged within the limits of 41° to 42° will produce all the spectral colors between violet at the bottom and red at the top according to the size of angle DEX. Likewise, the individual raindrops within the range of 51° to 52° will produce the spectral colors between red at the bottom and violet at the top according to the size of angle KEX. Thus when we rotate the entire system about EX as an axis, two spectral arcs, each 1° thick, will be produced, the lower one being the primary bow, the upper one the secondary bow. Since the size of the composite bow is determined by where it is cut by the horizon, it will be largest when the sun is just at the horizon and will diminish as it rises.[57]

However ingenious, mathematically exact, and forward looking it may be, Descartes's account of the rainbow in no way depends on or vindicates his association of color with the rotation of Ether particles along any given light ray. Why the light-bearing Ether emerging along DE at an angle of 42° should spin

57. *Météores*, 8, in AT, vol. 6, 325–30. Descartes's figures of 42° for the top of the primary bow and 51° for the bottom of the secondary one are only approximations; later, in AT, vol. 6, 340, he pinpoints the respective values at 41°47′ and 51°37′.

most swiftly, and thus yield red, while the light-bearing Ether emerging at an angle of 41° should spin most slowly, and thus yield violet, is left unexplained by the model. Yet Descartes's rainbow analysis does accomplish several important things. First, it leaves no doubt about what the rainbow colors are not. They are not tiny, fragmented images of the sun in the raindrops, as Aristotle would have it. Nor are they the product of the relative darkness or brightness of the surrounding cloud. Nor are they different from the colors of opaque objects. "I have been unable to appreciate the distinction drawn by the [scholastic] philosophers," Descartes sums it up, "when they claim that there are some colors that are true and others that are only false and apparent." The very nature of color, he insists, is "to appear," so it is "a contradiction to say that [colors] are false and that they appear."[58] Color is therefore not objectively in anything, be it a raindrop or an opaque object. Furthermore, Descartes's rainbow analysis reinforces the link between color and the angle of refraction at which the light emerges out of the raindrop, a link established earlier by Theodoric of Freiberg and, independently, by Kamāl al-Dīn al-Fārisī. The association of particular colors with particular angles is now explained by the physical alterations the light undergoes in being refracted into and out of the raindrop. Finally, and perhaps most important, Descartes's overall analysis of color establishes that it is not ontologically distinct from light, as had been commonly assumed for at least two millennia. It is simply a mechanical modification of white light, which is itself a mechanical modification of Ether.

The idea of color as a mere mechanical modification of white light was taken up by Robert Hooke in the *Micrographia* of 1665. A devotee of Descartes, but not an entirely blind one, he was unconvinced by the Ether-spin model of color and strove to find a more compelling theory within the general framework of Cartesian mechanics. The model of light upon which Hooke based that theory foreshadows Huygens's longitudinal wave model in certain key respects, a point that Huygens himself acknowledged.[59] For a start, like

58. Ibid., 335.

59. See Thompson, *Treatise*, 20 (*Traité*, 18). Huygens also acknowledged the Jesuit Ignace Pardies (1636–1673) as one who, along with Hooke, "hitherto [has] begun to consider the waves of light." Thomas Hobbes should perhaps be included in this group. Francesco Grimaldi (1618–1663) might also be included as something of an outlier insofar as his study of what he called (and we still do) "diffraction" led him to think of light as a fluid that undulates in waves analogous to transverse waves in water when it passes by sharp edges. For a helpful account of these protowave theories, see Darrigol, *History*, 49–64; see also Alan E. Shapiro, "Kinematic Optics: A Study in the Wave Theory of Light in the Seventeenth Century," *Archive for History of Exact Sciences* 11 (1973): 134–218.

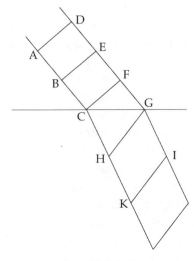

FIGURE 9.5

Huygens, Hooke avoided Descartes's vortex account with its emphasis on centrifugal tendency. Instead, he traced the origin of light to an extremely quick and "very short vibrating motion" of the particles in light sources. These vibrating particles, Hooke argued, create pulses in the transparent medium in contact with the luminous source, and those pulses are propagated spherically through the medium in "the least imaginable time," but not in an instant.[60]

The theory of refraction that follows from this model is illustrated in figure 9.5. Let AD, BE, and CF represent small, virtually straight segments of spherical light pulses, and let them be perpendicular to parallel lines AC and DG, which form a thin shaft of light in air. Let that shaft reach surface CG of a medium one-third denser than the air (that is, water). When the light impulse along "ray" AC enters that denser medium, it will speed up commensurately as it refracts, just as Descartes claimed for its force of movement. If, therefore, the light traveled three units of distance along BC, it will traverse four units along refracted ray CH. Moreover, in the same time that the refracted light travels four units from C to H in the denser medium, the light along ray FG will travel three units from F to G in the rarer one. From then on, the light throughout the refracted shaft will travel at the same speed in the denser medium, and

60. See *Micrographia*, 54–57 for Hooke's basic theory of light.

its angular deviation along that shaft will be determined by the relative speed after and before refraction: that is, 4 to 3, which is the sine relation in refraction from air to water.

Although the light accelerates when it is refracted into the denser medium, it is nonetheless deadened by the resistance that medium poses. Thus in figure 9.5, of all the individual light pulses that combine to form front CF, the one that travels along ray AC encounters that medium's resistance first. By taking the brunt of that resistance, this leading light pulse opens a somewhat easier, less resistant path for the one arriving just after it along the neighboring ray, and so on down the line as the light pulses from succeeding points along front CF enter the denser medium. Thus as Hooke describes it somewhat circuitously, "that part or end of the pulse which precedes the other, must necessarily be somewhat more obtunded, or impeded by the resistance of the transparent medium, than the other part or end of it which is subsequent, whose way is, as it were, prepared by the other." As a result, the light along HG and KI will be ordered spectrally according to relative strength or vividness. The light along ray CHK, which is most weakened by resistance, will therefore appear "dead blue" (Descartes's violet), while that along GI, which is least weakened by resistance, will appear red. All the other colors will appear at appropriate points between these extremes.[61]

In 1666, just a year after Hooke published his color theory in the *Micrographia*, young Isaac Newton was applying himself "to the grinding of Optick glasses of other figure than Spherical." While so engaged, he "procured . . . a Triangular glass-Prisme, to try therewith the celebrated Phenomenon of Colours." At least that is the story he tells us at the beginning of his letter "containing a New Theory about Light and Colors," which appeared in the 1671/72 issue of the *Philosophical Transactions*.[62] The "celebrated Phenomenon" Newton had in mind was the prismatic generation of spectral colors. In order to test it he closed his chamber off to darken it, drilled a small opening in the window shutter, and placed the prism in front of it to allow a beam of sunlight to refract through it and form a spectrum on the opposite wall. After some "very pleasing divertisement," he noticed something both Descartes and Hooke ignored: instead of forming a circle "according to the received laws of Refraction," the spectrum was oblong to such an extent that it "excited [him] to a more then [*sic*] ordinary curiosity of examining, from whence it

61. See ibid., 57–63, especially 62–63.

62. For the full text of Newton's letter, see *Philosophical Transactions* 6 (1671/72): 3075–87.

might proceed."[63] The first step in that examination was to carefully measure the relative dimensions of the spectrum, from which he determined that it subtended an angle of 2°49′ rather than the 31′ it should according to the sine law. He then conducted several experiments to confirm that this anomaly was not due to such things as irregularities in the glass or a "curvity" of the rays after refraction.[64]

With these possibilities excluded, Newton tells us, he lit upon the idea of conducting an "Experimentum Crucis." Taking two boards and drilling a small hole in each, he placed one in front of the prism so that light from the opening in the shutter could pass through the hole in the board to the prism and then be refracted into a spectrum. At a distance of 12 feet he placed the second board so that the spectrum was projected onto it. Then he posed another prism behind that second board so that it could capture the light passing through the board's hole and, by refracting it, project it onto the wall. What he found was that if he isolated one of the colors within the spectrum on the second board and allowed its light to pass through the hole to the second prism, the resulting colored light on the wall would remain integral rather than being dispersed after refraction. Turning the first prism on its axis in order to project different colors through the hole in the second board, he "saw by the variation of those places, that the light, tending to the end of the Image, towards which the refraction of the first Prisme was made, did in the second Prisme suffer a refraction considerably greater then the light tending to the other end." In other words, the light was refracted to a greater or lesser extent depending on its color, violet undergoing the greatest refraction and red the least. This, of course, explains the relationship between colors and angles in the rainbow.

Therefore, Newton realized, the true cause of the spectrum's elongation is "no other, then that Light consists of Rays differently refrangible."[65] This realization led him to conclude that chromatic aberration could never be overcome in refracting telescopes because it is due not to the lenses or their shape but to the light refracted through it—hence his turn toward reflecting telescopes. It was also this realization that led him to conclude that "Colours are not Qualifications of Light, derived from Refractions, or Reflections of natural Bodies (as 'tis generally believed,) but Original and connate properties, which

63. Ibid., 3076. Although neither Descartes nor Hooke mentioned the spectrum's elongation, their respective diagrams actually do show some slight elongation; see *Météores*, 8, in AT, vol. 8, 330, and *Micrographia*, Schema VI, fig. 4. Newton determined that in fact the spectrum was roughly five times as long as it was wide.

64. See *Philosophical Transactions* 6 (1671/72): 3076–78.

65. See ibid., 3078–79.

in divers Rays are divers."[66] Color, in short, is not a modification of white light, as Descartes and Hooke assumed. Quite the contrary, white light is essentially a modification of colors in that it consists of all of them.

A tour de force of reasoning from the phenomena, Newton's account of color in the "letter" of 1672 was also a rhetorical masterpiece because, by couching it in terms of "Rays," Newton avoided any speculation about the physical nature of whatever those rays consist of or represent. His new theory was therefore not a "hypothesis," hypotheses being a lifelong bugaboo for him. Under constant pressure from critics, however, he was forced far enough out of the closet in his "Hypothesis" of 1675 to suggest for those whose "heads . . . run much upon hypotheses" (Hooke in particular) that light might be thought to consist of particles of various "bignesses" shot off promiscuously and at great speed from luminous sources. Color would thus be a function of the particular size of these particles, the largest yielding red, the smallest violet. But that is not all. "It is to be supposed," as well, "that there is an aethereal medium much of the same constitution with air, but far rarer, subtler, and more strongly elastic." Like air, moreover, this "is a vibrating medium, . . . only the vibrations far more swift and minute." And one more thing: "It is [also] to be supposed, that light and aether mutually act upon one another, aether in refracting light, and light in warming aether; and that the densest aether acts most strongly."[67]

Taken together, these suppositions yield a hybrid particle-wave model according to which all optical phenomena should be explicable in terms of tiny, swiftly moving projectiles of different sizes, correlated to specific colors that interact with a highly elastic aethereal fluid. On the basis of this model Newton accounted for the periodic nature of "Newton's rings," the alternating colored and dark bands that appear in thin glass plates, bubbles, and films. On striking the surfaces of such objects, Newton conjectured, light creates wavelike "vibrations" in the elastic aether pervading them. These vibrations take form as successive condensations and rarefactions, much as a stone dropped into still water creates transverse waves in it. The "peaks" of these aether waves are so condensed that they force particles of certain sizes to reflect, thus yielding the colored bands. The "troughs," on the other hand, are so rarefied as to allow virtually all light particles to pass through, hence the appearance of darkness for lack of reflected light.[68]

66. Ibid., 3081.

67. See "An Hypothesis explaining the Properties of Light," in *The History of the Royal Society*, vol. 3, ed. Thomas Birch (London, 1757), 248–305, especially 249–51 and 255.

68. See ibid., 263–68.

Encountering unexpected opposition to his new theory and embittered by the experience, Newton withdrew from the fray soon after submitting his "Hypothesis" to the Royal Society and busied himself with other things, including the composition and publication of his *Principia* in 1686. In the very early 1690s, however, he picked up where he had left off in 1675 and, using the particle-wave model as a heuristic, revisited his account of color formation in thin glass plates and opaque bodies. In the process of rethinking these phenomena, Newton abandoned aether as the source of the "vibrations" that cause light's periodicity and looked directly to the rays themselves, which he supposed to vibrate appropriately. By the time he published his *Opticks* in 1704, he had moved from the idea of periodically vibrating rays to that of "Fits of easy Reflexion and easy Transmission" as a means of explaining the formation of Newton's rings, the rings in thick glass plates, and the alternating light and dark bands that appear when light passes sharp edges, that is, the phenomenon of diffraction or "inflexion."[69] Notwithstanding the problematic nature of this explanatory model, the experimental and mathematical precision with which Newton analyzed all these phenomena lent his *Opticks* an authority not unlike that enjoyed by the *Principia*. In fact, it is not much of an exaggeration to say that the *Opticks* did for the physics of light and color what the *Principia* did for the physics of motion. For that reason it remained canonical, though not uncontested, throughout the eighteenth century.[70]

5 . THE EPISTEMOLOGICAL CONSEQUENCES

Nothing succeeds like success, so the adage goes, and it certainly applies to seventeenth-century physical optics. The mechanistic approach to that science had proved to be extraordinarily fruitful as it evolved toward greater sophistication, analytic rigor, and comprehensiveness between the early 1600s and the beginning of the eighteenth century. On its basis an astonishing number of optical phenomena had been more or less satisfactorily accounted for, all of them unknown before 1600. The sine law of refraction had been validated on both theoretical and empirical grounds. The relationship between magnification and focal length was well understood and had been put to practical

69. It is worth noting that Newton still appealed to aether gradients to account for refraction (*Opticks*, 349–50) and that he postulated "Sides" to the rays in order to explain double refraction.

70. See Alan E. Shapiro, *Fits, Passions, and Paroxysms* (Cambridge: Cambridge University Press, 1993), 3–207. See also Sabra, *Theories*, 251–342, and Darrigol, *History*, 78–108.

use in telescopy and microscopy. Light was shown to propagate not only at finite speed but also slowly enough to require several minutes to reach from Sun to Earth. Refractive dispersion had been thoroughly analyzed, its role in the formation of rainbows, Newton's rings, and chromatic aberration now clear. Double refraction in transparent crystals had been measured and made comprehensible on mechanical grounds, and so had diffraction or "inflexion." Each new success added plausibility to both the approach and the presuppositions underlying it. For if all these physical manifestations of light and color are so readily explained in mechanical terms, then light and color must actually *be* mechanical effects in an inherently mechanical universe. With its subject matter, methodology, and analytic principles so defined, optics became increasingly autonomous and self-referential over the seventeenth century.

But the concept of light and color resulting from this mechanistic approach is clearly—and completely—incompatible with that of the perspectivists. For the perspectivists, light and color were real, ontologically distinct, and inherent properties of objects that conspire to make them visible. They were thus "primary" insofar as all the other physical properties of those objects, which consist of such common sensibles (or visible intentions) as shape, size, motion, and the like, must ride on their coattails in order to be perceived. These common sensibles were therefore "secondary," not because they lacked objective existence but because they had to be inferred from the primal impressions of luminous color formed on the eye's lens. By the seventeenth century, the order had been reversed. Strictly limited to spatial qualities, the common sensibles had become "primary" and light and color "secondary," the latter now riding on the coattails of the former—particularly those qualities pertaining to motion—in order to be perceived. If that was not dramatic enough, many leading thinkers had stripped light and color, as qualities, of all objective existence; they were neither real nor actually in the objects from which they seem to emanate. How, then, do we see them, and when we do, what do we actually see? Or, to put it in Keplerian terms, how do the mechanical effects constituting light and color penetrate "the opaque wall" of the retina to be transformed into sensations and perceptions in the optic nerves and brain? And what are they sensations and perception of?

Kepler had left it to the natural philosophers to solve this mystery, and Descartes was glad to accept the challenge. To start with, he counseled in Discourse One of the *Dioptrique*, we should think of seeing by analogy (*comparaison*) to how a blind person discerns physical objects by means of his cane, which constitutes a sixth sense of sorts. As soon as the tip of his cane encounters an object, the shock of the encounter is transmitted immediately to his hand, where the impulse is felt. By continually tapping the cane against the object, he gets an ever-clearer notion of its size, shape, distance, texture, and so forth. Now think of Ether as comparable to the cane and the light-impulses through it as the shocks of encounter. Accordingly, just as a blind person feels "the differences . . . between trees, stones, water, and the like by means of his cane," his sighted counterpart "feels" the differences "between red, yellow, green, and all the other colors" by means of the light impulses and spins reaching his retina through the Ether. In both cases, it is not the stone or the red but, rather, the shock or impulse that is felt. Consequently, there is no need "to suppose . . . that there is anything in those objects resembling the ideas or feelings we have of them" because those ideas and feelings are nothing like the impulses and spins that give rise to them. They are just responses of one kind to stimuli of another. We can therefore wipe the slate clean "of all those little images fluttering through the air, called intentional species, that so exercise the imagination of [scholastic] philosophers." Nothing passes between object and eye but impulses and spins transmitted through Ether particles.[71]

On reaching the back of the eye through the pupil, these impulses and spins in combination form a "painting [on the retina] that will represent quite artlessly in perspective all the objects lying outside." Not just an opaque screen upon which this painting is limned, though, the retina "is composed of all the endpoints of the optic nerve."[72] These endpoints belong to individual filaments that are even finer than the threads of a spider web, and a myriad of them extend from the retina through the hollow optic nerve to the brain. Thus when a particular light impulse, with its particular spin, paints a specific point on the retina, both the impulse and the spin are transmitted instantaneously through the length of the nerve filament associated with that point. Furthermore, because the hollow optic nerve is inflated with extremely fine, air-like animal spirit "arriving from the chambers or cavities that are in the brain," all the filaments passing through the nerve are kept apart rather than being squashed together. As a result, the impulse affecting any single filament affects none of the others. Transformed into individual impulses and spins in countless individual filaments, the "painting" on the retina passes instantaneously through the optic nerve in point-to-point but reversed order to the brain. Nevertheless, Descartes warned, "even though this painting, while passing thus into our head, still bears some resemblance to the objects from which it

71. For Descartes's exploitation of the cane analogy, see *Dioptrique*, 1, in AT, vol. 6, 83–86; all translated quotations from p. 85.

72. Translated quotations from *Dioptrique*, 5, in AT, vol. 6, 115.

derives," we must not think "that it is by means of this resemblance that [we] sense [those objects] as if [by] other eyes in the brain." It is the complex of movements transmitted from the retinal painting to the brain, not what the painting portrays, that arouses visual sensation, and it does so by "acting immediately on our soul."[73]

Although he barely adverted to it in the *Dioptrique*, Descartes addressed the issue of how these movements arouse sensation and perception in the soul fairly extensively in two major works published after the *Dioptrique*: *Les passions de l'âme* ("The Passions of the Soul," Paris, 1649), and *L'Homme* ("Man," Paris, 1664), which was part of *Le monde* as originally planned in 1633. The account articulated in these two treatises presupposes that the human body is an extraordinarily complex, hydraulically driven machine whose operations center is the pineal gland poised upright on a nest of arteries in the center of the brain's ventricles. Veins, arteries, and nerves are the pipes of this machine, and the physiological model Descartes adopted to explain their function is decidedly Galenic in inspiration. This is particularly evident in his account of how animal spirit is produced from "the most energetic, strongest, and finest parts" of arterial blood forced by the bellows-like heart through the carotid arteries to the brain. These arteries branch off into countless smaller tributaries surrounding the cerebral ventricles and then reconverge on the pineal gland to infuse it with animal spirit. So copious is the production of this spirit that the pineal gland expels it forcefully enough through countless minute pores to inflate not only the cerebral ventricles but also the nerves plugged into them from all parts of the body.[74]

Each of the optic nerves enters a corresponding lobe of the two lobes comprising the brain's frontmost ventricle, and the filaments within each optic nerve are connected to tiny tubes that debouch into that ventricle but are all but closed off by "valves" so as to keep the ventricle pressurized with animal spirit while also inflating the nerves. When light impulses and color spins strike the retina and create the appropriate mechanical effects in the individual filaments, the filaments respond by pulling their valves open, thus creating a disequilibrium in pressure. Each of these tiny, individual pressure changes will cause the spirit "to issue from the corresponding points (*les endroits . . . qui les*

73. On the structure of the nerves, the filaments contained by them, and their inflation with animal spirit, see *Dioptrique*, 4, in AT, vol. 6, 109–12; on Descartes's warning against thinking of the transmitted "image" as a likeness, see *Dioptrique*, 6, in AT, vol. 6, 130.

74. For Descartes's account of this physiological model, see *L'Homme*, in AT, vol. 11, 119–30 and 170–74, and *Passions*, ibid., 334–35; English quotation from Stephen Gaukroger, *Descartes: The World and Other Writings* (Cambridge: Cambridge University Press, 1998), 104.

regardent) on the [pineal] gland more freely and more rapidly than they otherwise would." Conjointly, all of these tiny pressure releases will trace a figure on the surface of the pineal gland that correlates to the set of impulses giving rise to it from the retinal painting. These and "only these [traced figures] should be taken as the forms or images which, when united to this machine, the rational soul (*l'âme raisonnable*) will consider directly when it imagines some object or senses it." In effect, then, Descartes located the faculties of common sense and imagination, as well as the point of image fusion, in the pineal gland (*le siège de l'imagination, et du sens commun*) while transforming the sensible species into an encoded tracing on its surface that represents but in no way replicates or portrays its objective source. Like words, Descartes explained at the beginning of the *Traité de la lumière*, these tracings stand for, or "signify," things without resembling them in any way.[75] In that capacity, they evoke the mental images that we assign to those things.

Since it is not my object to explicate Descartes's perceptual model in full, I will ignore his accounts of memory and bodily movement, which he also reduced to mechanical interactions in a hydraulic system. I have said enough, I think, to make it clear that although he was still operating within the framework of scholastic faculties psychology, Descartes found a coherent way to get the retinal image into the brain in the form of mechanical actions conveyed through neurological filaments. He also found a way to get that image represented in what amounts to the common sense and imagination in the form of mechanically encoded traces configured on the pineal gland, and all without recourse to intentional species or to an animate substrate that "realizes" them. Descartes's animal spirit is no more animate than fire or water. With the full array of traces from all five senses, those configurations "can give the soul occasion to sense movement, size, distance, colours, sounds, smells, and other qualities; and even things that can make it sense pleasure, pain, hunger, thirst, joy, sadness, and other such passions." But that still leaves unexplained how such encoded configurations are converted into the "forms or images" the rational soul considers when thinking. The qualities of "red," "smooth," "sweet," and "crisp" in the "apple" considered by Descartes's rational soul are neither explicitly nor implicitly in the mechanically produced traces etched on the pineal gland. Lacking any meaningful cause-effect relationship to those traces, these qualities, as well as the apple bearing them, are quite literally fig-

75. On this perceptual model, see *L'Homme* and *Passions*, in AT, vol. 11, 174–77 and 343–357, respectively; English quotations from Gaukroger, *Descartes: World*, 149. On Descartes's likening mental images to words, see *Traité*, chapter 1, in Gaukroger, *Descartes: World*, 4.

ments of the imagination. As such, they have subjective existence only. This is not to deny that there is some correlative object; it is only to deny that the object actually is as it is portrayed in the imagination. In a sense, then, both the encoded traces and their cognitive interpretations form "veils" between the perceiving subject, that is, the rational soul, and the perceived objects.

By disavowing any resemblance between external reality and its qualitative representation in the rational soul, Descartes dove headlong into the "epistemic gap" between sense perception and cognition that the perspectivists had papered over with the theory of intentional species. This gap is reflected ontologically in the distinction between incorporeal thinking substance (*res cogitans*) and corporeal extended substance (*res extensa*) so carefully drawn by Descartes in the *Meditationes de prima philosophia* ("Meditations on First Philosophy") of 1641 and recapitulated in the first part of his *Principia philosophiae* ("Principles of Philosophy") of 1644.[76] Rational soul, with its qualitative apprehension, is thinking substance; physical body, with its mechanistic functions, is extended substance, and the two coexist uneasily in all living humans. This, of course, is the basis for the mind-body dualism associated with Cartesian philosophy, but such dualism, as well as the epistemic gap resulting from it, was hardly unique to him.[77] Lurking beneath the scholastic distinction between sensitive soul and intellect, it can in fact be traced back to Plato's and Aristotle's efforts to link sense perception and cognition through the ensoulment of the body. Yet whereas scholastic theorists had made concerted efforts to render both the ontological and epistemic gaps paper thin and to bridge them with intentional species, animate spirit, and illuminative insight, Descartes did nothing to hide them, beyond perhaps adducing the pineal gland as a tertium quid between body and soul. In doing so, he effectively pushed Kepler's "opaque wall" back from the retina to the surface of that gland.

Which came first, Descartes's conception of light and color as mechanical effects in an ethereal medium or his conception of the body as a machine, is the sort of chicken-or-egg question that renders any attempt to answer it a futile exercise in intellectual tail chasing. It is fairly certain, though, that Descartes did not undertake a systematic study of anatomy and physiology until

76. On the relationship between metaphysics and physics in Descartes's *Principa philosophiae*, see Stephen Gaukroger, *Descartes' System of Natural Philosophy* (Cambridge: Cambridge University Press, 2002).

77. For a brief account of Descartes's dualism and its implications, see John Cottingham, "Cartesian Dualism: Theology, Metaphysics, and Science," in *The Cambridge Companion to Descartes*, ed. Cottingham (Cambridge: Cambridge University Press, 1992), 236–57. See also Gaukroger, *Descartes: Biography*, 269–90.

late 1629, long after he had arrived at his mature theory of light and color. It is therefore reasonable to assume that the physiological model Descartes constructed in *L'Homme*, which was originally intended as a companion piece to the *Traité de la lumière*, was tailored to his theory of light and color rather than the reverse. The particular way in which Descartes mechanized the process of sensation and perception was therefore a response to the particular way in which he mechanized light and color. Descartes, in short, followed Kepler's lead by subordinating sight theory to light theory. That helps explain his aversion to species theory, with its emphasis on abstract, formal replicas impressed in an animate, spiritual medium pervading the sense organs, nerves, and brain. It also helps explain why he adopted a theory of mental imaging that precludes any meaningful, cause-effect correspondence between objective reality and its subjective portrayal in the rational soul.

Like his theory of light and color, Descartes's theory of sensation and perception found its share of critics. Scholastic faculties psychology was by no means dead in the 1630s and 1640s, and for those who still subscribed to it, Descartes's repudiation of intentional species in favor of mechanical interactions among purely material entities was unacceptable.[78] Equally unacceptable was the lack of correspondence between physical cause and psychic effect that follows from his mechanistic approach to perception. On the other hand, Descartes's account was not radical enough for Hobbes. Committed to "a theory of perception and of primary knowledge exclusively in terms of space, time, body, and motion," he saw no need whatever to posit an incorporeal soul to perceive or think for the body.[79] In addition, Descartes was wrong on various anatomical details, most notably in locating the pineal gland in the middle of the cerebral ventricles and connecting it to a network of arteries through which it could be continually infused with inert animal spirit. Details aside, Descartes's effort to link the physics of light and color and the psychology of visual perception at the physiological level by means of Hooke's "small Machines of Nature" provided the basic model for subsequent attempts to establish this linkage. At the end of his *Opticks* of 1704, for example, Newton clearly adjusted the Cartesian physiological model to his theory of light and color. "Do not the rays of Light in falling upon the body of the Eye excite vibrations in the *Tunica Retina*?" he asks rhetorically in Query 12. And does it not follow

78. On the persistence and adaptation of species theory during the early seventeenth century, see Leen Spruit, Species Intelligibilis: *From Perception to Knowledge*, vol. 2 (Leiden: Brill, 1994–95), especially 267–351.

79. Ibid., 391.

that those "vibrations, being propagated along the solid fibres of the optick Nerves into the Brain, cause the sense of seeing?" And "do not several sort of rays make vibrations of several bignesses," he continues in Query 13, "which, according to their bignesses excite sensations of several Colours, much after the manner that the vibrations of the Air, according to their several bignesses excite sensation of several sounds?"[80]

6. CONCLUSION

By the time Newton posed these rhetorical questions, the Keplerian turn was virtually complete. The rupture between light theory and sight theory implicit in Kepler's account of retinal imaging was now absolutely explicit, and Descartes was pivotal in making it so by locating it at the juncture between body and mind. In physical optics, it was he who set things afoot by developing a relatively coherent mechanical model of light and color that reduced them to impulses and spins transmitted through ether. This move effectively stripped light and color, as well as transparency, of all qualitative existence outside the perceiver. That, in turn, foreclosed any hope of bridging the gap between physical cause and visual effect through similitude, as the perspectivists had done. Suitably modified, moreover, Descartes's mechanistic approach to light and color was applied successfully to a wide variety of optical phenomena that would have resisted the qualitative approach of perspectivist optics. It also brought to light a number of optical phenomena that had gone unnoticed before 1600. Self-reinforcing, the continued success of this approach encouraged a range of optical thinkers from Descartes onward to concentrate on physical questions with little or no regard for the perceptual consequences of their theoretical responses. Not that the eye was dropped out of the account entirely, especially in the explanation of how telescopes and microscopes work. But in that context the eye served only as a reference point subsidiary to the focal points determining the real or virtual images presented to it. With or without the eye to see them, those images are somehow "there" in the optical functioning of the instruments.

Along with the mechanization of light and color, the mechanization of sensation and perception that followed suit at the physiological level left the perspectivist paradigm hanging by nothing but a few terminological threads. Descartes still appealed to the faculties of common sense and imagination, but his

80. *Opticks*, Queries 12 and 13, pp. 135–36.

conception of the two was radically different from that of the perspectivists.[81] The same holds for his inanimate version of animal spirit and its hydraulic functions. Newton, as well, used the vocabulary of "species" to denominate what passes through the optic nerves to the brain in sensation, but he certainly did not think of them as formal entities portraying their objects in true qualitative detail.[82] Still, no matter how far into the brain the mechanisms of sensation and perception are made to reach, there comes a point at which any account based on such mechanisms fails to explain the intuitive experience of perception. When we perceive "red," "smooth," "sweet," or "crisp," we perceive them not as mechanical effects but as real, inherent properties of the specific objects they qualify and define. Descartes addressed this problem by lodging such perceptual experience in an incorporeal mind somehow conjoined with the cerebral machine and thus able to read its encodings. In adducing this entity, however, Descartes and nearly all of his seventeenth-century successors managed only to transpose Kepler's opaque wall from the retina to that mysterious place where mind and brain supposedly meet.

81. For a discussion of the development of psychological theory after Descartes, see Fernando Vidal, *The Sciences of the Soul* (Chicago: University of Chicago Press, 2011); see also Gary Hatfield, "The Cognitive Faculties," in *The Cambridge History of Seventeenth-Century Philosophy*, ed. Daniel Garber and Michael Ayers (Cambridge: Cambridge University Press, 1998), 953–1002.

82. See *Opticks*, Query 15, pp. 136–37.

BIBLIOGRAPHY

Acerbi, Fabio. "Damianus of Larissa." *Complete Dictionary of Scientific Biography*. 2008. Encyclopedia.com. http://www.encyclopedia.com/doc/1G2-2830905607.html (accessed May 30, 2011).

Ackerman, James. "Leonardo's Eye." *Journal of the Warburg and Courtauld Institutes* 41 (1978): 108–46.

Adam, Charles, and Paul Tannery, eds. *Oeuvres de Descartes*. 13 vols. Paris: Vrin, 1897–1913.

Adamson, Peter. *Alkindi*. Oxford: Oxford University Press, 2007.

———. "Vision, Light, and Color in al-Kindī, Ptolemy, and the Ancient Commentators." *Arabic Sciences and Philosophy* 16 (2006): 207–35.

Adamson, Peter, and Richard Taylor, eds. *The Cambridge Companion to Arabic Philosophy*. Cambridge: Cambridge University Press, 2005.

Aertsen, Jan A., Kent Emory Jr., and Andreas Speer, eds. *Nach der Verurteilung von 1277*. Berlin: Walter de Gruyter, 2001.

Aït-Touati, Frédérique. *Fictions of the Cosmos*. Chicago: University of Chicago Press, 2011.

Akbari, Suzanne Conklin. *Seeing through the Veil*. Toronto: University of Toronto Press, 2004.

Akdogan, Cemil, ed. and trans. *Albert's Refutation of the Extramission Theory of Vision and His Defence of the Intromission Theory*. Kuala Lumpur: International Institute of Islamic Thought and Civilization, 1998.

Albert the Great. *De sensu*. See Akdogan.

Alberti, Leon Battista. *On Painting*. See Sinisgalli.

Alexander of Aphrodisias. *De anima*. See Fotinis.

al-Farabi. *On the Intellect*. See McGinnis.

———. *The Perfect State*. See Walzer.

Al-Khalili, Jim. *The House of Wisdom: How Arabic Science Saved Ancient Knowledge and Gave Us the Renaissance*. London: Penguin, 2011.

Alonso, Manuel. "Notas sobre los traductores Toledanos Domingo Gundisalvo y Juan Hispano." In *Dominicus Gundissalinus (12th c.) and the Transmission of Arabic Philosophical Thought to the West*, edited by Fuat Sezgin, 23–56. Frankfurt: Institute for the History of Arabic-Islamic Science at the Johann Wolfgang Goethe University, 2000.

Alpers, Svetlana. *The Art of Describing*. Chicago: University of Chicago Press, 1983.

Alverny, Marie-Thérèse d'. *Avicenne en Occident*. Paris: Vrin, 1993.

———. "Translations and Translators." In *Renaissance and Renewal in the Twelfth Century*, edited by Robert Benson and Giles Constable, 421–62. Cambridge, MA: Harvard University Press, 1982.

Alverny, Marie-Thérèse d', and Françoise Hudry. "Al-Kindi: *De radiis.*" *Archives d'histoire doctrinale et littéraire du Moyen Âge* 41 (1974): 139–260.

Anselm of Canterbury. *On Truth*. See McInerny. See also Schmitt.

Apuleius. *Apologia*. See Butler and Hunink.

Aristotle. *Complete Works*. See Barnes.

———. *De anima*. See J. A. Smith.

———. *De memoria et reminiscentia*. See Beare.

———. *De sensu*. See Beare.

———. *Meteorology*. See Webster.

———. *Physics*. See Hardie.

Armstrong, A. H., trans. *Plotinus*. 5 vols. Cambridge, MA: Harvard University Press, 1966–84.

Arnzen, Rüdiger, and Jörn Thielmann, eds. *Words, Texts, and Concepts Cruising the Mediterranean Sea*. Louvain; Dudley, MA: Peeters, 2004.

Augustine. *Against the Academicians*. See King.

———. *The Greatness of the Soul (De quantitate animae)*. See Colleran.

———. *On Free Will*. See Williams.

———. *On Genesis Interpreted Literally*. See Hill.

———. *On the Magnitude of the Soul*. See Colleran.

———. *On the Teacher*. See King.

———. *On the Trinity*. See McKenna.

———. *The Teacher (De Magistro)*. See Colleran.

Averroes. *Long Commentary on the De Anima*. See Taylor.

Avicenna. *De anima seu Sextus de naturalibus*. See Riet.

———. *Healing (Shifā)*. See Bakoš.

———. *Physics of the Healing*. See McGinnis.

———. *Salvation (Najat)*. See Rahman.

Bacon, Roger. *De multiplicatione specierum*. See Lindberg.

———. *De speculis comburentibus*. See Lindberg.

———. *Opus Majus*. See Burke.

———. *Perspectiva*. See Lindberg.

Bakoš, Ján, trans. *Psychologie d'Ibn Sīnā (Avicenne) d'après son oeuvre Aš-šifāʾ*. Prague: Czech Academy of Sciences, 1956.

Bala, Arun. *The Dialogue of Civilizations in the Birth of Modern Science*. New York: Palgrave Macmillan, 2006.

Baldwin, John W. *The Scholastic Culture of the Middle Ages*. Lexington, MA: D. C. Heath, 1971.

Barnes, Jonathan, ed. *The Complete Works of Aristotle*. 2 vols. Princeton, NJ: Princeton University Press, 1984.

Barney, Stephen A., and others, trans. *The* Etymologies *of Isidore of Seville*. Cambridge: Cambridge University Press, 2006.

Barrow, Issac. *Lectiones Opticae XVIII*. See Fay.

Bartholomaeus Anglicus. *De proprietatibus rerum*. See Long.

Baur, Ludwig, ed. *Die Philosophischen Werke des Robert Grosseteste, Bischofs von Lincoln*. Münster: Aschendorff, 1912.

Beare, J. I., trans. *De memoria et reminiscentia*. In *The Complete Works of Aristotle*. Vol. 1, edited by Jonathan Barnes, 714–20. Princeton, NJ: Princeton University Press, 1984.

———. *De sensu*. In *The Complete Works of Aristotle*. Vol. 1, edited by Jonathan Barnes, 693–713. Princeton, NJ: Princeton University Press, 1984.

Beierwaltes, Werner, ed. *Eriugena Redivivus*. Heidelberg: Carl Winter, 1987.

Belting, Hans. *Baghdad and Florence: Renaissance Art and Arab Science*. Translated by Deborah Lukas Schneider. Cambridge, MA: Harvard University Press, 2011.

Benson, J. L. *Greek Color Theory and the Four Elements*. Amherst, MA: ScholarWorks@UMassAmherst. 2000. http://scholarworks.umass.edu/art_jbgc/1/.

Benson, Robert, and Giles Constable, eds. *Renaissance and Renewal in the Twelfth Century*. Cambridge, MA: Harvard University Press, 1982.

Beretta, Marco. *When Glass Matters*. Florence: Leo S. Olschki, 2004.

Beretta, Marco, Antonio Clericuzio, and Lawrence M. Principe, eds. *The Accademia del Cimento and Its European Context*. Sagamore Beach, MA: Science History Publications, 2009.

Berger, Albert Maria, ed. and trans. *Die Ophthalmologie* (Liber de oculo) *des Petrus Hispanus*. Munich: J. F Lehmann, 1899.

Bernard of Chartres. *Glosae super Platonem*. See Dutton.

Bernard Silvestris. *Cosmographia*. See Dronke.

Birch, Thomas, ed. *The History of the Royal Society*. 4 vols. New York: Johnson Reprint, 1968.

Björnbo, Axel, and Sebastian Vogl. *Alkindi, Tideus und Pseudo-Euclid: Drei optische Werke*. Leipzig: Teubner, 1912.

Black, Deborah. "Avicenna's 'Vague Individual' and Its Impact on Medieval Latin Philosophy." In *Vehicles of Transmission, Translation, and Transformation in Medieval Textual Culture*, edited by Robert Wisnovsky, Faith Wallis, Jamie Furno, and Carlos Fraenkel, 259–92. Turnhout: Brepols, 2012.

———. "Logic in Islamic Philosophy." In *Routledge Encyclopedia of Philosophy*. Vol. 5, edited by Edward Craig, 706–13. London; New York: Routledge, 1998.

Blair, Ann M. *Too Much to Know*. New Haven, CT: Yale University Press, 2010.

Bloch, Herbert. *Monte Cassino in the Middle Ages*. Cambridge, MA: Harvard University Press, 1986.

Bloom, Jonathan. *Paper before Print: The History and Impact of Paper in the Islamic World*. New Haven, CT: Yale University Press, 2001.

Blumenthal, Henry J. *Aristotle and Neoplatonism in Late Antiquity*. Ithaca, NY: Cornell University Press, 1990.

———. "Neoplatonic Elements in the *De Anima* Commentaries." In *Aristotle Transformed*, edited by Richard Sorabji, 305–24. Ithaca, NY: Cornell University Press, 1990.

———, trans. *"Simplicius": On Aristotle's* "On the Soul." Ithaca, NY: Cornell University Press, 2000.

———. "Themistius: The Last Peripatetic Commentator on Aristotle?" In *Aristotle Transformed*, edited by Richard Sorabji, 113–23. Ithaca, NY: Cornell University Press, 1990.

Bondanella, Julia, and Peter Bondanella, trans. *Giorgio Vasari*: The Lives of the Artists. Oxford: Oxford University Press, 1991.

Bowen, Alan C., and Robert B. Todd, trans. *Cleomedes' Lectures on Astronomy: A Translation of* The Heavens. Berkeley: University of California Press, 2004.

Boyer, Carl B. *The Rainbow: From Myth to Mathematics*. New York: Yoseloff, 1959.

Bradbury, Savile. *The Evolution of the Microscope*. Oxford: Pergamon Press, 1976.

Briggs, Charles F. "Literacy, Reading, and Writing in the Medieval West." *Journal of Medieval History* 26 (2000): 397–420.

Brown, Peter. *Chaucer and the Making of Optical Space*. Oxford: Peter Lang, 2007.

Brown, Peter Robert Lamont. *Augustine of Hippo: A Biography*. 2nd ed. Berkeley: University of California Press, 2000.

Buringh, Eltjo. *Medieval Manuscript Production in the Latin West: Explorations with a Global Database*. Leiden: Brill, 2011.

Burke, Robert B., trans. *The* Opus Majus *of Roger Bacon*. Philadelphia: University of Pennsylvania Press, 1928.

Burnett, Charles. "Arabic Philosophical Works Translated into Latin." In *The Cambridge History of Medieval Philosophy*. Vol. 2, edited by Robert Pasnau and Christina Van Dyke, 814–22. Cambridge: Cambridge University Press, 2010.

———. "The Coherence of the Arabic-Latin Translation Program in Toledo in the Twelfth Century." *Science in Context* 14 (2001): 249–88.

Burnett, Charles, and Danielle Jacquart, eds. *Constantine the African and 'Alī ibn al-'Abbās al-Maǧūsī: The* Pantegni *and Related Texts*. Leiden: Brill, 1994.

Burnett, D. Graham. *Descartes and the Hyperbolic Quest*. Transactions of the American Philosophical Society 95.3. Philadelphia: American Philosophical Society Press, 2005.

Burnyeat, Miles. "Archytas and Optics." *Science in Context* 18 (2005): 35–53.

Burton, Dan, ed. and trans. *Nicole Oresme's* De visione stellarum. Leiden: Brill, 2007.

Burton, Harry E., trans. "The Optics of Euclid." *Journal of the Optical Society of America* 35 (1945): 357–72.

Butler, H. E., trans. *The* Apologia *and* Florida *of Apuleius of Madaura*. Oxford: Clarendon Press, 1909.

Buttimer, Charles. *Hugonis de Sancto Victore* Didascalicon. Washington, DC: Catholic University Press, 1939.

Calcidius. *Timaeus*. See Waszink.

Callus, D. A., and R. W. Hunt, eds. *Johannes Blund*, Tractatus de anima. London: Oxford University Press, 1970.

Carabine, Deirdre. *John Scottus Eriugena*. New York: Oxford University Press, 2000.

Caston, Victor. "Aristotle's Psychology." In *A Companion to Ancient Philosophy*, edited by Mary Louise Gill and Pierre Pellegrin, 316–46. Malden, MA; Oxford: Blackwell, 2006.

————. "Aristotle's Two Intellects: A Modest Proposal." *Phronesis* 44 (1999): 199–227.

Charlton, William, trans. *Philoponus: On Aristotle's "On the Soul 2.7–12."* Ithaca, NY: Cornell University Press, 2005.

Clagett, Marshall. "William of Moerbeke: Translator of Archimedes." *Proceedings of the American Philosophical Society* 126 (1982): 356–66.

Clanchy, M. T. *From Memory to Written Record: England, 1066–1307.* Cambridge, MA: Harvard University Press, 1979.

Clark, David. "Optics for Preachers: The *De oculo morali* of Peter of Limoges." *Michigan Academician* 9 (1977): 329–43.

Clark, Gordon H. "Plotinus' Theory of Sensation." *Philosophical Review* 51 (1942): 357–82.

Clark, Kenneth. *The Drawings of Leonardo da Vinci in the Collection of Her Majesty the Queen at Windsor Castle.* 3 vols. 2nd ed. London: Phaidon, 1968–69.

Clark, Stuart. *Vanities of the Eye: Vision in Early Modern European Culture.* Oxford: Oxford University Press, 2007.

Clarke, Edwin, and Kenneth Dewhurst. *An Illustrated History of Brain Function.* Berkeley: University of California Press, 1972.

Cleomedes. *The Heavens.* See Bowen.

Colish, Marsha. *The Stoic Tradition from Antiquity to the Early Middle Ages.* Leiden: Brill, 1985.

Colleran, Joseph, trans. *Saint Augustine*: The Greatness of the Soul; The Teacher. Westminster, MD: Newman Press, 1950.

Colombo, Realdo. *De re anatomica libri XV.* Venice, 1559.

Corcoran, Thomas H., trans. *Seneca VII*: Naturales Quaestiones I. Cambridge, MA: Harvard University Press, 1971.

Cornford, Francis, trans. *The* Timaeus *of Plato.* New York: Liberal Arts Press, 1957.

Cottingham, John, ed. *The Cambridge Companion to Descartes.* Cambridge: Cambridge University Press, 1992.

————. "Cartesian Dualism: Theology, Metaphysics, and Science." In *The Cambridge Companion to Descartes*, edited by John Cottingham, 236–57. Cambridge: Cambridge University Press, 1992.

Courtenay, William. *Covenant and Causality in Medieval Thought.* London: Variorum Reprints, 1984.

Craig, Edward, ed. *Routledge Encyclopedia of Philosophy.* 10 vols. London: Routledge, 1998.

Crew, Henry, trans. *The* Photismi De Lumine *of Maurolycus: A Chapter in Late Medieval Optics.* New York: Macmillan, 1940.

Crombie, Alistair. *Science, Optics, and Music in Medieval and Early Modern Thought.* London: Hambledon Press, 1990.

Cunningham, Andrew, and Sachiko Kusukawa, trans. *Natural Philosophy Epitomised: Books 8–11 of Gregor Reisch's* Philosophical Pearl (1503). Farnham, UK: Ashgate, 2010.

Curr, Matthew, and Italo Ronca, trans. *A Dialogue on Natural Philosophy* (Dragmaticon Philosophiae), *William of Conches.* Notre Dame, IN: University of Notre Dame Press, 1997.

D'Ancona, Cristina. "Greek into Arabic: Neoplatonism in Translation." In *The Cambridge Companion to Arabic Philosophy*, edited by Peter Adamson and Richard Taylor, 10–31. Cambridge: Cambridge University Press, 2005.

Dahlberg, Charles, trans. *The Romance of the Rose*. Hanover, NH: University Press of New England, 1983.

Daley, Walter J., and Robert D. Yee, trans. "The Eye Book of Master Peter of Spain—A Glimpse of Diagnosis and Treatment of Eye Disease in the Middle Ages." *Documenta Ophthalmologica* 103 (2001): 119–53.

Damianos. *Optics*. See Schöne.

Dante Alighieri. *The Divine Comedy: Purgatorio*. See Singleton.

Darrigol, Olivier. *A History of Optics from Greek Antiquity to the Nineteenth Century*. Oxford: Oxford University Press, 2012.

Davidson, Herbert. "Averroes and Narboni on the Material Intellect." *Association for Jewish Studies Review* (1984): 175–84.

Davies, Brian, and Gillian Evans, eds. *Anselm of Canterbury: The Major Works*. Oxford: Oxford University Press, 1998.

Davies, Sir John. *Nosce teipsum*. In *The Complete Poems of Sir John Davies*, edited by Alexander Grosart, vol. 1, 15–116. London: Chatto and Windus, 1876.

Dear, Peter. *Revolutionizing the Sciences*. Princeton, NJ: Princeton University Press, 2001.

De Groot, Jean. *Aristotle and Philoponus on Light*. New York: Garland, 1991.

De Lacy, Phillip, trans. *Galen: On the Doctrines of Hippocrates and Plato*. Berlin: Akademie Verlag, 1980.

Delia, Diana. "From Romance to Rhetoric: The Alexandrian Library in Classical and Islamic Traditions." *American Historical Review* 97 (1992): 1449–67.

Della Porta, Giambattista. *De refractione*. Naples, 1593.

———. *Magiae Naturalis Libri XX*. Naples, 1589.

———. *Magiae Naturalis sive de Miraculis Naturalium Libri IIII*. Naples, 1558.

———. *Natural Magick*. London: Thomas Young and Samuel Speed, 1658.

Denery, Dallas. *Seeing and Being Seen in the Later Medieval World*. New York: Cambridge University Press, 2005.

Descartes. *Dioptrique*. In *Oeuvres de Descartes*. Vol. 6, edited by Charles Adam and Paul Tannery, 81–227. Paris: Leopold Cerf, 1902.

———. *L'Homme*. In *Oeuvres de Descartes*. Vol. 11, edited by Charles Adam and Paul Tannery, 119–202. Paris: Leopold Cerf, 1909.

———. *Météores*. In *Oeuvres de Descartes*. Vol. 6, edited by Charles Adam and Paul Tannery, 231–366. Paris: Leopold Cerf, 1902.

———. *Le Monde; ou, Traité de la lumière* (1664). In *Oeuvres de Descartes*. Vol. 11, edited by Charles Adam and Paul Tannery, 3–118. Paris: Leopold Cerf, 1902.

———. *Passions de l'âme*. In *Oeuvres de Descartes*. Vol. 11, edited by Charles Adam and Paul, 327–488. Paris: Leopold Cerf, 1909.

Dijksterhuis, E. J. *Archimedes*. Princeton, NJ: Princeton University Press, 1987.

Dijksterhuis, Fokko Jan. *Lenses and Waves*. Dordrecht: Kluwer, 2004.

Diocles. *On Burning Mirrors*. See Toomer.

Diogenes Laertius. *Lives of the Eminent Philosophers*. See Hicks.

Dodge, Bayard, ed. and trans. *The* Fihrist *of al-Nadim*. New York: Columbia University Press, 1970.

Donahue, William, trans. *Johannes Kepler, Optics: Paralipomena to Witelo and Optical Part of Astronomy.* Santa Fe, NM: Green Lion Press, 2000.

Drake, Stillman, trans. *Galileo*: Dialogue concerning the Two Chief World Systems. Berkeley: University of California Press, 1967.

Drake, Stillman, and C. D. O'Malley, trans. *The Controversy on the Comets of 1618.* Philadelphia: University of Pennsylvania Press, 1960.

Dronke, Peter, ed. *Bernardus Silvestris* Cosmographia. Leiden: Brill, 1978.

Dupré, Sven. "Ausonio's Mirrors and Galileo's Lenses: The Telescope and Sixteenth-Century Practical Optical Knowledge." *Galileiana* 2 (2005): 145–80.

———. "Kepler's Optics without Hypotheses." *Synthese* 185 (2012): 501–25.

———. "Mathematical Instruments and the Theory of the Concave Spherical Mirror: Galileo's Optics beyond Art and Science." *Nuncius* 15 (2000): 551–88.

———. "Optics, Pictures, and Evidence: Leonardo's Drawings of Mirrors and Machines." *Early Science and Medicine* 10 (2005): 211–36.

———. "Playing with Images in a Dark Room: Kepler's *Ludi* inside the Camera Obscura." *Early Science and Medicine* 13 (2009): 219–44.

———. "William Bournes' Invention: Projecting a Telescope and Optical Speculation in Elizabethan England." In *The Origins of the Telescope*, edited by Albert Van Helden and others, 129–45. Amsterdam: KNAW Press, 2010.

Dürer, Albrecht. See Kurth.

Dutton, Paul Edward, ed. *The* Glosae super Platonem *of Bernard of Chartres*. Toronto: Pontifical Institute of Mediaeval Studies, 1991.

Eamon, William. *Science and the Secrets of Nature.* Princeton, NJ: Princeton University Press, 1994.

Eastwood, Bruce. "Al-Farabi on Extramission, Intromission, and the Use of Platonic Visual Theory." *Isis* 70 (1979): 423–25.

———. *The Elements of Vision: The Micro-Cosmology of Galenic Visual Theory According to Ḥunayn Ibn 'Isḥāq.* Transactions of the American Philosophical Society 72.5. Philadelphia: American Philosophical Society Press, 1982.

———. "Grosseteste's 'Quantitative' Law of Refraction: A Chapter in the History of Non-Experimental Science." *Journal of the History of Ideas* 28 (1967): 404–14.

Edgerton, Samuel Y. *The Heritage of Giotto's Geometry.* Ithaca, NY: Cornell University Press, 1991.

———. *The Mirror, the Window, and the Telescope.* Ithaca, NY: Cornell University Press, 2009.

Eijk, Philip J. van der, trans. *Philoponus*: On Aristotle's "On the Soul 1.3–5." Ithaca, NY: Cornell University Press, 2006.

Eisenstein, Elizabeth. *The Printing Press as an Agent of Change: Communications and Cultural Transformations in Early Modern Europe.* Cambridge: Cambridge University Press, 1979.

———. *The Printing Revolution in Early Modern Europe.* Cambridge: Cambridge University Press, 1983.

Elkins, James. *The Poetics of Perspective.* Ithaca, NY: Cornell University Press, 1994.

Elshakry, Marwa. "When Science Became Western: Historiographical Reflections." *Isis* 101 (2010): 98–109.

Endress, Gerhard. "The Circle of al-Kindī: Early Arabic Translations from the Greek and the Rise of Islamic Philosophy." In *The Ancient Tradition in Christian and Islamic Hellenism*, edited by Gerhard Endress and Remke Kruk, 43–76. Leiden: Research School CNWS, 1997.

Endress, Gerhard, and Remke Kruk, eds. *The Ancient Tradition in Christian and Islamic Hellenism*. Leiden: Research School CNWS, 1997.

Enoch, Jay M. "The Enigma of Early Lens Use." *Technology and Culture* 39 (1988): 273–91.

Erigena, Johannes Scotus. *De Divisione Naturae*. See Jeauneau.

Euclid. *Elements*. See Heath.

———. *Opera Omnia*. See Heiberg.

———. *Optics*. See Burton and Ver Eecke.

Eyjólfur Kjalar Emilsson. *Plotinus on Intellect*. Oxford: Clarendon Press, 2007.

———. *Plotinus on Sense-Perception: A Philosophical Study*. Cambridge: Cambridge University Press, 1988.

Fay, H. C., trans. *Isaac Barrow's* Optical Lectures. London: Worshipful Company of Spectacle Makers, 1987.

Finamore, John, and John M. Dillon, trans. *Iamblichus* De anima. Leiden: Brill, 2002.

Fiori, Cesare, and Mariangela Vandini. "Chemical Composition of Glass and Its Raw Materials: Chronological and Geographical Development in the First Millennium A.D." In *When Glass Matters*, edited by Marco Beretta, 151–94. Florence: Leo S. Olschki, 2004.

Fisher, Saul. *Pierre Gassendi's Philosophy and Science: Atomism for Empiricists*. Leiden: Brill, 2005.

Fotinis, Athanasius, trans. *The* De anima *of Alexander of Aphrodisias*. Washington, DC: University Press of America, 1979.

Frymire, John. *The Primacy of the Postils: Catholics, Protestants, and the Dissemination of Ideas in Early Modern Germany*. Leiden: Brill, 2010.

Gal, Ofer, and Raz Chen-Morris. *Baroque Science*. Chicago: University of Chicago Press, 2013.

Galen. *De anatomicis administrationibus*. See Singer.

———. *De placitis Hippocratis et Platonis*. See De Lacy.

———. *De usu partium*. See May.

Galilei, Galileo. *Assayer*. See Drake.

———. *Dialogue*. See Drake.

———. *On Sunspots*. See Reeves.

———. *Sidereus Nuncius*. See Van Helden.

Garber, Daniel, and Michael Ayers, eds. *The Cambridge History of Seventeenth-Century Philosophy*. Cambridge: Cambridge University Press, 1998.

Gascoigne, John. "A Reappraisal of the Role of the Universities in the Scientific Revolution." In *Reappraisals of the Scientific Revolution*, edited by David Lindberg and Robert Westman, 207–60. Cambridge: Cambridge University Press, 1990.

Gaukroger, Stephen. *Descartes: An Intellectual Biography*. Oxford: Clarendon Press, 1995.

———. *Descartes' System of Natural Philosophy*. Cambridge: Cambridge University Press, 2002.

———. *Descartes*: The World and Other Writings. Cambridge: Cambridge University Press, 1998.

Gauthier, René, ed. *Sentencia libri de anima*. Paris: Vrin, 1984.

———. "Le Traité *De anima et de potenciis eius* d'un maître ès Arts (vers 1225)." *Revue des sciences philosophiques at théologiques* 66 (1982): 3–55.

Gersh, Stephen, and Maarten Hoenen, eds. *The Platonic Tradition in the Middle Ages: A Doxographic Approach*. Berlin: Walter de Gruyter, 2002.

Gerson, Lloyd. *Aristotle and Other Platonists*. Ithaca, NY: Cornell University Press, 2005.

———, ed. *The Cambridge History of Philosophy in Late Antiquity*. 2 vols. Cambridge: Cambridge University Press, 2010.

Gill, Mary Louise, and Pierre Pellegrin, eds. *A Companion to Ancient Philosophy*. Malden, MA; Oxford: Blackwell, 2006.

Gilson, Étienne. *The Christian Philosophy of Saint Augustine*. Translated by L. E. M. Lynch. New York: Random House, 1960.

———. *Reason and Revelation in the Middle Ages*. New York: Scribner's Sons, 1938.

———. "Les sources gréco-arabes de l'augustinisme avicennisant." *Archives d'histoire doctrinale et littéraire du Moyen-Âge* 4 (1930): 5–149.

Gilson, Simon. "Dante and the Science of 'Perspective': A Reappraisal." *Dante Studies* 115 (1997): 185–219.

———. *Medieval Optics and Theories of Light in the Works of Dante*. Lewiston, NY: Edward Mellen, 2000.

Gingerich, Owen. *The Book Nobody Read: Chasing the Revolutions of Nicolaus Copernicus*. 2nd ed. London: Penguin, 2005.

———. "Review of *Islamic Science and the Making of the European Renaissance*, by George Saliba." *Journal of Interdisciplinary History* 39 (2008): 310–11.

Glick, Thomas. "Science in Medieval Spain: The Jewish Contribution in the Context of *Convivencia*." In *Convivencia: Jews, Muslims, and Christians in Medieval Spain*, edited by Vivian Mann, Thomas Glick, and Jerrilynn Dodds, 83–111. New York: Georges Braziller and the Jewish Museum, 1992.

Gouguenheim, Sylvain. *Aristote au Mont-Saint-Michel*. Paris: Seuil, 2008.

Goulding, Robert. "Thomas Harriot's Optics, between Experiment and Imagination: The Case of Mr. Bulkeley's Glass." *Archive for History of Exact Science* 67 (2013): DOI: 10.1007/s00407-013-0125-1.

Grant, Edward. "Reflections of a Troglodyte Historian of Science." *Osiris* 27 (2012): 133–55.

———. "Science and Theology in the Middle Ages." In *God and Nature: Historical Essays on the Encounter between Christianity and Science*, edited by David Lindberg and Ronald Numbers, 49–75. Berkeley: University of California Press, 1986.

———, ed. *A Source Book in Medieval Science*. Cambridge, MA: Harvard University Press, 1974.

Grant, Edward, and John Murdoch, eds. *Mathematics and Its Applications to Science and Natural Philosophy in the Middle Ages*. Cambridge, New York: Cambridge University Press, 1987.

Grassus, Benvenutus. *De oculis eorumque egritudinibus et curis*. See Wood.

Grosart, Alexander, ed. *The Complete Poems of Sir John Davies*. 2 vols. London: Chatto and Windus, 1876.

Guillaume de Lorris. *Romance of the Rose*. See Dahlberg.

Gutas, Dimitri. "Avicenna's Marginal Glosses on *De anima* and the Greek Commentatorial Tradition." *Bulletin of the Institute of Classical Studies* 47 (2004): 77–88.

———. "Geometry and the Rebirth of Philosophy in Arabic with al-Kindī." In *Words, Texts, and Concepts Cruising the Mediterranean Sea*, edited by Arnzen Rüdiger and Jörn Thielmann, 195–209. Louvain; Dudley, MA: Peeters, 2004.

———. "Greek Philosophical Works Translated in Arabic." In *The Cambridge History of Medieval Philosophy*. Vol. 2, edited by Robert Pasnau and Christina Van Dyke, 802–14. Cambridge: Cambridge University Press, 2010.

———. *Greek Thought, Arabic Culture*. London: Routledge, 1998.

———. "Origins in Baghdad." In *The Cambridge History of Medieval Philosophy*. Vol. 1, edited by Robert Pasnau and Christina Van Dyke, 11–25. Cambridge: Cambridge University Press, 2010.

Hackett, Jeremiah. *Roger Bacon and the Sciences*. Leiden: Brill, 1997.

———. "Roger Bacon: His Life, Career, and Works." In *Roger Bacon and the Sciences*, edited by Jeremiah Hackett, 9–23. Leiden: Brill, 1997.

———. "Roger Bacon on *Scientia Experimentalis*." In *Roger Bacon and the Sciences*, edited by Jeremiah Hackett, 277–315. Leiden: Brill, 1997.

Hall, Robert E. "Intellect, Soul, and Body in Ibn Sīnā." In *Interpreting Avicenna: Science and Philosophy in Islam*, edited by Jon McGinnis, 62–86. Leiden: Brill, 2004.

Hankins, James. "The Myth of the Platonic Academy of Florence." *Renaissance Quarterly* 44 (1991): 429–75.

Hardie, R. P., and R. K. Gaye, trans. *Physics*. In *The Complete Works of Aristotle*. Vol. 1, edited by Jonathan Barnes, 315–446. Princeton, NJ: Princeton University Press, 1984.

Haskins, Charles Homer. *The Renaissance of the Twelfth Century*. Cambridge, MA: Harvard University Press, 1927.

———. *Studies in the History of Mediaeval Science*. Cambridge, MA: Harvard University Press, 1927.

Hasse, Dag. *Avicenna's* De anima *in the Latin West*. London: Warburg Institute, 2000.

Hatch, Robert A. "Between Erudition and Science: The Archive and Correspondence Network of Ismaël Boulliau." In *Archives of the Scientific Revolution: The Formation and Exchange of Ideas in Seventeenth-Century Europe*, edited by Michael Hunter, 49–71. Woodbridge, UK: Boydell, 1998.

———. "Coherence, Correspondence, and Choice: Gassendi and Boulliau on Light and Vision." In *Actes du Colloque International Pierre Gassendi*, 365–85. Digne-les-Bains: Société Scientifique et Littéraire des Alpes de Haute-Provence, 1994.

———. "The Republic of Letters: Boulliau, Leopoldo, and the Accademia del Cimento." In *The Accademia del Cimento and Its European Context*, edited by Marco Beretta, Antonio Clericuzio, and Lawrence M. Principe, 165–80. Sagamore Beach, MA: Science History Publications, 2009.

Hatfield, Gary. "The Cognitive Faculties." In *The Cambridge History of Seventeenth-Century*

Philosophy, edited by Daniel Garber and Michael Ayers, 953–1002. Cambridge: Cambridge University Press, 1998.

Heath, Thomas L., trans. *The Thirteen Books of Euclid's Elements*. 2nd ed. New York: Dover, 1956.

Heiberg, I. L., ed. and trans. *Euclidis Opera Omnia*. 9 vols. Leipzig: Teubner, 1985.

Heiberg, I. L., and Emil Wiedemann. "Ibn al-Haitams Schrift über parabolische Hohlspiegel." *Bibliotheca Mathematica* 10, series 3 (1909–10): 201–37.

Henderson, Julian. "Tradition and Experiment in First Millennium A.D. Glass Production— The Emergence of Early Islamic Glass Technology in Late Antiquity." *Accounts of Chemical Research* 35 (2002): 594–602.

Henry, John. *The Scientific Revolution and the Origins of Modern Science*. New York: Palgrave Macmillan, 2008.

Hero of Alexandria. *Opera quae supersunt omnia*. See Schmidt.

Herrin, Judith. *The Formation of Christendom*. Princeton, NJ: Princeton University Press, 1987.

Hicks, R. D., trans. *Diogenes Laertius: Lives of the Eminent Philosophers*. 2 vols. Cambridge, MA: Harvard University Press, 1959.

Hill, Edmond, trans. *On Genesis*. New York: New City Press, 2002.

Hockney, David. *Secret Knowledge: Rediscovering the Lost Techniques of Old Masters*. New York: Viking Studio, 2001.

Hogendijk, Jan, and Abdelhamid Sabra, eds. *The Enterprise of Science in Islam*. Cambridge, MA: MIT Press, 2005.

Hon, Giora, and Yaakov Zik. "Galileo's Knowledge of Optics and the Functioning of the Telescope." arXiv.org. http://arxiv.org/abs/1307.4963, July 18, 2013 (accessed July 23, 2013).

———. "Kepler's *Optical Part of Astronomy* (1604): Introducing the Ecliptic Instrument." *Perspectives on Science* 17 (2009): 307–45.

———. "Magnification: How to Turn a Spyglass into an Astronomical Telescope." *Archive for History of Exact Sciences* 66 (2012): 439–64.

Hooke, Robert. *Micrographia*. London, 1665.

Hudson, Anne, and Michael Wilks, eds. *From Ockham to Wyclif*. Oxford: Blackwell, 1987.

Huff, Toby. "Review of *Islamic Science and the Making of the European Renaissance*, by George Saliba." *Middle East Quarterly* 15 (2008): 77–79.

———. *The Rise of Early Modern Science: Islam, China, and the West*. 2nd ed. Cambridge: Cambridge University Press, 2003.

Hugh of St. Victor. *Didascalicon philosophiae*. See Buttimer.

Ḥunayn ibn 'Isḥāq. *The Ten Treatises On the Eye*. See Meyerhof.

Hunink, Vincent, trans. *Apuleius of Madauros* Pro Se De Magia (Apologia). Amsterdam: J. C. Gieben, 1997.

Hunter, Michael, ed. *Archives of the Scientific Revolution: The Formation and Exchange of Ideas in Seventeenth-Century Europe*. Woodbridge, UK: Boydell Press, 1998.

Huxley, George L. *Anthemius of Tralles: A Study in Later Greek Geometry*. Cambridge, MA: Easton Press, 1959.

Huygens, Christiaan. *Traité de la lumière*. Leiden, 1690. See also Thompson.

Iamblichus. *De anima*. See Finamore.

Ibn al-Haytham. *De aspectibus.* See Pietquin and Smith.

———. *De speculis comburentibus.* See Heiberg.

———. *On the Configuration of the World.* See Langermann.

———. *Optics.* See Sabra.

Ibn al-Nadīm. *Fihrist.* See Dodge.

Ibn Khaldūn. *Muqaddimah.* See Rosenthal.

Ilardi, Vincent. "Renaissance Florence: The Optical Capital of the World." *Journal of European Economic History* 22, no. 3 (1993): 507–41.

———. *Renaissance Vision from Spectacles to Telescopes.* Philadelphia: American Philosophical Society Press, 2007.

Iribarren, Isabel, and Martin Lenz, eds. *Angels in Medieval Philosophical Inquiry.* Aldershot, UK; Burlington, VT: Ashgate, 2008.

Isidore of Seville. *Etymologies.* See Barney.

Ivry, Alfred. "The Ontological Entailments of Averroes' Understanding of Perception." In *Theories of Perception in Medieval and Early Modern Philosophy*, edited by Simo Knuutila and Pekka Kärkkäinen, 73–87. Dordrecht: Springer, 2008.

Jacquart, Danielle. *L'Épopée de la science arabe.* Paris: Gallimard, 2005.

———. *La médicine dans le cadre parisien, XIVᵉ–XVᵉ siècle.* Paris: Fayard, 1998.

Jean de Meun. *Romance of the Rose.* See Dahlberg.

Jeauneau, Edouard, ed. *Iohannis Scotti Euriugenae* Periphyseon (De Divisione Naturae) *Liber Quartus.* Translated by John J. O'Meara. Dublin: Dublin Institute for Advanced Studies, 1995.

Johannes Blund. *Tractatus de anima.* See Callus.

Johns, Adrian. *The Nature of the Book.* Chicago: University of Chicago Press, 1998.

Jones, Alexander. "Pseudo-Ptolemy *De Speculis.*" *SCIAMVS* 2 (2001): 145–86.

Karamanolis, George. *Plato and Aristotle in Agreement?* Oxford: Clarendon Press, 2006.

Karger, Elizabeth. "Ockham's Misunderstood Theory of Intuitive and Abstractive Cognition." In *The Cambridge Companion to Ockham*, edited by Paul V. Spade, 168–226. Cambridge: Cambridge University Press, 1999.

Karnes, Michelle. *Imagination, Meditation, and Cognition in the Middle Ages.* Chicago: University of Chicago Press, 2011.

Kaster, Robert A., trans. *Macrobius*: Saturnalia *Books 6–7.* Cambridge, MA: Harvard University Press, 2011.

Kelso, Carl, ed. and trans. "Witelonis *Perspectivae* liber quartus." PhD diss., University of Missouri, Columbia, 1994.

Kemp, Martin. "Imitations, Optics, and Photography: Some Gross Hypotheses." In *Inside the Camera Obscura—Optics and Art under the Spell of the Projected Image*, preprint 333, edited by Wolfgang Lefèvre, 243–64. Berlin: Max Planck Institute for the History of Science, 2007.

———. *The Science of Art.* New Haven, CT: Yale University Press, 1990.

———. "Science, Non-Science, and Nonsense: The Interpretation of Brunelleschi's Perspective." *Art History* 1 (1978): 134–61.

Kemp, Martin, and Margaret Walker, eds. and trans. *Leonardo on Painting.* New Haven, CT: Yale University Press, 1989.

Kepler, Johannes. *Ad Vitellionem paralipomena*. Frankfurt, 1604. See also Donahue.

———. *Dioptrice*. Augsburg, 1611.

———. *Dissertatio cum Nuncio siderio*. Prague, 1610. See also Rosen.

Kessler, Herbert. "Speculum." *Speculum* 86 (2011): 1–41.

Kheirandish, Elaheh. *The Arabic Version of Euclid's* Optics. New York: Springer, 1999.

———. "The Many Aspects of 'Appearances': Arabic Optics to 950 AD." In *The Enterprise of Science in Islam*, edited by Jan Hogendijk and Abdelhamid Sabra, 55–83. Cambridge, MA: MIT Press, 2003.

Kindī, al. *De radiis stellarum*. See Alverny.

———. *On the Intellect*. See McGinnis.

King, Henry C. *The History of the Telescope*. Cambridge, MA: Sky Publications, 1955.

King, Peter, trans. *Augustine*: Against the Academicians, *and* The Teacher. Indianapolis: Hackett, 1995.

Kirchner, Eric, and Mohammed Bagheri. "Color Theory in Islamic Lapidaries: Nīshābūrī, Tūsī, and Kāshānī." *Centaurus* 55 (2013): 1–19.

Knorr, Wilbur. "Archimedes and the Pseudo-Euclidean *Catoptrics*: Stages in the Ancient Geometrical Theory of Mirrors." *Archives internationales d'histoire des sciences* 35 (1985): 28–104.

Knuutila, Simo, and Pekka Kärkkäinen, eds. *Theories of Perception in Medieval and Early Modern Philosophy*. Dordrecht: Springer, 2008.

Kuhn, Thomas S. *The Structure of Scientific Revolutions*. 2nd ed. Chicago: University of Chicago Press, 1970.

Kurth, Willi, ed. *The Complete Woodcuts of Albrecht Dürer*. New York: Crown Publishers, 1946.

Langermann, Tzvi, trans. *Ibn al-Haytham's* On the Configuration of the World. New York: Garland, 1990.

Lechler, Gotthard, ed. *Johannis Wiclif*: Trialogus cum supplemento Trialogi. Oxford: Clarendon Press, 1869.

Lee, Raymond, and Alistair Fraser. *The Rainbow Bridge: Rainbows in Art, Myth, and Science*. University Park: Pennsylvania State University Press, 2001.

Lefèvre, Wolfgang, ed. *Inside the Camera Obscura—Optics and Art under the Spell of the Projected Image*, preprint 333. Berlin: Max Planck Institute for the History of Science, 2007.

Lejeune, Albert, ed. and trans. *L'Optique de Claude Ptolémée dans la version latine d'après l'arabe de l'émir Eugène de Sicile*. Leiden: Brill, 1989.

———. *Recherches sur la catoptrique grecque*. Mémoires de l'Académie Royale de Belgique: Classe des lettres et des sciences morales et politiques 52.2. Brussels: Palais des Académies, 1957.

Leonardo da Vinci. *On Painting*. See Kemp.

———. "Treatise on the Eye." See Strong.

———. See MacCurdy and Richter.

Lindberg, David C. "Alkindi's Critique of Euclid's Theory of Vision." *Isis* 62 (1971): 469–89.

———. *A Catalogue of Medieval and Renaissance Optical Manuscripts*. Toronto: University of Toronto Press, 1975.

———. "The Genesis of Kepler's Theory of Light: Light Metaphysics from Plotinus to Kepler." *Osiris* 2 (1986): 5–42.

———. Introduction to *Opticae Thesaurus*. Edited by Friedrich Risner. New York: Johnson Reprint, 1972.

———, ed. and trans. *John Pecham and the Science of Optics*. Madison: University of Wisconsin Press, 1970.

———. "Lines of Influence in Thirteenth-Century Optics: Bacon, Witelo, and Pecham." *Speculum* 46 (1971): 66–83.

———. "Optics in Sixteenth-Century Italy." *Annali dell'Istituto e museo di storia della scienza* 2 (1983): 31–148.

———, ed. and trans. *Roger Bacon and the Origins of* Perspectiva *in the Middle Ages: A Critical Edition and English Translation of Bacon's* Perspectiva *with Introduction and Notes*. Oxford: Clarendon Press, 1996.

———, ed. and trans. *Roger Bacon's Philosophy of Nature: A Critical Edition, with English Translation, Introduction, and Notes, of* De multiplicatione specierum *and* De speculis comburentibus. Oxford: Clarendon Press, 1983.

———. *Theories of Vision from al-Kindi to Kepler*. Chicago: University of Chicago Press, 1976.

———. "The Theory of Pinhole Images from Antiquity to the Thirteenth Century." *Archive for History of Exact Sciences* 5 (1968): 154–76.

———. "The Theory of Pinhole Images in the Fourteenth Century." *Archive for History of Exact Sciences* 6 (1970): 299–325.

Lindberg, David C., and Ronald Numbers, eds. *God and Nature: Historical Essays on the Encounter between Christianity and Science*. Berkeley: University of California Press, 1986.

Lindberg, David C., and Robert Westman, eds. *Reappraisals of the Scientific Revolution*. Cambridge: Cambridge University Press, 1990.

Lindsay, Wallace M., trans. *Isidori Hispalensis Episcopi*: Etymologiarum sive Originum. 2 vols. Oxford: Clarendon Press, 1911.

Long, R. James, ed. *Bartholomaeus Anglicus on the Properties of Soul and Body:* De proprietatibus rerum libri III et IV. Toronto: Pontifical Institute of Mediaeval Studies, 1979.

Loveday and Forester. See Pseudo-Aristotle.

Lucretius. *De rerum natura*. See Rouse.

Lyons, Jonathan. *The House of Wisdom: How the Arabs Transformed Western Civilization*. New York: Bloomsbury, 2009.

MacCurdy, Edward, ed. and trans. *The Notebooks of Leonardo da Vinci*. 2 vols. New York: Reynal and Hitchcock, 1938.

Macrobius. *Saturnalia*. See Kaster.

Mahoney, Michael. *The Mathematical Career of Pierre de Fermat*. Princeton, NJ: Princeton University Press, 1994.

Manetti, Antonio. *The Life of Brunelleschi*. See Saalman.

Mann, Vivian, Thomas Glick, and Jerrilynn Dodds, eds. *Convivencia: Jews, Muslims, and Christians in Medieval Spain*. New York: Georges Braziller and the Jewish Museum, 1992.

Marenbon, John. "Platonism—A Doxographic Approach: The Early Middle Ages." In *The Platonic Tradition in the Middle Ages: A Doxographic Approach*, edited by Stephen Gersh and Maarten Hoenen, 67–89. Berlin: Walter de Gruyter, 2002.

Marshall, Peter. "Parisian Psychology in the Mid-Fourteenth Century." *Archives d'histoire doctrinale et littéraire du Moyen Âge* 50 (1983): 101–93.

Martin, Thomas F., and Allan D. Fitzgerald. *Augustine of Hippo: Faithful Servant, Spiritual Leader*. Boston: Prentice Hall, 2011.

Maurolico, Francesco. *Photismi De Lumine, et umbra ad perspectivam, et radiorum incidentiam facientes*. Naples, 1611. See also Crew.

May, Margaret Tallmadge, trans. *Galen on the Usefulness of the Parts of the Body*. Ithaca, NY: Cornell University Press, 1968.

McCray, W. Patrick. *Glassmaking in Renaissance Venice: The Fragile Craft*. Aldershot, UK; Burlington, VT: Ashgate, 1999.

McEvoy, James. "Ioannes Scottus Eriugena and Robert Grosseteste: An Ambiguous Influence." In *Eriugena Redivivus*, edited by Werner Beierwaltes, 192–223. Heidelberg: Carl Winter, 1987.

———. *The Philosophy of Robert Grosseteste*. Oxford: Clarendon Press, 1982.

McGinnis, Jon. *Avicenna*. Oxford: Oxford University Press, 2010.

———, trans. *Avicenna: The Physics of* The Healing. Provo, UT: Brigham Young University, 2009.

———. "Avicenna's Naturalized Epistemology and Scientific Method." In *The Unity of Science in the Arabic Tradition*, edited by Shahid Rahman, Tony Street, and Hassan Tahiri, 129–52. Berlin: Springer, 2008.

———, ed. *Interpreting Avicenna: Science and Philosophy in Islam*. Leiden: Brill, 2004.

McGinnis, Jon, and David Reisman, eds. and trans. *Classical Arabic Philosophy*. Indianapolis: Hackett, 2007.

McInerny, Ralph, trans. *On Truth*. In *Anselm of Canterbury: The Major Works*, edited by Brian Davies and Gillian Evans, 158–59. Oxford: Oxford University Press, 1998.

McKenna, Stephen, trans. *Augustine*: On the Trinity, *Books 8–15*. Cambridge: Cambridge University Press, 2002.

Meyerhof, Max. "New Light on Hunain Ibn Ishaq and His Period." *Isis* 8 (1926): 685–724.

———, trans. *The Ten Treatises on the Eye*. Cairo: Government Press, 1928.

Micheau, Françoise. "'Alī ibn al-'Abbās al-Maǧūsī et son milieu." In *Constantine the African and 'Alī ibn al-'Abbās al-Maǧūsī: The* Pantegni *and Related Texts*, edited by Charles Burnett and Danielle Jacquart, 1–15. Leiden: Brill, 1994.

Minio-Paluello, Lorenzo. "Jacobus Veneticus Grecus, Canonist and Translator of Aristotle." *Traditio* 8 (1952): 265–304.

Modrak, Deborah K. W. "The Nous-Body Problem in Aristotle." *Review of Metaphysics* 44 (1991): 755–74.

Molesini, Guiseppe. "Testing Telescope Optics of Seventeenth-Century Italy." In *The Origins of the Telescope*, edited by Albert Van Helden and others, 271–80. Amsterdam: KNAW Press, 2010.

Morelon, Régis, and Ahmad Haznawi, eds. *De Zenon d'Élée à Poincaré*. Louvain: Peeters, 2004.

Morrow, Glenn R., trans. *Proclus: A Commentary on the First Book of Euclid's* Elements. 2nd ed. Princeton, NJ: Princeton University Press, 1992.

Muckle, J. T., ed. "The Treatise De Anima of Dominicus Gundissalinus." *Medieval Studies* 2 (1940): 23–103.

Nemesius, Bishop of Emesa. *On the Nature of Man*. See Sharples.

Neri, Antonio. *L'Arte vetraria*. Florence, 1612.

Neugebauer, Otto. *The Exact Sciences in Antiquity*. 2nd ed. New York, Dover: 1969.

Newhauser, Richard. "*Inter scientiam et populum*: Roger Bacon, Peter of Limoges, and the 'Tractatus moralis de oculo.'" In *Nach der Verurteilung von 1277*, edited by Jan A. Aertsen and others, 682–703. Berlin: Walter de Gruyter, 2001.

———, trans. *Peter of Limoges*: The Moral Treatise on the Eye. Toronto: Pontifical Institute of Mediaeval Studies, 2012.

Newton, Francis. "Constantine the African and Monte Cassino: New Elements and the Text of the *Isagoge*." In *Constantine the African and 'Alī ibn al-'Abbās al-Maǧūsī: The* Pantegni *and Related Texts*, edited by Charles Burnett and Danielle Jacquart, 16–47. Leiden: Brill, 1994.

Newton, Issac. "An Hypothesis explaining the Properties of Light." In *The History of the Royal Society*. Vol. 3, edited by Thomas Birch, 248–305. London, 1757.

———. "Letter concerning a New Theory of Light and Color." *Philosophical Transactions of the Royal Society* 6 (1671/72): 3075–87.

———. *Opticks*. London, 1704.

Newton, Robert R. *The Crime of Claudius Ptolemy*. Baltimore: Johns Hopkins Press, 1977.

O'Daly, Gerard. *Augustine's Philosophy of Mind*. London: Duckworth, 1987.

O'Donnell, James. *Augustine: A New Biography*. New York: Ecco, 2005.

Omar, Saleh. *Ibn al-Haytham's Optics: A Study in the Origins of Experimental Science*. Minneapolis: Bibliotheca Islamica, 1977.

Oresme, Nicole. *De visione stellarum*. See Burton.

Panofsky, Erwin, ed. and trans. *Abbot Suger on the Abbey Church of St.-Denis and Its Art Treasures*. Princeton, NJ: Princeton University Press, 1946.

Pappus of Alexandria. *Mathematical Collections*. See Ver Eecke.

Park, David. *The Fire within the Eye: A Historical Essay on the Nature and Meaning of Light*. Princeton, NJ: Princeton University Press, 1997.

Parronchi, Alessandro. "La perspettiva dantesca." In *Studi su la "dolce" prospettiva*, edited by Alessandro Parronchi, 4–90. Milan: Aldo Martello, 1964.

———. *Studi su la "dolce" prospettiva*. Milan: Aldo Martello, 1964.

Pasnau, Robert. *Theories of Cognition in the Later Middle Ages*. Cambridge: Cambridge University Press, 1997.

———, trans. *Thomas Aquinas: A Commentary on Aristotle's* De anima. New Haven, CT: Yale University Press, 1999.

———. *Thomas Aquinas on Human Nature*. Cambridge: Cambridge University Press, 2002.

Pasnau, Robert, and Christina Van Dyke, eds. *The Cambridge History of Medieval Philosophy*. 2 vols. Cambridge: Cambridge University Press, 2010.

Pecham, John. *Perspectiva communis*. See Lindberg.

Pendergrast, Mark. *Mirror Mirror: A History of the Human Love Affair with Reflection*. New York: Basic Books, 2003.

Perler, Dominik. "Thought Experiments: The Methodological Function of Angels in Late Medieval Epistemology." In *Angels in Medieval Philosophical Inquiry*, edited by Isabel Iribarren and Martin Lenz, 143–53. Aldershot, UK; Burlington, VT: Ashgate, 2008.

Peter of Limoges. *Tractatus moralis de oculo*. See Newhauser.

Peters, Francis E. *Aristoteles Arabus.* Leiden: Brill, 1968.

Petrus Hispanus. *Tractatus de oculis.* See Berger, Daley, and Smith.

Phillips, Heather. "John Wyclif and the Optics of the Eucharist." In *From Ockham to Wyclif,* edited by Anne Hudson and Michael Wilks, 245–58. Oxford: Blackwell, 1987.

Philoponus, John. *On Aristotle's* De anima. See Charlton and Eijk.

Piché, David. *La condamnation Parisienne de 1277.* Paris: Vrin, 1999.

Pietquin, Paul, ed. and trans. *Le septième livre du traité* De aspectibus *d'Alhazen, traduction latine médiévale de* l'Optique *d'Ibn al-Haytham.* Louvain: Académie royale de Belgique, 2010.

Plato. *Timaeus.* See Cornford, Wasink, and Zeyl.

Platter, Felix. *De corporis humani structura et usu.* Basel, 1583.

Porphyry. *Life of Plotinus.* See Armstrong.

Poster, Carol, and Linda Mitchell, eds. *Letter-Writing Manuals and Instruction from Antiquity to the Present.* Columbia: University of South Carolina Press, 2007.

Proclus. *Commentary on the First Book of Euclid's* Elements. See Morrow.

Pseudo-Aristotle. *On Colours.* Translated by T. Loveday and E. S. Forster. In *The Complete Works of Aristotle.* Vol. 1, edited by Jonathan Barnes, 1219–28. Princeton, NJ: Princeton University Press, 1984.

Ptolemy. *Almagest.* See Toomer.

———. *Optics.* See Lejeune and Smith.

Quinlan-McGrath, Mary. *Influence: Art, Optics, and Astrology in the Italian Renaissance.* Chicago: University of Chicago Press, 2013.

Rahman, Fazlur, trans. *Avicenna's Psychology.* Westport, CT: Hyperion Press, 1952.

———. *Islam.* 2nd ed. Chicago: University of Chicago Press, 1979.

Rahman, Shahid, Tony Street, and Hassan Tahiri, eds. *The Unity of Science in the Arabic Tradition.* Dordrecht: Springer, 2008.

Rashdall, Hastings. *The Universities of Europe in the Middle Ages.* 3 vols. Oxford: Clarendon Press, 1895.

Rashed, Roshdi. *Les catoptriciens grecs I: Les mirroirs ardents.* Paris: Les Belles Lettres, 2000.

———. *The Development of Arabic Mathematics: Between Arithmetic and Algebra.* Dordrecht: Kluwer Academic, 1994.

———. *Géométrie et dioptrique au Xe siècle.* Paris: Les Belles Lettres, 1993.

———. *Les mathématiques infinitésimales du IXe au XIe siècle.* Vol. 2. London: Al-Furqan, 1993.

———. *Oeuvres philosophiques et scientifiques d'al-Kindī.* Vol. 1. Leiden: Brill, 1997.

———. "A Pioneer in Anaclastics: Ibn Sahl on Burning Mirrors and Lenses." *Isis* 81 (1990): 464–91.

———. "Problems of the Transmission of Greek Scientific Thought into Arabic: Examples from Mathematics and Optics." *History of Science* 27 (1989): 199–209.

Reeves, Eileen. *Galileo's Glassworks.* Cambridge, MA: Harvard University Press, 2008.

Reeves, Eileen, and Albert Van Helden, trans. *Galileo Galilei and Christoph Scheiner on Sunspots.* Chicago: University of Chicago Press, 2010.

Reisch, Gregor. *Margarita Philosophica.* Freiburg im Breisgau, 1503. See also Cunningham.

Richardson, William Frank, and John Burd Carman, trans. *Vesalius* On the Fabric of the Human Body, *Books III and IV*. Novato, CA: Norman Publishing, 2002.

Richter, Jean Paul, ed. and trans. *The Literary Works of Leonardo da Vinci.* 2 vols. 3rd ed. London: Phaidon, 1970.

Richter, Jean Paul, and Irma Richter, ed. and trans. *The Literary Works of Leonardo da Vinci.* 2 vols. 2nd ed. London: Oxford University Press, 1939.

Richter, Jean Paul, ed. and trans., and Carlo Pedretti, commentary. *The Literary Works of Leonardo da Vinci.* 2 vols. Berkeley: University of California Press, 1977.

Riedl, Clare, trans. *Robert Grosseteste*: On Light. Milwaukee: Marquette University Press, 1942.

Riet, Simone van, ed. *Avicenna Latinus*: Liber de anima *seu* Sextus de naturalibus. 2 vols. Leiden: Brill, 1968–72.

Risner, Friedrich, ed. *Opticae Thesaurus*. Basel, 1572.

Robert Grosseteste. *De iride.* See Baur.

———. *De lineis, angulis, et figuris.* See Baur.

———. *De luce.* See Baur and Riedl.

Rocca, Julius. *Galen on the Brain: Anatomical Knowledge and Physiological Speculation in the Second Century AD*. Leiden: Brill, 2003.

Ronca, Italo, ed. *Guillelmo de Conchis*, Dragmaticon Philosophiae. Turnhout: Brepols, 1997.

———. "The Influence of the *Pantegni* on William of Conches's *Dragmaticon*." In *Constantine the African and 'Alī ibn al-'Abbās al-Maǧūsī: The* Pantegni *and Related Texts*, edited by Charles Burnett and Danielle Jacquart, 266–85. Leiden: Brill, 1994.

Ronchi, Vasco. *The Nature of Light.* Translated by V. Barocas. Cambridge, MA: Harvard University Press, 1970.

Rorem, Paul. *Pseudo Dionysius.* New York: Oxford University Press, 1993.

Rosen, Edward, trans. *Kepler's Conversation with Galileo's Sidereal Messenger*. New York: Johnson Reprint, 1965.

Rosenthal, Franz, trans. *Ibn Khaldun*: The Muqaddimah. Princeton, NJ: Princeton University Press, 1969.

Ross, Helen E. "Cleomedes (c. 1st century AD) on the Celestial Illusion, Atmospheric Refraction, and Size-Distance Invariance." *Perception* 29 (2000): 863–71.

Ross, Helen E., and Cornelis Plug. *The Mystery of the Moon Illusion.* Oxford: Oxford University Press, 2002.

Ross, Helen E., and George M. Ross. "Did Ptolemy Understand the Moon Illusion?" *Perception* 5 (1976): 377–85.

Rouse, Richard H., and Mary A. Rouse, eds. *Authentic Witnesses: Approaches to Medieval Texts and Manuscripts*. Notre Dame, IN: Notre Dame University Press, 1991.

———. "The Book Trade at the University of Paris, ca. 1250–ca. 1350." In *Authentic Witnesses: Approaches to Medieval Texts and Manuscripts*, edited by Richard H. Rouse and Mary A. Rouse, 259–338. Notre Dame, IN: Notre Dame University Press, 1991.

Rouse, William H. D., and M. F. Smith, trans. *Lucretius*: De rerum natura. Cambridge, MA: Harvard University Press, 1982.

Rowland, Ingrid D., trans. *Vitruvius*: Ten Books on Architecture. New York: Cambridge University Press, 1999.

Rudd, M. Eugene. "Chromatic Aberration of Eyepieces in Early Telescopes." *Annals of Science* 64 (2007): 2–18.

Rüegg, Walter, ed. *A History of the University in Europe.* 4 vols. Cambridge: Cambridge University Press, 1992–2011.

Russell, Gül. "The Anatomy of the Eye in ʿAli ibn al-ʿAbbas al-Maǧūsī: A Textbook Case." In *Constantine the African and ʿAli ibn al-ʿAbbas al-Maǧūsī: The* Pantegni *and Related Texts,* edited by Charles Burnett and Danielle Jacquart, 247–65. Leiden, New York: Brill, 1994.

Saalman, Howard, and Catherine Engass, ed. and trans. *The* Life of Brunelleschi *by Antonio di Tuccio Manetti.* University Park: Pennsylvania State University Press, 1970.

Sabra, Abdelhamid. "The 'Commentary' That Saved the Text: The Hazardous Journey of Ibn al-Haytham's Arabic 'Optics.'" *Early Science and Medicine* 12 (2007): 117–33.

———. "One Ibn al-Haytham or Two? An Exercise in Reading the Bio-Bibliographic Sources." *Zeitschrift für Geschichte der Arabisch-Islamischen Wissenschaften* 12 (1998): 1–40.

———, ed. *The Optics of Ibn al-Haytham. An Edition of the Arabic Text of Books IV–V: On Reflection and Images Seen by Reflection.* Kuwait: National Council for Culture, Arts, and Letters, 2002.

———, ed. *The Optics of Ibn al-Haytham. Books I–II–III: On Direct Vision.* Kuwait: National Council for Culture, Arts, and Letters, 1983.

———, trans. *The Optics of Ibn al-Haytham: Books I–III on Direct Vision.* London: Warburg Institute, 1989.

———. "Psychology versus Mathematics: Ptolemy and Alhazen on the Moon Illusion." In *Mathematics and Its Applications to Science and Natural Philosophy in the Middle Ages,* edited by Edward Grant and John Murdoch, 217–47. Cambridge: Cambridge University Press, 1987.

———. "Science and Philosophy in Medieval Islamic Theology." *Zeitschrift für Geschichte der Arabisch-Islamischen Wissenschaften* 9 (1994): 1–42.

———. *Theories of Light from Descartes to Newton.* Cambridge: Cambridge University Press, 1981.

Said, Edward. *Orientalism.* New York: Parthenon, 1978.

Saliba, George. *Islamic Science and the Making of the European Renaissance.* Cambridge, MA: MIT Press, 2007.

———. "Whose Science Is Arabic Science in the Renaissance?" Columbia University. http://www.columbia.edu/%7Egas1/project/visions/case1/sci.1.html (accessed July 14, 2010).

Schechner, Sarah J. "Between Knowing and Doing: Mirrors and Their Imperfections in the Renaissance." *Early Science and Medicine* 10 (2005): 137–62.

Scheiner, Christoph. *Oculus.* Innsbruck, 1619.

———. *On Sunspots.* See Reeves.

———. *Rosa Ursina.* Bracciano, 1630.

Schmidt, W., ed. and trans. *Catoptrics.* In *Heronis Alexandrini Opera Quae Supersunt Omnia.* Vol. 2.1. Leipzig: Teubner, 1900.

Schmidt, W., L. Nix, H. Schoene, and I. L. Heiberg, eds. *Heronis Alexandrini Opera Quae Supersunt Omnia.* 5 vols. Leipzig: Teubner, 1899–1914.

Schmitt, Franciscus, ed. *Sancti Anselmi, Cantuariensis Archiepiscopi, Opera Omnia.* 6 vols. Edinburgh: Thomas Nelson and Sons, 1946–61.

Schöne, Richard, ed. and trans. *Damianos Schrift über Optik*. Berlin: Reichsdrukerei, 1897.

Schwinges, Rainer. "Student Education, Student Life." In *A History of the University in Europe*, edited by Walter Rüegg. Vol. 1, 195–243. Cambridge: Cambridge University Press, 1992.

Seneca. *Naturales Quaestiones*. See Corcoran.

Sentencia libri de anima. See Gauthier.

Sezgin, Fuat, ed. *Dominicus Gundissalinus (12th c.) and the Transmission of Arabic Philosophical Thought to the West*. Frankfurt: Institute for the History of Arabic-Islamic Science at the Johann Wolfgang Goethe University, 2000.

Shapin, Stephen. *The Scientific Revolution*. Chicago: University of Chicago Press, 1996.

———. *A Social History of Truth*. Chicago: University of Chicago Press, 1994.

———, and Simon Schaffer. *Leviathan and the Air Pump*. Princeton, NJ: Princeton University Press, 1989.

Shapiro, Alan E. "Artists' Colors and Newton's Colors." *Isis* 85 (1994): 600–630.

———. *Fits, Passions, and Paroxysms*. Cambridge: Cambridge University Press, 1993.

———. "Kinematic Optics: A Study in the Wave Theory of Light in the Seventeenth Century." *Archive for History of Exact Sciences* 11 (1973): 134–218.

Sharples, R. W., and Philip J. van der Eijk, trans. *Nemesius: On the Nature of Man*. Liverpool: Liverpool University Press, 2008.

Sheldon-Williams, I. P., ed. and trans. *Iohannis Scotti Euriugenae* Periphyseon (De Divisione Naturae) *Liber Secundus*. Dublin: Dublin Institute for Advanced Studies, 1972.

———. *Iohannis Scotti Euriugenae* Periphyseon (De Divisione Naturae) *Liber Tertius*. Dublin: Dublin Institute for Advanced Studies, 1981.

Sheppard, Anne. "The Mirror of Imagination: The Influence of *Timaeus* 70e ff." *Bulletin of the Institute of Classical Studies* 46 (2011): 203–12.

Siegel, Rudolph E. *Galen on Sense Perception*. Basel; New York: Karger, 1970.

Simms, D. L. "Archimedes and the Burning Mirrors of Syracuse." *Technology and Culture* 18 (1977): 1–24.

Simplicius. *De anima*. See Blumenthal.

Simon, Gérard. *Archéologie de la vision*. Paris: Seuil, 2003.

———. "L'Expérimentation sur la réflexion et la réfraction chez Ptolémée et Ibn al-Haytham." In *De Zenon d'Élée à Poincaré*, edited by Régis Morelon and Ahmad Haznawi, 335–75. Louvain: Peeters, 2004.

———. "Optique et perspective: D'Ibn al-Haytham à Alberti." In Gérard Simon, *Archéologie de la vision*, 167–81. Paris: Seuil, 2003.

———. *Le regard, l'être et l'apparence dans l'optique de l'antiquité*. Paris: Seuil, 1988.

Singer, Charles, trans. *Galen on Anatomical Procedures*. London: Oxford University Press, 1956.

Singleton, Charles, trans. *Dante Alighieri*: The Divine Comedy: Purgatorio. 2 vols. Princeton, NJ: Princeton University Press, 1973.

Sinisgalli, Rocco, trans. *Leon Battista Alberti*: On Painting. Cambridge: Cambridge University Press, 2011.

———. *Perspective in the Visual Culture of Classical Antiquity*. Cambridge: Cambridge University Press, 2012.

Smith, A. Mark, ed. and trans. *Alhacen on Image-Formation and Distortion in Mirrors: A Critical Edition, with English Translation and Commentary, of Book 6 of Alhacen's* De aspectibus, *the Medieval Latin Version of Ibn al-Haytham's* Kitāb al-Manāẓir. Transactions of the American Philosophical Society 98.1 and 2. Philadelphia American Philosophical Society Press, 2008.

——, ed. and trans. *Alhacen on Refraction: A Critical Edition, with English Translation and Commentary, of Book 7 of Alhacen's* De aspectibus, *the Medieval Latin Version of Ibn al-Haytham's* Kitāb al-Manāẓir. Transactions of the American Philosophical Society 100.1 and 2. Philadelphia: American Philosophical Society Press, 2010.

——, ed. and trans. *Alhacen on the Principles of Reflection: A Critical Edition, with English Translation, Introduction, and Commentary, of Books 4 and 5 of Alhacen's* De aspectibus, *the Medieval Latin Version of Ibn al-Haytham's* Kitāb al-Manāẓir. Transactions of the American Philosophical Society 94.2 and 3. Philadelphia: American Philosophical Society Press, 2006.

——. "Alhacen's Approach to 'Alhazen's Problem.'" *Arabic Sciences and Philosophy* 18 (2008): 143–63.

——, ed. and trans. *Alhacen's Theory of Visual Perception: A Critical Edition, with English Translation and Commentary, of the First Three Books of Alhacen's* De Aspectibus, *the Medieval Latin Version of Ibn al-Haytham's* Kitāb al-Manāẓir. Transactions of the American Philosophical Society 91.4 and 5. Philadelphia: American Philosophical Society Press, 2001.

——. "Le *De aspectibus* d'Alhacen: Révolutionnaire ou réformiste?" *Revue d'histoire des sciences* 60 (2007): 65–81.

——. *Descartes's Theory of Light and Refraction: A Discourse on Method.* Transactions of the American Philosophical Society 77.3. Philadelphia: American Philosophical Society Press, 1987.

——. "Extremal Principles in Ancient and Medieval Optics." *Physis* 31 (1994): 113–40.

——. "Getting the Big Picture in Perspectivist Optics." *Isis* 72 (1981): 568–89.

——. "The Latin Source of the Fourteenth-Century Italian Translation of Alhacen's *De aspectibus* (Vat. Lat. 4595)." *Arabic Sciences and Philosophy* 11 (2001): 27–43.

——. "The Latin Version of Ibn Muʿādh's Treatise 'On Twilight and the Rising of Clouds.'" *Arabic Sciences and Philosophy* 2 (1992): 38–88.

——. "The Physiological and Psychological Grounds of Ptolemy's Visual Theory: Some Methodological Considerations." *Journal of the History of the Behavioral Sciences* 34 (1998): 231–34.

——. "Picturing the Mind: The Representation of Thought in the Middle Ages and Renaissance." *Philosophical Topics* 20 (1992): 149–70.

——. "Practice vs. Theory: The Background to Galileo's Telescopic Work." *Atti di Giorgio Ronchi* 54 (2001): 149–62.

——. "The Psychology of Visual Perception in Ptolemy's *Optics*." *Isis* 79 (1988): 189–207.

——. *Ptolemy and the Foundations of Ancient Mathematical Optics.* Transactions of the American Philosophical Society 89.3. Philadelphia: American Philosophical Society Press, 1999.

——. "Ptolemy's Search for a Law of Refraction: A Case-Study in the Classical Methodology of 'Saving the Appearances' and Its Limitations." *Archive for History of Exact Sciences* 26 (1982): 221–40.

———, trans. *Ptolemy's Theory of Visual Perception: An English Translation of the* Optics *with Introduction and Commentary*. Transactions of the American Philosophical Society 86.2. Philadelphia: American Philosophical Society Press, 1996.

———. "Reflections on the Hockney-Falco Thesis: Optical Theory and Artistic Practice in the Fifteenth and Sixteenth Centuries." *Early Science and Medicine* 10, no. 2 (2005): 163–85.

———. "Review of *The Fire within the Eye: A Historical Essay on the Nature and Meaning of Light*, by David Park." *Physics Today* 51 (1998): 62–63.

———. "Saving the Appearances of the Appearances: The Foundations of Classical Geometrical Optics." *Archive for History of Exact Sciences* 24 (1981): 73–100.

———, ed. and trans. *Witelonis* Perspectivae *liber quintus*. Studia Copernicana 23. Wroclaw: Polish Academy of Sciences, 1983.

Smith, A. Mark, and Arnaldo Pinto Cardoso. *The Treatise on the Eyes by Pedro Hispano*. Lisbon: Alêtheia Editores, 2008.

Smith, J. A., trans. *De anima*. In *The Complete Works of Aristotle*. Vol. 1, edited by Jonathan Barnes, 641–92. Princeton, NJ: Princeton University Press, 1984.

Somfai, Anna. "The Eleventh-Century Shift in the Reception of Plato's 'Timaeus' and Calcidius's 'Commentary.'" *Journal of the Warburg and Courtauld Institutes* 65 (2002): 1–21.

Sorabji, Richard, ed. *Aristotle Transformed*. Ithaca, NY: Cornell University Press, 1990.

Spade, Paul V., ed. *The Cambridge Companion to Ockham*. Cambridge: Cambridge University Press, 1999.

Spruit, Leen. Species Intelligibilis: *From Perception to Knowledge*. 2 vols. Leiden: Brill, 1994–95.

Squadrani, Ireneo. "*Tractatus de luce* fr. Bartholomaei de Bononia." *Antonianum* 7 (1932): 139–238, 337–76, 465–94.

Staden, Heinrich von. *Herophilus: The Art of Medicine in Early Alexandria*. Cambridge: Cambridge University Press, 1989.

Stahl, William. *Roman Science: Origins, Development, and Influence to the Later Middle Ages*. Madison: University of Wisconsin Press, 1962.

Steadman, Philip. "Allegory, Realism, and Vermeer's Use of the Camera Obscura." *Early Science and Medicine* 10 (2005): 287–313.

Steffens, Bradley. *Ibn al-Haytham: First Scientist*. Greensboro, NC: Morgan Reynolds, 2007.

Stewart, Devin. "The Structure of the *Fihrist*: Ibn al-Nadim as Historian of Islamic Legal and Theological Schools." *International Journal of Middle East Studies* 39 (2007): 369–87.

Stratton, George M., trans. *Theophrastus and the Greek Physiological Psychology before Aristotle*. London: Allen and Unwin, 1917.

Strong, Donald S. "Leonardo da Vinci on the Eye: The MS D in the Bibliothèque de l'Institut de France, Paris, Translated into English and Annotated with a Study of Leonardo's Theories of Optics." PhD diss., University of California, Los Angeles, 1967. Also New York: Garland, 1979.

Stump, Eleonore. "The Mechanism of Cognition: Ockham on Mediating Species." In *The Cambridge Companion to Ockham*, edited by Paul V. Spade, 168–226. Cambridge: Cambridge University Press, 1999.

Summers, David. *The Judgement of Sense: Renaissance Naturalism and the Rise of Aesthetics.* Cambridge: Cambridge University Press, 1987.

———. *Vision, Reflection, and Desire in Western Painting.* Chapel Hill: University of North Carolina Press, 2007.

Tachau, Katherine. *Vision and Certitude in the Age of Ockham.* Leiden: Brill, 1988.

Takahashi, Ken'ichi. *The Medieval Latin Traditions of Euclid's Catoptrica.* Kyushu: Kyushu University Press, 1992.

Taton, René, ed. *Roemer et la vitesse de la lumière.* Paris: Vrin, 1978.

Taylor, Jerome, trans. *The Didascalicon of Hugh of St. Victor.* New York: Columbia University Press, 1961.

Taylor, Richard C., trans. *Averroes (Ibn Rushd) of Cordoba: Long Commentary on the De Anima of Aristotle.* New Haven, CT: Yale University Press, 2009.

Theisen, Wilfrid. "*Liber de visu*: The Greco-Latin Translation of Euclid's *Optics.*" *Mediaeval Studies* 41 (1979): 44–105.

———. "The Medieval Tradition of Euclid's *Optics.*" PhD diss., University of Wisconsin, Madison, 1972.

Themistius. *De anima.* See Todd.

Theophrastus. *De sensibus.* See Stratton.

Thomas Aquinas. *Summa theologiae.* See Pasnau.

Thompson, Silvanus P., trans. *Treatise on Light.* Repr.; New York: Dover, 1962.

Todd, Robert. "Philosophy and Medicine in John Philoponus' Commentary on Aristotle's 'De anima.'" *Dumbarton Oaks Papers* 39 (1984): 103–10.

———, trans. *Themistius on Aristotle's On the Soul.* Ithaca, NY: Cornell University Press, 1996.

Toomer, Gerald J., trans. *Ptolemy's Almagest.* Princeton, NJ: Princeton University Press, 1998.

———, trans. *Diocles on Burning Mirrors.* New York: Springer, 1976.

Unguru, Sabetai, ed. and trans. *Witelonis* Perspectivae *liber primus.* Studia Copernicana 15. Wroclaw: Polish Academy of Sciences Press, 1977.

———, ed. and trans. *Witelonis* Perspectivae *liber secundus et liber tertius.* Studia Copernicana 28. Wroclaw: Polish Academy of Sciences, 1991.

Van Helden, Albert. "Christopher Wren's 'De Corpore Saturni.'" *Notes and Records of the Royal Society of London* 23 (1968): 213–29.

———. "Galileo's Telescope." In *The Origins of the Telescope*, edited by Albert Van Helden and others, 183–201. Amsterdam: KNAW Press, 2010.

———. "The Importance of the Transit of Mercury of 1631." *Journal of the History of Astronomy* 7 (1976): 1–10.

———. *The Invention of the Telescope.* Transactions of the American Philosophical Society 67.4. Philadelphia: American Philosophical Society Press, 1977.

———. *Measuring the Universe.* Chicago: University of Chicago Press, 1985.

———. "Saturn and His Anses." *Journal of the History of Astronomy* 5 (1974): 105–21.

———, trans. *Sidereus Nuncius.* Chicago: University of Chicago Press, 1989.

———. "The Telescope in the Seventeenth Century." *Isis* 65 (1974): 38–58.

———. "Telescopes and Authority from Galileo to Cassini." *Osiris* 9 (1994): 8–29.

Van Helden, Albert, and others, eds. *The Origins of the Telescope.* Amsterdam: KNAW Press, 2010.

Vasari, Giorgio. *Vite de' più eccellenti architetti, pittori et scultori italiani*. See Bondanella.

Ver Eecke, Paul, trans. *Euclide: L'*Optique *et la* Catoprique. Paris: Albert Blanchard, 1959.

———. *Pappus d'Alexandrie*. Paris; Bruges: Desclée de Brouwer, 1933.

Vesalius, Andreas. *De humani corpors fabrica libri septem*. Basel, 1543. See also Richardson.

Vescovini, Graziella. "Alhazen vulgarisé: Le De li aspecti d'un manuscrit du Vatican (moitié du XIVe siècle) et le troisième commentaire sur l'optique de Lorenzo Ghiberti." *Arabic Sciences and Philosophy* 8 (1998): 67–96.

———. "A New Origin of Perspective." *Res* 37 (2000): 73–81.

———. *Studi sulla prospettiva medievale*. Turin: Giappichelli, 1965.

Vidal, Fernando. *The Sciences of the Soul*. Chicago: University of Chicago Press, 2011.

Vitruvius. *De architectura*. See Rowland.

Walzer, Richard, ed. and trans. *Al-Farabi on the Perfect State*. Oxford: Clarendon Press, 1985.

Warren, Edward W. "Consciousness in Plotinus." *Phronesis* 9 (1964): 83–97.

———. "Imagination in Plotinus." *Classical Quarterly* 16 (1966): 277–85.

Waszink, Jan H., ed. Timaeus *a Calcidio Translatus*. London: Warburg Institute, 1962.

Webster, E. W., trans. *Meteorology*. In *The Complete Works of Aristotle*. Vol. 1, edited by Jonathan Barnes, 555–625. Princeton, NJ: Princeton University Press, 1984.

Wedin, Michael V. *Mind and Imagination in Aristotle*. New Haven, CT: Yale University Press, 1988.

Wetherebee, Winthrop, trans. *The* Cosmographia *of Bernardus Silvestris*. New York: Columbia University Press, 1973.

Whiston, William. *Astronomical Lectures*. London, 1715.

White, Hayden. *The Content of the Form*. Baltimore: Johns Hopkins Press, 1987.

Willach, Rolf. *The Long Route to the Invention of the Telescope*. Transactions of the American Philosophical Society 98.5. Philadelphia: American Philosophical Society Press, 2008.

William of Conches. *Dragmaticon Philosophiae*. See Curr and Ronca.

Williams, Thomas, trans. *Augustine*: On Free Choice of the Will. Indianapolis: Hackett, 1993.

Wilson, Catherine. *The Invisible World: Early Modern Philosophy and the Invention of the Microscope*. Princeton, NJ: Princeton University Press, 1995.

Wisnovsky, Robert, Faith Wallis, Jamie Furno, and Carlos Fraenkel, eds. *Vehicles of Transmission, Translation, and Transformation in Medieval Textual Culture*. Turnhout: Brepols, 2012.

Witelo. *Perspectiva*. See Kelso, Smith, Risner, and Unguru.

Wolfson, Harry A. "The Internal Senses in Latin, Arabic, and Hebrew Philosophical Texts." *Harvard Theological Review* 28 (1935): 69–133.

Wood, Casey, trans. *De oculis eorumque egritudinibus et curis*. Stanford, CA: Stanford University Press, 1929.

Wyclif, John. *Trialogus cum supplemento Trialogi*. See Lechler.

Zeyl, Donald, trans. *Plato*: Timaeus. Indianapolis: Hackett, 2000.

Zupko, Jack. *John Buridan: Portrait of a Fourteenth-Century Arts Master*. Notre Dame, IN: University of Notre Dame Press, 2003.

INDEX

ʿAbbāsid caliphate, 7, 155, 157

accidental (*accidentale*)/artificial perspective, 312

accidental light (*lux accidentalis/lumen accidentale*), 258–59

accidental sensibles (*sensibilia per accidens*), 264

Adamson, Peter, 165

aerial perspective, 92, 279, 299n47, 313

aether/aither, 32n22, 122, 125, 185–86, 218, 222, 393, 399–401, 407–8, 410; longitudinal waves in, 399, 400; transverse waves in, 407, 408.

Afflacius, 239n27

afterimage, 90, 188, 250

Aḥmad ibn ʿĪsā, visual theory of, 172–75

air pump, 379

Akbari, Suzanne Conklin, 291–95

Alberti, Leon Battista, 300–304, 312, 314, 317, 350–51

Albertus Magnus, 246–47, 256, 294

albugineous humor, 187–88

Alcmaeon, 232n6

Alexander of Aphrodisias, 132, 141, 243, 244

Alexandrian library, 11

Alfanus, 239n27

Alfonso VI of Castile, 242

Alhacen/Alhazen. *See* Ibn al-Haytham

Alhazen's Problem, 201–4, 226, 272, 375

Alighieri, Dante, 291, 293–94

allegory, 292

Almagest (*Mathēmatikē Suntaxis*). *See* Ptolemy, works of

anaklasis, 55, 78

anamorphosis, 315

angelic cognition, 288

animal, cerebral spirit, 164, 250, 254, 263, 276, 350, 410, 411–12, 413, 414, 416

anomaly, 54–55, 61–63, 127–28, 277, 320, 338, 371, 399–400, 406

Anselm of Canterbury, 236–37

Anthemius of Tralles, 71n106, 131, 166, 172–75, 206

Antonino. *See* Pierozzi, Antonino (Saint)

Apollonius of Perga, 15, 69n102, 71, 183n7, 204

apophatic (or negative) theology, 235

Apuleius of Madaura, 55n88, 72n107

aqueous humor (*ōoeidēs*), 39

Aquinas, St. Thomas, 246, 247, 250, 251

Arabic translation movement, 13–14, 19, 132, 155–58, 166n99

aranea (spider's web), 186, 270, 352

Archimedes, 71, 72n107, 109, 206, 244, 331, 337

Archytas of Tarentum, 28n10

Aristotle, cosmological model of, 14, 122

Aristotle, visual theory of, 72, 74, 80, 276; on color, 31–33, 35, 43, 45–46, 82, 127, 135, 246, 403; on the common sensibility, 140–41; compared to Augustine, 152; compared to Plato, 131–32, 138, 143; compared to Ptolemy, 127; on images as central to thinking, 34–35, 249; on imagination, 33–34, 135; on intellect, 136–37, 164; in Latin West, 232–33, 246–47, 264, 267, 294, 377; on light, 32, 73, 74, 135, 139, 148, 246; on the optical system, 36; on perceptual theory, 135–37; on psychological faculties, 43, 136–37, 178–79; on rainbows, 26–28, 29, 31, 43, 45–46, 52, 74, 232–33, 403; role of heart in, 141, 138; on sensibles, 33; on the soul, 134–37, 138, 140–41; use of geometry by, 25–26, 73; on visual illusions, 125; on visual ray theory, 28–31

Aristotle, works of: in Arabic East, 17–18; *De anima*, 10, 31, 131, 132, 143, 156, 242, 244–48, 278; *De sensu*, 29, 31, 45, 242, 244, 246, 256, 278; influence of, 14, 232–33, 241, 245–46; *Metaphysics*, 259; *Meteorology*, 26, 125, 148n169, 243, 244, 259; *Organon*, 155; *Parva naturalia*, 156; *Physics*, 25; translations of, 13–14, 155–58, 230, 242–45

armillary sphere, 122, 324n3

assimilation, 34, 135, 138, 146

astrology, 285, 322, 330, 331, 378

astronomy, 14n34, 18, 25, 54, 79, 117, 128–29, 282n10, 322, 324n2, 330, 383–85, 389–90

atmosphere, 122–23, 125, 127, 218, 222, 274n100, 285

atmospheric refraction. *See* refraction

atomism, 30, 232, 233, 377, 393, 398

Atto, 239n27

Augustine of Hippo: Christ as intellect for, 151, 153; Christian background for, 150; compared to Aristotle, 152; defense of extramission in, 154, 266; on form and

species, 152–53, 248; Greek language and, 230; on the human will, 153–54; in Latin West, 132, 235, 236, 245, 248, 261, 266; on light, 257; on psychological faculties, 152; on senses, 152; on the soul, 151, 154; theological authority of, 154; theory on divine illumination, 152–53, 250

Ausonio, Ettore, 344n41, 366

Averroes (Ibn Rushd): authority of, 165n209; in Latin West, 245, 247, 250–51, 263; *Long Commentary*, 244, 259; translations of, 244

Avicenna (Ibn Sīnā): on agent intellect, 164–65; arguments against extramission in, 160–61, 179, 255; authority of, 133; on color, 161; compared to Ibn al-Haytham, 183, 184, 194; on epistemology of, 164–65; on imagination, 164, 194; on intentionality, 163–64; on the internal senses model, 162–66, 194, 241, 243, 245, 247, 372; on intromission, 159, 225; in Latin West, 165–66, 241, 245, 247–48, 251, 253–55, 261–62, 265, 267, 275, 350, 372; on light, 161, 183; optical system according to, 161; two-part memory in, 163; on vague individuals, 191n19

Avicenna, works of: *Book of Salvation*, 160; *Canon of Medicine*, 160, 165, 240, 243, 350; *Healing (Shifā)*, 10, 160, 241–43, 260; influence of, 183, 240, 241; translations of, 240, 242–43

Bacon, Francis, 379, 380

Bacon, Roger: on burning mirrors, 334; defense of extramission by, 265–66; influence of, 16, 284, 293, 339; on light, 395; on the optical system, 261–62; on practical optics, 269–70, 337; sources relied on for, 11, 260–61; on species multiplication, 262–64; spherical aberration and, 336; on spiritual vision, 270–71, 288–89; on the three levels of visual process, 264–65; use of geometry by, 256, 268, 271; on visual illusions, 266–70

Bacon, Roger, works of: *De multiplicatione specierum*, 260; *De speculis comburentibus*,

261, 271, 278, 334; *Opus maius*, 261, 293, 336; *Perspectiva*, 10, 261, 271, 328; printing history of, 328
Baghdad, 7, 155–57, 169
Baghdad Peripatetics, 157
Barrow, Isaac, 375
Bartholin, Erasmus, 400
Bartholomaeus Anglicus, 247–48
Bartholomew of Bologna, 290
Bayt al-Ḥikmah (House of Wisdom), 156
Bernard of Chartres, 237–38
Bernard Silvestris, 238–39
Biagio Pelacani di Parma, 302
Bible, 12n29, 17–18n44, 270
bifocal construction, 301
binocular vision, 43, 87, 89, 92, 97, 127, 148n57, 159n82, 170, 179, 193n24, 312–13
Bīrūnī, al-, 19n49
Bloch, Herbert, 239n27
Blund, John, 245–46
Boethius, 230, 249n51, 259
Bonaventure, Saint, 293
Boulliau, Ismaël, 380
Boyle, Robert, 377–78, 379, 398
Brahe, Tycho, 322, 384–85
brain: role in perception, 37, 40, 170, 186–87, 236, 240–41, 306, 350, 369, 409–12; as seat of perception and intellect, 43, 83n26, 128, 142, 149, 159, 232, 234, 238, 241. *See also* cerebral ventricles
Braudel, Fernand, 10
Brunelleschi, Filippo, 299–302, 304, 326
Buridan, Jean, 253
Buringh, Eltjo, 17n43
Burnyeat, Miles, 28n10, 73–74
Būyids, 157
Byzantines, 155, 229

Calcidius, 232, 237
Callisthenes, 232n6
camera obscura, 271, 307, 309, 317, 320, 322–23, 344, 353–54, 366
Campani, Giuseppe, 387
Campanus of Novara, 282n10

Canobus (Canopus), 79
Caravaggio, 315
Carolingian Renaissance, 231, 235
Cassini, Giovanni, 389
Cassiodorus, 230
cathedral schools, 13, 230
cathetus rule, 59, 61–62, 67, 93, 95, 97, 104, 111, 123, 170–71, 175, 205, 267–68
catoptrics (*katoptrikē*), 55, 77, 78, 302
caustic curve, 343, 358–59
Cavendish, Margaret, 391
center of sight (*centrum visus*), 188; biconal theory and, 308; preservation of, 225–26, 370–71; reflection and, 200–202, 268, 375; refraction and, 215–18, 222–24; rejection of, 363, 371; Renaissance art and, 300, 305; species theory and, 264; visual disorders and, 341
cerebral ventricles, 37, 40, 146, 149, 152, 162, 163, 164, 179, 180, 238–41, 247–48, 250, 253–55, 262, 276–77, 411–12
certification (*certificatio/verificatio*), 190
Cesi, Federico, 380
Chaucer, Geoffrey, 281n7, 291, 294–95, 316
chiaroscuro, 313, 315
choroid or "afterbirth-like" tunic (*chitōn choroeidēs*), 38, 39
Christ, 151–52, 153, 288
Christian education, 13, 17n44, 20, 230
Christina of Sweden, 381n12
chromatic aberration, 386, 409
Clark, Stuart, 316
Clement IV, Pope, 261
Cleomedes, 125n108
cognition: by abstraction, 34, 141–42, 144, 163–65, 178–79, 248, 298; abstractive, 252; Arabic thought on, 158; as grounded in sensation, 143, 238, 251; human will as critical to, 153–54; role of the soul in, 136, 138, 143, 145, 154; species multiplication and, 248–50; understanding through, 35, 152–53, 165, 237, 241, 249, 250–52, 298, 368; visual connotations of, 133–34, 149–50, 153

cognitive faculty (*cogitativa*), 149, 163, 255, 264–65

Colombo, Realdo, 351

color perspective, 313

color spectrum, 45–46, 401–2, 405–7

color theory: color as primary/objective cause of vision, 31–32, 35, 40–41, 43, 45, 75, 82–83, 127, 161, 162, 193–94, 236, 240, 245–46, 264, 369; color as a special sensible, 135, 148; color as a subjective/secondary quality, 44, 75, 373–74, 400–401; color in a hybrid-particle-wave model, 407–8; illumination of color, 73, 74, 139–40, 147, 186; primary colors, 73; visual illusions associated with, 89–91; white light and, 400–407. *See also individual thinkers*

common axis, 88, 97, 107

common nerve (*nervus communis*), 264, 289. *See also* optic chiasma

common sense (*sensus communis*): in Cartesian philosophy, 412, 415–16; as distinct from imagination, 162; function of, 163, 247, 253, 262, 306, 308; location of, 162, 254, 262, 306, 412; proper objects of, 163, 248, 264; as similar to the final sensor, 194; species multiplication in, 276

common sensibility (*aisthēsis koinē*), 33, 35, 88, 135, 140–41, 144, 152

common sensibles, 33–34, 35, 43, 82, 127, 135, 136, 141, 152, 163, 189, 194, 262, 276, 409

compositive imagination, 162–63, 194, 247n48, 248, 319. *See also* phantasy (*fantasia*)

condemnation of 1277, 251

cone of radiation, 179, 188, 193, 215–18, 225–26, 234, 259, 264, 284n15, 311

conjunctiva, 39

Conon of Samos, 71

conservation. *See* principle of conservation

Constantine I, Emperor of Rome, 229

Constantine the African, 239–40

Constantinople, 229, 242, 244

Copernicus, Nicolas, 8n19, 380, 384–85

cornea (*keratoeidēs*), 32, 38–39, 40, 72, 81,

146, 147, 186, 187–88, 194, 261, 307n73, 308, 350, 374

corpuscularism, 393, 398

costruzione legittima, 312, 317

craft knowledge, 279, 323–27, 332

critical angle, 115–16, 213–14,

crystalline humor (*krustalloeidēs*), 39, 245, 247, 306

crystalline lens: as cause of visual disorders, 193, 339; corrective lenses and, 340–41; image apprehension on, 307–9, 312; as an image projector, 351–52, 354; location of, 38–40, 159, 186–88, 194, 308, 339, 351, 374; as seat of sensitivity, 41, 81, 159, 161, 170, 250, 254, 261, 263, 350, 354; selective sensitivity of, 188, 215, 217, 340–41; shape of, 351

Damianos (Heliodorus) of Larissa, 130

Davies, Sir John, 295–98, 320

ḍaw', 161, 183. *See also* nūr

Delambre, Jean Baptiste, 79

Della Porta, Giambattista, 323, 380; *Magiae Naturalis*, 331–32, 337; ray tracing in, 347–49; influence of, 366, 367, 371; on refraction, 344–49

Democritus, 30–31, 36, 233, 374n1

Denery, Dallas, 290

density (*densitas* or *soliditas*), 97, 110, 115, 122–23, 125, 185–86, 209–11, 289, 364, 396–97. *See also* opacity

derivative (*dirivativa*) shadows, 310–11

Descartes, René, visual theory of: background in Mechanical Philosophy, 377–78, 413–14; on color, 400–404; epistemological implications of, 376, 413; human body as a hydraulically driven machine in, 411, 415–16; on light, 392–96; patronage of, 381; perceptual model in, 409–13; pineal gland as seat of sensitivity in, 411–12; as pivotal to modern optics, 415–16; on rainbows, 403; shortcomings of, 414–15; on sine-law of refraction, 394–96; sources relied on for, 391–92, 395, 401n56

Despars, Jacques, 350–51

Didymus, 166

diffraction, 349, 403n59, 408, 409

Diocles, 69–72, 166, 206, 334

Diocletian, Emperor of Rome, 229

Diogenes of Apollonia, 36

Dionysius the Areopagite, 235. *See also* Pseudo-Dionysius

dioptra, 78n8, 93–95, 105, 111, 113, 115

dioptrics, 78n8, 398

diplopia/double vision, 87, 88–90, 91, 97, 148n57, 159n82, 170, 193n24

discursive reason (*logistica*), 135, 143–46, 149, 162, 179, 194, 240, 241–42

dispersion, 349

distance perception, 49–50, 53, 61, 62, 83–85, 167, 192, 218, 224, 226, 312

divine illumination, 153, 248, 251, 255, 265–66, 413. *See also* intellectual illumination

Divini, Eustachio, 387

Dositheus of Pelusium, 71

Dresden, 366

Dtrūms, 166n98, 206

dura mater, 37, 186

Dürer, Albrecht, 304

early medieval Europe: formation of, 228–29; Greek language in, 235; intellectual developments in, 229–31; in so-called Dark Ages, 228; survival of ancient philosophy in, 230–32

Eastwood, Bruce, 159

Edgerton, Samuel, 301–4

Eisenstein, Elizabeth, 327

Elkins, James, 314, 315

Empedocles, 29, 30, 36, 44

Epicurus, 30, 233

epistemic gap, 165, 250–51, 370, 376, 413

epistemology, 3, 4, 34–35, 152–53, 164–65, 241, 250–51, 276, 288, 290, 294–95, 370, 379–80, 413

epithumetikon, 137

equal-angles law of reflection, 28n9, 51n82, 52, 56–57, 66–68, 93–97, 111, 128, 130n1, 170, 172, 195–97, 267–68, 274, 294, 396

Erasistratus, 42–43

Eriugena, John Scottus, 235–37, 257

estimative faculty (*estimativa*), 162–65, 194, 248, 255, 262, 264

ether. *See* aether/aither

Eucharistic theology, 288

Euclid: on cathetus rule, 59, 61–62, 67; on color, 54; compared to Ptolemy, 91–92; on focal properties, 68; in Latin West, 267, 272, 281, 283, 333–34, 335; on light, 54, 73; on reflection, 55–61, 98, 107, 108; shortcomings of, 53–55, 62–63, 68–69, 75, 84–85, 128, 167, 334; on spatial perception, 47–53; use of geometry by, 11, 54–55, 62, 73–74, 79–80, 180; on visual illusions, 59, 61, 62, 64, 91–92, 98

Euclid, works of: in Arabic East, 171–72; authenticity of, 23–24, 55n89, 72n107; *Catoptrics*, 10, 24, 28–29, 51, 55, 62, 68, 72n107, 77–78, 109, 166, 170, 171–72, 179, 243, 260, 267–68, 283, 333–34; *Elements*, 11–12, 15, 47, 55, 100, 155, 259; influence of, 109, 166, 170, 179, 183n7, 204, 268, 334; *Optics*, 10, 24, 29, 47–53, 109, 131n3, 157, 166, 243, 259, 260, 283, 328–29; printing history of, 328–29; translations of, 155, 157, 243, 283, 334

Eudoxus, 47n76

Eugene of Sicily, 76–77, 242

Eurocentricism, 8–10, 13

Experimental Philosophy, 379–80, 391

experiments: bronze-disk, 93–96, 111–16, 198, 213–14; diplopia texted in, 88–90, 266; empirical ray tracing, 347–49, 358–60; Experimentum Crucis, 406–7; floating-coin, 78, 110–11; Ibn al-Haytham on reflection, 195–200; Ibn al-Haytham on refraction, 206–15; limitations of, 96, 199, 213–14, 225, 325; needle, 217, 222, 307, 309, 317; Ptolemy on refraction, 119–20

extended substance (*res extensa*), 413

external senses, 134, 140–41, 152, 163, 234, 235–37, 249, 254, 264, 295 extramission: disproved, 352, 366; intromissionist challenges and modifications to, 30, 74, 147–48, 160–61, 179, 184–85, 225, 245, 256, 265–66, 273, 293, 306, 350–51, 374–75; support for, 29–30, 154, 158, 233, 236, 237, 240–41, 260–61, 283–84, 289. *See also* visual ray theory

Eyck, Jan van, 315

eyeglasses, 270n93, 280, 291, 325, 327, 336–37, 363, 382, 383

faculties, psychological: Arabic thoughts on, 158; Aristotle on, 33–34, 140–41; Avicennian model of, 162–66; Cartesian philosophy and, 412–13, 415–16; as cause of visual illusion, 320; hierarchy of, 149, 162, 234, 237–38; implications of retinal imaging for, 368–69; location of, 43, 146, 152, 162, 247–48; multiplication of species through, 248–49, 276–77; order of processes in, 162–63, 178–80, 241, 253–55, 262; perspectivist visual theory on, 262; scholastic visual theory on, 247–48, 253–55; subordination of to agent intellect, 165; visible intentions in, 189–92, 226

faculty of discrimination *(virtus distinctiva)*, 189, 190–91, 265; faculty of judgment, 289, 369

Falco, Charles, 317–19

Fārābī, al-: authority of, 133, 165n97; biographical background for, 157; *Catalogue*, 178n128, 182; estimative faculty according to, 162; influence of, 182; intromission and, 159–60; purpose of optics according to, 178n128

farsightedness. *See* visual disorders: presbyopia

Fermat, Pierre de, 396–98, 399

field of view/visual field, 41, 48, 80, 82, 84–86, 94n63, 168, 170, 189, 193, 216, 217, 300, 307, 361, 362, 368, 383, 386, 388

final cause, 31, 32, 35

final sensor *(ultimus sentiens/sensator)*, 189–90, 194, 264

fiqh, 18n45

first principles *(propositiones prime)*, 189–90

Florence, 327

focal point, 68–70, 173, 205, 220–21, 334–35, 344n40, 346–47, 358, 361–62, 365, 415

Form *(eidos)*, 34, 133–34, 137–42, 144, 149, 151, 178, 250, 255

form *(forma)*, 151, 152–53, 237, 238, 296. *See also* particular form/form of a particular; species *(species)*; universal form *(forma universalis)*

formal cause, 31, 32

Fourth Crusade, 231

Frederick II, Emperor, 13n31, 244

Frymire, John, 330n19

Galen: appeal to *hēgemonikon* in, 83; on color, 40–41, 43, 127; compared to Aristotle, Ptolemy, 80, 127; on extramission, 43; in Latin West, 146, 232, 234, 236n21, 277, 350, 351–52, 411; on light, 73, 74; on the optical system, 36–40, 72, 87, 159, 187; originality of, 42–43; use of geometry by, 43, 74; visual cone according to, 41, 47

Galen, works of, 23; in Arabic East, 133, 158–59, 170, 194; *De placitis Hippocratis et Platonis*, 37; *De usu partium corporis humani*, 10, 37, 41, 170; translations of, 157, 239

Galilean telescope, 382–87

Galilei, Galileo, 337, 366, 373–74, 378, 380, 381, 383–86, 396n43

Gassendi, Pierre, 375, 377–78, 398

Geminus, 55

Gerard of Cremona, 15, 242, 243

Ghiberti, Lorenzo, 303

ghost images, 107, 204, 215n60

Gilson, Étienne, 248

glacial humor *(glacialis)*, 187, 188, 261, 264, 270

glass technology and quality, 10–11, 21, 63–64, 113, 214, 280, 324–25, 326–27, 336–37, 347n46, 349, 382

Goughenheim, Sylvain, 21, 228n2

governing faculty (*virtus regitiva*), 83, 86, 88, 107, 127. See also *hēgemonikon*

grace, 151, 236, 270–71, 289, 293

Grassus, Benvenutus, 350n49

Greek language, 13, 14n31, 230, 235

Grew, Nehemiah, 389

Grimaldi, Francesco, 403n59

Grosseteste, Robert, 256–60, 263, 266, 271, 275, 290, 392

guilds, 326

Guillaume de Lorris, 291, 292

Gundissalinus, Dominicus, 243, 245

Gutas, Dimitri, 155, 156n76

Gutenberg, Johannes, 12n29, 327

hadīth, 17

Hagia Sophia, 172

Haly Abbas. *See* Majūsī

Harriot, Thomas, 7n15, 367n75, 383, 393n40

Hartlib, Samuel, 380

Harvey, William, 389

Haskins, Charles Homer, 20n50

Hatch, Robert, 380n11

heart, 40, 240, 411; as seat of perception, 43, 83n26, 141, 236

hēgemonikon, 40, 83, 127

Heiberg, I. L., 55n89, 72n107

Hellenistic learning, 13, 79, 155

Henry VI, Emperor of Sicily, 76

Henry Aristippus, 76

Hero of Alexandria: authenticity of works, 23–24, 72n107; on burning mirrors, 71; *Catoptrics*, 64, 71, 80, 93, 109, 166; compared to Ptolemy, 97; on dioptrics, 78n8; influence of, 80, 93, 97, 130n1, 274; on least-distance principle, 66, 119, 396; on light, 73; on reflection, 65–68; use of geometry by, 73; on visual illusions, 64–68; on visual rays, 74

Herophilus, 42–43, 232n6

Hobbes, Thomas, 377–78, 414

Hockney, David, 317–19

Hodgson, Marshall, 7n13

Holbein, Hans, 315

Hon, Giora, 385

Hooke, Robert, 379, 387–89, 390, 397, 403–6, 414

Ḥubayash, 157

Huff, Toby, 17

Hugh of Saint Victor, 238n26

Hulagu, 8n16

human will, 151

humor. *See* albugineous humor; aqueous humor (*ōoeidēs*); crystalline humor (*krustalloeidēs*); glacial humor (*glacialis*); vitreous humor (*hualoeidēs*)

Ḥunayn ibn ʾIsḥāq, 179; authority of, 133; biographical background to, 156–57; cerebral ventricles according to, 162; compared to Galen, 159; on extramission, 158–59; in Latin West, 248, 277, 308, 351; on light, 159; optical system according to, 158–59, 161, 187, 350, 351; translations of, 239

Huygens, Christiaan, 375, 381, 386, 389, 398–400, 403–4

Huygens-Fresnel principle, 399

Iamblichus, 132, 143

Iberian Peninsula, 229

Ibn Abī Uṣaybīʿa, 157n190, 181

Ibn al-Haytham: atmospheric refraction in, 218, 222–24; biographical background to, 181–82; on color, 186; compared to Avicenna, 183, 184, 191n19, 194; compared to Ptolemy, 193, 200, 204, 224–26, 364; on distance perception, 226; epistemological implications of, 227; final sensor in, 194; on image distortion, 204–5, 218–20; on image formation, 221–22; on image fusion, 194; on imagination, 194; in Latin West, 6, 16, 183, 221, 227, 248, 256, 259–60, 265–76, 281, 284–85, 289, 292–93, 305, 306–7, 316, 322, 329, 365, 375–76, 391; lens theory and, 225; on light, 180, 184–85, 209–11, 224; mathematical drive behind, 11, 184, 199, 204–5; optical system according to, 186–88, 194; on peripheral vision, 215–18,

Ibn al-Haytham (*continued*)
226; as pioneer of modern optics, 198–99, 224–27; reflection analysed in, 108, 199–200, 201–4; refraction analyzed in, 216–24; shortcomings of, 206, 211, 213–14; sources relied on for, 98, 104, 179–80, 183–84, 202, 329; on spatial perception, 192; on spherical aberration, 220–22; on transparency, 185–86; visible intentions model in, 189–92, 194; on visual illusions, 192–93, 223–24; on visual ray theory, 184–85, 205, 225, 317n95

Ibn al-Haytham, experiments of: "Alhazen's Problem" in, 201–4, 226; compared with Ptolemy, 198, 213–14; feasibility of, 199, 214, 225; image formation tested in, 222; oblique radiation tested in, 217, 307; reflection tested in, 195–204; refraction tested in, 206–14; quantitative analysis of refraction in, 211–15

Ibn al-Haytham, works of: in Arabic East, 20; authenticity of, 181–82; *Book of Optics (Kitāb al-Manāẓir)*, 1, 14, 179–80, 182; critical editions of, 3; *De aspectibus*, 1, 10, 14–17, 21, 182–84, 195, 205, 206, 215, 223, 225–27, 241, 243, 248, 256, 259–60, 267–69, 271–75, 278, 279, 284–85, 322, 328, 329, 365, 375–76; *De speculis comburentibus*, 182, 205–6, 220–21, 243, 274, 278; misrepresentations of, 215, 220–22, 221, 225; printing history of, 328–29; translations of, 1, 14–17, 21, 182–83, 225, 241, 243, 279; *Treatise on the Burning Sphere*, 182, 220–21

Ibn al-Nadīm, 157n78

Ibn al-Qifṭī, 157n78, 181

Ibn Bājjah (Avempace), 165n97

Ibn Daud (Avendauth), 243

Ibn Khaldūn, 17n44, 18n45, 18n48, 165n209, 286

Ibn Muʿādh, 243, 274

Ibn Sahl: influence of, 176–78, 183, 206, 225, 385–86; *On Burning Instruments*, 176;

Ptolemaic influence on, 179; sine law of refraction in, 7, 9, 11, 176–78, 393n40

Ilardi, Vincent, 325

image (*phantasm*), 33, 34, 135, 141, 153, 162–63, 179, 234, 235, 253–54, 288

image fusion, 42, 43, 74, 89, 194, 412

image mapping, 300–302, 304

image projection, 302, 307–9, 318, 352, 354, 360–62, 375

images, mirror: distortion of, 204–5, 233, 268n91, 292, 382; formation of, 221–22, 292, 344, 369; inversion of, 59, 63, 205, 272, 307–8, 317n95, 367–68, 374, 375, 386; location of, 59, 93, 104–5, 108, 113, 119, 170, 200–201, 215n60, 220, 267; multiplicity of, 59, 102, 202–3; reversal of, 61–63

imagination (*phantasia*), 297; in Cartesian philosophy, 412–13, 415–16; compositive, 162–63, 194, 247n48, 248, 319; as distinct from common sense, 162; formation of images (*phantasmata*) in, 33, 135; inherently sensibles in, 264; location of, 238, 240, 247–48, 254, 262, 412–13; as mediator between sensation and cognition, 141–42, 143–44, 149, 152; as mirror, 250, 297–98; as mnemonic storehouse (retentive), 35, 162–64, 190, 194; pneuma shaped by, 146; secondary qualities in, 373–74; sensible species in, 141–42, 238, 241, 248–49, 253–55, 262, 276; vague particulars etched on, 265

imperceptible sensibles (*insensate forme*), 264

imprinting (*tupousthai*), 30–32, 135, 139, 140, 142, 146, 153, 161–63, 188, 217, 238, 241, 249, 263–64, 295, 354

incidental sensible, 136, 264

index of refraction, 113, 178, 347n46, 349, 357, 395

inflexion. *See* diffraction

inherently sensible (*sensibilia per se*), 264

Innocent III, Pope, 288n24

intellect (*mens*), 153, 194, 234, 235, 237–38,

241, 248, 265, 298, 413; active (productive), 136, 143–45, 149, 179, 251; agent, 149, 164–65, 250–51, 253, 255; as Christ, 151, 153; formation of species in, 262; passive (*pathetikos*) intellect, 136, 145, 149; potential, 144–45, 149, 164, 179, 251, 253, 255; as replicator of Forms, 138–39; as separable from the body, 136–37

intellectual illumination, 138n133, 142, 145, 149, 164, 179, 250. *See also* divine illumination

intentions, 163–64, 246, 255. *See also* visible intentions (*intentiones visibiles*)

interior sense (*sensus interior*), 152, 194, 235–37

internal senses model. *See* Avicenna (Ibn Sīnā)

intromission. *See* extramission: intromissionist challenges and modifications to

intuition (*intuitio*), 190

intuitive cognition, 252–53

iris, 36, 36n37, 81n16, 362

Ishāq ibn Hunayn, 157

Isidore of Seville, 234

Islamic law, 18–19

Italy, 244

James of Venice, 13n31, 242

Jean de Meun, 292–93, 316

Jesenský, Ján (Johannes Jessenius), 366

John of Toledo, Archbishop, 243

John Philoponus: influence of, 161, 167, 168; on intromission 179, 225; on light, 147–48; optical system according to, 146–47; philosophical background for, 132, 143

Johns, Adrian, 327n12

judgement of sense, 314

Julião, Pedro, 350n49

Jupiter, 384, 390, 398

Kamāl al-Dīn al-Fārisī, 7–8, 16n42; on rainbows, 285, 403

Kemp, Martin, 318–19

Kepler, Johannes: academies and scholarly networks of, 266–67, 333; *Ad Vitellionem Paralipomena*, 2, 10, 322, 353, 392; biographical background for, 322–23; on camera obscura, 353–54; empirical ray tracing, 358–59; on epistemology, 370, 376; on image inversion, 367–68; influence of, 391–92, 414; lens theory in, 344, 354–63, 385; on light, 392; on the optical system, 362–63; patronage for, 381; perceptual limitations of, 368–69, 374, 378–79, 413; perspectivist visual theory and, 1–4, 284, 321, 329, 363–72; as pivotal to modern optics, 5, 364, 370–72, 373, 375–76, 415; on retinal imaging model, 4, 5–6, 323, 333, 352, 354, 363, 367, 369–72, 374; on virtual images, 360

Keplerian (astronomical) telescope, 386

Kindī, al-: on burning mirrors, 172–74; on epistemology, 165; on intellect, 164; knowledge of Greek optics of, 166–67, 171–72, 175, 179; in Latin West, 167, 183–84, 259, 260, 274, 283; on light 185; sources available to, 132–33, 156–57; on visual ray theory, 158, 166–69, 171–72; on visual scanning, 167; works translated of, 243

Koyré, Alexandre, 2n2

Knorr, Wilbur, 79

knowledge (*scientia*), 151–53, 261

Kuhn, Thomas S., 2, 277

last limit (*finis ultimus*), 358, 360

late antique encyclopedias, 232

Latin Empire of Constantinople, 231, 244

Latin translation movement, 1, 13–16, 24, 76–77, 154, 215, 228, 231, 237, 242–45

least-distance principle of reflection, 66–67, 97, 119, 396

least-time principle of refraction, 396–98

Leeuwenhoek, Antonie van, 388

Lejeune, Albert, 77, 79

lenses: burning, 177–79, 206, 220–21, 335–36, 342–44; compared to mirrors, 344, 346–47; concave spherical, 220–21, 342–44,

lenses (*continued*)
  358, 367; convex spherical, 4, 5, 318, 367; in correcting visual disorders: 338–40, 363; ellipsoidal, 391–92; focal points of, 341–44; hemispherical glass, 336; mass production of, 325; in microscopes, 387, 388; in telescopes, 382, 385
lens grinding, 383, 387, 392
lens theory, 220–22, 225, 354–63, 385
Leonardo da Vinci: on biconal visual theory, 309; challenges to perspectivist theory in, 305–9, 320, 371; educatonal background of 317, 326, 341n37; on the eye as a camera obscura, 307–8, 367; on the optical system, 350; on shadow-casting, 309–12; use of geometry of, 312–15
Leucippus, 30
liberal arts, 13–14, 20, 230, 231, 253
light theory: corporeality of light, 200, 209–11, 270–71, 396–400; hybrid particle-wave model, 407–8, 414–15; impulse light theory, 393, 396, 399, 410; incorporeality of light, 139–40, 143, 147–48, 392; light as catalyst of transparency, 31–32, 72–73, 74, 135, 139, 245–46; light as *ḍaw'* and *nūr*, 161, 183; light as primary/objective cause of vision, 81, 128, 264; light as quasi-kinetic, 169; light as subjective/secondary quality, 400; light as vehicle of heat, 72, 159, 172–74; light rays compared to visual rays, 75, 80, 176, 179, 198, 224–46; longitudinal wave model, 403–4; *lux-lumen* distinction, 183, 185, 245–46, 253, 257, 262; speed of light, 398–99, 404, 409; subordination of sight theory to, 2, 5, 283–84, 364, 370–72, 414–15; wave fronts of light, 399–400. *See also individual thinkers*
likeness (*similitudo*), 133, 153, 235, 248–49. See also *simulacra*
Lindberg, David, 1–4, 261, 351, 363–64
linear perspective, 50, 299, 303–4, 312–13. See also *costruzione legittima*
Lipperhey, Hans, 381–83
Lombard, Peter: *Sentences,* 278, 288

Lucian, 71n106
Lucretius, 30
lumen, 183, 185, 246, 257, 258–59, 262, 392
lux, 183, 246, 257, 259, 262–63, 392
lux minima, 185

Macrobius, 233
magnification: with eyeglasses, 205, 269n92, 337–38, 363; in mirrors, 120–21, 218–20, 270; moon illusion and, 125–27, 224, 289; point of inversion and, 344; in spiritual vision, 291; in telescopy and microscopy, 382–87, 408–9; as unnatural, 391
Majūsī, al- (Haly Abbas), 239
Malpighi, Marcello, 389
Ma'mūn, al-, 156
ma'nā/ma'ānī. See intentions
Manetti, Antonio, 299
Manṣūr, al-, 155
Maragha observatory, 8n16, 8n29
Marius Victorinus, 150
Mars, 390
Martianus Capella, 230
material cause, 32
mathematical (geometrical) optics: disregard for other optical approaches in, 73–74, 147–48, 205; identified, 11, 184; interest in, 133, 166, 244–45, 378; level of sophistication in, 72, 131, 232, 268, 271, 273–76, 312, 378; perceptual challenges to, 54, 91, 127, 312–15; philosophical or psychological optics melded with, 75, 108–9, 127–28, 179, 225, 256, 260, 275–76
mathematics, development and study of, 18, 25, 69–70n102, 71, 178, 230, 280–83, 324n2, 378
matter, 134, 138, 145–46
matter-form analogy, 145
Maurolico, Francesco, 323, 338–44, 367, 371, 401n56
Maximilian I, Emperor, 253
Mecca, 18, 155
Mechanical Philosophy, 377–79, 392, 396, 398, 400–401

Medici, Cosimo II de', 384

memory (*memorativa*), 144–46, 149, 153, 162, 163, 179, 233, 234, 238, 240, 248, 253, 255, 262, 264, 276, 297

Menaechmus, 71

Mercury, 389–90

Mersenne, Marin, 380

metaphysics of light, light metaphysics, 256–57, 290–91, 392

Meun, Jean de, 291

microscope/microscopy, 376, 379, 387–91, 409, 415; compound microscope, 387, 388; simple microscope/flea glass, 388

Milan, 328

mind-body problem, 134, 136, 138, 413, 416

minimal light. See *lux minima*

mirrors: ancient, 63–64, 97n69; burning, 68–71, 73, 131, 166, 172–74, 176–78, 205, 341; composite, 108; Eucharist theology explained through, 288; God as, 290, 293, 319–20; imagination as, 250, 292, 298; iron, 94, 96, 195; in literary allegories, 292–93; parabolic/paraboloidal, 70–71, 205–6, 274; used in Renaissance art, 299, 302, 315, 318

mirrors, concave conical and cylindrical, 195, 204

mirrors, concave spherical, 58–59, 71, 358; Alhazen's Problem with, 202–3; burning with, 68–70, 131, 274, 341, 344; compared to lenses, 344–46; image anomalies seen in, 59, 62–63, 98–104, 171, 272; rainbows and, 233; reflection in, 4, 56–62, 67, 92–93, 103–4, 170, 195, 268, 334; refraction in, 273; in telescopes, 386–87; visual illusions seen in, 107; widespread availability of, 336–37

mirrors, convex conical and cylindrical, 195, 204

mirrors, convex spherical: Alhazen's Problem with, 201–2, 204; image anomalies seen in, 59, 61, 170–71; reflection in, 56–58, 67, 92–93, 195, 197–98, 200; visual illusions seen in, 90, 98

mirrors, plane: burning with, 174; image

anomalies seen in, 59, 61, 204; reflection according to, 56–58, 67, 92–93, 95, 97, 170, 195–97; visual illusions seen in, 64, 90, 98

monastic education, 13, 230–31

Monte Cassino, 239

moon illusion, 125–28, 223–24, 289n26, 312

motor nerves, 37

Muʿāwīyah, 155

Muḥammed, 17, 155

Murano glassworks, 325

Muslim conquests, 229, 242

Muslim education, 17–21

Muʿtaṣim, al-, 156

Naṣīr al-Dīn al-Ṭūsī, 8n16

natron, 325n4

natural (*naturale*) perspective, 299, 302, 312

naturalistic art, 279, 299, 304–5, 311–12, 313, 324

natural philosophy: applied to optics, 11, 184, 226, 258, 260, 368, 371, 398, 409; in education, 14, 17–18, 20, 231, 232, 253, 287, 324n2; seventeenth-century changes in, 376–78

natural spirit (*naturalis spiritus*), 232

nature: as perfectly efficient, 67–68, 119–211; as providential, 37, 86–87

nearer limit (*quod terminet citime*), 355, 360, 365

nearsightedness. *See* visual disorders: myopia

Neoplatonism (Late Platonism), 131–32, 235, 257

Neri, Antonio, 325n5

nerve filament, 410

Netherlandish art, 376n6

networks of correspondence, 12, 286, 366, 380–81

Neugebauer, Otto, 117

Newhauser, Richard, 279n2, 289n25

Newton, Isaac, 386–87, 396, 405–8, 414–15, 416

Newton's rings, 407–8, 409

nūr, 161, 183

Nuremberg, 327, 328

objective cause, 379

oblique radiation, 217, 307, 341

occult/occultism, 331–32, 378, 381, 388

ocular anatomy and physiology, 36, 38–43, 81, 146, 159, 161, 186–88, 234, 236–40, 261–62, 289, 339, 350–52, 362–63, 374

oculogyral illusion, 90, 267n88

Oldenburg, Henry, 380

Olivi, Peter John, 252, 319

opacity, 45, 81, 128, 185–86, 189, 209

optical literacy, 277; evident in literature, 279, 281n7, 291–98, 316, 319; evident in Renaissance art, 298–316; evident in theology, 278–79, 287–91, 319

optic chiasma, 37, 42, 43, 88, 159n82, 161, 187, 188, 193n24, 194, 264n82, 368, 369

optic nerves: anatomical placement of, 36, 37–38, 186, 232, 239, 351n51, 374; binocular vision and image fusion with, 42, 188; as cause of visual disorders, 228; as conduit of cerebral spirit, 161, 187, 194, 234, 240, 254, 261; as conduit of forms and species, 188, 194, 241, 245, 247, 264, 308, 339, 416; as location of visualty faculty, 245; role in impulse light theory, 409–11; role in retinal imaging model, 368, 371, 409

optics, teaching of, 21, 275, 280–87, 328

Oresme, Nicole, 253, 285

painting (*pictura*), 359–60, 367–68, 410–11

Panofsky, Erwin, 256n66

paper, 12, 21, 156n76

Pappus of Alexandria, 131

papyrus, 12

paradigm. *See* scientific/optical paradigm

parchment, 12, 286, 327

Pardies, Ignace, 403n59

Parmigianino (Francesco Mazzola), 315

Parronchi, Alessandro, 294n42

particular form/form of a particular, 140, 190, 194

patents, 326, 382, 383

patronage, 13n31, 19, 156, 242, 243, 381

Paul, Saint, 250, 319

Pecham, John, 271–73, 275, 307n73; influence of, 16, 256, 284, 339, 353n57; printings of *Perspectiva communis*, 328; on shadow-casting in, 310; spherical aberration and, 335–36; university education featuring, 21, 281–83

Peiresc, Nicolas-Claude Fabri de, 375, 380

penumbral shadow, 311

perception: through recognition/precognition (*comprehensio per scientiam antecedentem*), 191–93, 233, 265; by sense alone (*comprehensio solo sensu*), 188, 193; by syllogism (*comprehensio per syllogismum*), 189–93, 265

perception of motion, 85–86, 90

perceptual judgment, 33–35

Peripatetics, 143, 232

peripheral vision, 215–18, 225, 226, 341, 368

perspective. *See* accidental (*accidentale*)/artificial perspective; aerial perspective; color perspective; linear perspective; natural (*naturale*) perspective

perspectivist visual theory: epistemological implications of, 3, 287–88, 413; foundation for, 256–60; historical significance of, 1, 2, 16, 275–77; on light, 262–63, 392, 395; literary use of, 281n7, 291–98, 316; mirror and lens analysis in, 333–36; on the optical system, 352; on psychological faculties, 262; among Renaissance artists, 4, 302–14, 318; shortcomings of, 319–21, 334–36, 365; on species theory, 262–66; subversion of, 277, 323, 340–41, 354, 360, 363–72, 374–76, 392, 409–10, 415; theological use of, 278–79, 287–91; theological warrant for optics in, 257, 261, 270–71, 275; in university education, 21, 280–87; on virtual images, 309

Peter of Limoges, 278, 288–90, 292, 303

phantasies (*phantasiae*), 235–36

phantasy (*fantasia*), 248, 253, 254–55; as Wit's looking-glass, 297, 319–20

physical optics, 62, 67, 73–74, 80, 127, 131, 205, 371, 374, 408–9, 415–16

pia mater, 37, 186, 261

Pierozzi, Antonino (Saint), 290, 303

pineal gland, 411–14

Pinelli, Gian Vincenzo, 366n73, 367

plane of: reflection, 93, 197, 200, 375; refraction, 209, 212, 364

Plato, 276; challenge to mathematics in, 74; on color, 44–45, 83; compared to Aristotle, 131–32, 132, 138, 143; in Latin West, 232, 251, 294; on light, 73; on the optical system, 36; on sense perception, 137–38, 142, 178–79; on the soul, 134, 137–39, 151; use of geometry by, 73; on visual ray theory, 29–30

Plato, works of: authenticity of, 23; Christianization of, 230; influence of, 234, 236; in Latin West, 232, 234, 236, 237; *Phaedo*, 157; *Republic*, 137, 234, 324n1; *Timaeus*, 29–30, 44, 74, 80, 83, 132, 157, 232, 236, 237, 292; translations of, 13, 157, 232

Platter, Felix, 366, 374

Plotinus: on color, 139–40; compared to Aristotle, 139; de-materialization of the perceptual process, 142; on imagination, 142, 144, 152, 250n52; influence of, 132, 143, 156–57; on intellect, 138–39, 142; on light, 139–40; on memory, 163; on psychological faculties, 146; on sense perception in, 139–40, 252; shortcomings of, 147

pneuma (*pneuma*), 40, 42–43, 80, 125n108, 127, 140n26, 142, 146, 149, 158, 159, 180, 234. *See also* psychic pneuma (*pneuma psuchikon*); vital pneuma (*pneuma zōtikon*)

point of inversion (*punctum inversionis*), 344

polarization, 400

Pontormo, Jacopo, 315

Porphyry, 131–32, 134, 143, 230, 249n51, 250n52

Posidonius, 125, 233

potential intellect, 250

pre-Socratics, 23, 29, 72

primary colors, 45–46, 73

primary qualities, 373–74, 378, 392, 409

primary visibile characteristics, 82

primitive (*primitiva*) shadows, 310–11, 313

principle of conservation: conservation of power, 119; horizontal conservation, 394, 396

principle of natural economy, 67–68, 396

principle of reciprocity, 118, 211, 220, 348, 356–57, 364

print culture, 327

printing industry, 21, 280, 327–31

printing revolution, 327

Priscianus, 143

prisms, 386, 401, 405–7

Proclus, 12n26, 47n76, 55n88, 77, 78n8, 157, 250n52

projectile motion, 65, 97, 393

Pseudo-Aristotelian works, 46, 73, 155n75

Pseudo-Dionysius, 235, 257n67

Pseudo-Euclidean works, 172, 243, 259, 260

Pseudo-Simplicius, 143

psychic pneuma (*pneuma psuchikon*), 40, 73, 236n21

psychological optics: appeal to, 74, 79–80, 91, 107, 126–28, 131, 160, 178–80, 184, 205, 223–24, 248, 275–76, 305, 312; dismissal of, 309, 368, 370–71, 414, 416n81; interest in, 165–66, 244–45

Ptolemy: appeal to psychology in, 75, 79–80, 91, 107–8, 126–28, 131; astronomy and, 79, 117, 128–29; on atmospheric refraction, 121–25; on color, 82–83, 127; compared to Aristotle, 127–28; compared to Euclid, 83, 91–92, 128; compared to Galen, 80, 128; compared to Hero, 97, 119; compared to Ibn al-Haytham, 193–94, 198, 200, 213–14, 225–26, 364; data tabulation in, 79, 116–18, 274, 364, 365; on ghost images, 204; on image formation, 104–8; on image location, 119–21; in Latin West, 14n34, 260, 266, 274, 283; on light, 80, 81–82, 128; on the optical system, 81, 86–87, 88, 307n73; on reflection, 92–104, 128; on refraction, 109–19; sources relied on for, 97, 108–9; on spatial perception, 83–86; use of geometry by, 75, 79–80, 91–92, 108, 128; on visual acuity, 86–87; on the visual faculty, 107, 126, 226;

Ptolemy (*continued*)
  on visual illusions, 87, 88–92, 98, 102, 104, 125–28, 224; on visual rays, 80, 84–85, 92, 97, 109, 126
Ptolemy, works of: *Almagest* (*Mathēmatikē Suntaxis*), 11–12, 76–79, 121–22, 128–29, 282n10, 324n3; in Arabic East, 166, 170, 171–72, 175–76, 178–80, 182, 183, 200; authenticity of, 23–24, 72n107, 79; influences of, 130–31, 364; *Optics*, 24, 54, 76–79, 80n14, 91, 121, 126–31, 166, 170, 172, 175, 178, 179, 182, 183, 204, 243, 259, 260, 283, 299n47, 364; translations of, 13n31, 76–77, 80n14, 242–43, 283
punctiform light radiation, 183, 185, 261, 263
pupil (*korē*): anatomical placement of, 39, 187; as cornea, 36, 81n16; in extramission, 41, 158, 236, 237, 240–41; limitations of in peripheral vision, 215–17; as similar to the aperture of a camera obscura, 306–9, 344, 410; in spiritual vision, 270, 289; as visually sensitive, 148, 166–67, 169–70, 234
Pythagoras, 83, 143
Pythagorean theory, 80

qibla, 18
quadrivium, 13, 324n2, 332
Qur'ān, 17
Qusṭā ibn Lūqā, 133, 169–70; knowledge of Greek optics, 171–72; on light and visual rays, 170–71, 179, 184
Quṭb al-Dīn al-Shīrāzī, 8n16

rainbow, 26–28, 29, 31, 43, 45–46, 74, 232–33, 259, 261–62n76, 268, 273, 274, 285, 292–93, 316, 400–403, 406, 409; primary, secondary, 27, 401
Rashed, Roshdi, 3, 7, 8n18, 166, 174n123, 176n126, 181, 182, 206, 220–21
rational apprehension (*scientia*), 151, 152, 261
ray geometry. *See* mathematical (geometrical) optics; mathematics, development and study of

Raymond of Toledo, Archbishop, 242, 243
ray tracing: empirical, 347–49, 358–60, 365; theoretical, 347, 357–58, 365, 367n75
real images, 205, 307–8, 317–18, 360
reason (*ratio*), 137, 138, 141, 144, 148–49, 152, 154, 234, 235–38, 248, 373
recollection (*anamnēsis*), 35, 135, 162, 163, 194, 249, 253
Reconquista, 231
Reeves, Eileen, 337
reflection, 4–5, 55, 97, 171–72, 256, 258; Alhazen's Problem and, 201–4, 375; compared to refraction, 56, 78, 92, 109–11, 119, 396; as or like physical rebound, 65, 96, 97, 110, 200, 209, 258; principles of, 28, 56–59, 65–68, 92–97, 98, 110, 128, 195, 197, 199–200, 399–400; in rainbows, 401–2; in Renaissance art, 279; in spiritual vision, 289; in telescopes, 386–87, 406. *See also* cathetus rule; equal-angles law of reflection
refraction, 4–5, 97, 258, 268–69, 284, 344; atmospheric, 122–23, 125, 176, 218, 222–24, 285, 322; compared to reflection, 56, 78, 92, 109–11, 119, 396; in corrective eyewear, 341; double, 406, 409; impulse light theory and, 404–5; Kepler's lens theory on, 354–63; principles of, 110, 118–19, 128, 131n3, 176–77, 209, 211, 213, 214–15, 219–20, 260, 358, 364, 399–400; in rainbows, 285, 401–3, 409; in Renaissance art, 279; in spiritual vision, 289, 291; in telescopes, 406. *See also* principle of reciprocity; sine law of refraction
Reisch, Gregor, 253–55, 284, 320
Renaissance academies, 21, 333, 366, 380–81
Renaissance art, 4, 46, 298–320
rete mirabile, 40, 236n21, 240
retina (*amphiblēstroeidēs*): anatomical placement of, 38, 186–87, 362–63; complications of image inversion and, 367–68, 371, 375; as conduit for cerebral spirit, 159; with corrective lenses, 363; impulse light theory and, 409–11, 413, 416; as opaque

division between mind and body, 369; as primary component of vision, 352; projection of images onto, 354
retinal imaging, 5, 352, 362–63, 369–70, 374, 410–11, 412, 415
Risner, Friedrich, 222, 322, 328, 329, 365, 375
Roemer, Ole, 398, 399
Roman de la Rose, 292–93
Ronchi, Vasco, 2n4
Rudolph II, Emperor, 322
Ryle, Gilbert, 127–28

Sabra, Abdelhamid I., 3, 181, 182
Sacrobosco, Johannes de, 14n34
Said, Edward, 8
Saliba, George, 8–9
sapientia. See wisdom (*sapientia*)
Saturn, 384, 390
Savonarola, Girolamo, 290–91, 338
Sbaraglia, Girolamo, 390–91
Scheiner, Christoph, 374–75, 384, 386, 391–92
scholastic visual theory: agent intellect in, 250–51; Avicennian foundations for, 245; on color and light, 245–47; on intentionality, 246–47, 249–50; on psychological faculties, 247–48, 253–55, 414; refutations of Aristotle in, 247; sixteenth-century, 253–55; on species theory, 248–50, 252–53, 255. *See also* perspectivist visual theory
scientia. See rational apprehension (*scientia*)
scientific/optical paradigm, 2, 277, 280, 283, 295, 316, 320, 321, 364, 370–72, 374, 375, 415
scientific professionalization, 381
Scientific Revolution, 376–77
sclera (*consolidativa*), 186
scleral or "hard" tunic (*chitōn sklēros*), 38
Scot, Michael, 244
scriptorium, 230
seal and wax analogy, 32, 135, 139, 153, 161, 295
secondarily visible characteristics, 82

secondary illumination/light. *See* accidental light (*lux accidentalis/lumen accidentale*)
secondary qualities, 373–74, 400, 409
Seleucid period, 79, 117
self-reflection/introspection, 290, 292, 295–98, 319
Seneca, 232–33, 267
sense perception (*cognitio solo sensu*), 265
sensible form, 179, 194
sensory nerves, 37, 149, 180
sermons, 279, 290–91, 330
sfumato, 313
shadow-casting, 50–52, 168, 274, 309–11. *See also* derivative (*dirivativa*) shadows; primitive (*primativa*) shadows
Simon, Gérard, 88n41, 213
Simplicius, 132, 146
simulacra, 30, 232, 233, 238
sine law of refraction, 7, 9, 113, 177–78, 206, 367n75, 387, 393–96, 397, 401, 406, 408
size-distance invariance hypothesis, 192, 218, 269
size perception, 5, 48–54, 61–62, 83–85, 90, 125–26, 161, 170–71, 175, 189, 192, 193, 218, 222–24, 269, 312–13, 338
Snel, Willibrord, 7, 9, 11, 393n40
social construction theory, 9
Somfai, Anna, 237
Sosigenes, 72n107
soul: Aristotle's theory of, 32, 35, 134–37; Augustine's theory of, 151–54; immortality of, 133, 250–51, 289; irrational soul, 139, 142, 143–44, 235; as material, 145–46, 414; as object of introspection, 290; Plato's theory of, 132n7, 137–38; Plotinus's theory of, 133, 138–42; rational soul, 135, 139, 142, 143–44, 235, 248, 262, 412–14; sensitive soul, 135, 248n49, 250, 255, 276, 291, 413; soul and intellect, 234, 236, 237, 248, 250, 251; vegetative soul, 134–35; World Soul, 137–39
Spain, 12n29, 13, 15, 231
spatial perception, 47–53, 83–86, 90, 91, 284n15

special form, 140

special sensibles, 33–34, 35, 135–36, 148, 276

species (*species*), 152–53, 234, 291–92, 352, 416; intelligible (*species intelligibiles*), 248, 249, 252, 253, 255, 276; intentional, 249–50, 368, 369, 370, 392, 410, 413; memorative (*species memoriales*), 248–49; multiplication of, 253, 262–66, 272, 276, 293; sensible (*species sensibiles*), 249, 250, 252–55, 262, 276, 288, 412; skepticism about, 252, 370, 414; visible, 249, 264, 339, 340, 374

spherical aberration, 70, 220–21, 334–36, 341–43, 346, 347, 355–56, 365, 385–86

spirit. *See* animal, cerebral spirit; natural spirit (*naturalis spiritus*); visual spirit

spiritual existence (*esse spirituale*), 246–47, 288

spiritual vision, 270–71, 288–91, 303, 319–20

spyglasses, 382–83

Stahl, William, 130–31n2

Stoics, 42–43, 78n9, 80, 83, 109, 125n108, 134, 140n26, 142, 143, 232, 234. *See also* sympathy

Strato of Lampsacus, 46

Suger, abbot of Saint-Denis, 256n66

Summers, David, 314, 315

sunnah, 17

sunspots, 374, 384

Swammerdam, Jan, 389

sympathy, 40, 43, 80, 140, 142, 147

Syriac language, 155, 156n76, 157

Syrus, 12n28

Tachau, Katherine, 3, 252–53, 287

Takahashi, Ken'ichi, 55n89

Targ, William, 329

telescopes/telescopy, 325, 337, 376, 379, 381–87, 389–91, 398, 406, 408–9, 415. *See also* reflection; refraction

Thagaste, 150

Theaetetus, 47n76

Themistius, 132, 144–45, 244, 251

Theodoric of Freiberg, 285, 401, 403

Theon of Alexandria, 55n89, 72n107, 131n3

theophany, 235, 257

Theophilus Presbyter, 330

Theophrastus, 30–31, 36, 46

thinking substance (*res cogitans*), 413

thumoeides, 137

Tideus, 243, 260

Toledo, 242–44

transparency: density as tied to, 185–86, 211; effects of *lux* and *lumen* on, 245–46; impulse light theory on, 393, 415; as an inherent property, 161, 183–84; as potential, 31–32, 73, 147–48

trigonometry, 18, 347

trivium, 13

trompe l'oeil, 128, 315

twelfth-century renaissance, 20–21n50, 228

ultimus sentiens. See final sensor (*ultimus sentiens/sensator*)

'Umayyad caliphate, 155

universal form (*forma universalis*), 191, 194

Universals, 34, 136–37, 144, 149, 153, 163, 164, 178, 252, 253, 265, 276

universities, 13, 14, 16, 20–21, 280–87, 328, 331

University of: Oxford, 281; Padua, 328, 383; Pavia, 328; Prague, 281; Tübingen, 353–54n57; Vienna, 281

uvea, 186, 187, 362, 363

vague individual/ vague particular (*particulare vagum*), 191n19, 265, 276

Vasari, Giorgio, 313–14

Venice, 325, 326–27, 328

Venus, 384, 389–90

vernacular, 21, 279, 280, 326n7, 329, 331

Vesalius, Andreas, 339n35, 351

Vescovini, Graziella, 279n3, 302

virtual images, 205, 309, 323, 360, 415

virtual motion, 393, 395–98

virtual rays, 217–18, 225, 341

virtus distinctiva. *See* faculty of discrimination (*virtus distinctiva*)

visible intentions (*intentiones visibiles*), 189–92, 250, 264, 409

vision, preconditions for, 91, 127, 192–93, 266, 296

visual acuity, 54, 86–87, 167, 168–69, 190, 217, 338

visual axis, 86–88, 91, 97, 186, 190, 261, 374

visual cone, 169–70, 179, 225; Alberti's cone of projection and, 304; doubled, 307–8, 317; extramission and, 148, 160, 168; formation of, 41, 47, 80, 81, 166–67, 300; perception and, 48, 53, 84–86, 161; reflection and, 93; rejection of, 127, 363, 370–71; visual illusions and, 87. *See also* cone of radiation

visual disorders: cataracts, 338; corneal lesions, 338; myopia, 286, 325n6, 336, 337–41; presbyopia, 89, 286, 325n6, 336, 337–41

visual faculty (*vis vivendi*), 82, 83, 84–85, 89–90, 107, 119, 126, 127, 216–17, 220, 226, 234, 245

visual field. *See* field of view/visual field

visual flux, 128; abnormalities of, 89, 90, 91, 127; complications for extramission with, 160–61; interaction with light, 77; perception initiated by, 82–83, 107, 158–59, 234, 238; radiation of, 80–81; reflection and refraction of, 90, 92, 97; visual acuity and, 86

visual illusions (*deceptiones visus*): corrected by reason, 319; corrective eyewear as creating, 338; physical and psychological conditions for, 87–91, 126–28, 192–93, 223, 266–67; reflective/refractive radiation and, 61, 64–65, 90, 92, 98, 104, 107, 267–70. *See also* diplopia/double vision; moon illusion

visual ray theory, 26n6, 28–30, 56, 166–69, 172, 184–85, 233, 256, 259–60, 374–75; color perception according to, 44–46;

criticisms of, 53–55, 147–48, 160, 245, 265–66, 273, 364; materiality of visual rays, 44–45, 74, 147–48; optical system according to, 41; projectile motion of visual rays, 65, 97; Renaissance art and, 300–301; replacement of visual rays with light rays in, 72, 180, 198, 224–26, 364; spatial perception according to, 47–53, 83–87, 92; species according to, 268; in university education, 283–84; visual acuity according to, 54, 86–87, 167, 168; visual rays compared to light rays, 130n113, 167–68, 176, 179. *See also* extramission

visual scanning, 48, 54, 80, 167, 190

visual spirit, 159, 161, 188, 193, 194, 250, 261, 270, 338

vital pneuma (*pneuma zōtikon*), 40, 236n21

vitreous humor (*hualoeidēs*), 39, 187, 188, 194, 261

Vitruvius, 324n2

vortex theory, 393, 404

Whiston, William, 390

William I, King of Sicily, 76

William of Conches, 240–41, 292

William of Moerbeke, 64n92, 244, 274

William of Ockham, 252–53, 319

Wilson, Catherine, 389, 391

wisdom (*sapientia*), 151, 153, 237, 261

Witelo, 273–74; influence of, 16, 256, 284, 294, 306–7, 322, 329, 357, 364, 365, 368, 375, 391; printing history of *Perspectiva*, 328; on shadow-casting, 310; significance of the *Perspectiva*, 274–75; sources relied on for, 222, 273–75; spherical aberration and, 335–36; university education featuring, 21, 281, 282–83

Wolfson, Harry, 162

Wyclif, John, 288–89, 292

Zik, Yaakov, 385